Springer Series on Environmental Management

Bruce N. Anderson
Robert W. Howarth
Lawrence R. Walker
Series Editors

Springer Series on Environmental Management
Volumes published since 1989

The Professional Practice of Environmental Management (1989)
R.S. Dorney and L. Dorney (eds.)

Chemicals in the Aquatic Environment: Advanced Hazard Assessment (1989)
L. Landner (ed.)

Inorganic Contaminants of Surface Water: Research and Monitoring Priorities (1991)
J.W. Moore

Chernobyl: A Policy Response Study (1991)
B. Segerståhl (ed.)

Long-Term Consequences of Disasters: The Reconstruction of Friuli, Italy, in its International Context, 1976–1988 (1991)
R. Geipel

Food Web Management: A Case Study of Lake Mendota (1992)
J.F. Kitchell (ed.)

Restoration and Recovery of an Industrial Region: Progress in Restoring the Smelter-Damaged Landscape near Sudbury, Canada (1995)
J.M. Gunn (ed.)

Limnological and Engineering Analysis of a Polluted Urban Lake: Prelude to Environmental Management of Onondaga Lake, New York (1996)
S.W. Effler (ed.)

Assessment and Management of Plant Invasions (1997)
J.O. Luken and J.W. Thieret (eds.)

Marine Debris: Sources, Impacts, and Solutions (1997)
J.M. Coe and D.B. Rogers (eds.)

Environmental Problem Solving: Psychosocial Barriers to Adaptive Change (1999)
A. Miller

Rural Planning from an Environmental Systems Perspective (1999)
F.B. Golley and J. Bellot (eds.)

Wildlife Study Design (2001)
M.L. Morrison, W.M. Block, M.D. Strickland, and W.L. Kendall

Selenium Assessment in Aquatic Ecosystems: A Guide for Hazard Evaluation and Water Quality Criteria (2002)
A.D. Lemly

Quantifying Environmental Impact Assessments Using Fuzzy Logic (2005)
R.B. Shepard

Changing Land Use Patterns in the Coastal Zone: Managing Environmental Quality in Rapidly Developing Regions (2006)
G.S. Kleppel, M.R. DeVoe, and M.V. Rawson (eds.)

The Longleaf Pine Ecosystem: Ecology, Silviculture, and Restoration (2006)
S. Jose, E.J. Jokela, and D.L. Miller (eds.)

Shibu Jose
Eric J. Jokela
Deborah L. Miller
(Editors)

The Longleaf Pine Ecosystem
Ecology, Silviculture, and Restoration

With 92 Illustrations

Springer

Shibu Jose
School of Forest Resources and
  Conservation
University of Florida
Gainesville, FL 32611
USA
sjose@ufl.edu

Eric J. Jokela
School of Forest Resources and
  Conservation
University of Florida
Gainesville, FL 32611
USA
ejokela@ufl.edu

Deborah L. Miller
Department of Wildlife Ecology
  and Conservation
University of Florida
Milton, FL 32583
USA
dlmi@ufl.edu

Series Editors:
Bruce N. Anderson
Planreal Australasia
Keilor, Victoria 3036
Australia
bnanderson@compuserve.com

Robert W. Howarth
Program in Biogeochemistry and
  Environmental Change
Cornell University
Corson Hall
Ithaca, NY 14853
rwh2@cornell.edu

Lawrence R. Walker
Department of Biological
  Sciences
University of Nevada
Las Vegas, NV 89154
walker@unlv.nevada.edu

Library of Congress Control Number: 2005936717

ISBN-10: 0-387-29655-7
ISBN-13: 978-0387-29655-5
eISBN: 0-387-30687-0

Printed on acid-free paper.

© 2006 Springer Science+Business Media, LLC
All rights reserved. This work may not be translated or copied in whole or in part without the written permission of the publisher (Springer Science+Business Media, Inc., 233 Spring Street, New York, NY 10013, USA), except for brief excerpts in connection with reviews or scholarly analysis. Use in connection with any form of information storage and retrieval, electronic adaptation, computer software, or by similar or dissimilar methodology now known or hereafter developed is forbidden.
The use in this publication of trade names, trademarks, service marks, and similar terms, even if they are not identified as such, is not to be taken as an expression of opinion as to whether or not they are subject to proprietary rights.

Printed in the United States of America.

9 8 7 6 5 4 3 2

springer.com

Contributors

Janaki R.R. Alavalapati, School of Forest Resources and Conservation, University of Florida, Gainesville, Florida 32611

John Blake, Savannah River Site, USDA Forest Service, New Ellenton, South Carolina 29809

William D. Boyer, Southern Research Station, USDA Forest Service, Auburn, Alabama 36849

Dale G. Brockway, Southern Research Station, USDA Forest Service, Auburn, Alabama 36849

J. Bachant Brown, The Nature Conservancy, Jay Florida Office, The Gulf Coastal Plain Ecosystem Partnership of Jay, Florida 32565

Douglas R. Carter, School of Forest Resources and Conservation, University of Florida, Gainesville, Florida 32611

Vernon Compton, The Nature Conservancy, Jay Florida Office, The Gulf Coastal Plain Ecosystem Partnership of Jay, Florida 32565

Ralph Costa, U.S. Fish and Wildlife Service, Clemson Field Office, Department of Forestry and Natural Resources, Clemson University, Clemson, South Carolina 29634

Roy S. DeLotelle, DeLotelle and Guthrie, Inc., 1220 SW 96th Street, Gainesville, Florida 32607

Cecil Frost, Adjunct Faculty, Curriculum in Ecology, University of North Carolina, 119 Pot Luck Farm Road, Rougemont, North Carolina 27572.

Dean Gjerstad, School of Forestry and Wildlife Sciences, Auburn University, Auburn, Alabama 36849

J.C.G. Goelz, USDA Forest Service, Southern Research Station, Pineville, Louisiana 71360

James M. Guldin, Arkansas Forestry Sciences Laboratory, Southern Research Station, USDA Forest Service, Monticello, Arkansas 71656

Timothy B. Harrington, USDA Forest Service, Pacific Northwest Research Station, Olympia, Washington 98512

Larry D. Harris, Department of Wildlife Ecology and Conservation, University of Florida, Gainesville, Florida 32611

M. Hicks, The Nature Conservancy, Jay Florida Office, The Gulf Coastal Plain Ecosystem Partnership Jay, Florida 32565

Thomas S. Hoctor, Department of Landscape Architecture, University of Florida, Gainesville, Florida 32611

Alan W. Hodges, Institute of Food and Agricultural Sciences, University of Florida, Gainesville, Florida 32611

Don Imm, Savannah River Site, USDA Forest Service, New Ellenton, South Carolina 29809

Steven B. Jack, Joseph W. Jones Ecological Research Center, Newton, Georgia 39870

Rhett Johnson, Solon Dixon Forestry Education Center, School of Forestry and Wildlife Sciences, Auburn University, Andalusia, Alabama 36420

Eric J. Jokela, School of Forest Resources and Conservation, University of Florida, Gainesville, Florida 32611

Shibu Jose, School of Forest Resources and Conservation, University of Florida, Gainesville, Florida 32611

John S. Kush, School of Forestry and Wildlife Sciences, Auburn University, Auburn, Alabama 36849

Peter E. Linehan, The Pennsylvania State University, Mont Alto Campus, Mont Alto, Pennsylvania 17237

Jagannadha R. Matta, School of Forest Resources and Conservation, University of Florida, Gainesville, Florida 32611

D. Bruce Means, Coastal Plains Institute and Land Conservancy, Tallahassee, Florida 32303

Deborah L. Miller, Department of Wildlife Ecology and Conservation, University of Florida, Milton, Florida 32583

Robert J. Mitchell, Joseph W. Jones Ecological Research Center, Newton, Georgia 39870

W. Keith Moser, USDA, Forest Service, North Central Research Station, Forest Inventory and Analysis, St. Paul, Minnesota 55108

Contributors

W. Leon Neel, Joseph W. Jones Ecological Research Center, Newton, Georgia 39870

Reed F. Noss, Department of Biology, University of Central Florida, Orlando, Florida 32816

Kenneth W. Outcalt, Southern Research Station, USDA Forest Service, Athens, Georgia 30602

Robert K. Peet, Department of Biology, University of North Carolina, Chapel Hill, North Carolina 27599-3280

P. Penniman, The Nature Conservancy, Jay Florida Office, The Gulf Coastal Plain Ecosystem Partnership of Jay, Florida 32565

Andrea M. Silletti, USDA Forest Service, Southern Research Station, Clemson, South Carolina 29634

G. Andrew Stainback, School of Forest Resources and Conservation, University of Florida, Gainesville, Florida 32611

Joan L. Walker, USDA Forest Service, Southern Research Station, Clemson, South Carolina 29634

K. A. Whitney, Department of Urban and Regional Planning, University of Florida, Gainesville, Florida 32611

Richard A. Williams, School of Forest Resources and Conservation, University of Florida, Milton, Florida 32572.

Preface

The history and development of the longleaf pine (*Pinus palustris* Mill.) ecosystem in the southeastern United States has intrigued natural resource professionals, researchers, and the general public for many decades. Prior to European settlement, longleaf pine forests were one of the most extensive ecosystems in North America. Most recent estimates suggest that only about 2.2% of the original area remains today, making it one of the most threatened ecosystems in North America.

The reduction in land area of the longleaf pine ecosystem has been attributed to a number of factors, including: (i) extensive harvesting in the early 1900s that significantly reduced growing stock levels; (ii) an inadequate understanding of the biophysical factors influencing regeneration dynamics such as seeding habits and fire management; (iii) general intolerance of longleaf pine to shade and understory competition; and (iv) conversion of longleaf pine sites to other commercially important species such as loblolly (*P. taeda* L.) and slash pine (*P. elliottii* Engelm.).

Over the last decade, considerable interest has grown in conserving and restoring the longleaf pine ecosystem. For example, it provides habitat for a wide variety of wildlife species including the endangered red-cockaded woodpecker (*Picoides borealis* Vieillot) and gopher tortoise (*Gopherus polyphemus* Daudin). Similarly, interest in longleaf pine regeneration and management systems has been high among land managers, ecologists, the forest products industry, and the general public. One example of this is the formation of the Longleaf Alliance based in Andalusia, AL, which is a partnership of private landowners, forest products industries, state and federal agencies, university researchers, and others interested in promoting a regionwide recovery of longleaf pine forests for their ecological and economic benefits. A variety of conventional and alternative management systems are being studied (e.g., single and multiple cohort stands) with regard to achieving these goals.

Restoration efforts in the longleaf pine ecosystem have focused on expanding areas of critical habitat. Ecosystem restoration efforts, however, require effective training of natural resource practitioners. For example, knowledge regarding past history of the southeastern landscape, current status of the longleaf pine ecosystem, its potential economic and associated biodiversity values, and the role of fire in maintaining the system is of critical importance. The idea for this book, therefore, was conceived originally as a textbook for undergraduate and graduate students because the time-tested classic of Wahlenberg (1946; *Longleaf Pine: Its Use, Ecology, Regeneration, Protection, Growth and Management*) was out of print. To achieve that aim we desired a text with ecosystem-level coverage on topics related to the ecology, management, and restoration of longleaf pine. In addition to the biophysical aspects, we desired coverage on the historical, social, and political aspects as well.

It quickly became apparent that a book serving not only students, but also practitioners, scientists, policymakers, and the general public was needed. The skills required to effectively manage natural resources have changed considerably over the past two decades. In addition to managing ecosystems for products and services, increasing emphasis has been placed on ecosystem restoration. This has become particularly important in promoting the recovery, management, and ecological integrity of disturbed and degraded ecosystems.

The authors who contributed to this multidisciplinary book have diverse backgrounds. As editors, we endeavored to accommodate their ideas, experiences, and interpretations over a broad range of topics. We wanted to treat each chapter as a standalone manuscript. As a result, a certain degree of overlap between some of the chapters was inevitable. However, each chapter addresses unique aspects of the longleaf pine ecosystem. The book is not intended to be viewed as a practical guide or prescription handbook for students and managers. The focus, rather, is on providing a foundation to relate information on processes to field problems and their solutions using innovative management approaches. We hope that this book will be particularly useful to students, practitioners, and scientists seeking a broader perspective on the biophysical and social dimensions of managing and restoring the various components of the longleaf pine ecosystem.

We are grateful to a large number of individuals for assistance in accomplishing this task, particularly the authors for their commitment to the project and their synthesis of the current knowledge. Also, the invaluable comments and suggestions made by the referees significantly improved the clarity and content of the chapters. In addition to many of the chapter authors who served as reviewers for other chapters, we thank: Robert Abt, Larry Bishop, Lindsay Boring, Andre Clewell, Kenn Dodd, Kevin Enge, Dennis Hardin, Nancy Herbert, Katherine Kirkman, David Maehr, Michael Messina, Jaroslaw Nowak, Scott Roberts, Kevin Robertson, Linda Roth, Wayne Smith, George Tanner, Morgan Varner, and Jeff Walters. We are grateful to Larry Schnell who served as our copy editor during this project and wish to extend our sincere thanks to Janet Slobodien and her staff at Springer Science for their timely efforts in publishing this book.

Shibu Jose
Eric J. Jokela
Deborah L. Miller

Contents

Contributors v

Preface ix

Section I Introduction

1. The Longleaf Pine Ecosystem: An Overview 3
   Shibu Jose, Eric J. Jokela, and Deborah L. Miller

2. History and Future of the Longleaf Pine Ecosystem 9
   Cecil Frost

   Box 2.1: The Naval Stores Industry 43
   Alan W. Hodges

Section II Ecology

3. Ecological Classification of Longleaf Pine Woodlands 51
   Robert K. Peet

4. Longleaf Pine Regeneration Ecology and Methods 95
   Dale G. Brockway, Kenneth W. Outcalt, and William D. Boyer

5. Plant Competition, Facilitation, and Other Overstory–Understory Interactions in
   Longleaf Pine Ecosystems 135
   Timothy B. Harrington

6. Vertebrate Faunal Diversity of Longleaf Pine Ecosystems 157
   D. Bruce Means

Section III Silviculture

7. Uneven-Aged Silviculture of Longleaf Pine 217
    James M. Guldin

   Box 7.1: The Stoddard–Neel Approach 242
      Steven B. Jack, W. Leon Neel, and Robert J. Mitchell

   Box 7.2: The Stoddard–Neel System—Case Studies 246
      W. Keith Moser

8. Longleaf Pine Growth and Yield 251
    John S. Kush, J. C. G Goelz, Richard A. Williams, Douglas R. Carter,
    and Peter E. Linehan

Section IV Restoration

9. Restoring the Overstory of Longleaf Pine Ecosystems 271
    Rhett Johnson and Dean Gjerstad

10. Restoring the Ground Layer of Longleaf Pine Ecosystems 297
    Joan L. Walker and Andrea M. Silletti

    Box 10.1: Prescribed Burning for Understory Restoration 326
       Kenneth W. Outcalt

    Box 10.2: Restoring the Savanna to the Savannah River Site 330
       Don Imm and John Blake

11. Reintroduction of Fauna to Longleaf Pine Ecosystems: Opportunities
    and Challenges 335
       Ralph Costa and Roy S. DeLotelle

12. Spatial Ecology and Restoration of the Longleaf Pine Evosystem 377
       Thomas S. Hoctor, Reed F. Noss, Larry D. Harris, and K. A. Whitney

13. Longleaf Pine Restoration: Economics and Policy 403
       Janaki R. R. Alavalapati, G. Andrew Stainback, and Jagannadha R. Matta

14. Role of Public–Private Partnership in Restoration: A Case Study 413
       Vernon Compton, J. Bachant Brown, M. Hicks, and P. Penniman

Index 431

Section I
Introduction

Chapter 1

The Longleaf Pine Ecosystem
An Overview

Shibu Jose, Eric J. Jokela, and Deborah L. Miller

An Ecosystem in Peril?

The longleaf pine (*Pinus palustris* Mill.) ecosystem once occupied an estimated 37 million hectares in the southeastern United States (Frost this volume). These forests dominated the Coastal Plain areas ranging from Virginia to Texas through central Florida, occupying a variety of sites ranging from xeric sandhills to wet poorly drained flatwoods to the montane areas in northern Alabama. The extent of the longleaf pine ecosystem has greatly declined since European settlement. At present, it occupies less than 1 million hectares, making it one of the most threatened ecosystems in the United States. Will this ecosystem always be in peril? Maybe not! The objective of this chapter is to provide an overview of the book's content that will examine the historical, ecological, silvicultural, and restoration aspects of longleaf pine ecosystems.

In the second chapter in Section I, Frost describes the historic context of the decline of the longleaf pine ecosystem and examines the current status and future outlook. Longleaf pine was exploited from first settlement; however, before 1700 travel and trade limited impacts to coastal regions along navigable streams. Land clearing and open range cattle and feral hogs that fed on longleaf pine seedlings in nearby woods were characteristic features of these early domesticated landscapes. Commercial logging had little impact until introduction of the water-powered sawmill in 1714, but by the 1760s hundreds of these mills were turning out sawn lumber. Still, deforestation was limited to narrow dendritic patterns defined by streams and rivers. By this time much of the eastern Piedmont was fully settled and the frontier had passed on toward the Appalachians.

By the Civil War, all of the best land on the Atlantic slope was in fields and pasture, but much virgin forest remained along the Gulf Coast. The naval stores industry that caused further decline in the area of longleaf pine stands is also discussed in detail by Hodges in Box 2.1. This crude turpentine industry, which began in Virginia in 1608, was practiced through the Colonial Period. By that time, there had been little impact farther to the south, with exception of stands found along rivers in North Carolina. Then, in 1834, adaptation of the copper whiskey still for turpentine

---

Shibu Jose and Eric J. Jokela • School of Forest Resources and Conservation, University of Florida, Gainesville, Florida 32611.   **Deborah L. Miller** • Department of Wildlife Ecology and Conservation, University of Florida, Milton, Florida 32583.

distillation made the fledgling forest industry vastly more efficient and profitable. Turpentining, along with the communities and jobs it supported, moved south into Georgia and then west along the Gulf Coast. Eventually, the turpentine industry reached virgin stands in Texas by around 1900. Steam technology mushroomed by 1870, with proliferation of logging railroads, steam log skidders, and steam sawmills. An intensive era of logging activities occurred in the South from 1870 to 1920. The 1920s also saw the beginning of commercial pine plantations, now more than 20% of southern uplands.

The presettlement range of longleaf pine was estimated at 37 million hectares, of which 23 million were longleaf dominant and 14 million had longleaf in mixtures with other pines and hardwoods. By 1946, longleaf pine had dwindled to one-sixth its original area. This decline has continued, such that only about 2.2% of the original area remains today. Of the original range, only about 0.2% of the land in 2000 was being managed with fire sufficient to perpetuate the open structure and species diversity represented by the hundreds of fire-dependent plant and animal species of the longleaf pine ecosystem.

Ecological Significance

The longleaf pine ecosystem plays a prominent role in the ecology and economy of the southeastern United States. These ecosystems have one of the richest species diversities outside the tropics. Although the overstory is dominated by one species, the understory is host to a plethora of plant species. The diversity among the herbaceous plants is the main contributor to its high biodiversity. In general, the composition of the understory is site specific, but is mainly dominated by grass species. In the western Gulf Coastal Plain, the understory is comprised mainly of bluestem (*Andropogon* and *Schizachyrium* spp.) grasses. In Florida and along the Atlantic Coast wiregrass (*Aristida beyrichiana*) is dominant, with *Aristida stricta* occurring from central South Carolina through North Carolina.

The first chapter in Section II (Chapter 3) by Peet illustrates how complex the plant associations can be in longleaf pine forests. Based on data from his own work and other published sources, Peet has classified the seemingly homogenous expanse of longleaf pine woodlands into 135 vegetation associations. Recognizing the considerable variation that occurs in longleaf pine communities with simple geographic distance and subtle environmental changes is of particular importance in making management decisions. The vegetation associations described in Chapter 3 could serve as a benchmark for classifying longleaf pine forests for conservation and providing targets for restoration.

One of the significant reasons for the reduction of longleaf pine regeneration was the interruption of natural fire cycles in the understory. Understanding the role of fire and the autecology of longleaf pine is vital for the restoration of this ecosystem. The chapter by Brockway et al. (Chapter 4) discusses the ecology of longleaf pine and the silvicultural reproduction methods commonly used for this species. Longleaf pine is a very intolerant pioneer species (Landers et al. 1995) and does not compete well for site resources with other more aggressive species (Brockway and Lewis 1997; Harrington this volume). Compared to other pine species, longleaf pine is not a prolific seed producer. Longleaf pine seeds require over 3 years for their physiological development. Thus, good seed crops are infrequent and may arise only once every 6–8 years. The seeds are large and heavy and do not disperse great distances. The short dissemination distances of the seeds prevent longleaf pine from colonizing and establishing in areas far from the seed source. Longleaf pine requires an exposed mineral soil seedbed that is free of surface litter. Fire exclusion results in accumulation of forest litter that hinders proper germination of longleaf pine seeds (Croker 1975).

With the removal of fire, the less fire adapted shrub species can spread into the understory. The encroaching hardwoods compete for site resources and light with the longleaf seedlings and hinder their growth and regeneration. Longleaf pine seedlings undergo an extended stemless phase without height initiation under competition from surrounding vegetation. This phase, also known as the "grass stage," varies in length depending on site resources and competition and may last as long as 10–25 years. These competitive interactions are the subject of Chapter 5 by Harrington.

Chapter 6 by Means explores the past and present vertebrate faunal diversity of the longleaf pine ecosystem. The highest species richness of turtles, frogs, and snakes in the United States and Canada (Kiester 1971), as well as a large salamander fauna (Means this volume), occurs on the Coastal Plain of the southeastern United States. However, bird species richness (Stout and Marion 1993) is not particularly high and mammal fauna is depauperate. With a number of threatened and endangered species and loss of over 97% of their habitat, these vertebrates still represent one of the largest vertebrate faunas in temperate North America. There are 212 resident vertebrate species in longleaf pine savannas of which 38 are specialists occurring exclusively or primarily in longleaf pine savannas.

The gopher tortoise and red-cockaded woodpecker are keystone species in this ecosystem that enable increased species richness by providing shelter for many species through their specialized activities. The gopher tortoise is a longleaf pine specialist, which excavates extensive underground burrows used by more than 300 species of other vertebrates and invertebrates (Jackson and Milstrey 1989). The red-cockaded woodpecker is the only woodpecker to make cavities in living trees. Because the longleaf pine trees are alive when cavities are excavated, the latter persist for up to 400 years and are used by many other animals over the lifetime of the tree.

Silvicultural Considerations

Uneven-aged silviculture of longleaf pine has received considerable attention in the recent past. This reproduction method and management system has been successfully applied in other southern pine stands such as mixed loblolly (*P. taeda* L.)–shortleaf (*P. echinata* Mill.) pine in the upper west Gulf Coastal Plain. In the first chapter in Section III (Chapter 7), Guldin presents an overview of lessons learned from loblolly–shortleaf uneven-aged management and explains the underlying principles of applying the same approach in longleaf pine ecosystems. Described in detail are reproduction methods, stand-level regulation, and developmental dynamics. The Stoddard–Neel approach to uneven-aged management is also described in detail by Jack et al. and Moser in Boxes 7.1 and 7.2, respectively. Available literature on the growth and yield of both plantation and natural stands of longleaf pine is summarized in Chapter 8 by Kush et al.

Ecological Restoration

The Society for Ecological Restoration (SER) defines restoration as an intentional activity that initiates or accelerates the recovery of an ecosystem with respect to its health, integrity, and sustainability (SER 2004). The ecosystem that requires restoration may be degraded, damaged, transformed, or entirely destroyed as the direct or indirect result of anthropogenic activities. The vast majority of the remaining longleaf pine ecosystems fall into one of the above-mentioned categories. Most have been altered beyond their resiliency; therefore, it is nearly impossible for them to revert back to the predisturbance state or historic developmental trajectory without human intervention.

Ecological restoration attempts to return sites formerly occupied by longleaf pine ecosystems to their historic trajectory. Historic conditions are therefore the ideal starting point for restoration design. Restoration of longleaf pine ecosystems requires identifying important reference communities that have conditions characteristic of a "historic" state. However, using a static image for restoring a dynamic forest ecosystem, is not only difficult to achieve, but may not be an appropriate goal (Hobbs and Harris 2001). There is a need to discuss in detail ecological indicators for restoration assessments. These indicators should be identified for their influence on determining the dynamics of plant community succession and soil productivity (Burger and Kelting 1999).

In the past, ecological restoration has been practiced using a retrospective approach, trying to capture the properties of an ecosystem that existed during some designated period of the past (Hobbs and Harris 2001). Current planning augments historical information by characterizing ecosystem composition, structure, function, biodiversity, and resilience from an existing system that is free of degradation and located within a reasonable distance (Harris 1999). This neighboring system is used as a model or reference for comparison. The advantage is that these reference systems can be studied over time and space. Sources of information that can be used in describing the reference ecosystem include (SER 2004):

1. Ecological descriptions, species lists, and maps of the project site prior to damage
2. Historical and recent aerial and ground-level photographs
3. Remnants of the site to be restored, indicating previous physical conditions and biota
4. Ecological descriptions and species lists of similar intact ecosystems; herbarium and museum specimens
5. Historical accounts and oral histories by persons familiar with the project site prior to damage
6. Paleoecological evidence, e.g., fossil pollen, charcoal, tree ring history, rodent middens

Based on the lessons learned from several operational restoration projects, Section IV explores the current status of restoration of the longleaf pine ecosystem. Restoring the overstory is the focus of the first chapter (Chapter 9) by Johnson and Gjerstad. The authors outline restoration strategies for 10 scenarios, representing 10 degraded conditions commonly encountered within the natural range of longleaf pine. Walker and Silletti (Chapter 10) discuss the techniques employed in restoring the understory community. The importance of fire for understory restoration is further explained by Outcalt in Box 10.1. Imm and Blake narrate a success story of putting savanna back to the Savanna River Site in Box 10.2.

Costa and DeLotelle discuss the reintroduction and augmentation, via translocation, of native fauna into longleaf pine ecosystems in Chapter 11. The focus is on rare species, including those considered "sensitive," "of special concern," or "candidates" for listing by conservation groups, or state or federal agencies. Their discussion also includes federally listed species as either "threatened" or "endangered" under the Endangered Species Act. Special emphasis is also placed on the red-cockaded woodpecker.

The importance of a landscape approach in restoring the longleaf pine ecosystem is the topic covered in Chapter 12 by Hoctor et al. Given the distinctive ecology and current condition of longleaf pine communities, landscape ecology and regional reserve design principles are crucial for guiding restoration efforts. Chapter 13 by Alavalapati et al. explores the socioeconomic and policy aspects of restoration. Incentive programs in place to promote restoration activities are also discussed. An example regional approach is presented in Chapter 14 by Compton et al. The successful Gulf Coastal Plain Ecosystem Partnership is emerging as a model for restoring longleaf pine across its former range.

Are We There Yet?

Restoration Ecology, the art and science behind ecological restoration, is not an exact science. Because ecosystems are dynamic, it is difficult to identify exact values to determine restoration success (van Diggelen et al. 2001). Instead, a range of values are used to identify restoration trajectories and "thresholds" (SER 2004; Suding et al. 2004). An ecosystem is considered to have reached a restored state when the system has been shifted across recovery thresholds and has returned to the general direction and boundaries of the historic trajectory. Exceeding recovery thresholds becomes an important goal in the restoration process. An ecosystem is restored when it contains sufficient biotic and abiotic resources to continue its development (trajectory) without further assistance. It will sustain itself structurally and functionally. The Society for Ecological Restoration has identified nine attributes for determining when restoration has been accomplished (SER 2004). They are:

1. The restored ecosystem contains a characteristic assemblage of the species that occur in the reference ecosystem so that it provides an appropriate community structure.
2. The restored ecosystem consists of indigenous species to the greatest extent possible. In restored cultural ecosystems, allowances can be made for domesticated alien species and for noninvasive ruderal (plants that colonize disturbed sites) and segetal (plants that grow intermixed with crop species) species that presumably co-evolved with them.
3. All functional groups necessary for the continued development and/or stability of the restored ecosystem are present or, if they are not, the missing groups have the potential to colonize by natural means.
4. The physical environment of the restored ecosystem is capable of sustaining viable reproducing populations of the species necessary for its continued stability or development along the desired trajectory.
5. The restored ecosystem functions normally for its ecological stage of development, and signs of dysfunction are absent.
6. The restored ecosystem is integrated into a larger ecological matrix or landscape, with which it interacts through abiotic and biotic flows and exchanges.
7. Potential threats to the health and integrity of the restored ecosystem from the surrounding landscape have been eliminated or reduced as much as possible.
8. The restored ecosystem is sufficiently resilient to endure the normal periodic stress events in the local environment that serve to maintain the integrity of the ecosystem.
9. The restored ecosystem is self-sustaining to the same degree as its reference ecosystem, and has the potential to persist indefinitely under existing environmental conditions. Nevertheless, aspects of its biodiversity, structure, and functioning may change as part of normal ecosystem development, and may fluctuate in response to normal periodic stress and occasional disturbance events of greater consequence. The species composition and other attributes of a restored ecosystem may evolve as abiotic conditions change.

A monitoring and evaluation program should be in place to track the success of the restoration efforts. A good monitoring program should be focused on a few key indicators in order to provide for statistically sound information (Lindenmayer 1999). Monitoring should be conducted in a systematic manner, designed to provide the needed information. The following steps have been recommended to ensure a functional monitoring plan (Block et al. 2001): (a) Set monitoring goals, (b) identify the resources to monitor, (c) establish threshold points, (d) develop a sampling design, (e) collect and analyze data, and (f) evaluate results (Fig. 1).

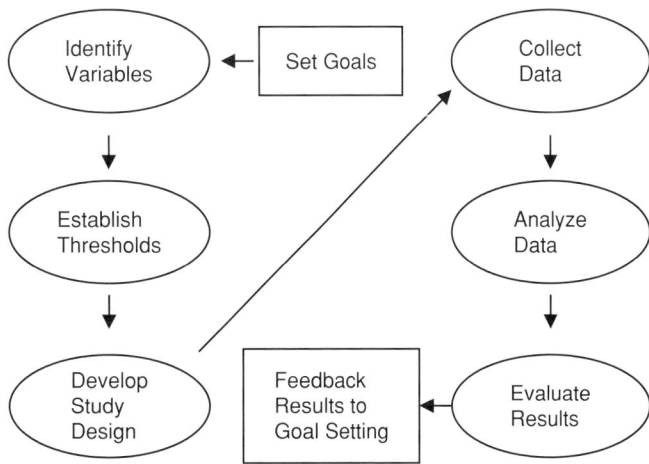

FIGURE 1. Flow diagram of monitoring process. Modified from Block et al. 2001.

Of the listed steps, identifying the ecosystem variables to monitor can be the most difficult. A small group of interrelated properties of a community can be used to develop a range of values instead of any single attribute such as an indicator species or species richness index. This will help avoid identification of a false threshold based on a single community attribute or a single threshold point (Block et al. 2001). Finally, monitoring should provide a feedback mechanism whereby the researcher or manager can make adjustments to the monitoring program based on the analyzed data. Since monitoring provides data about the dynamics of a community over time, a model can be developed from the results of monitoring the preselected group of community properties (indicators), which can illustrate how the community functions on a continuum.

Longleaf pine still occurs over most of its former natural range. By restoring degraded, destroyed, damaged, or transformed tracts and by expanding these pockets, it should be feasible to gradually increase longleaf pine acreage in the Southeast (Landers et al. 1995). As pointed out by Van Lear et al. (2005), restoring the longleaf pine ecosystem is a daunting task that raises many questions. Identification and removal of critical constraints to moving the system across recovery thresholds is the most important step. However, once the desired condition is achieved, it can be maintained with adaptive management using proven silvicultural practices (Van Lear et al. 2005).

References

Block, W.M., Franklin, A.B., Ward, J.P., Ganey, J.L., and White, C. 2001. Design and implementation of monitoring studies to evaluate the success of ecological restoration on Wildlife. *Restor Ecol* 9:293–303.

Brockway, D.G., and Lewis, C.E. 1997. Long-term effects of dormant-season prescribed fire on plant community diversity, structure and productivity in a longleaf pine wiregrass ecosystem. *For Ecol Manage* 96:167–183.

Burger, J.A., and Kelting, D. 1999. Using soil quality indicators to assess forest stand management. *For Ecol Manage* 122:155–166.

Croker, T.C. 1975. Seedbed preparation aids natural regeneration of longleaf pine. USDA Forest Service, Southern Forest Experiment Station, Research Paper SO-112, New Orleans, LA.

Harris, R.R. 1999. Defining reference conditions for restoration of riparian plant communities: Examples from California, USA. *Environ Manage* 24:55–63.

Hobbs, R.J., and Harris, J.A. 2001. Restoration ecology: Repairing the Earth's ecosystems in the new millennium. *Restor Ecol* 9:239–246.

Jackson, D.L., and Milstrey, E.G. 1989. The fauna of gopher tortoise burrows. In *Proceedings of the Gopher Tortoise Relocation Symposium*, eds. J. E. Diemer, D.R.Jackson, J.L. Landers, J.N. Layne, and D.A. Woods, pp. 86–98. Florida Game and Fresh Water Fish Commission Nongame Wildlife Program Technical Report 5. Tallahassee, FL.

Kiester, A.R. 1971. Species density of North American amphibians and reptiles. *Syst Zool* 20:127–137.

Landers, J.L., Van Lear, D. H., and Boyer, W.D. 1995. The longleaf pine forests of the southeast: Requiem or renaissance? *J For* 93:39–44.

Lindenmayer, D.B. 1999. Future directions for biodiversity conservation in managed forests: Indicator species, impact studies and monitoring programs. *For Ecol Manage* 115:277–287.

SER 2004. The SER International Primer on Ecological Restoration. Society for Ecological Restoration International, Tucson, AZ.

Stout, I.J., and Marion, W.R. 1993. Pine flatwoods and xeric pine forests of the southern (lower) Coastal Plain. In *Biodiversity of the Southeastern United States: Lowland Terrestrial Communities*, eds. W.H. Martin, S.G. Boyce, and A.C. Echternacht, pp. 373–446. New York: John Wiley & Sons.

Suding, K.N., Gross, K.L., and Houseman, G.R. 2004. Alternative states and positive feedbacks in restoration ecology. *Trends Ecol Evol* 19:46–53.

Van Diggelen, R., Grootjans, A.P., and Harris, J.A. 2001. Ecological restoration: State of the art or state of the science? *Restor Ecol* 9:115–118.

Van Lear, D.H., Carroll, W.D., Kapeluck, P.R., and Johnson, R. 2005. History and restoration of the longleaf pine–grassland ecosystem: Implications for species at risk. *For Ecol Manage* 211:150–165.

Chapter 2

History and Future of the Longleaf Pine Ecosystem[1]

Cecil Frost

Introduction

From Virginia to Texas, much of the coastal plain landscape was once covered by a "vast forest of the most stately pine trees that can be imagined..." (Bartram 1791 [1955]). Longleaf pine could be found from sea level, on the margins of brackish marshes, to around 2000 feet on the Talladega National Forest in Alabama (Harper 1905; Stowe et al. 2002). The spectacular failure of the primeval longleaf pine forest (Fig. 1) to reproduce itself after exploitation is a milestone event in the natural history of the eastern United States, even greater in scale and impact than the elimination of chestnut (*Castanea dentata*) from Appalachian forests by blight. This chapter discusses presettlement extent and summarizes major events in the decline of the longleaf pine ecosystem and its displacement from more than 97% of the lands it once occupied.

Land uses ranging from 100 to 400 years of agriculture; open range grazing by hogs and other livestock; logging; production of turpentine, and elimination of naturally occurring wildfires have left less than 3% of the upland landscape in entirely natural vegetation. While much has been made of the loss of some 10% to 30% of wetlands in the region (Hefner and Brown 1985), the elimination of natural vegetation on 97% of uplands (Table 1) has gone largely unnoticed.

Presettlement Vegetation of the Longleaf Pine Region

The presettlement range of longleaf pine has been estimated at 37 million hectares, of which 23 million were longleaf dominant and 14 million had longleaf in mixtures with other pines and hardwoods (Frost 1993). States bordering the Atlantic, and some of the Gulf Coast region, lack the systematic database of witness trees that were recorded when lands were surveyed after 1790 under the township, range, and section system in the rest of the country. Thus, there can be no easy reconstruction of virgin forests from such data. Even where historical survey records are available, interpretation is compromised because surveyors routinely failed to distinguish the various species of pine, just lumping them as "pine" on records and survey plats. There is, however, an exceptional

---

Cecil Frost • Adjunct Faculty, Curriculum in Ecology, University of North Carolina, 119 Pot Luck Farm Road, Rougemont, North Carolina 27572.

FIGURE 1. Virgin longleaf pine savanna 10 miles east of Fairhope, Baldwin County, Alabama, August 13, 1902. Note the absence of woody understory and the classic bilayered structure of fire-resistant canopy over a rich herbaceous layer under a natural fire regime (estimated at 1–3 years at this site). Roland Harper commented that "...it may never be possible to take such a picture in Alabama again." Photo from Harper (1913).

2. History and Future of the Longleaf Pine Ecosystem

TABLE 1. Distribution of natural vegetation and land use categories in presettlement forests, in 1900, and in 2000 for the 412 counties of the original longleaf pine ecosystem.

		Percent of uplands	Percent of region	ha × 1000	a × 1000
Presettlement					
1.	**Longleaf pine (dominant)**	**52.0**	**36.0**	**22,852**	**56,430**
2.	**Longleaf (mixed)**[a]	**33.2**	**23.0**	**14,606**	**36,064**
3.	Mixed (w/o longleaf)	9.0	6.3	4,001	9,878
4.	Upland slash pine	3.3	2.3	1,440	3,555
5.	Beech-magnolia	2.5	1.7	1,108	2,735
11.	Wetlands	0	30.7	19,496	48,137
		100.0	100.0	63,503	156,799
1900					
1.	Longleaf pine (natural)	24.2	17.5	11,109	27,430
2 + 3.	Mixed pyrophytic spp.	20.7	15.1	9,581	23,657
4.	Upland slash pine	1.7	1.2	775	1,914
5.	Beech-magnolia	0.4	0.3	166	410
6.	Successional forests	25.0	18.1	11,501	28,399
7.	Pine plantation	0	0	0	0
8 + 9.	Pasture and cropland	27.0	19.6	12,448	30,733
10.	Developed	1.0	0.7	460	1,137
11.	Wetlands	0	27.5	17,463	43,119
		100.0	100.0	63,503	156,799
2000					
1.	Longleaf pine (natural)	2.1	1.7	1,017	2,510
2 + 3.	Mixed pine-hardwood	0.5	<0.4	250	618
4.	Upland slash pine	0.4	0.3	222	547
5.	Beech-magnolia	0.4	0.3	222	547
6.	Successional forests	44.0	34.6	20,104	49,639
7.	Pine plantation (all species)	15.2	12.0	11,077	27,350
8.	Pasture	6.4	5.0	3,456	8,534
9.	Cropland	20.8	16.3	6,027	14,882
10.	Developed	10.2	8.0	7,538	18,616
11.	Wetlands	0	21.4	13,590	33,556
		100.0	100.0	63,503	156,799

Vegetation and Land Use Categories
1. Natural, fire-maintained communities dominated by longleaf pine
2. Longleaf-dominant patches and longleaf pine in fire-maintained mixed species savanna and woodland having longleaf, shortleaf, loblolly, pond pine, and sometimes hardwoods in various combinations
3. Pyrophytic woodlands without longleaf pine
4. Natural, fire-maintained slash pine on uplands
5. Southern mixed hardwood forest (nonpyrophytic, fire-refugial beech-magnolia)
6. Successional mixed pine-hardwood forests resulting from logging, old field abandonment, and fire exclusion
7. Pine plantation (all species)
8. Pasture
9. Cropland
10. Cities, towns, roads, industry
11. All wetlands: types wetter than hydric longleaf pine savanna

[a] Of the combined area of longleaf-dominant and longleaf-mixed species stands with patches of pure longleaf, I estimated the total original area of longleaf-dominant stands at 30 million hectares.

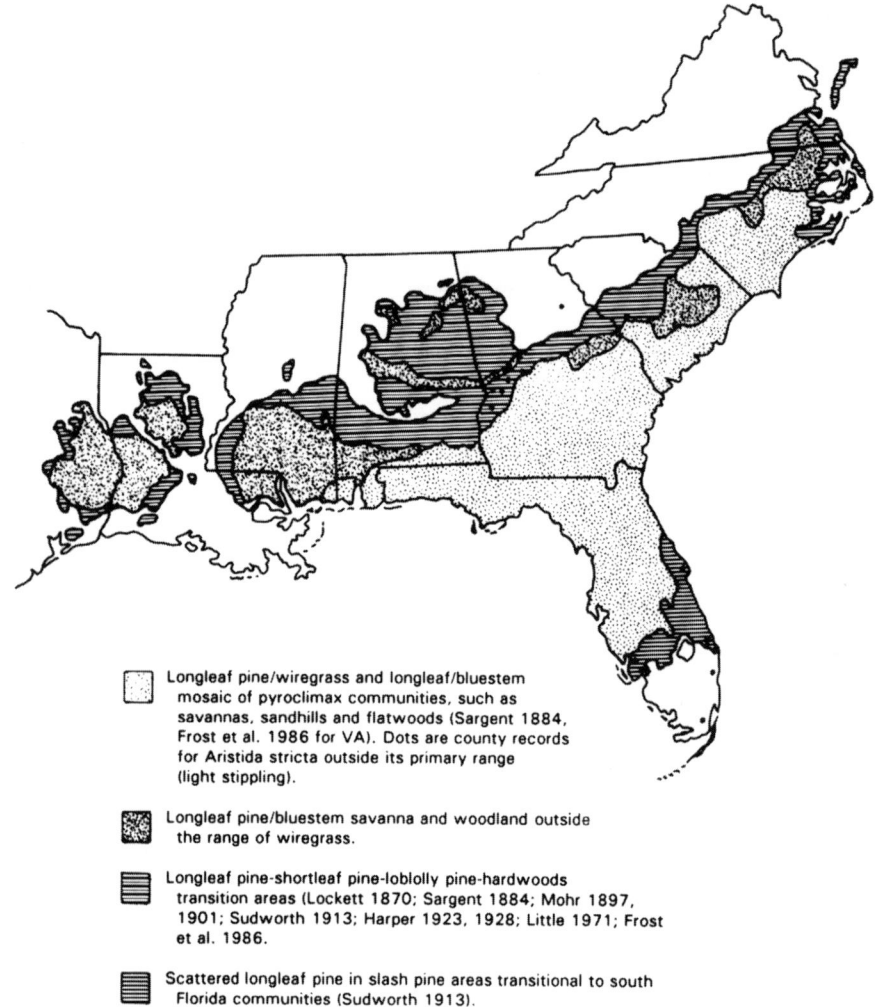

FIGURE 2. Presettlement range and major divisions of the longleaf pine ecosystem, showing the transition region between frequent fire communities of the Coastal Plain and the fire communities of the Piedmont described by Sargent (1884). Reprinted from Frost 1993 with permission from the Tall Timbers Research Station.

narrative literature on the longleaf pine forests, dating from 1608 when Captain John Smith exported the first barrels of pitch and tar made from pines near the new settlement at Jamestown, Virginia (Smith 1624).

Because of its primacy as the commercial tree of the South, longleaf pine became in the 1880s the first forest species to be studied in detail by botanists and early professional foresters. Major studies by Sargent (1884), Mohr (1896), Ashe (1894a), and Harper (1913, 1928) include literally hundreds of locations of longleaf pine as well as maps, lumbering records, and calculations of acreage and board feet by state, allowing a reasonable approximation of its original range and abundance. Figure 2 is a reconstruction of the original range of longleaf pine, using as a base a compilation of the state maps prepared by Sargent. Range maps and numerous locations provided by Ruffin (1861), Lockett (1870), Hale (1883), State Board of Agriculture (1883), Ashe (1894a,b), Harper (1905, 1906, 1911, 1913, 1914, 1923, 1928), Sudworth (1913), Mattoon

(1922), Wakeley (1935), Wahlenburg (1946), and Little (1971) were also useful. In addition, numerous historical references and remnant locations for longleaf were used to fill in areas unknown to Sargent and reconstruct its original northern range in North Carolina and Virginia. The resulting map includes all areas known to have once supported longleaf pine. In all, in the presettlement range of longleaf pine there were 412 counties in nine states. Sources of statistics and methods for reconstructing the original range are discussed further in Frost (1993). Figures for pine plantation (all species) were updated using a projection for 2000 by McWilliams (1987), and corrected for the area of each state lying outside the original range of longleaf pine.

Amount of Longleaf Pine Remaining in 2000

According to data of the 1995 Forest Inventory and Analysis (FIA), there were some 1.02 million hectares of longleaf pine remaining at that time. About 15% of this, or 178,200 hectares, consisted of pine plantation, mostly on old field or mechanically prepared sites, so about 85%, or 841,800 hectares, of naturally regenerated longleaf pine having some degree of understory integrity persist (Outcalt and Sheffield 1996). There are a variety of factors of uncertainty in the estimate of remaining longleaf pine. The FIA data are based only on stands with at least 50% longleaf pine canopy cover, so will be an underestimate of the total remaining. On the other hand, longleaf pine in FIA permanent sample plots declined by 22% from 1985 to 1995 (Kelly and Bechtold 1990; Outcalt and Sheffield 1996): the data were already 9 years out of date as of January 2004 and so will be an overestimate of the longleaf dominant natural stands remaining in 2005. We would expect these under- and overestimates to partially cancel each other, making the figure of 841,800 hectares a reasonable estimate of naturally regenerated longleaf in all stands in 2000. This is about 2.2% of the presettlement extent of longleaf pine.

Fire Relations of the Original Forests

In the pastoral landscapes of Britain, domesticated since Roman times, wildfire was an alien concept. A British traveler in South Carolina in 1829 was astonished to discover a recently burned stand of longleaf pine:

There was no underwood properly so-called, while the shrubs had all been destroyed a week or two before by a great fire. The pine-trees, the bark of which was scorched to a height of about 20 feet, stood on ground as dark as if it had rained Matchless Blacking for the last month. Our companions assured us that although these fires were frequent in the forest, the large trees did not suffer. This may be true, but certainly they did look very wretched, though their tops were green as if nothing had happened. (Hall 1829, p. 137)

Historically, agents of fire included lightning, Native Americans, and European settlers. Agents of fire suppression were bodies of water, topography (steep slopes, islands, peninsulas [Harper 1911]), a few plantation owners (Gamble 1921, p. 27), and government agencies (Sherrard 1903). Varying effects of fire in the landscape mosaic have been attributed to fire frequency, fire intensity, and season of burn (Garren 1943; Komarek 1974). Given that lightning fires would mostly have been growing season fires, fire frequency must have been the most important fire variable in presettlement vegetation.

Mattoon (1922) commented that longleaf lands experienced fire at an average of every 2–3 years over millions of hectares. There is evidence that fire frequency is proportional to fire compartment size: the larger the fire compartment the higher the fire frequency, and in the largest fire compartments (over 1000 km^2), the original fire frequency averaged 1–3 years (Frost 2000). On the Pamlico Terrace and other terraces of the lower Coastal Plain from Virginia to Texas, there were numerous tracts of land from several hundred to over a thousand square kilometers in size without a single natural firebreak. In Florida, Komarek (1965) reported that 99 wildfires were started by lightning on a single summer day. On the

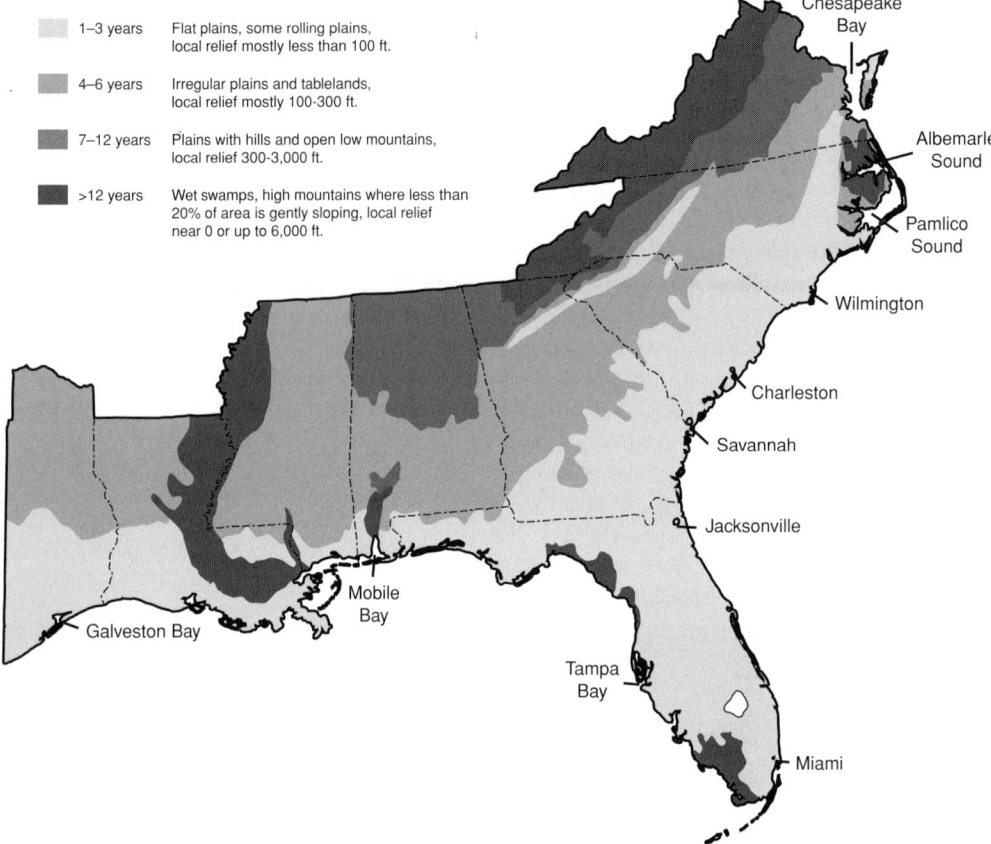

FIGURE 3. Presettlement fire regimes of the southeastern United States. Frequencies are for the most fire-exposed parts of the landscape. Each region contains variously fire-protected areas with lower incidences of fire (revised from Frost 1995, 2000). Revised from Frost 1995 with permission from the Tall Timbers Research Station.

Pamlico Terrace, where a single ignition might burn 1000 km², a few ignitions in each state might be sufficient to burn most of the landscape. On the other hand, fire frequency should decrease inland on the more dissected upper Coastal Plain and Piedmont, where numerous separate ignitions would be required to burn the decreasingly smaller fire compartments. The resulting decrease in fire frequency, along with clayey soils, colder winter temperatures, and increased topographic variation should explain the admixtures of other pine species and hardwoods with longleaf in the transition regions (Sargent 1884).

Figure 3 shows generalized presettlement fire frequencies of the longleaf pine region. Before immigration of Indians into the Southeast near the end of the Wisconsin glaciation some 12,000 to 13,000 years ago, essentially all fires would have been caused by lightning. E V. Komarek marshaled evidence to support the idea that lightning alone is adequate to account for evolution of pyrophytic vegetation, the antiquity of which far exceeds the appearance of aboriginal peoples on the scene. This provided a basis for thinking about fire as a ubiquitous environmental parameter, as influential as slope, aspect, rainfall, and temperature on shaping vegetation structure and the species composition of plant communities (Komarek 1964, 1965, 1966, 1967, 1968, 1972, 1974). His paper on ancient wildfires (1972) seems to have had particular impact on paleoecologists, and opened a door into inquiries concerning the role of fire and vegetation through geologic and evolutionary time (Cloud 1976; Cope and Chaloner 1980; Scott 1989; Scott and Jones 1994).

The emerging picture suggests that terrestrial vegetation has evolved with fire from its very beginning in the early Devonian Era, some 400 million years ago. Some species, such as Venus flytrap (*Dionaea muscipula*), have been shown to require a mean fire frequency of at least 3 years to survive (Frost 2000). Venus flytrap is a highly evolved species with a suite of adaptations far too complex to have evolved in the short time since Native Americans appeared on the scene. Further, the requirement for fire for reproduction in species such as longleaf pine, and even a fire frequency of 1–3 years for species such as Venus flytrap, may be millions of years old. Such adaptations may have developed along with evolution of the species themselves, rather than representing adaptations to fire in the mere 12,000 or 13,000 years that humans have been using fire in the Western Hemisphere. Such species indicate that some parts of the landscape—the largest fire compartments—experienced a natural fire frequency of 2–3 years long before immigration of man into the Western Hemisphere, and before man, the only agent that could have provided a frequent ignition source was lightning.

On the other hand, burning by Native Americans did transform vegetation in many parts of the southeastern landscape. Accounts from the Colonial Period describing Indian burning practices indicate that use of wildfire by Indians in the Southeast peaked in fall and winter when fires were set to drive game (Smith 1624; Lawson 1709; Byrd 1728; Martin 1973). On the outer Coastal Plain, where annual spring and summer lightning fires pre-empted fuel, the effect of any Indian burning may have been only a slight increase in burn area resulting from the inclusion of peninsulas and isolated patches of uplands that otherwise were naturally protected from fire. On the other hand, Indian influence may have been much more significant on dissected inland terraces and the Piedmont, where their primary effect, in compartments missed by lightning, would have been a net increase in fire frequency. Early explorers described some regions of the Piedmont that were dotted with prairies and open woodlands maintained by fire. These open landscapes were almost certainly the result of burning by Native Americans (Barden 1997).

Distribution of Major Vegetation Types in Presettlement Forests

Sargent (1884) divided the range of longleaf pine into two regions, the larger having longleaf as the most common dominant tree, and a second region around the margins of the first, in which longleaf occurred in patches or in mixed stands transitional to other types outside its range. Each of these two was further divided in Fig. 2. In the flat-to-gently rolling lands Sargent described longleaf as the "prevailing growth" on the uplands and F. A. Michaux reported that "Seven-tenths of the country are covered with pines of one species, or *Pinus palustris* . . ." (Michaux, 1805 [1966]). This longleaf-dominated landscape included a diverse mosaic of pine savannas, sandhills, and flatwoods, with variants in other habitats, such as riparian sand ridges, Carolina bay sand rims, coastal scarps, and dunes (Peet and Allard 1993; Harcombe et al. 1995).

Boundaries of the primary region were compiled almost exactly as drawn on Sargent's individual state maps. In Fig 2, I divided this first region into two, depending on presence or absence of wiregrass. Wiregrass in North Carolina and the northern third of South Carolina is *Aristida stricta*, that from southern South Carolina to Mississippi is *Aristida beyrichiana* (Peet 1993). Vegetation type 1 indicates the portion of the known historical range of wiregrass that occurs within the longleaf pine ecosystem, based on herbarium records (Parrott 1967; Peet 1993).

Transitional Communities

Sargent's second major assemblage of communities included the mosaic of forest types transitional between coastal plain regions dominated by nearly pure stands of longleaf, and the oak–hickory–shortleaf pine pyrophytic woodlands of the Piedmont. Sargent described the transition regions as "long leaved pine (*Pinus palustris*) with hardwoods in about equal proportion" in the Gulf states and "short leaved (*Pinus echinata*) and loblolly pine (*P. taeda*) intermixed with hardwoods and scattered long leaved pine" in the Atlantic states. I added the transitional woodlands around the northern and eastern sides of the primary longleaf range in Virginia and North Carolina. Not described by Sargent, these stands included variants in which pond pine (*Pinus serotina*) was added to the mixture (Ashe 1894a).

Mixed Patches versus Mixed Species

The importance of natural mixtures of longleaf pine with other fire-resistant trees has been generally overlooked. In Sargent's transition regions we can further distinguish the difference between mixed longleaf-dominant patches in a landscape with other forest types, and true mixed-species stands. The first was a patch mosaic having nearly pure stands of longleaf pine on south slopes and upland ridges. Both Mohr (1896) and Harper (1905, 1923, 1928) described pure stands as well as mixed stands. In the second group, they pictured the mixed pyrophytic types as open woodland with a geographically varying mixture of the dominant trees, which were longleaf, shortleaf pine, loblolly pine, post oak, white oak, southern red oak, hickories, and various scrub oaks. From historical photos, these were bilayered communities, having a tree canopy and a savannalike grass–forb understory, indicative of a frequent fire regime. The existence of natural mixed species stands has been overshadowed by the remarkable pure longleaf stands that dominated most of the southern uplands, and by the fact that the mixed stands occurred on the moister and finer textured, more fertile soils, the preponderance of which were cleared for farming long ago (Williams 1989). These diverse communities, with all their geographic variation, have never been adequately described. With rising interest in restoring longleaf pine, well-intentioned individuals have in some cases eliminated natural mixed longleaf–shortleaf savanna in the transition regions and replaced them with pure longleaf.

Hardwoods in Presettlement Forests

Several types of natural hardwood communities occur interspersed in the longleaf pine uplands. Besides longleaf pine stands with understory turkey oak (*Quercus laevis*), there are stands of mixed scrub oaks (*Quercus laevis, Quercus marilandica, Quercus incana* and *Quercus margaretta*); pyrophytic woodland with mixed longleaf, post oak, southern red oak, and mockernut hickory (*Carya tomentosa*); and patches of post oak savanna (*Quercus stellata*), the importance of which has been mostly overlooked.

In contrast to the dominant fire communities, small areas of nonpyrophytic types such as Southern Mixed Hardwood Forest, dominated by beech, magnolia, semievergreen oaks, and other hardwoods, may have been confined to naturally fire-sheltered sites within the range of longleaf pine (Harper 1911). Old-growth stands of beech and other mesophytic hardwoods can be found on steep slopes, islands in swamps, and a few upland flats on peninsulas. In many places, species such as beech (*Fagus grandifolia*) are now escaping from these fire refugia onto the uplands (Ware 1978). Studies by Delcourt and Delcourt (1977) in the Apalachicola bluffs region of the Florida Panhandle suggest that fire-refugial Southern Mixed Hardwood Forest occupied less than 1% of the presettlement landscape.

Landscape Changes 1565 to 1900

Ecosystem Changes in the Early Colonial Period

While the landscape that greeted the first two major groups of European settlers held astonishing forest resources, neither the English nor the Spanish were well equipped to exploit them, and the two cultures used radically

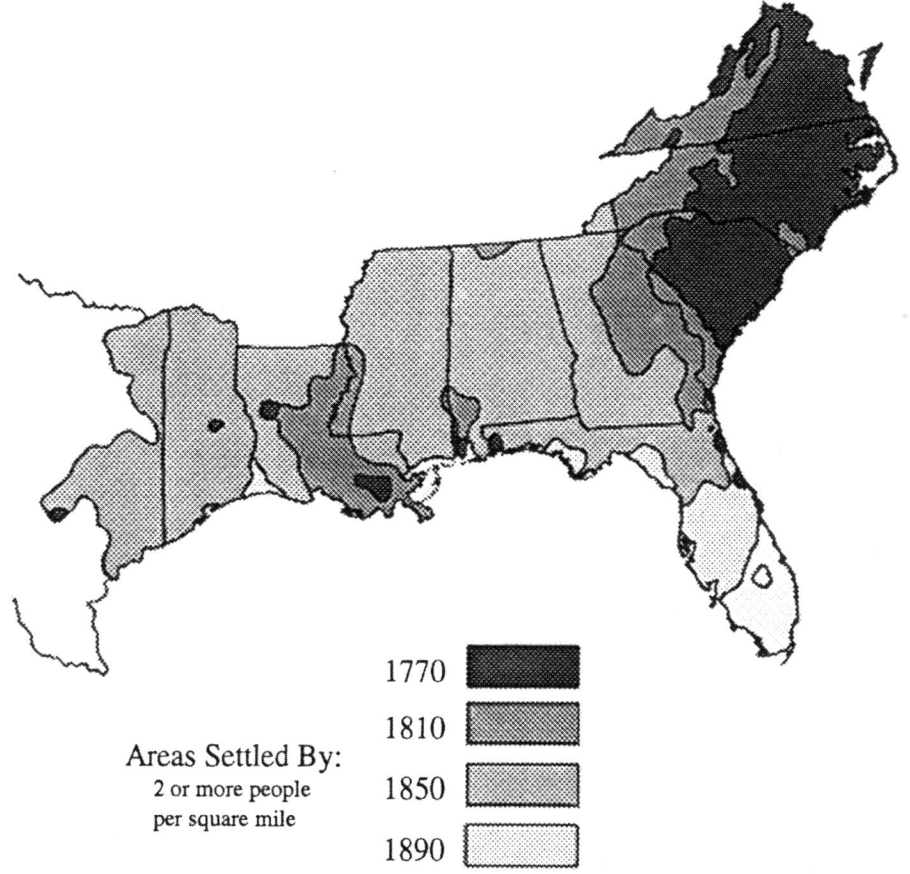

FIGURE 4. Pattern of settlement in the Southeast to 1890. Note the three small centers of population in Florida, which comprised most of its sparse population until 1821. With exception of the new cotton plantation regions, most virgin forest of the interior of the six Gulf states remained intact in 1850. Map redrawn from Hammond Inc., Maplewood, NJ. Reprinted from Frost 1993 with permission from the Tall Timbers Research Station.

different approaches in exploitation of the New World.

DeSoto set out in 1539 to explore the Gulf Coast interior, an epic overland journey complete with army, horses, and droves of hogs, that took him as far inland as the Cherokee towns of North Carolina and west beyond the Mississippi River (Bakeless 1961). While the Spanish, disappointed with the scarcity of interesting targets for conquest and pillage, lost interest in the north Gulf interior, they continued to control access to much of that vast region from Florida to Texas. What is significant for landscape history is that during their 256-year tenure—from establishment of St. Augustine in 1565 until cession of Florida to the United States in 1821—the Spanish blocked settlement of the Gulf Coast interior, leaving longleaf pine forests of much of the region in pristine condition well into the nineteenth century. Curiously, with the exception of a handful of coastal villages such as St. Augustine and Pensacola, they never pursued immigration and settlement of the land. In 1821, at the end of their occupation, the entire European population of Florida was barely more than 20,000 people, scarcely enough for a reputable town. Note the contrast in settlement patterns between Spanish lands and English settlements along the Atlantic in Fig. 4.

Unlike the Spanish military outposts, English settlements were commercial ventures financed by corporations of wealthy stockholders. Backers of the 1607 Jamestown, VA, expedition under John Smith promoted settlement and domestication of the land in order to establish a productive populace from which they could harvest taxes, agricultural produce, and whatever natural products the land could supply (Smith 1624).

For the first 150 years, dependence on water for travel and trade limited settlement to the nearest high lands along coastal sounds, bays, and the tidal portions of major and minor streams (Hart 1979). The tidewater area included at least 10,000 miles of shoreline from Virginia to Texas, and until the coastal zone was thoroughly populated there was little incentive to push inland. Domestication of this easily accessible landscape resulted in land clearing and establishment of saturation densities of open-range hogs and other livestock that fed on longleaf pine seedlings in nearby woods.

At that time, in the absence of machinery, timber was worthless except for local use in fencing and log cabin construction. The only milled boards were laboriously pit sawed by hand with crosscut saws, using one man in a pit and another above (Hindle 1975). A very early exception, a water-powered sawmill built at Henrico on the James River in Virginia in 1611, was destroyed by the Indians a few years later (Hindle 1975). Port records from the British Public Records Office from the early 1600s show that while lumber was a frequent item in ship's cargoes, the quantities were small. Cooperage stock—barrel staves and wooden water pipes made from oak and white cedar—supplied practically the only manufactured items for export for the first hundred years (British Public Records 1607–1783).

At the onset of agriculture, timber was little more than an obstruction. Settlers simply killed trees by girdling them, and the land was then burned and grazed, or planted in corn and other crops beneath the dead timber (Beverley 1705 [1947]). Since most livestock were allowed to graze on open range in the woods, it was necessary to fence them out of the small crop patches (Beverley 1705 [1947]). As a result, the principal early demand for timber was for fencing. Of great importance to natural savanna and woodland communities, though little remarked historically, was the introduction of swarms of hogs, cattle, horses, mules, sheep, and goats onto open range in all of the settled areas. Of these ravening herds, hogs in particular would play a major part in the decline of longleaf pine.

Naval Stores and the Original Northern Range of Longleaf Pine to the Virginia/Maryland Border

Tar, pitch, rosin, and turpentine were collectively called naval stores (Ashe 1894a; Mohr 1896) and were produced in the Southeast almost exclusively from longleaf pine, although smaller amounts were made from slash pine, shortleaf, and sometimes even loblolly pine (Michaux 1871) (see box by Hodges in this chapter). There were five substances commonly produced from longleaf pine gum: crude turpentine, spirits of turpentine, tar, pitch, and rosin. Crude turpentine was just the fresh gum exuded from the tree when a section of bark was removed. Spirits of turpentine was the aromatic fraction produced by distilling crude gum, and rosin was the dense, waxy residue left over from distillation. These materials were produced from the living tree. Tar was the product of distillation of dead "lightwood," the resin-rich heartwood from old stumps, or gathered from partly decayed trunks on the forest floor and distilled in tar kilns. The black, much thicker pitch was simply tar that had been burned down in iron "pitch kettles" to about one-third its original volume.

The early history of naval stores and longleaf pine has been all but lost, since the species was commercially extirpated from much of its northern range by 1850. Even Mohr (1896) states that the naval stores industry began in North Carolina. Such was not the case, however; it had been carried on earlier for over 200 years in Virginia. Longleaf once extended to within a mile of the Maryland border (Fig. 5), and likely continued into that state. I examined a herbarium specimen of longleaf pine collected near Sinnickson, VA, in 1925. I also visited the site and interviewed the collector

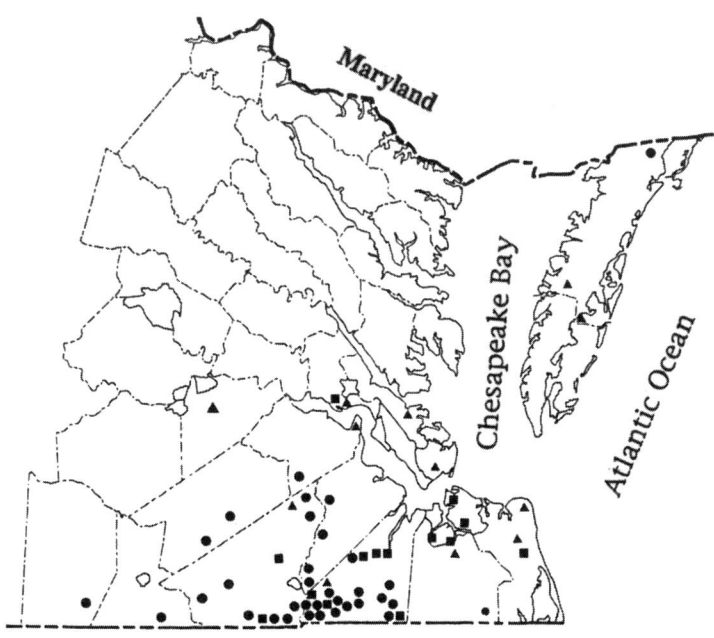

FIGURE 5. Documentation of the original range of longleaf pine in Virginia. Circles indicate herbarium specimens or living trees seen from 1960 to 2004, or reported to me by local foresters (also includes two tar kilns visited in Suffolk and Chesapeake). Squares denote clear historical records, some as early as 1608, but lacking herbarium specimens. Triangles are used for naval stores place names like Pitch Kettle Road, Lightwood Swamp, Tar Pit Swamp, and Tar Bay. Reprinted from Frost 1993 with permission from the Tall Timbers Research Station.

before his death (Moldenke 1979, personal communication). He reported that he collected the specimen from a natural stand growing on the ridges of forested coastal sandhills that lie on the scarp that forms the eastern uplands before dropping down into the coastal marshes on the Atlantic side of the Eastern Shore. These low sandhills continue into Maryland only 2 miles from this site. This area is part of a large, unbroken fire compartment, and it is almost certain that longleaf pine once extended at least into Worcester County, MD. This state, however, was not included in the presettlement range map for lack of a verifiable record.

Tar and pitch were produced in Virginia for over 200 years before the boom in North Carolina that gave the Tarheel State its nickname. We know of the early trade, the extent of which has never been thoroughly investigated, only through disparate and widely scattered records. The southern naval stores industry began in 1608 when John Smith exported the first "tryalls of Pitch and Tarre" (Smith 1624). The settlement was founded in 1607 and the next year the Jamestown, VA, colony exported some three or four dozen barrels to England. To all indications, longleaf was sparse on the north side of the James River, where Smith reported finding only a tree here and there "fit for the purpose" [of making naval stores].

Tar and pitch were absolutely essential commodities until the development of petroleum-based substitutes in the mid-1800s. Wagons could not move without tar to grease the axles. Ships could not sail without tar and pitch for waterproofing cordage and sails, for caulking leaks, and for coating hulls to prevent destruction by shipworms (Wertenberger 1931). During the Revolutionary War, Captain H. Young wrote to his colonel "...let me entreat you once more to lay before the Council my distressed situation for the want of two Barrels of Tar." "I have offer'd Brown (who is the only one that has Tar) his price in specie, or two barrels of Tar for one, both of which[2] offers he has refused. Our waggons can't run for the want of tar" (Young 1781 [Calendar of State Papers (Virginia) 1881], 2:619). Colonel Davies had his own problems with the recalcitrant Mr. Brown, while trying to ship 30 cannon to prevent their capture by the British: "Our own vessels are all in readiness, except for some slight repairs, for the finishing of which some small quantity of tar is necessary, tho' not more than a barrel at the utmost—we cannot procure this quantity under some time unless we obtain it from Mr. Brown, who will not part with it upon any other terms than for specie,[2] of which the State has none to pay" (Davies 1781 [Calendar of State Papers (Virginia) 1881], 2:599).

Early naval stores production concentrated on burning tar kilns for tar and pitch. Tar kilns were earth-covered mounds of several cords of collected dead pine "lightwood" that were burned under controlled conditions by carefully regulating the amount of air let into the mound. This sometimes dangerous process took up to 2 weeks of continuous management—from the first drops which might not appear for several days, until the tar ceased to flow into the barrels placed below (Catesby 1731, 1743). The second, more destructive practice involved boxing of live trees for the crude gum that was then shipped to New England or Europe for distillation of spirits of turpentine in crude iron retorts. While boxing was practiced as early as 1608 (Smith 1624), the necessity of shipping the bulky crude gum long distances limited the price and demand for the first hundred years.

While tar and pitch were made from 1608 on, most seem to have been consumed locally until around 1700. In 1697, Governor Sir Edmund Andros said that Virginia produced no naval stores for sale except along the Elizabeth River [Norfolk County], where about 1,200 barrels of tar and pitch were made annually (Pierce 1953). This would have had ready market at the port of Norfolk just a few miles downstream. The industry was carried on by poor men who built their kilns unassisted by servants or slaves, and considered a few dozen barrels a year an excellent output (Wertenberger 1931). F. A. Michaux, writing about his own observations made around 1802, notes that "toward the north, the Long-leaved Pine first makes its appearance near Norfolk, in Virginia, where the pine-barrens begin" (Michaux 1871).

In 1704 Jenings (1704 [1923]) reported some 3000 barrels of tar produced in Princess Anne County and part of Norfolk County. The disposition was split three ways: local consumption, sale to ship's masters, and export to the West Indies. Customs records on file for ports from around the Chesapeake Bay list barrels of naval stores as one of the most common exports from the colony from the late 1600s until the Revolution (British Public Records).[3] In a typical entry, the customs official at Hampton, VA, noted on April 12, 1745, "Cleared at Hampton, the snow John and Mary, Thomas Bradley, for Liverpool with 106 hhd. tobacco, 500 bbl tar, 60 walnut stocks and 5600 staves" (a snow [pronounced like "now"] was a square-rigged sailing vessel, one of the most frequently mentioned trading-ship designs in early eighteenth century). The exact point of origin of the goods is seldom determinable since ships often stopped at plantations up and down the rivers to pick up cargo, sailing on to be cleared through customs at the ports of Accomack, Hampton, or Norfolk.

Twenty-five years later, the export trade had increased such that, from March 25 to September 29, 1726, 17 vessels were cleared from Hampton, only one of the ports, with 1194 barrels of pitch and 6004 barrels of tar. One ship alone carried 1580 barrels of tar and 130 of pitch (British Public Records 1726). By 1791 the port at Norfolk exported 29,376 tons of naval stores (La Rochefoucauld 1799). By 1803, the number of ships cleared for foreign ports from Norfolk and Portsmouth reached 484, and it was reported that Virginia was no longer able to meet the export demand for yellow pine (Wertenberger 1931). The designation "yellow pine" most often meant lumber from longleaf pine in the early trade.

Early channels of trade in tar and pitch in Virginia were the Elizabeth and Nansemond Rivers, with their tidal tributaries interpenetrating the lands in the interiors of Norfolk and Nansemond counties. Not a single longleaf pine remains within the watersheds of these two stream systems today, and not a single tree remains in the former longleaf counties of Norfolk and Princess Ann. The only evidence remaining in the three counties east of the Nansemond River are a few remnant tar kilns and a handful of isolated trees in Suffolk. Most of the remainder of Colonial production of tar, pitch, and turpentine originated from counties along the south side of the James River, where there is evidence of once-extensive longleaf pine forests (Frost and Musselman 1987).

There were as much as 600,000 hectares in the original range of longleaf pine in Virginia, based on the extent of suitable soils in the original range defined in Fig. 5. Longleaf pine forests in Virginia appear to have been largely exhausted by 1840, after which no further naval stores production was listed (U.S. Census Office 1841). The Census of Manufactures for that year listed 5012 barrels produced from five counties. The species no longer occurs in two of these and I was able to find fewer than 200 mature native trees left in this state—enough to stock perhaps 5 hectares—where once there were more than 4000 km^2 dominated by longleaf pine. In 1893, forester B. E. Fernow concluded that "[i]n Virginia the longleaf pine is, for all practical purposes, extinct."

In Southampton County, Virginia, I met a farmer, 84 years old—born around 1896—whose recollection went back to the days of "longstraw" pine as it was known there in the past. Perhaps the last person in the state to remember that term from daily use, he took me to see three trees that he had ordered to be left when his land was logged. Longleaf pine has been completely extirpated from 11 of the original 15 counties of its range in Virginia. Remnant trees can now be found only in Isle of Wight, Southampton, Suffolk, and Greensville counties.

Southward Migration of the Naval Stores Industry, North Carolina to Texas

In 1622, John Pory traveled overland from Jamestown to the Indian town of Chowanoc, passing through a "great forest of Pynes 15. or 16. myle broad and above 60. mile long, which will serve well for Masts for Shipping, and for pitch and tarre, when we shall come to extend our plantations to those borders" (Powell 1977, p. 101). These were the great pine barrens of western Isle of Wight and Nansemond counties, Virginia, and Gates and Chowan counties, North Carolina. The first record of naval stores produced in North Carolina was in 1636, 17 years before the first settler set up a house and trading post in 1653. A visitor from Bermuda sailing up the Chowan River was surprised to discover a large number of men there busily producing "sperrits of rosin" (Clay et al. 1975). This was in the vicinity of the "Sand Banks" of western Gates County. The crew had apparently come south, overland from the settlements, only a few years old, along the James River in Virginia. From 1980 to 1990 I was only able to locate about 25 old longleaf trees in the Sand Banks region. I counted annual rings when some of these were logged around 1980: the largest was 308 years old and only 23 inches (60 cm) in diameter on the stump when cut.

Schoepf (1788 [1911]) traveling down the coastal plain from Virginia to South Carolina observed that "...the greatest and most important part of the immense forests of this fore-county consists of pine...", and commented on "...the opportunity for considerable gain from turpentine, tar, pitch, resin and turpentine-oil." In the northern tier of North Carolina counties, as mentioned above, some 20 mature trees remain in Gates County, only 2 trees are known in Hertford County, and a single tree in Perquimans County. The last stand of longleaf in Northampton County was logged about 1980, and longleaf pine has also been extirpated from Currituck, Pasquotank, Washington, and Tyrrell counties.

Fernow (1893) observed that "in North Carolina, in the division of mixed growth and in the plain between the Albemarle and Pamlico Sound, the long-leaf pine has likewise been almost entirely removed and is replaced with the loblolly." In the central part of the state, there was considerable turpentining activity along the Tar River in the central Coastal Plain by 1732, and by 1850 the state was the world's leading supplier of naval stores (U.S. Censuses of Agricultural and of Manufactures 1841, 1853, 1864, 1872, 1883, 1895). Agriculturalists complained that the entire labor force of the Coastal Plain was employed in the turpentine orchards, to the neglect of agriculture

FIGURE 6. Boxing trees for turpentine. Bark and cambium were removed and large boxes were chopped into the base to collect the crude gum. Photo courtesy of U.S. National Archives.

by Merrens (1964). Gamble (1921), Croker (1987), and Earley (2004) have reviewed the history of naval stores for the rest of the South.

Few mature trees escaped the turpentine boxing procedure. Large trees were boxed on three or even four sides (Schoepf 1788), with deep wedges cut into the base to collect the resin (Fig. 6). Crude gum was dipped from the box six to eight times a season and transported by cart or boat to the nearest still (Figs. 7 to 9). Casks of distilled spirits of turpentine and barrels of rosin, the residue after distillation, then were shipped downstream to the nearest port (Fig. 10). Using nineteenth-century methods, virgin stands often produced for only about 4 years (Mohr 1896). Weakened trees in abandoned turpentine orchards often were blown over or killed when the

(Ruffin 1861). By 1900 longleaf had been decimated in North Carolina and the industry had passed on to the south. Ashe (1894b) commented: "In North Carolina most of the trees which now bear seed are boxed and have been in this condition for 50–100 years...."

Introduction of the copper still in 1834 allowed concentration of the final product into distilled "spirits of turpentine" making the process highly efficient, slashing shipping costs, and touching off a wave of commercial exploitation which swept south from North Carolina to Texas decade by decade, decimating the longleaf pine region within 80 years (Mohr 1896). Sargent's state maps (1884) for Louisiana and Texas show the extent of turpentine orcharding being carried into the virgin pine forests. The history of naval stores in North Carolina has been reviewed

FIGURE 7. Gum was collected every few weeks by dipping with large spoons. Barrels were crafted locally from white oak. Photo courtesy of U.S. National Archives.

2. History and Future of the Longleaf Pine Ecosystem 23

FIGURE 8. Barrels of crude gum were taken by boat or wagon to the nearest still. Photo courtesy of Forest History Society.

next ground fire set the residue ablaze in the boxes (Fig. 11). Much of the virgin timber thus was wasted until around 1870, when narrow-gauge logging railroads were extended into upland forests. As forests of each state were exhausted the industry moved south and by 1890 foresters raised the alarm that without provision for reforestation the turpentine industry would soon come to an end (Ashe 1894b).

Thomas Gamble (1921, p. 35) summarized the wave of turpentining that decimated the virgin longleaf forests:

The exhaustion of the South Carolina pine forests so far as heavy supplies of naval stores were concerned, was astoundingly rapid. Such a thing as conservation was undreamed of. The vast forests of Georgia and Alabama and Florida were too inviting to promote the thought of care in the use of what remained of the Carolina pine forests that had

FIGURE 9. Introduction of the copper still into the woods in 1834 permitted reduction of crude gum to spirits of turpentine, saving shipping costs and making the process immensely more PROFITABLE. Photo courtesy of U.S. National Archives.

FIGURE 10. The rosin yards at Savannah, GA, in 1893. Every 50-gallon barrel of distilled turpentine contained the entire life's production of 33 virgin longleaf pine trees, with a by-product of 4 barrels of rosin. Net profit per tree was about 32 cents (Mohr 1893). Photo courtesy of U.S. National Archives.

evoked the admiration of the early discoverers and explorers. No section of the primeval longleaf pine forests was more quickly or more effectively obliterated than that through which the "Tar Heelers" pressed on their way from North Carolina to Georgia. A very few years and they had cut their last boxes, hacked their last trees, gathered their last crops of crude gum, and, like an army of locusts leaving a Kansas wheat farm, moved on to fields new and pastures green.

Mohr (1896) described the situation in most of the South by 1896: "... the forests invaded by turpentine orcharding present, in five or six years after they have been abandoned, a picture of ruin and desolation painful to behold, and in view of the destruction of the seedlings and the younger growth all hope of the reforestation of these magnificent forests is excluded." This grim prediction was largely fulfilled when the last of the virgin forests were depleted in the 1920s.

The Spread of Agriculture in the Longleaf Pine Region

Indians were the first farmers, and the full extent of Indian agriculture in the South has

FIGURE 11. This virgin longleaf stand in Beaufort County, SC, had been boxed for turpentine. Fires further weakened the trees by setting the boxes ablaze and in coastal areas, hurricanes often finished the job. Photo, Sherrard 1903.

never been delimited. Bartram (1791) described "tallahassees" or abandoned Indian old fields in north Florida. To the north, the hunter-gatherer cultures of North Carolina and Virginia farmed on a smaller scale in patches adjacent to villages, while much of the diet came from fishing and hunting (Harriott 1590 [1972]; Smith 1624). In the Creek country of Alabama, however, Bartram traversed a region of Indian farmland broken only by small tracts of woods between the outlying agricultural lands of one village and the next (Bartram 1791 [1955]). Clearly a portion of the longleaf pine region had already been domesticated long before arrival of the first Europeans.

Along the Atlantic slope, settlers finally began expanding out of the tidewater region in the 1730s (Clay et al. 1975) and, with later waves of immigrants, settled the Piedmont, reaching the foothills of the Appalachians by the 1790s (Fig. 4). During the period 1750–1850 virtually all longleaf communities of the more fertile soils were converted to farmland and pasture (Williams 1989). Both the American Revolution and the Civil War interrupted agriculture for a number of years and in 1795 it was reported that "all Tidewater Virginia was full of 'old fields' reverting to timber" (Wertenberger 1922).

The longleaf pine region was fully settled by 1750 with the exception of Florida, Texas, and the interiors of Alabama and Mississippi (Fig. 12). As late as 1820 the vast longleaf forests of the interior of Alabama, Mississippi, Louisiana, and east Texas remained untouched. In 1821, however, cession of Florida to the United States by Spain, along with major land purchases from the Creek and Choctaw Indians, opened this region to settlement. By 1850 the fertile Black Belt region of central Alabama and Mississippi had been plowed and converted to cotton plantations by large slave-holding planters. A map compiled from the Census of 1840 (Williams 1980) shows the distribution of major cotton plantations in three dense regions: coastal South Carolina and Georgia, the lower Mississippi River valley, and the Black Belt.

By the Civil War, nearly all lands optimally suitable for agriculture were in production. By

FIGURE 12. Virgin longleaf stands of the interior hills of the Piedmont and southern tip of the Appalachians were nearly as open as those of the Coastal Plain. Fresh boxes had just been chopped into the bases of these trees for the turpentine process, which had just reached the hills in 1905. Bibb or Coosa Co., Alabama. Photo, Reed 1905.

1900, 12.5 million hectares, or about 27% of the uplands in the former range of longleaf pine upland was listed as "improved" farmland, a category that included pasture, roads and buildings as well as cropland (U.S. Census Office 1902). While there were no separate figures for land in pasture in 1900, it was necessary to maintain pasture or range on every farm for horses, mules, and oxen used for plowing and transportation, and until around 1880 much livestock was still maintained on open range in the woods.

History of Logging: Hand Power, Waterpower, and Steam

Effects of timbering were minor through the early Colonial Period (beginning in 1607 in Virginia, 1565 in Florida) to the mid-1730s, when logging was done by hand, using horses, mules, and oxen to drag the logs. Commercial logging was limited to the vicinity of streams where the harvest could be transported. While waterpower was tried as early as 1611 in Virginia, this technology did not take hold until around a century later, with introduction of water-powered sawmills in Louisiana about 1714 (Hindle 1975) and the Cape Fear region of North Carolina in the 1730s. In 1732, Governor Burrington reported that an abundance of sawmills was being constructed along the Cape Fear River. In 1764 Governor Dobbs reported that 40 sawmills had been completed on branches of the Cape Fear, and Governor Tryon reported that the number had risen to 50 two years later in 1766 (Merrens 1964).

Waterpower opened up the first possibility of a commercial lumber industry. Steel saw blades were imported from Holland where the technology had been developed, and sawmills proliferated rapidly along streams in settled areas. Still, these were slow, straight-bladed reciprocating saws (slash saws), with an up and down action, mimicking the human-powered

FIGURE 13. A "carry-log" drawn by mules. Economical range of this kind of transport was less than 4 miles (Croker 1987). Photo courtesy of U.S. National Archives.

pit saws: the circular saw and band saw were still 100 years away, not coming into general use until after the Civil War (Hindle 1975). Many of these small mills operated only part time—when there was enough water in the mill pond in winter and spring to turn the wheel. Many were plantation-owned, producing boards for local use, with only a small surplus shipped downstream to coastal towns. As late as 1826, a few decades before the appearance of steam-powered sawmills, Mills (1826) commented that the pine timber was still used mostly for local construction.

While waterpower helped the clapboard house replace the log cabin, lumber production remained a minor industry from 1730 to around 1850. Most logging occurred along streams where logs were skidded out by horses, mules, and oxen. The giant wheeled "carry-log" (or "caralog," Fig. 13) was important from this time until the late nineteenth century when it was supplanted by logging railroads and steam skidders. Logs were dragged this way to the nearest water and then rafted downstream to mills. The maximum effective distance for this kind of overland transport was only 3 or 4 miles (Croker 1987) and so commercial exploitation was limited to narrow zones along navigable streams.

Prosperous South Carolinians were fascinated by steam power and in 1833 constructed the first railroad in the United States, connecting Charleston on the coast to the vicinity of Augusta on the Savannah River. The entire route lay through longleaf pine country, and on some of the first runs the engine slowed to a crawl from lack of steam and had to stop while hands ran to chop longleaf pine lightwood for fuel (Derrick 1930). In 1856, the first steam-powered dredges were used in Norfolk County, VA, to build the Albemarle and Chesapeake Canal (Ruffin 1861), and the period 1850–1870 saw explosive proliferation of steam technology for logging railroads, steam skidders, and steam-powered sawmills (Anon. 1907). By the end of the Civil War, with resumption of intensive turpentining throughout the longleaf forests of North and South Carolina, and with steam logging methods perfected, the stage was set for cataclysmic decimation of the longleaf ecosystem.

After the war, huge tracts of southern lands were bought by railroad companies (Fig. 14). After construction the railroads sold surplus

FIGURE 14. Clearing right-of-way through virgin longleaf forest in Mississippi for the Natchez, Columbia and Mobile Railroad in 1907. All timber was soon cut within several miles of railroads and more distant lands were sold to logging companies. Photo, American Lumberman 1907.

lands to logging companies. Lands sometimes changed hands at the rate of 40,000 hectares or more, at prices of $3 per hectare (Napier 1985). The decade 1880 to 1890 saw standardization of track sizes and concatenation of isolated railroad lines, making overland transport of lumber cheap and efficient (Hale 1883; Anon. 1907). By 1880, all commercial timber had been removed from lands within a few miles of streams and railroads. Tapping of virgin forests of the interior had just begun, but huge volumes of lumber were being produced. Sargent reported an annual cut of over a billion board feet in 1884 (Table 2), increasing to 3.7 billion

TABLE 2. Virgin longleaf pine remaining in 1880 and annual cut in 1880 (board feet)[a]

	Merchantable longleaf pine	Annual cut
Virginia	No reported commercial production	
North Carolina	5,229,000,000	108,411,000
South Carolina	5,316,000,000	124,492,000
Georgia	16,778,000,000	272,743,000
Florida	6,615,000,000	208,054,000
Alabama	18,885,000,000	245,396,000
Mississippi	18,200,000,000	108,000,000
Louisiana	26,588,000,000	61,882,000
Texas	20,508,000,000	66,450,000
Totals	118,119,000,000	1,194,428,000

[a] Figures are only for major longleaf pine regions and major logging companies. While virgin growth had been depleted in Virginia and exhaustion in the Carolinas was imminent, stands in Louisiana and Texas still were largely untouched (Sargent 1884).

board feet by 1896 (Mohr 1896). This phase of intensive logging, from 1870 to 1930, saw removal of virtually all remaining virgin forests in the South. By 1900, it was apparent that many cutover longleaf areas, particularly those on better soils, were being occupied by scrubby second growth of other species, while some remained open and nearly treeless. To the growing concern of foresters, longleaf pine replaced itself only sporadically in a small percentage of its former landscape (Mohr 1896). Given the vast extent of longleaf once reproducing naturally in primeval forests, what could explain its failure to do so now?

The Disappearance of Longleaf Pine

Failure of Longleaf Pine Regeneration after Logging

Historical records suggest that two factors combined to explain the final disappearance of longleaf pine after initial exploitation for turpentine and lumber. First was the fondness of feral livestock, especially hogs, for the seedlings (Mohr 1896; Hopkins 1947a,b,c). Unlike other pines, longleaf seedlings have a non-resinous, carbohydrate-rich meristem, which, while in the grass stage, is vulnerable to grazing for 5 to 7 years or more. Hogs have been observed to feed heavily on longleaf seedlings, consuming up to 400 each in a day (Hopkins 1947a,c). The second and final nail in the coffin was twentieth-century fire suppression.

By the 1890s foresters saw clearly that, over large expanses of the landscape, longleaf was not replacing itself after logging (Ashe 1894a,b; Mohr 1884, 1896). On the road on the ridge between the Cooper and Ashley rivers out of Charleston, Edmund Ruffin observed changes in the forest, on lands long settled:

The trees are nearly all pine, & generally of second growth, the land having been formerly cultivated & afterwards turned out.

The pines of original forest are mostly of the 'long leaf' species, & many of the great size & beauty for which that kind is distinguished. But whenever of second growth, whether after culture, after mere cutting down the first growth for fuel, the second growth pines are of the "loblolly" or "old-field" kind, of mean sized appearance. (Ruffin 1843, p. 60)

Mohr (p. 64) commented, "on the lowlands of the Atlantic coast toward its northern limit this pine is almost invariably replaced by the Loblolly Pine." "In the stronger soil of the upper division of the maritime pine belt, the region of mixed growth, where seedlings of the Longleaf Pine spring up simultaneously with the hard wood trees and the seedlings of the Shortleaf Pine, these latter will eventually gain the supremacy and suppress those of the Longleaf Pine." "It is evident that the offspring of the Longleaf Pine is rarely seen to occupy the place of the parent tree, even in the region most favorable to its natural renewal, and that final extinction of the forests of the Longleaf Pine is inevitable unless proper forest management is applied." To Mohr's mind proper management meant eliminating all fire, encouraging 15 to 20 years later, shade-tolerant tree species below the longleaf to build up a humus layer "to secure improvement and permanency of favorable soil conditions." These sentiments were echoed by Sherrard (1903). Unfortunately, this was a prescription for extirpation of longleaf pine.

The question that dogged foresters was, why did longleaf not reproduce, at least on those lands where nothing else was done other than logging of the virgin timber? Contemporary with Mohr, one of the first foresters to wrestle with this problem was W. W. Ashe, who noted that not only was the longleaf seed crop produced in irregular mast years, but also that the seeds were descended upon by a variety of predators: "... its large and sweet seeds are eaten in large quantities by fowls of various kinds, rats, squirrels, and by swine, which prefer them to all other kinds of mast, and when there is enough long leaf pine mast become very fat on it" (Ashe 1894b, p. 57). This had been noticed as early as 1728 by William Byrd during the survey of the Virginia–North Carolina line, and Ruffin (1861) commented that "[t]hey are so eagerly sought for by hogs that scarcely any are left on the ground to germinate" (p. 255). Ashe was one of the first to report the fondness of hogs for the larger seedlings. "No sooner, however, has the young pine gotten a foot high and its root an inch in diameter than the hog attacks it, this time eating out the roots, which until two inches in diameter, are very tender and juicy, pleasantly flavored and free of resinous matter" (Ashe 1894b, p. 57).

Like most foresters of his time, Ashe regarded fire as the unrelenting enemy of forest regeneration, even going so far as to insist that in North Carolina "... the burnings of the present and future, if not soon discontinued, will mean the final extinction of the long leaf pine in this state" (Ashe 1894b). This opinion echoed that of Mohr (1884) and others on the destructive nature of fire. The groundwork for the field of fire ecology had clearly not yet been laid.

Ashe concluded that the chief agencies preventing regrowth of longleaf pine were fire and hogs. In contrast, later authors asserted the actual dependence of the species upon fire to prevent site appropriation by shade-tolerant pines and hardwoods (Harper 1913). When some of the early assertions were tested in southeastern South Carolina, longleaf pine was found to be replaced by slash pine when both fire and hogs were excluded (Sherrard 1903), and studies in 1935 showed only 8% fire mortality in 2-year-old longleaf plantations in Louisiana, versus 53% for 7-year-old loblolly (Wakeley 1935). If fire is excused as one of the two principle culprits, that leaves hogs conspicuously in need of closer scrutiny.

In 1539, DeSoto made the first introduction of swine to the South (Bakeless 1961). Later, English settlements brought with them starter livestock (Strachey 1610 [1964]; Smith 1624). Hogs showed an astounding reproductive potential, and demonstrated an ability to fend entirely for themselves in the woods with no attention from their owners (Beverley 1705 [1947]; Blakeley 1812 [1910]). The capacity of the landscape to support open range hogs has never been investigated, but the evidence suggests that they quickly reached saturation density within a few decades after settlement. By 1617 the log palisades with which the town was walled off were not sufficient to keep the hogs out of the streets of Jamestown, Virginia. Capt. Samuel Argall and James Rolfe on landing there in May of that year, only 10 years after settlement, commented on the "innumerable numbers of swine" (Smith 1624).

Evidence for Early Saturation of the Landscape by Hogs in Coastal Regions

Both the Spanish and English experiences demonstrated the potential of hogs to increase from a handful to thousands in a few years under conditions of complete neglect on open range. By 1702 a Swiss visitor to coastal Virginia declared that "pigs are found there in such numbers that I was astonished" (Michel 1702 [1916]). This was corroborated by Beverley (1705 [1947]) who stated that "hogs swarm like Vermine upon the Earth.... The Hogs run where they list, and find their own Support in the Woods, without any Care of the Owner; and in many Plantations it is well, if the Proprietor can find and catch the Pigs, or any part of a Farrow when they are young, to mark them...." A few years later, Brickell (1737 [1968]) reported similar conditions in northeastern North Carolina where he saw "... swine, breeding in vast numbers...."

A considerable meat packing business had sprung up in Norfolk, VA, the major seaport in the mid-Atlantic region, to supply salt pork to sailing ships. The first direct evidence that hogs had reached saturation density in North Carolina is provided by the report of Governor Barrington in 1733, that about 50,000 hogs were driven annually to the Norfolk market from the Albemarle region of North Carolina (Wertenberger 1931). The first livestock census figures from these six small counties showed no increase in hog numbers from 1840 to the Civil War, indicating that saturation density had been reached, with an average of 14,800 hogs on open range in each of the six counties south of the state line within hog driving range of Norfolk. This gives an average of 4.3 hectares per hog (U.S. Census 1841). For the 1890 census only, supplementary figures were kept for hogs consumed or hogs that died. In Alabama, which still had hogs on open range, an annual number equal to 45% of the total hogs alive were consumed and 23% died. This gives us an approximation for surplus hogs that could be harvested when populations were near capacity (U.S. Census Office 1902: U.S. Census of Agriculture for 1902). If the total number of hogs in the six North Carolina counties mentioned above were at carrying capacity in 1750, the numbers should be nearly the same as in 1840 (88,850 hogs), then the surplus should have been 45%, or 40,000 hogs. The fact that the reported surplus of 50,000 fully grown hogs driven to Virginia exceeds our estimate of 40,000 strongly suggests that carrying capacity had been reached in this region sometime before 1733. These counties were settled between the years of 1655 and 1700 so there had been from 35 to 78 years, easily sufficient for hogs to reach saturation density.

While hogs spread inland from southeastern Virginia and northeastern North Carolina, other introductions were made along the Atlantic and Gulf coasts. In 1663, explorers stepping ashore on the barrier island at Cape Fear, NC, were astonished at being offered pork for sale by the Indians, livestock having been placed on the islands a few years earlier by stockmen from New England (Lawson 1709 [1967]). On the Gulf Coast, Mobile, founded in 1711, was the first permanent city (Hamilton 1910 [1976]), and in 1812, free-ranging hogs were kept on three islands of about 1600 hectares each at the head of Mobile Bay. Josiah Blakeley, the owner, wrote that "[c]attle and hogs do well upon them, and no expense. Upon them I have about 30 head of cattle and hundreds of hogs, the hogs wild. I shoot or catch them with a dog" (Blakeley 1812 [1910], p. 405). There is no evidence, however, that hogs spread very far beyond the frontier, where Indians and other predators likely kept them under control.

From the descriptions above, it seems likely that tidewater Virginia was saturated with hogs by around 1700, and the whole coastal plain of Virginia and the portion of North Carolina north of Albemarle Sound by 1730. The first

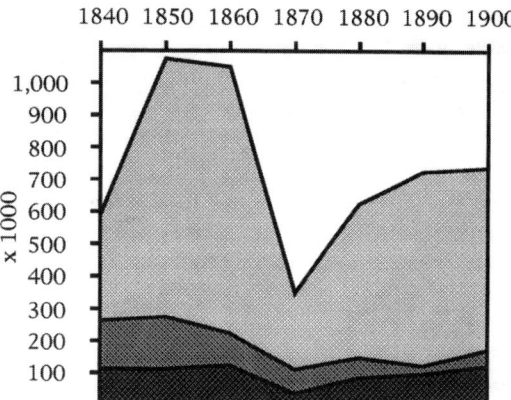

FIGURE 15. Evidence for saturation of the landscape by feral hogs. The lower curves represent stable hog populations in coastal regions long-settled by 1840—more than 200 years for coastal Virginia (bottom line) and over 100 years for coastal Alabama (middle line). The vast regions in central Alabama, only opened to settlement in 1821, had just reached carrying capacity in 1850, with over a million hogs on open range. The dip from 1860 to 1870 was the result of overconsumption of all livestock during the starvation that accompanied the Civil War. Data from U.S. Censuses of Agriculture, 1840–1890. Reprinted from Frost 1993 with permission from the Tall Timbers Research Station.

regularly kept figures, however, were not available until a century later with the 1840 Census of Agriculture. The lower line in Fig. 15 shows the total number of hogs from the 15 Virginia counties within the original range of longleaf pine from 1840 to 1900. The plunge in numbers occasioned by famine during the Civil War is characteristic of all the southern states and is closely paralleled by figures for cattle and other livestock (U.S. Census Office 1841, 1853, 1864, 1872, 1883, 1895, 1902). Note that the population curve for the decades preceding the Civil War is flat, and recovers to a relatively flat slope within two or three decades afterward. This supports the notion that carrying capacity had been reached some time before such records were kept.

In contrast, figures for Alabama indicate that only the coastal region was saturated by 1840. The middle line in Fig. 15, which parallels that for Virginia, represents the number of hogs in the seven old, long-settled coastal counties near Mobile. The upper line represents the middle counties. The interior remained Spanish territory until 1821, when settlers from Georgia and the coast were poised for entry (see Fig. 4). By 1840, only 19 years after opening of the territory, immigration was in full swing but the country was still sparsely settled. Figure 15 shows increasing numbers of hogs in the central counties, but leveling off after 1850, within 19 years of the 1821 opening of the land to settlement (the large numbers for the central counties reflect the much greater land area). The flattening of the curve again suggests carrying capacity had been reached, and demonstrates the capacity of hogs to saturate a vast landscape in less than 20 years.

Hogs were not the only competitor for forage on open range, however. While hogs were consistently the most abundant livestock species reported in the agricultural censuses, the range was shared by cattle, horses, mules, sheep, and goats, whose numbers collectively equaled those of hogs (U.S. Censuses 1841, 1853, 1864, 1872, 1883, 1895, 1902).

One writer estimated that 5–10 hectares of unmanaged southern woodland was required to support one cow, while 0.8 hectare of good pasture would suffice (Gardner 1979). No figures were ever determined for carrying capacity of southern range for hogs (Grelen 1980). As noted above, the apparent saturation density of hogs in 1840 in northeastern North Carolina was 4.3 hectares. While this might seem an abundance of land per hog, keep in mind that there was an equal number of other open range livestock competing for food, the county acreage included water, areas from which hogs were fenced out, and large areas of upland forests where there may have been little forage except for the fall mast crop of acorns and pine seeds. There was also stiff competition for the mast crop from birds and native animals (Ashe 1894b; Wahlenburg 1946). Longleaf pine seedlings, on the other hand, were available and vulnerable all year round.

While birds have been observed to consume from 8% to up to 42% of the longleaf seed crop (Wahlenburg 1946), they do not molest the seedlings, and this much predation must have been tolerable, since birds were a natural part of the landscape in which longleaf pine flourished. Wakeley (1954, p. 151) considered hogs by far the most serious threat to longleaf: "Where there are many hogs it is foolhardy to plant longleaf pine without fencing.... To this species hogs are infinitely more destructive than fire."

There are several hog-and-fire exclusion studies to back up this assertion, two of which reported complete failure of stand regeneration on tracts where feral hogs were present. Two experimental tracts at Urania, LA, after 5 years of protection against hogs, contained an average of 16000 longleaf saplings per ha, as compared with an average of only 20 per ha on two unprotected tracts (Mattoon 1922). In an area with free-ranging hogs in Georgetown County, SC, hogs were fenced out of 32 0.04 ha plots. After two growing seasons the fenced areas contained 1200 large seedlings (those with root collar diameters of 1.3 cm or larger) per ha, while unfenced areas contained only 20 per ha (Lipscomb 1989). The hogs largely ignored small first-year seedlings but focused on those large enough to have accumulated starchy root content. Density of hogs was not controlled but was estimated to be about three to six animals on the 24-hectare study area, or 4–8 hectares per hog. This is comparable to the hog densities of 4.3 hectares per hog reported above, on open range in colonial North Carolina, which we have suggested may represent carrying capacity.

Ashe (1894b) and Mohr (1896) both commented on the palatability of longleaf pine seedling roots in the 1.5 to 5 centimeter diameter range. Wakeley (1954) reported hog consumption of 200 to 1000 longleaf seedlings per day, at rates of up to 6 per minute. Hopkins (1947a,b,c), after observing hogs rooting up hundreds of seedlings a day, analyzed the root starch content and found them to be as nutritious as corn. Little wonder then that hogs would be drawn to longleaf seedlings, which, in the grass stage, are highly conspicuous and vulnerable for 3 to 7 years. With 10,000 to 40,000 hogs on open range in every settled county in the longleaf region (U.S. Censuses 1841, 1853, 1864, 1872, 1883, 1895, 1902), all that would be required to eliminate reproduction would be for a drove of hogs to happen upon a regenerating plot once every 3 or 4 years to largely eliminate the species from the landscape.

Hogs on open range were completely dependent on natural forage. If carrying capacity had been reached, survival would be tenuous and occasional disasters could be expected when mast crops or other wild foods failed. A curious example occurred in Illinois when hogs starved in winter after passenger pigeons unexpectedly descended on a local area and ate all the fall mast of acorns, beechnuts, and chestnuts (Bakeless 1961). This raises the question about the reverse situation, that saturation of the landscape with hogs contributed to the extinction of the passenger pigeon. Their summer breeding range extended only as far south as Virginia but from late September to early November the flocks migrated to the winter range from South Carolina to Florida (Bent 1932 [1963]). This coincided with longleaf seed fall, and it has been observed that related birds like mourning doves and quail have their crops "crammed" with longleaf seeds during this time (Wahlenburg 1946). The distinct parallel between the decline of longleaf pine, a major winter food source, and that of the passenger pigeon may not be a coincidence. In the South, memory of the species persists only in place names like "Passager Swamp" in Isle of Wight County, VA.[4]

The End of Open Range

The effects of hogs on longleaf pine were not noticed until the massive wave of logging that followed the Civil War physically removed the forest. Most of the timber cut in the period 1870–1900 was still virgin forest (Mohr 1896), where the effects of hogs in eliminating seedlings could be overlooked as long as the trees stood. Note that longleaf had indeed been extirpated from much of the northern range a hundred years before, but the process had taken 200 years, while decimation of the forest using steam-logging technology seemed to occur overnight. This precipitated an immediate shortage of lumber for fencing (Hale 1883), and forced landowners to look at the problem of livestock on open range. For the first three centuries, crops had been fenced in to protect them from livestock, which had free run of the land. Even if a farmer had little stock of his own, he had no choice but to fence his crops against the animals of his neighbors. As more land came into agriculture, demands for fencing increased until the timber shortage made it apparent that it would more economical to fence in the livestock rather than the crops.

In response, fence laws (stock laws) were passed throughout the South, beginning in the 1870s. In 1883 a statewide law was passed in South Carolina making it incumbent upon the owners of livestock to see that they do not trespass on the lands of others. A respondent to an 1880 timber survey, from Anson County, NC, commented that "every man who owns cattle, hogs, sheep, goats or horses in Anson County is now compelled to pasture them on his own land. None are allowed to run at large on the range. This system came into effect in our county about two years ago, and so much is it esteemed already that a return to the old style of fencing the crops against the incursions of stock is next to impossible. This is regarded as the most important single step taken in this county in the last twenty years" (Hale 1883). The process took decades to become effective over the whole South and there are still some areas where hogs run wild (Lipscomb 1989).

Landscape Changes from 1900 to 2000

Fire Suppression and the Decline of Fire as a Natural Determinant of Vegetation

The end of open range should have been a boon to longleaf pine, but while three centuries of open range were drawing to a close, a new threat was in the making. Fire was still widespread, but by the Civil War, much of the landscape had been fragmented by agriculture, reducing the size of fire compartments. In central South Carolina there were an average of 20.3 hectares per farm cleared and tilled (State Board of Agriculture 1883). As long as raising stock was the primary source of income the remaining woodlands were burned by the residents to green up forage for livestock. This practice may have perpetuated longleaf pine and its associated flora of wiregrass and savanna herbs, in a landscape where roads, plowed fields, and other man-made firebreaks fragmented the fire landscape and eliminated landscape-scale fires ignited by lightning. When cattle grazing declined in importance after the Civil War, the practice of spring burning was abandoned in major agricultural areas. Describing the resultant vegetation changes in South Carolina, one writer noted that "the uplands were covered, as they still are, with a large growth of yellow pine, but a deer might then have been seen, in the vistas made by their smooth stems, a distance of half a mile, where now, since the discontinuance of the spring and autumn fires, it could not be seen fifteen paces, for the thick growth of oak and hickory that has taken the land" (State Board of Agriculture 1883, p. 79).

On all but the drier lands, longleaf reproduction is completely eliminated by other pines and hardwood, and shrub invasion within a few years after fire exclusion (Sherrard 1903). Nowhere in the South can longleaf be seen reinvading the mesophytic mixed pine–hardwood succession that has replaced it.

Modern fire laws and the state apparatus for prevention and suppression of wildfire did not come into being in most of the South until the period 1910–1930. This left a window of some 50 years, between the end of open range around 1880 and the beginning of twentieth-century fire suppression, in which longleaf pine had a safe opportunity to reproduce. Many of the stands that did result have now been logged and the oldest of those naturally regenerated stands still remaining, date to the end of this window of opportunity.

Fernow (1893) was one of the first to argue for governmental involvement in forestry: "there exist some legislative provisions regarding forest fires in almost every State, but they are rarely if ever carried into execution for lack of proper machinery." Most states remedied this condition with a vengeance in the next 30 years. In 1919, Virginia passed laws creating the position of State Forester and provided for forest wardens. The act also imposed fines and a minimum penalty of a year in prison for maliciously starting a forest fire, a far cry from the days when burning was a casual management practice.

Few of the early foresters cared to acknowledge the role of lightning as an ignition source. In South Carolina, Sherrard (1903) blamed all fires on humans, stating that fires were "carelessly set to improve grazing, to clear land, and to protect woods where turpentine is being gathered." Burning in this case was done after first raking pine straw away from the flammable boxes in the bases of the trees. Ashe even believed that one of the reasons longleaf pine was being replaced by loblolly was that it was more sensitive to fire:

The loblolly pine is less injured by fire because its bark is thicker and so offers more protection to the growing wood, –the bark, too, lying closer to the wood in firmly appressed layers, does not so easily take fire.

The chief agencies, then, which prevent a regrowth of long leaf pine on the high sandy lands, are the hogs and the fires...the burnings of the present and future, if not soon discontinued, will mean the final extinction of the long leaf pine in this State. (Ashe 1894, p. 58, writing about North Carolina)

In contrast, Sherrard observed that "the Longleaf Pine may rightly be called a fireproof species in so far as the survival of scattered groups and patches of second growth and individuals is concerned." Still, he was one of the first to call for a public campaign: "the people must be educated to a sentiment against fires."

The first voice to clearly distinguish the natural role of fire was Roland Harper, who stated, "it can be safely asserted that there is not and never has been a long-leaf pine forest in the United States...which did not show evidences of fire, such as charred bark near the bases of the trees; and furthermore, if it were possible to prevent forest fires absolutely the long-leaf

FIGURE 16. The first documented study showing the effects of exclusion of fire and hogs from longleaf pine. A dense forest of slash pine is regenerating in a fenced plot after exclusion of fire and hogs for several years. Old boxed longleaf survivors and scattered slash pine make up the canopy (Sherrard 1903). Sherrard aspired to produce a similar forest on all pine lands in the two counties being studied and fire exclusion became the general forest prescription for the South. Southeastern South Carolina. Photo, Sherrard 1903.

pine—our most useful tree—would soon become extinct" (Harper 1913, p. 16).

If not recognized by early foresters, it was well known to inhabitants of the longleaf pine region as early as the 1830s that lightning was often responsible for fires in the "turpentine orchards." On a large estate in Onslow County, NC, damage to the turpentine crop was prevented by providing log cabins free of rent to poor white families, whose duties included fighting summer lightning fires:

These men are required to do three things: first, they are to guard the orchards from fire, and if a small fire occur, as it often does in the summer time by lightning striking and igniting a resinous pine tree, they and their families must extinguish it. If it gets beyond their control they are to blow horns, summoning the neighboring tenants, sending all around for help, fight the fire until it is put out ... (Gamble 1921, p. 27)

The slow and patchy reproduction characteristic of unmanaged longleaf under conditions of frequent growing season fires was a legitimate concern, and foresters were hungry for solutions. While most were convinced that both hogs and fire were inimical to longleaf regeneration, the first real demonstration was conducted in 1903. Sherrard (1903) examined a fenced plot from which fire and hogs were excluded. Within a few years a dense stand of slash pine had established itself beneath the longleaf (Fig. 16). Sherrard was pleased with the result. Never mind that the new forest would be composed of a new dominant species and of entirely different structure than the open longleaf forests. And curiously, neither he nor Ashe nor Mohr ever questioned that if fire were the enemy of longleaf, why did its exclusion lead to an entirely different forest type? While it must have been apparent that this kind of succession would eventually lead to replacement of longleaf, it was sufficiently good news in a landscape recently denuded of its primeval forest cover, that within a few years, fire exclusion and a program of educating the public "to a sentiment against fires" became the general forest prescription for the South.

TABLE 3. The first pine plantations: 1892–1931[a]

	1892–1928	1928	1929	1930	1931	Total
Virginia	136	19	141	128	162	587
North Carolina	618	124	220	109	190	1,261
South Carolina	1,308	—	45	195	301	1,850
Georgia	608	2	324	1,030	62	2,026
Florida	391	0	14	595	756	1,756
Alabama	36	20	133	108	14	311
Mississippi	—	—	—	217	241	457
Louisiana	7,914	3,756	4,286	4,298	1,002	20,018
Texas	—	—	—	105	—	105
						28,371 ha

[a] Figures are from Wakeley (1935), with exception of the area planted before 1928 in South Carolina, from Boyce (1979).

Pine Plantation

Pine plantations scarcely existed in 1900. The earliest plantations of record in the South were three small plots established by farmers in 1892, 1896, and 1907 (Wakeley 1935). The first large attempt at plantations by the U.S. Forest Service, 365 hectares on the Choctawhatchee and Ocala National Forests in 1911, proved a failure. Wakeley knew of only 200 hectares successfully established by 1919. Problems with technique were soon worked out, however, and Table 3 shows the extent of pine plantation in the nine states within the range of longleaf pine by 1931.[5] By this time more than 20 lumber and paper companies were involved and they accounted for at least 78% of the area planted.

Fire was a threat to pine plantations, but establishment of increasingly large areas protected from fire in the 1930s and 1940s made it seem feasible to plant loblolly and slash pine as commercial crops. Pine planting was expanded by large timber corporations in the 1940s and 1950s, and there were 12,460,000 acres (5,046,300 hectares) established in the years 1965 to 1967 (Boyce 1979). Forced into more marginal lands by development pressures, timber companies found it increasingly desirable to produce pine pulpwood and sawtimber using intensive management. In the former longleaf region, there are at present about 11 million hectares of pine plantations, primarily loblolly and slash pine, but also small amounts of shortleaf and longleaf (based on figures and projections in Boyce 1979; McWilliams 1987; Outcalt and Sheffield 1996).

Expansion of Agriculture and Developed Land

While much mixed pine–hardwood is now converted to plantation after logging, some is also cleared and converted to cropland or pasture. While commercial dairy operations have proliferated since 1900, total pasture and cropland have declined. After World War II, mules and horses were retired by tractors, and surplus pasture lands went into cropland or succeeded to loblolly pine and hardwoods (Boyce and Knight 1980). The relative percentages of land in cropland and forest are the net result of a complexity of changes that include forest succession of abandoned cropland on small uncompetitive farms between 1940 and 1965, and clearing of new cropland from woodland by large farming operations. Agricultural land area peaked in 1930 and has been reverting to forest and other land uses ever since (Williams 1989). The 1997 Census of Agriculture reported 3,456,000 hectares in pasture (7% of the uplands) and 6,027,000 hectares in cropland (12% of the uplands) in the portions of the 412 counties included in the former longleaf pine region (Table 1).

TABLE 4. Fire regime conditions in 785 stands of longleaf pine in the northern range of the species[a]

	CC1	CC2	CC3	CC4	CC5
Virginia	0	1	2	2	18
North Carolina	42	60	60	77	116
South Carolina	17	16	114	83	104
Northeast Georgia	12	4	17	17	23
Totals	71	81	193	179	261

[a] Condition Class 1 contains stands that had been burned often enough to have retained at least 70% of their original plant species diversity. Condition Class 2 indicates longleaf dominant stands with loss of more than 30% of species but that had been burned recently enough to retain conspicuous fire char on their trunks. CC3 stands were also dominated by longleaf but fire had been excluded long enough for the understory to fill in with dense woody vegetation: with no other treatment, these stands, when next logged for longleaf pine, will largely convert to hardwoods and early successional pines such as loblolly, with a few residual stems of longleaf. This successional process also accounts for most of the CC4 stands which consisted of scattered longleaf pines in fire-suppressed stands dominated by other species. CC5 were former stands in which fewer than 10 longleaf pines were found, in some cases consisting of a single ancient boundary line tree in a forest completely converted to other types.

Fire Regimes Today and the Condition of Remnant Longleaf Pine Communities

The few substantial, well-maintained remnants of longleaf pine communities are now found primarily on military bases whose managers have sufficient fire staff to maintain effective fire regimes. Smaller, fire-maintained examples can be found locally on national forests and other public lands and private preserves, and fire programs are now gearing up for restoration of natural stands suffering from various stages of fire regime alteration.

Over the 25-year period 1978–2003, I examined 785 stands of longleaf pine ranging from the northernmost remaining tree in Isle of Wight County, VA, to stands in northeast Georgia (Table 4). I evaluated each in terms of its departure from the natural fire regime. Stand investigation ranged in intensity from detailed 1/10-hectare study plots to 100-square-meter plots or quick visual evaluations.

By 2000, only 19% of remnant stands in the northern range of longleaf pine were being maintained with fire (Classes 1 and 2), and even this interpretation is optimistic. Only 9% of stands retained something approaching the full complement of plant species that they once supported under the natural fire regimes indicated in Fig. 3. The stands in Condition Class 2 had experienced some reduction in fire frequency and many were stands to which fire had been reintroduced after a long period of fire exclusion in the mid to late twentieth century. Of these, most had lost more than 50% of their understory species diversity. In most natural longleaf pine communities, more than 90% of the plant species diversity is found in the herb layer, as a rich assortment of native grasses and forbs. Most of the rare species are also found in this layer (Walker 1995). In the worst case, during the initial stages of reintroduction of fire, I saw several stands during this survey with not a single herbaceous species in a study plot.

Longleaf pine has been extirpated from all but about 2.2% of its original range (excluding recent plantations), or about 1,050,000 hectares. Of that fraction, only about 19%, or 193,000 hectares, is currently being maintained with fire, and only 9% has escaped significant loss of species diversity resulting from episodes of fire suppression. Fire-suppressed stands typically were invaded by hardwoods, loblolly pine, sweetgum, or slash pine. Instead of the two-layered structure typical of natural longleaf communities, there were heavy shrub and midstory layers. The resulting shade, along with deep pine needle litter and duff accumulation, had completely eliminated wiregrass and most of the rest of the herb layer on many sites. With only 9% of remnant stands in Condition Class 1, that means that less than 96,572 hectares, or less than 0.2% of the original extent of the longleaf pine ecosystem, remains in condition good enough to support most of its native plants and animals.

The logging boom of the late nineteenth century left in its wake cutover lands and dense, scrubby second growth, and efforts of crusading fire exclusionists guaranteed that over much of the region, the sunny, open, fire-maintained woodlands would be seen no more. For the inhabitants who lived during the first decades, seeing the forest of centuries fall around them was often a disheartening experience that transformed their world. One respondent to a timber survey in 1882 in Currituck County, NC, noted bitterly:

The avaricious and insatiable saw mills, together with the desire of every man who could buy a pair of oxen and "Carry-Log", have demolished and transported nearly all of our pine.... This certainly looks like a gloomy report, but more truth than poetry. (Hale 1883, p. 222)

The Future of the Longleaf Pine Ecosystem

If less than 0.2% of the original extent of longleaf pine remains in condition good enough to support a significant diversity of their native plants and animals, then the few areas that have been burned often enough to have retained their full complement of species are exceedingly valuable—as refugia for species, and as reference communities for setting restoration targets for the rest of the longleaf pine landscape. With so few remnants, we are now compelled to make every effort to get fire back into all remaining longleaf stands.

Encouraging signs for the future are now appearing on public lands. Remnant stands are being bought or protected, the dense thickets resulting from decades of fire exclusion are being subjected to midstory thinning, and fire is being restored to the land. The newest efforts include reintroduction of grasses and other herb layer species. Recent government actions mandate the determination of original fire regimes and Fire Regime Condition Class (FRCC)—the degree from which current fire regimes and stand conditions have departed from that in nature (Hann 2002). As FRCC is determined for lands across the country we will then have targets for restoration of the fire regimes that thousands of species rely upon. This gives us some cause to hope that 2000 represented the low point for the longleaf pine ecosystem.

Within the 2.2% (1.01 million hectares) of the landscape that still supports natural longleaf pine today, there is a remarkable galaxy of sites large and small, only one generation away from logging and turpentining, some of which have recovered nicely. These we may still be able to maintain, and perhaps we can restore more of Bartram's "... expansive, airy pine forests...of the great long-leaved pine...the earth covered with grass, interspersed with an infinite variety of herbaceous plants, and embellished with extensive savannas, always green...."

Endnotes

1. An earlier version of this chapter was first published in *Proceedings of the Tall Timbers Fire Ecology Conference*, No. 18, The Longleaf Pine Ecosystem: ecology, restoration and management, ed. Sharon Hermann, Tall Timbers Research Station, Tallahassee, FL, 1993.
2. Gold or other coin, "hard cash."
3. I am grateful to the staff of the Colonial Williamsburg Foundation Library, Williamsburg, VA, for access to the original records on microfilm.
4. The term "passenger" pigeon is a pejoration of the original word "passager," as used in Colonial times. Most people only saw them as birds of passage since the great flocks, except for a few weeks while nesting, were constantly on the move in search of food. As a consequence they were called "passager pigeons."

5. A small amount of planted trees were hardwood, something under 5%.

References

Anon. 1907. A trip through the varied and extensive operations of the John L. Roper Lumber Company. *Am Lumberman* [n.v.n.], April: 51–114.

Ashe, W.W. 1894a. The long leaf pine and its struggle for existence. *J Elisha Mitchell Soc* 11:1–16.

Ashe, W.W. 1894b. *The Forests, Forest Lands and Forest Products of Eastern North Carolina*. Raleigh, NC: Josephus Daniels, State Printer.

Bakeless, J. 1961. *The Eyes of Discovery*. New York: Dover Publications.

Barden, L.S. 1997. Historic prairies in the piedmont of North and South Carolina, USA. *Nat Areas J* 17:149–152.

Bartram, W. 1791 [1955]. *Travels Through North and South Carolina, Georgia, East and West Florida*. Reprint. New York: Dover Publications.

Bent, A.C. 1932 [1963]. *Life Histories of North American Gallinaceous Birds*. Reprint. New York: Dover Publications.

Beverley, R. 1705 [1947]. *The History and Present State of Virginia*. Chapel Hill: University of North Carolina Press.

Blakeley, J., dated February 28, 1812, at Mobile. In Hamilton, P.J. 1910 [1976]. *Colonial Mobile*. University: University of Alabama Press.

Boyce, S.G. 1979. Prospective ingrowth of southern pine beyond 1980. USDA Forest Service, Southeastern Forest Experiment Station Research Paper SE-200.

Boyce, S.G., and Knight, H.A. 1980. Prospective ingrowth of southern hardwoods beyond 1980. USDA Forest Service, Southeastern Forest Experiment Station, Research Paper SE-203.

Brickell, J. 1737 [1968]. *The Natural History of North Carolina*. Reprint. Murfreesboro, NC: Johnson Publishing Company.

British Public Records, Colonial Office. 1607–1783. Williamsburg, VA: Original records on microfilm at Colonial Williamsburg Foundation Library.

Byrd, W. 1728 [1967]. *Histories of the Dividing Line Betwixt Virginia and North Carolina*. Reprint. New York: Dover Press.

Catesby, M. 1731. *The Natural History of Carolina, Florida and the Bahama Islands. Vol. I.* London: M. Catesby.

Catesby, M. 1743. *The Natural History of Carolina, Florida and the Bahama Islands. Vol. II.* London: M. Catesby.

Clay, J.W., Orr, D.M., Jr., and Stuart, A.W. 1975. *North Carolina Atlas*. Chapel Hill: University of North Carolina Press.

Cloud, P. 1976. Beginnings of biospheric evolution and their biogeographical consequences. *Palaeobiology* 2:351–387.

Cope, M.J., and Chaloner, W.G. 1980. Fossil charcoal as evidence of past atmospheric composition. *Nature* 283:647–649.

Croker, T.C., Jr. 1987. Longleaf pine, a history of man and a forest. USDA Forest Service, Southern Forest Experiment Station, Forestry Report R8-FR 7.

Davies, W. 1781 [1881]. Letter to D. Jamieson. Calendar of state papers [Virginia] 2:599.

Delcourt, H.R., and Delcourt, P.A. 1977. Presettlement magnolia-beech climax of the Gulf Coastal Plain: Qualitative evidence from the Apalachicola River bluffs, north-central Florida. *Ecology* 58:1085–1093.

Derrick, S.M. 1930. *Centennial History of the South Carolina Railroad*. Columbia, SC: The State Co.

Earley, L. 2004. *Looking for Longleaf: The Fall and Rise of an American Forest*. Chapel Hill: University of North Carolina Press.

Fernow, B.E. 1893. Results on investigations on long-leaf pine. In Timber physics, Part II. Forestry Division, Bulletin No. 8. Washington, DC: US Department of Agriculture.

Frost, C.C. 1993. Four centuries of changing landscape patterns in the longleaf pine ecosystem. In *The Longleaf Pine Ecosystem: Ecology, Restoration and Management*, ed. S.M. Hermann, pp. 17–43. Proceedings Tall Timbers Fire Ecology Conference No. 18.

Frost, C.C. 1995. Presettlement fire regimes in southeastern marshes, peatlands and swamps. In *Fire in Wetlands: A Management Perspective*, eds. S.I. Cerulean and R.T. Engstrom, pp. 39–60. Proceedings Tall Timbers Fire Ecology Conference No. 19.

Frost, C.C. 2000. Studies in landscape fire ecology and presettlement vegetation of the southeastern United States. Doctoral dissertation. Chapel Hill: University of North Carolina.

Frost, C.C., and Musselman, L.J. 1987. History and vegetation of the Blackwater Ecological Preserve. *Castanea* 52:15–46.

Frost, C.C., Legrand, H.E., Jr., and Snyder, R.E. 1990. Regional inventory for critical natural areas, wetland ecosystems, and endangered species habitats of the Albemarle–Pamlico estuarine region: Phase 1. Raleigh, North Carolina: U.S. Environmental Protection Agency and N.C. Dept. of Environment, Health and Natural Resources.

Albemarle–Pamlico Estuarine Study, Project No. 90-01.

Gamble, T., ed. 1921. *Naval Stores: History, Production, Distribution and Consumption*. Savannah, GA: Review Publishing & Printing Co.

Gardner, A. 1979. A cow in a woodlot is as welcome as a bullfrog in a punchbowl. *Westvaco CFM News* [n.v.n.] summer 1979.

Garren, K.H. 1943. Effects of fire on vegetation of the southeastern United States. *Bot Rev* 9:617–654.

Grelen, H.E. 1980. Letter dated January 14, 1980, in response to my inquiry about carrying capacity of southern forest range for feral hogs.

Hale, P.M. 1883. *The Woods and Timbers of Eastern North Carolina*. New York: E.J. Hale & Son.

Hall, B. 1829. *Travels in North America in the Years 1827 and 1828, Vol. III*. Edinburgh, UK: Cadell and Co.

Hamilton, P.J. 1910 [1976]. *Colonial Mobile*. University: University of Alabama Press.

Hammond. 1980. *United States Atlas*. Maplewood, NJ.

Hann, W.J. 2002. Mapping fire regime condition class: A method for watershed and project scale analysis. In *Proceedings of the 22nd Tall Timbers Fire Ecology Conference: Fire in Temperate, Boreal and Montane Ecosystems*. Tallahassee, FL: Tall Timbers Research Station.

Harcombe, P.A., Glitzenstein, J.S., Knox, R.G., Orzell, S.L., and Bridges, E.L. 1995. Western Gulf coastal plain communities. In *The Longleaf Pine Ecosystem: Ecology, Restoration and Management*, ed. S.M. Hermann. Proceedings Tall Timbers Fire Ecology Conference No. 18.

Harper, R.M. 1905. Some noteworthy stations for Pinus palustris. *Torreya* 5:55–60.

Harper, R.M. 1906. A phytogeographical sketch of the Altamaha Grit Region of the Coastal Plain of Georgia. *Ann NY Acad Sci* 17:1–415.

Harper, R.M. 1911. The relation of climax vegetation to islands and peninsulas. *Bull Torrey Bot Club* 38:515–525.

Harper, R.M. 1913. Economic botany of Alabama. Part 1. University, Alabama: Geological Survey of Alabama, Monograph 8.

Harper, R.M. 1914. Geography and vegetation of northern Florida. Sixth Annual Report, pp. 163–437, Florida Geological Survey.

Harper, R.M. 1923. Some recent extensions of the known range of Pinus palustris. *Torreya* 23:49–51.

Harper, R.M. 1928. Economic botany of Alabama. Part 2. University, Alabama: Geological Survey of Alabama, Monograph 9.

Harriott, T. 1590 [1972]. *A Briefe and True Report of the New Found Land of Virginia*. Reprint. New York: Dover Press.

Hart, J.F. 1979. The role of the plantation in southern agriculture. *Proc Tall Timbers Ecol Manage Conf* 16:1–19.

Hefner, J.M., and Brown, J.D. 1985. Wetland trends in the southeastern United States. *Wetlands* 4:1–11.

Hindle, B. 1975. *America's Wooden Age—Aspects of Its Early Technology*. Tarrytown, NY: Sleepy Hollow Restorations.

Hopkins, W. 1947a. Perhaps the hog is hungry. USDA Forest Service, Southern Forest Experiment Station, Southern Forestry Notes No. 50.

Hopkins, W. 1947b. Pigs in the pines. *Forest Farmer* 7:3,8.

Hopkins, W. 1947c. Hogs or logs? *South Lumberman* 175:151–153.

Jenings, E. 1704 [1923]. Letter to Her Majesties Lords Commissioners for Trade & Plantations. *William Mary Quart* 3:209–210.

Kelly, J.F., and Bechtold, W.A. 1990. The longleaf pine resource. In *Proceedings of Symposium on the Management of Longleaf Pine, Long Beach, MS 1989*, pp. 11–22. USDA Forest Service Gen. Tech. Rep. SO-75. Southern Forest Experiment Station, New Orleans, LA.

Komarek, E.V. 1964. The natural history of lightning. *Proc Tall Timbers Fire Ecol Conf* 3:139–183.

Komarek, E.V. 1965. Fire ecology—Grasslands and man. *Proc Tall Timbers Fire Ecol Conf* 4:169–220.

Komarek, E.V. 1966. The meteorological basis for fire ecology. *Proc Tall Timbers Fire Ecol Conf* 5:85–125.

Komarek, E.V. 1967. The nature of lightning fires. *Proc Tall Timbers Fire Ecol Conf* 6:5–41.

Komarek, E.V. 1968. Lightning and lightning fires as ecological forces. *Proc Tall Timbers Fire Ecol Conf* 8:169–197.

Komarek, E.V. 1972. Ancient fires. *Proc Tall Timbers Fire Ecol Conf* 12:219–240.

Komarek, E.V. 1974. Effects of fire in temperate forests and related ecosystems: Southeastern United States. In *Fire and Ecosystems*, eds. T.T. Kozlowski and C. Ahlgren, pp. 251–277. New York: Academic Press.

La Rochefoucauld. 1799. *Voyages dans les Etats-Unis. Vol. IV*. Paris.

Lawson, J. 1709 [1967]. *A New Voyage to Carolina*. Reprint. Chapel Hill: University of North Carolina Press.

Lipscomb, D.J. 1989. Impacts of feral hogs on longleaf pine regeneration. *South J Appl For* 13:177–181.

Little, E.L. 1971. *Atlas of United States Trees. Vol. 1. Conifers and Important Hardwoods*. Miscellaneous Publication 1146. Washington, DC: USDA Forest Service.

Lockett, S.H. 1870. *Louisiana as it is*. Baton Rouge: Louisiana State University Press.

Martin, C. 1973. Fire and forest structure in the aboriginal eastern forest. *The Indian Historian* 6:38–42, 54.

Mattoon, W.R. 1922. Longleaf pine. USDA Bulletin No. 1061. Washington, DC: US Government Printing Office.

McWilliams, W.H. 1987. Developing timber-supply projections for the South's fourth forest. In *Proceedings: Current Challenges to Traditional Wood Procurement Practices. Atlanta, GA, September 20–30*, pp. 16–21. Madison, WI: Forest Products Research Society.

Merrens, H.R. 1964. *Colonial North Carolina in the Eighteenth Century*. Chapel Hill: University of North Carolina Press.

Michaux, F.A. 1805 [1966]. Travels to the west of the Alleghany Mountains. In *Early Western Travels, Vol. III*, ed. R. Thwaites. Reprint. New York: AMS Press.

Michaux, F.A. 1871. *North American Silva*. Philadelphia: W. Rutter & Co.

Michel, F.L. 1702 [1916]. Report of the journey of Francis Louis Michel from Berne, Switzerland, to Virginia, October 2, 1701–December 1, 1702. Wm. J. Hinkle, trans. *Virginia Mag Hist Biog* 24:1–43.

Mills, R. 1826. *Statistics of South Carolina, Including a View of its Natural, Civil and Military History*. Charles, SC: Hurlbut and Lloyd.

Mohr, C. 1884. Quoted in Report on the forests of North America, ed. C.S. Sargent. Washington, DC: USDI Census Office.

Mohr, C. 1893. Turpentine orcharding in America. In *Report of the Chief of the Division of Forestry for 1892*, ed. B.E. Fernow, pp. 342–346. Washington, DC: US Government Printing Office.

Mohr, C. 1896. The timber pines of the southern United States. USDA Division of Forestry, Bulletin No. 13. Washington, DC.

Mohr, C. 1901. Plant life of Alabama. USDA Division of Botany, Contrib. U.S. National Herbarium, Vol. VI. Washington, DC.

Moldenke, H. 1979 (letter and personal communication). Observations and field notes from 1925, with longleaf pine herbarium specimen collected at Sinnickson, VA, 1 mile south of the Maryland line.

Napier, J.H., III. 1985. *Lower Pearl River's Pineywoods*. Center for the Study of Southern Culture. University: University of Mississippi.

Outcalt, K.W., and Sheffield, R.M. 1996. The longleaf pine forest: Trends and current conditions. USDA Forest Service, Southern Research Station, Resource Bulletin SRS-9.

Parrott, R.T. 1967. A study of wiregrass (*Aristida stricta* Michx.) with particular reference to fire. Master's thesis, Duke University, Durham, NC.

Peet, R.K. 1993. A taxonomic study of *Aristida stricta* and *A. beyrichiana*. *Rhodora* 95:25–37.

Peet, R.K., and Allard, D.J. 1995. Longleaf pine vegetation of the southern Atlantic and eastern Gulf Coast regions: A preliminary classification. In *The Longleaf Pine Ecosystem: Ecology, Restoration and Management*, ed. S.M. Hermann. Proceedings Tall Timbers Fire Ecology Conference 18.

Pierce, A.M. 1953. *Tobacco Coast. A Maritime history of Chesapeake Bay in the Colonial Era*. Newport News, VA: The Mariners' Museum.

Powell, W.S. 1977. *John Pory, 1572–1636*. Chapel Hill: University of North Carolina Press.

Reed, F.W. 1905. *A Working Plan for Forest Lands in Central Alabama*. Washington, DC: US Government Printing Office.

Ruffin, E. 1843, 1844. *Report of Survey of South Carolina*. Charleston, SC: Southern Agriculturalist.

Ruffin, E. 1861. *Sketches of Lower North Carolina*. Raleigh: North Carolina State Printer.

Sargent, C.S. 1884. *Report on the Forests of North America*. Washington, DC: USDI Census Office.

Schoepf, J.D. 1788 [1911]. *Travels in the Confederation*. Philadelphia: William J. Campbell, Pub.

Scott, A.C. 1989. Observations on the nature of fusain. *Int J Coal Geol* 12: 443–475.

Scott, A.C., and Jones, T.J. 1994. The nature and influence of fires in Carbonaceous ecosystems. *Palaeogeogr Palaeoclimatol Palaeoecol* 106:91–112.

Sherrard, T.H. 1903. *A Working Plan for Forest Lands in Hampton and Beaufort Counties, South Carolina*. Washington, DC: US Government Printing Office.

Smith, J., Capt. 1624. *The general historie of Virginia, New England and the Summer Isles*. Reprint, n.d. (circa 1970). Murfreesboro, NC: Johnson Publishing Co.

State Board of Agriculture. 1883. *South Carolina: Resources and Population, Institutions and Industries*. Charleston, SC: Walker, Evans and Cogswell, Publishers.

Stowe, J.P., Jr., Varner, J.M., III, and McGuire, J.P. 2002. Montane longleaf pinelands. *Tipularia* 17:9–14.

Strachey, W. 1610 [1964]. A true repertory of the wracke, and redemption of Sir Thomas Gates Knight; upon, and from the Ilands of the Bermudas: his comming to Virginia, and the estate of that Colonie then, and after, under the government of the Lord La Warre, July 15, 1610. In *A Voyage to Virginia*, ed. L.B. Wright. Charlottesville: University Press of Virginia.

Sudworth, G.B. 1913. *Forest Atlas: Geographic Distribution of North American Trees, Part 1. Pines*. Washington, DC: USDA Forest Service.

U.S. Census Bureau. 1982, 1984, 1997. Census of agriculture. Vol. 1, Geographic area series. Part 1, Alabama; Part 9, Florida; Part 10, Georgia; Part 18, Louisiana; Part 24, Mississippi; Part 33, North Carolina; Part 40, South Carolina; Part 43, Texas; Part 46, Virginia. Washington, DC: US Government Printing Office.

U.S. Census Office 1841. Compendium of the sixth census of the United States. U.S. Census of Manufactures. Washington, DC: Department of State.

U.S. Census Office. 1853. The seventh census of the United States: 1850. Washington, DC: Robert Armstrong, Public Printer.

U.S. Census Office. 1864. Agriculture of the United States in 1860 compiled from the original returns of the eighth census. Washington, DC: US Government Printing Office.

U.S. Census Office. 1872. A compendium of the ninth census (June 1, 1870). Washington, DC: US Government Printing Office.

U.S. Census Office. 1883. Report on the productions of agriculture as returned at the tenth census (June 1, 1880). Washington, DC: US Government Printing Office.

U.S. Census Office. 1895. Report on the statistics of agriculture in the United States at the eleventh census: 1890. Washington, DC: US Government Printing Office.

U.S. Census Office. 1902. Twelfth census of the U.S. Vol. 5. Agriculture. Part 1. Washington, DC: US Government Printing Office.

Wahlenburg, W.G. 1946. *Longleaf Pine*. Washington, DC: Charles Lathrop Pack Forestry Foundation.

Wakeley, P.C. 1935. Artificial reforestation in the southern pine region. US Department of Agriculture Technical Bulletin No. 492. Washington, DC.

Wakeley, P.C. 1954. Planting the southern pines. Agriculture Monograph No. 18. Washington, DC: USDA Forest Service.

Walker, J. 1995. Regional patterns in the vascular flora and rare plant component of longleaf pine communities. In *The Longleaf Pine Ecosystem: Ecology, Restoration and Management*, ed. S.M. Hermann. Proceedings Tall Timbers Fire Ecology Conference No. 18.

Ware, S. 1978. Vegetational role of beech in the Southern Mixed Hardwood Forest and the Virginia Coastal Plain. *Va J Sci* 29:231–235.

Ware, S., Frost, C.C., and Doerr, P.D. 1993. Southern mixed hardwood forest: The former longleaf pine forest. In *Biodiversity of the Southern United States: Lowland Terrestrial Communities*, eds. W.H. Martin, S.G. Boyce, and A.C. Echternacht, pp. 447–493. New York: John Wiley & Sons.

Wertenberger, T.J. 1922. *Planters of Colonial Virginia*. Princeton, NJ: Princeton University Press.

Wertenberger, T.J. 1931. *Norfolk: Historic Southern port*. Durham, NC: Duke University Press.

Williams, M. 1980. Products of the forest: Mapping the census of 1840. *J For Hist* 24:4–23.

Williams, M. 1989. *Americans and Their Forests: A Historical Geography*. New York: Cambridge University Press.

Young, H. 1781 [1881]. Letter to Col. Davies. Calendar of state papers [Virginia] 2:619.

BOX 2.1
The Naval Stores Industry

Alan W. Hodges
*Institute of Food and Agricultural Sciences,
University of Florida, Gainesville,
Florida 32611*

The vast forests of longleaf pine (*Pinus palustris*) in the southeastern United States were once the basis of a very large industry for producing pine tar, rosin, turpentine, pine oil, and other products derived from the natural oleoresin of the tree. The term "naval stores" came about from the use of these products for building and maintaining wooden ships during the seventeenth through nineteenth centuries. Pine tar was used together with various fibrous materials to seal the seams between the wooden planking. It was an important strategic commodity for the naval-based European empires of Britain, Holland, Spain, and Portugal. Although the term "naval stores" is now antiquated, it is still frequently used in historical contexts. However, the preferred contemporary terminology in the industry is "pine chemicals" or "gum and wood chemicals."

Naval stores production is based upon exploiting the terpene chemical defense system of the pine tree, which protects against wood-decaying fungi and insect pests such as bark beetles (*Dendroctonus, Ips sp.*). When a tree suffers injury to the bark and cambium layer, oleoresin is secreted to prevent the establishment and spread of pathogens, acting as a natural biocide and preservative. Pine oleoresin is a complex mixture of about 30 to 50 different terpene molecules, comprised chiefly of diterpene resin acids and monoterpene essential oils, which impart different physical and biotic properties (Zinkel and Russell 1989). The terpene chemical defense system came about very early in the evolution of higher plants on earth, with some ambers from fossilized tree resins having been dated to the Carboniferous Era, over 200 million years before present (Langenheim 1969). Terpenes are synthesized through the basic malevonic–pyruvate biochemical pathway, and occur in many different groups of plants, as well as all conifers. Longleaf pine, slash pine (*P. elliottii*), and other subtropical pine species are especially rich in terpenes because of high year-round pest pressures.

Terpenes are produced in pine trees by specialized cells that form a network of microscopic ducts with interconnected longitudinal and radial segments, as illustrated in Box Fig. 1. The epithelial cells of the resin ducts arise from the parenchyma tissue, and lack the rigid cell wall of normal wood fibers. As terpenes are secreted into the lumen of the resin ducts, where they are stored, the elastic membrane of the epithelial cells maintains a relatively high pressure (300 psi) on the fluid oleoresin such that it can be mobilized in case of an injury. Oleoresin is present in all parts of the tree—leaves, branches, stem, roots, bark— and typically represents about 3 to 5% of total tree biomass (dryweight basis). In older trees, oleoresin accumulates in the stumps and heartwood, and in the wood around the base of major branches. Oleoresin is not to be confused with sap, the nutrient solution that is carried in a separate system of vascular tissues.

Methods for harvesting or extraction of oleoresin from pines have evolved significantly over the past 400 years, due to changes in technology and the forest resource. Beginning in the early 1600s, colonists in North America made pine tar for export to Europe by a pyrolysis process, i.e., by slowly burning resinous wood in earthen kilns. This activity reached its peak in North Carolina, where tar makers, known as "tarheels," exploited the abundant longleaf pine forests (Butler 1998). They gathered naturally occurring resinous wood from old-growth stumps, heartwood, and branch knots, and also deliberately

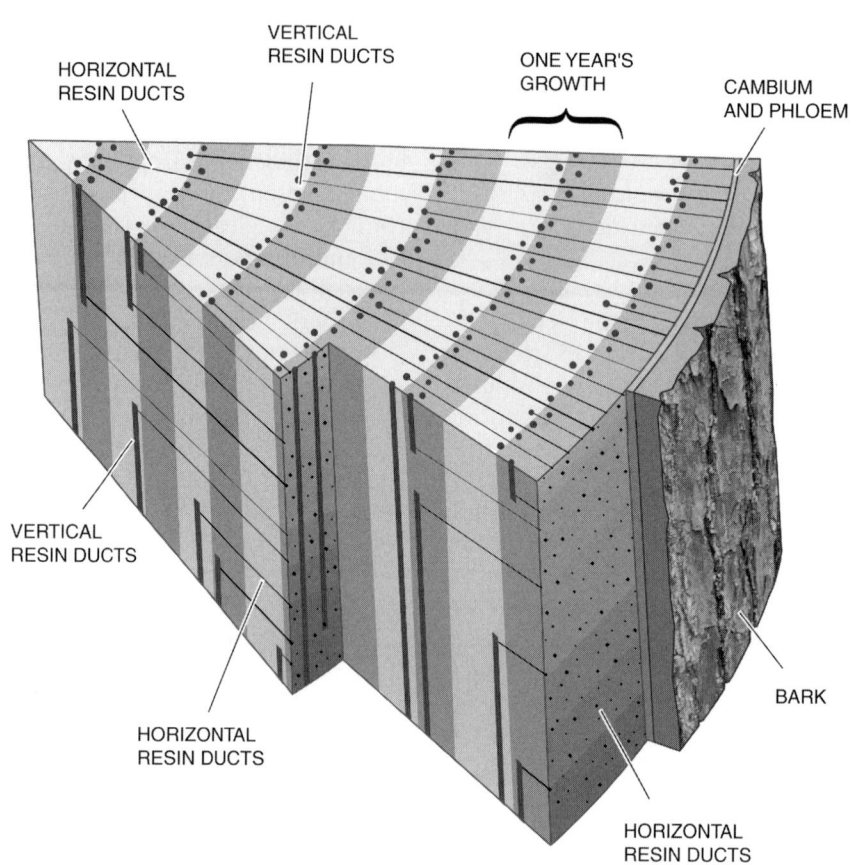

FIGURE 1. Anatomy of the oleoresin duct system of longleaf pine. Courtesy of USDA Forest Service. Illustration by Susan Trammell, Willisiton, FL.

injured trees to cause resinosis by removing bark and applying fire. Resin-saturated pine is known as "lightwood," because the settlers used the wood for torches.

During the 1800s and early 1900s, as settlers moved into the southeast U.S. coastal plain region, the gum naval stores industry developed for tapping of living longleaf and slash pine trees (Box Fig. 2). Trees were repeatedly wounded using a hook-shaped cutting tool known as a "hack" to cause the natural defensive response and bring about oleoresin exudation. The oleoresin was collected in cavities ("boxes") chopped into the base of the tree, and special "dipping" tools were used to periodically remove the accumulated oleoresin. This destructive practice often killed or weakened the trees to other mortality factors. The exploitative resin harvesting was usually followed by clear-cut logging, resulting in widespread deforestation. The laborious process of hacking and dipping was done mostly by black workers, many of whom were slaves, descendants of slave families, or prisoners, who lived in isolated camps. Most gum naval stores operators practiced annual controlled burning of the forest stands to improve accessibility, and carefully prepared the stand by raking away litterfall around each tree to avoid scorching the tapped face.

In the twentieth century, better methods were developed for collecting gum oleoresin in manufactured containers attached to the tree. Clay cups introduced by

2. History and Future of the Longleaf Pine Ecosystem

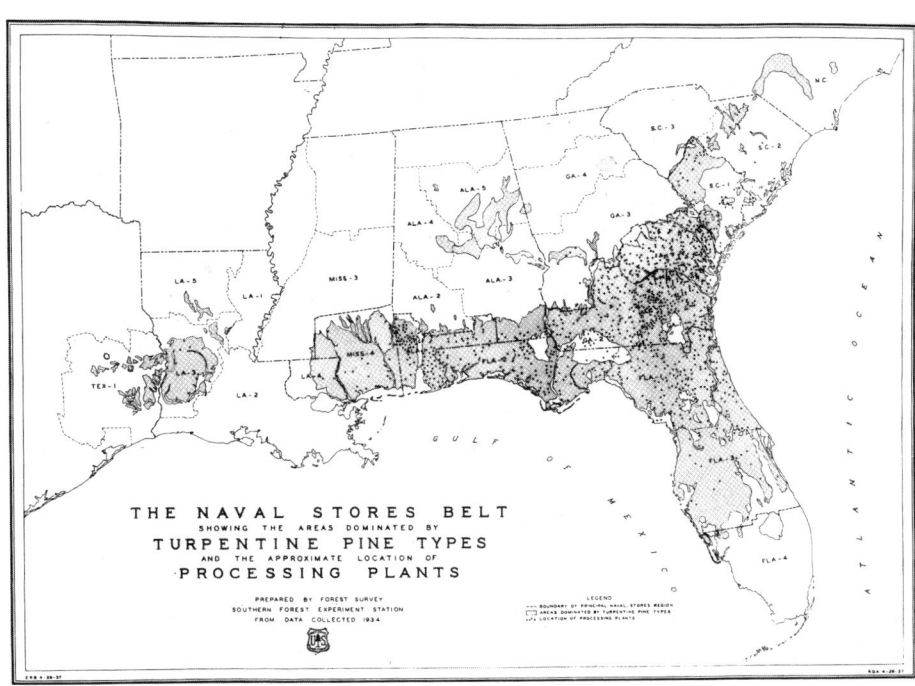

FIGURE 2. Map of the naval stores belt, and location of processing plants (1934). Courtesy of USDA Forest Service. Southern Forest Experiment Station, forest survey (1934).

Dr. Charles Herty in 1906 led to a resurgence of the industry. Typically, the cups were hung from a nail underneath a pair of sheet metal gutters that channeled the oleoresin into the cup (Box Fig. 3). Many of these old clay cups and tins can still be found in the woods today. The standard contemporary method of oleoresin tapping, known as bark chipping, was developed by U.S. Forest Service researchers during the 1940s. Strips of bark and cambium approximately 2 inches (5 cm) wide are removed across one-third of the tree's circumference, at intervals of 3 to 4 weeks throughout the March through October period in the southeastern United States. This method causes significantly less damage to trees than the previous practice of deeply chipping into the wood. It was also discovered that chipping stimulates a roughly a sevenfold increase in the number of resin ducts in new wood formed above the chipped face, which enables higher yields if light chipping is practiced (Gerry 1935). With appropriate conservative practices, trees may be tapped for about 12 years on two sides or "faces," and can then normally be used for the full range of wood products (Box Fig. 4).

Chemical treatments are also used for increasing oleoresin yields from tapping pines. Sulfuric acid is applied as a spray or paste solution to the freshly exposed cambium to destroy cells surrounding the opening of severed resin ducts, preventing premature occlusion of the ducts, and prolonging oleoresin flow for several weeks. A new generation of chemical stimulation has been developed using plant regulators such as ethylene, which acts as a general stressor, stimulating biosynthesis of oleoresin. With the best available method, the expected annual oleoresin yield from a 35-year-old longleaf pine, 10 inches (25 cm) DBH, is approximately 11.2 pounds (5.1 kg) (McReynolds and Kossuth 1984). For an 8-year production period, in a typical stand the 1930s, the primitive fire stills were replaced by about 30 large central steam processing plants.

Rosin is an amber-colored crystalline solid material at room temperature, and is used for making adhesives, sealants, coatings, fluxes, printing inks, emulsifiers, and food products such as chewing gum. Turpentine is used in solvents, cleaners, antiseptics, insecticides, flavors and fragrances, and synthetic resins. The rosin fraction typically represents about 70% of the original crude oleoresin, turpentine about 15%, and foreign material such as dirt, litter, and water the remaining 15%. Rosin is traded commercially based upon color and chemical composition, which is determined by the pine species, and methods of production and processing. "American" rosin from longleaf and slash pines makes an excellent grade that is recognized worldwide as the standard of quality. A characteristic of the rosin from longleaf pine is that it crystallizes very rapidly, due to the particular mix of diterpene resin acids. A significant portion of the oleoresin becomes crystallized and dried and on the face, and it is necessary to remove this material with a special scraping tool.

At its peak in 1910, the United States produced nearly 600,000 metric tonnes of rosin and turpentine, which is the highest of any country in history, and is unlikely to ever be achieved again. At this time, there were about 27,000 workers employed in the industry, with over 10,000 independent producers operating on about 8 million acres. Since the 1930s, the gum naval stores industry in the United States has steadily declined due to the high cost of labor, scarcity of suitable timber, and competition from foreign producers and substitute materials. The last remaining processing plant in the United States closed in 2001. However, gum naval stores is still an important industry in many developing countries, such as China, Indonesia, India, Brazil, and Mexico, with global production of gum

FIGURE 3. The Herty cup and gutter system for resin collection, circa 1906. Courtesy of USDA Forest Service. Image scanned from *A Pictoral Album of the Naval Stores Industry* (1937), University of Florida library.

of 50 tappable trees per acre, the revenue from a gum naval stores operation would be over $1500 per acre. Oleoresin yields are strongly influenced by the age, size, and vigor of trees, and stand density and canopy development.

The crude oleoresin extracted from pines is distilled to separate the principal constituents of diterpene resins and monoterpene essential oils, known commercially as rosin and turpentine. Originally, this was done at small farm-scale distilleries using fire-heated copper kettles adapted from Scottish whiskey stills, and at one time there were over 1800 such stills in the southeastern United States (Box Fig. 5). A few of these old stills have been preserved for public demonstrations of the process. Beginning in

2. History and Future of the Longleaf Pine Ecosystem

FIGURE 4. Typical woods scene of gum naval stores harvesting operation, circa 1900. Courtesy of USDA Forest Service. State archives of Florida, call number Rc 02612.

FIGURE 5. Turpentine still building and loading ramp. State archives of Florida, call number PR 12612.

rosin and turpentine in excess of 700,000 metric tons annually (Hodges 2002). About 10 major pine species are used around the world, including *Pinus elliottii* established from seedstock in the United States.

Wood naval stores is another facet of the industry based upon extraction of rosin and oils from the resin-saturated stumps of first-growth longleaf pine trees that died or were harvested decades ago. This process was developed in the 1920s, and continues today at a single plant in Brunswick, GA, operated by Hercules Inc. Weathered stumpwood contains about 40% extractives, by weight, and is highly decay-resistant. Stumps are recovered from cut-over forest lands with heavy equipment, then transported to the plant by truck or railcar, where they are finely shredded before extraction with solvents such as gasoline or hexane. The process produces an excellent grade of pale rosin, wood turpentine, and pine oil. Longleaf pine stumpwood is essentially a non-renewable resource, since trees are generally not allowed to grow to the age when heartwood formation occurs, after about 80 years. Nevertheless, the remaining supply of stumps is expected to last many more years at the current rate of recovery, to produce about 20,000 metric tons of wood rosin annually. A system for artificially inducing resinosis in young southern pines by treatment with herbicides such as paraquat was developed during the 1970s (Stubbs et al. 1984). This process achieved a 15-fold increase in total whole-tree extractives over a period of 2 years, under the optimal treatment. However, it was not commercialized because of difficulties with high tree mortality and insect pests.

The most important source of naval stores products in the United States today is a by-product of the sulfate or "Kraft" pulping process (Zinkel and Russell 1989). Turpentine volatilized by cooking of wood chips is recovered from pulp digestors, while nonvolatiles extracted from wood pulp are recovered from the black liquor stream as crude tall oil, then fractionated into rosin, fatty acids, and a variety of other compounds. Both the turpentine and rosin are contaminated by sulfates, but industrial users have adapted these materials to their needs at lower cost than other sources, and they compete with petroleum hydrocarbon chemical feedstocks. There is over 800,000 metric tons of crude tall oil produced in the United States and Canada (Hodges 2002). However, longleaf pine does not contribute substantially to this industry because it is seldom used for making paper.

References

Blount, R. 1995. *Spirits of Turpentine: A History of Florida Naval Stores, 1528–1950.* Florida Heritage Journal Monograph No. 3. Tallahassee: Florida Department of Agriculture and Consumer Services.

Butler, C.B. 1998. *Treasures of the Longleaf Pine: Naval Stores.* Shalimar, FL: Tarkel Press.

Gerry, E. 1935. A Naval Stores Handbook Dealing with the Production of Pine Gum or Oleoresin. Miscellaneous Publication 209, US Forest Service, Washington, DC.

Hodges, A., ed. 2002. *Forest Chemicals Review International Yearbook,* 2001. New Orleans, LA: Kreidt Publications.

Langenheim, J. 1969. Amber: A botanical inquiry. *Science* 163:1157–1169.

McReynolds, R.D., and Kossuth, S.V. 1985. CEPA in liquid sulfuric acid increases oleoresin yields. *South J Appl For* 9(3):170–173.

Stubbs, J., Roberts, D., and Outcalt, K. 1984. Chemical stimulation of lightwood in southern pines. General Technical Report SE-25. USDA Forest Service, Southeast Forest Experiment Station, Asheville, NC.

Zinkel, D. and Russell, J., eds. 1989. *Naval Stores: Production, Chemistry, Utilization.* Atlanta, GA: Pulp (Pine) Chemicals Association.

Section II
Ecology

Chapter 3

Ecological Classification of Longleaf Pine Woodlands

Robert K. Peet

Introduction

When Europeans first settled in southeastern North America and began to explore their new homeland, they found a landscape that was to a large extent dominated by open, savannalike longleaf pine woodlands. The pines were typically widely spaced, affording the traveler opportunities to see for long distances without obstruction by undergrowth. The ground layer was dominated by grasses with a great diversity of showy forbs. Vegetation of this character occurred from southeastern Virginia southward deep into peninsular Florida and west to western Louisiana and eastern Texas (Frost et al. 1986; Harcombe et al. 1993; Peet and Allard 1993; Ware et al. 1993; Platt 1999; Christensen 2000; Frost this volume).

Early descriptions and superficial treatment in textbooks have created and maintained the inaccurate perception, widespread outside the Southeast, that the vegetation of the original longleaf pine ecosystem was a homogeneous and monotonous bilayer community with pines above and grass below. While this is clearly an oversimplification, the original longleaf ecosystems were generally bilayered communities with the physiognomy maintained by frequent, low-intensity surface fires that removed most small woody plants and thereby kept the canopy open. However, this caricature obscures the remarkable floristic diversity of these systems. At a within-community scale, longleaf vegetation can be among the most diverse in North America with some examples having 40 or more species of higher plants per square meter (Walker and Peet 1983) or 170 per 1000 m^2 (Peet, Carr and Gramling 2006; W. J. Platt personal communication). But even more impressive is the diversity reflected in the change in composition of longleaf vegetation with subtle changes in environmental conditions, or with geographic distance. This diversity is particularly conspicuous in the floristic richness and endemism of the region. There are on the order of 6000 vascular plant taxa that occur on the southeastern Coastal Plain, which represents almost a quarter of all plant species that occur in North America north of Mexico. Moreover, 1630 taxa are endemic to the Coastal Plain, and with 1306 full species included (Sorrie and Weakley 2001, 2006). The region falls just short of qualifying as one of the top 25 biodiversity hotspots on the globe (see Myers et al. 2000).

Robert K. Peet • Department of Biology, University of North Carolina, Chapel Hill, North Carolina 27599-3280.

A large proportion of the endemics occur in the longleaf-dominated vegetation (Sorrie and Weakley 2005).

Although longleaf ecosystems once dominated the southeastern Coastal Plain, most of this vegetation was gone by 1920 and today less than 3% of the original extent of longleaf vegetation remains in natural conditions (Frost 1993; Ware et al. 1993; Frost this volume). Today, natural longleaf stands occur primarily on lands characterized by soils too wet or too dry for agriculture (Ashe 1897; Mohr 1901; Harper 1906; Frost 1993). From North Carolina to Mississippi the inner coastal plain with its predominantly finer-textured soils has lost essentially all of its original, vast extent of longleaf vegetation through conversion to agriculture. Gone also are nearly all longleaf populations from Virginia and from North Carolina north of the Neuse River, the region where longleaf was first exploited for naval stores. What remains of the original longleaf ecosystem is a small and biased sample of what was once one of the most extensive and diverse biomes of North America.

Simple removal of longleaf is far from the whole story of the demise of the longleaf ecosystem. Just as important has been the loss of the original fire regime. On all but the most sterile sites, a significant decrease in fire frequency quickly leads to a dense growth of woody plants, followed by a competitive failure of the original ground cover and thus most of the original biodiversity of the longleaf system (see Frost 2000). Although conservationists working in other ecological systems often identify critical tracts for preservation by the persistence of old-growth trees, this approach is generally inappropriate for longleaf systems. For longleaf ecosystems where the biodiversity is concentrated in the ground layer, tree age is relatively unimportant compared to the integrity of the ground-layer vegetation, which in turn depends on the long-term persistence of a regime of frequent, low-intensity surface fires.

Conservation and restoration of the natural longleaf ecosystem is remarkably difficult for several reasons (Walker and Silletti this volume). For many types of sites, nothing is left that might be used as a template for restoration efforts. Many sites where longleaf does persist have been significantly altered by fire suppression. Further, the great diversity, endemism, and spatial heterogeneity of the longleaf system means that any reserve system that aspires to preserve biodiversity needs to incorporate the range of environmental conditions at numerous places scattered over the original range of the species.

Vegetation description and classification play a key role in many areas of conservation, land management, and scientific research. A primary role of vegetation classification is to delimit natural communities so as to provide a framework for identifying, understanding, managing, and restoring the natural vegetation. Managers of conservation lands require accurate and detailed descriptions of the vegetation attributes they need to preserve or re-create. Without a well-formulated vegetation classification and description, a quality template for management or restoration is often impossible. In short, future conservation and restoration must be based on knowledge of vegetation composition across a broad range of sites selected to represent that range of natural conditions and geographic variation. Unfortunately, the diversity of longleaf vegetation has been relatively little studied. Documentation of compositional variation can be found in the scientific literature for small portions of this system over limited ranges of soil conditions. Vast areas of the longleaf region have not been subjected to rigorous ecological study. Although for some regions it is now too late, efforts to document the compositional variation of the remnants of this remarkable ecosystem are underway. My goal in this chapter is to combine information in the nascent U.S. National Vegetation Classification (NatureServe 2005; see Anderson et al. 1998) with available quantitative data to create a preliminary classification of natural, fire-maintained, longleaf-dominated vegetation types to guide future conservation and restoration efforts.

Physiography and Ecoregions of the Longleaf Ecosystem

The relatively gentle to imperceptible topography that characterizes much of the range of longleaf belies the considerable geographic variation and complexity in its environment and biota. One approach to understanding and managing this subtle complexity is to break the longleaf system into ecoregions that are relatively consistent in their climate, soils, and physiography. Comparison of vegetation within and between ecoregions provides a framework within which variation in the longleaf ecosystem can be understood.

Several alternative ecoregion systems have been published (e.g., Omernik 1987; Bailey 1995; Brown et al. 1998; Ricketts et al. 1999). However, the system that seems to match best the natural variation of longleaf vegetation is that of the U.S. Environmental Protection Agency (EPA; see Omernik 1987, and updates). EPA ecoregions are hierarchical with 4 levels. Although I have found level-4 ecoregions to correlate well with compositional variation in longleaf ecosystems, there are far too many types to provide a useful context for examining large-scale, range-wide patterns. I here describe six ecoregions largely based on the nine EPA level-3 ecoregions that span the natural range of longleaf pine, plus I treat separately one level-4 segregate, the Fall-line Sandhills of the Carolinas and Georgia (Fig. 1).

The Atlantic Coastal Plain Ecoregion

The Atlantic Coastal Plain Ecoregion includes the coastal flatlands of the EPA Middle Atlantic Coastal Plain region and that portion of the EPA Southeastern Plains region occurring from Virginia southwest to and including the Altamaha Grit region of Georgia, stopping at the Flint River as an arbitrary division between the Atlantic and Gulf Coastal Plains. The outer portion of the Atlantic Coastal Plain Ecoregion is derived from marine sediments of Miocene age or younger. This region tends to be extremely flat with significant topography restricted to banks and bluffs of major rivers. Local topographic relief is consistently under 20 meters. These Atlantic coastal flatlands are perhaps best visualized as a series of old barrier dunes and shorelines. Soils of the barrier dunes and shorelines per se tend to be extremely sandy and dry due to rapid percolation of water, whereas the soils of the once embayed regions tend to be seasonally saturated as a result of their low relief and finer soil texture, which make for poor drainage (DuBar et al. 1974; Daniels et al. 1984; Soller and Mills 1991). Old marine terraces are also prominent and tend to be flat with fine-textured, poorly drained soils. In almost all cases, the soils tend to be highly phosphorus deficient. Coarse, siliceous sands have formed dune systems on the northeast sides of all major rivers draining into the Atlantic as well as on the northeast sides of Carolina bay depression wetlands, in both cases the sands having been blown out of the river valleys and bay depressions prior to the last glacial advance.

The inland half of the Atlantic Coastal Plain Ecoregion is dominated by landforms that are generally older and the topography more hilly and complex than on the outer coastal plain. The overall aspect is of low, rolling hills, often with loamy rather than sandy soils. A distinctive region of clay hills occurs in South Carolina (Myers et al. 1986). The pronounced topographic relief of the rolling inner coastal plain results in generally well drained soils with the consequence that seasonally wet sites are less common than on the outer Coastal Plain, mostly of local occurrence, and associated with outcrops of impermeable soil horizons.

The Fall-line Sandhills Ecoregion

The Fall-line Sandhills Ecoregion is identical to the same-named level-4 segregate of the EPA Southeastern Plains ecoregion and forms its inland fringe from central North Carolina

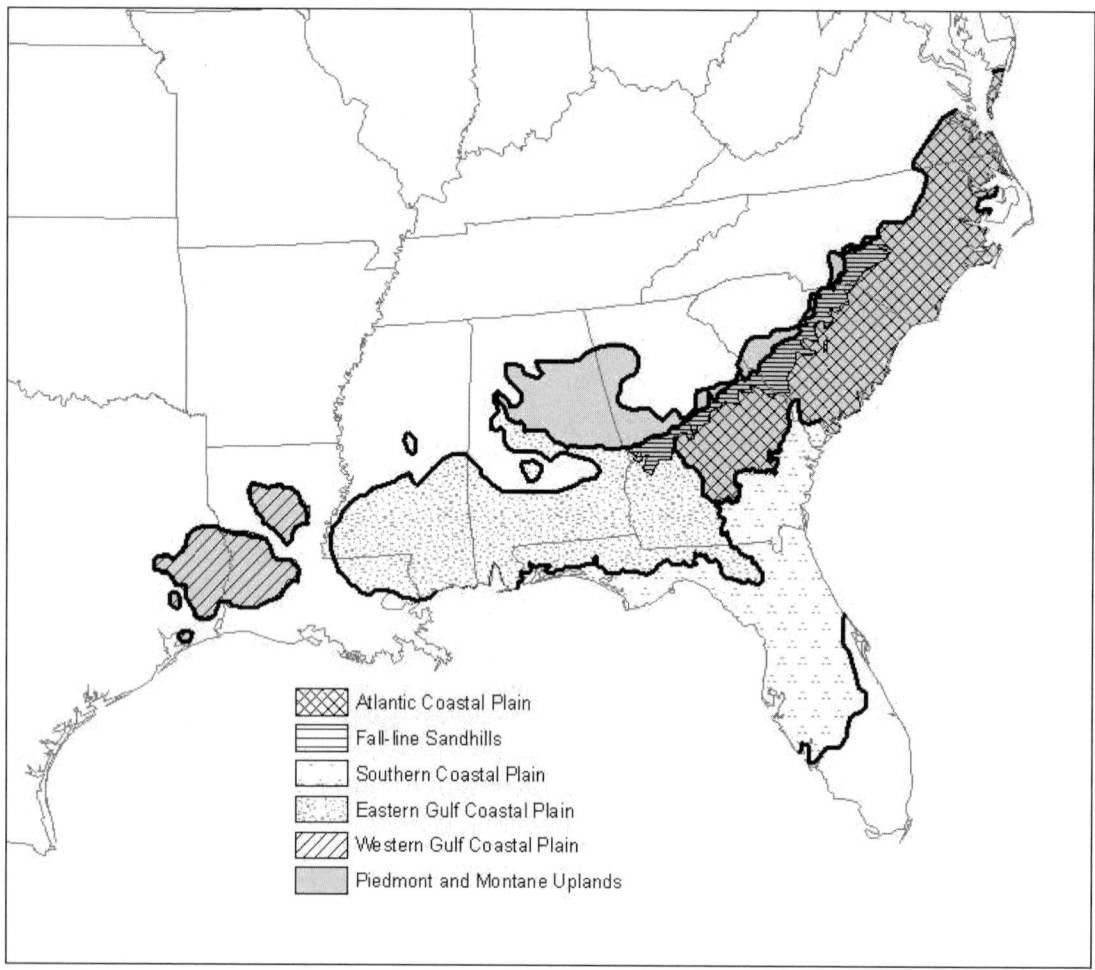

FIGURE 1. Six longleaf pine vegetation ecoregions, largely derived from EPA Ecoregions (see text).

to easternmost Alabama. This is a region of primarily coarse, Cretaceous-age sediments, often with a veneer of Miocene-age eolian dune sands. The sandhill sediments appear to have eroded from the ancient uplands that today form the Piedmont. Intermixed with the fall-line sands are layers of clay that impede drainage and result in seepage wetlands where the clay layers appear near or at the surface. The sandhills are also characterized by scattered subtle depressions, perhaps similar in origin to the Carolina bay depression wetlands of the flatlands south and east of the sandhills (James 2000). These sandhill depressions often contain finer-textured soils than the adjacent sandy landscape, perhaps deposited by winds. The area is both ecologically and edaphically distinct from the rest of Southeastern Plains, which tend to be dominated by fine-textured soils rather than coarse sands.

The Southern Coastal Plain Ecoregion

The Southern Coastal Plain Ecoregion includes most of the EPA Southern Coastal Plain Ecoregion including the range of longleaf in Florida south of the Cody Scarp, the lower coastal plain of Georgia, and a narrow fringe along the outermost coastal plain of southern South Carolina. Soils of this ecoregion, like those of the Atlantic Coastal Plain, are derived almost exclusively from marine sediments of Miocene age or younger, and the region tends to be extremely flat with significant topography restricted to banks and bluffs of major rivers. The Southern Coastal Plain differs from the Atlantic and Gulf Coastal Plains in the general absence of fine-textured soils, nearly the whole region being characterized by a veneer of marine sands. In addition, the influence of the underlying limestone of the Florida peninsula not infrequently is expressed in the form of higher soil phosphorus content than generally encountered elsewhere in the sandy soils of the longleaf region (Peet, Carr and Gramling 2006).

The Eastern Gulf Coastal Plain Ecoregion

The Eastern Gulf Coastal Plain Ecoregion includes the EPA Southeastern Plains ecoregion from the Flint River in Georgia southward to the Cody Scarp including the Tallahassee Red Hills of north Florida, and west to include the EPA Loess Plains ecoregion of Louisiana and Mississippi. Also included is the narrow fringe of EPA Southern Coastal Plain ecoregion west of the Florida–Alabama border. As with the inner Atlantic Coastal Plain, landforms here are generally older and the topography more hilly and complex than on most of the Coastal Plain. The overall aspect is of low, rolling hills, often with loamy rather than sandy soils. A distinctive region of clay hills occurs in Alabama and Mississippi (Hodgkins 1965; Hodgkins et al. 1979), which extend into southern Georgia (Harper 1930). The soils are generally well drained soils with wet sites primarily confined to the coastal fringe. The region near the Mississippi River is recognized by EPA as a separate ecoregion owing to its distinctive thick cap of loess, blown out of the river during the late glacial period. The loess cap gives the region distinctive, fine-textured, highly fertile soils that set it off from all other longleaf landscapes.

The West Gulf Coastal Plain Ecoregion

The West Gulf Coastal Plain Ecoregion includes all longleaf lands west of the Mississippi River and is included in the EPA Western Gulf Coastal Plain and South Central Plains ecoregions. The more southerly portion of this ecoregion is similar to its southeastern counterparts in being strikingly flat with extensive areas of poorly drained soils. The more inland portions are characterized by rolling hills and even occasional outcrops of the underlying sandstone. The Pleistocene terraces of this region are dominated by silty and even clayey soils, some exhibiting strong vertic tendencies.

The Piedmont and Montane Uplands Ecoregion

The Piedmont and Montane Uplands Ecoregion includes the eastern fringe of the igneous EPA Piedmont ecoregion in the Carolinas as well as the Piedmont and adjacent EPA Southwestern Appalachians and Blue Ridge ecoregions of northwestern Georgia and adjacent eastern Alabama. This ecoregion is the most atypical longleaf ecoregion in both topography and substrate. Here are found mature landscapes with well-developed drainage networks and complex topography which, except for the Ridge and Valley region, are characterized by kaolinitic clay soils largely derived from igneous rocks. Generally, longleaf sites of this region tend to be well drained, though seepage areas with longleaf vegetation do occur occasionally.

A Representative Longleaf Landscape

Over much of the range of longleaf pine, local variation in vegetation can be interpreted in terms of two primary gradients: soil moisture and soil texture. This typical pattern serves as a general model for factors that affect longleaf pine vegetation (Fig. 2). Although variation in soil nutrients can also be important, for a given moisture and texture regime the soils in a region are relatively predictable, as is the overall character of the associated vegetation. Only when one leaves the Coastal Plain for the Piedmont and Montane longleaf types

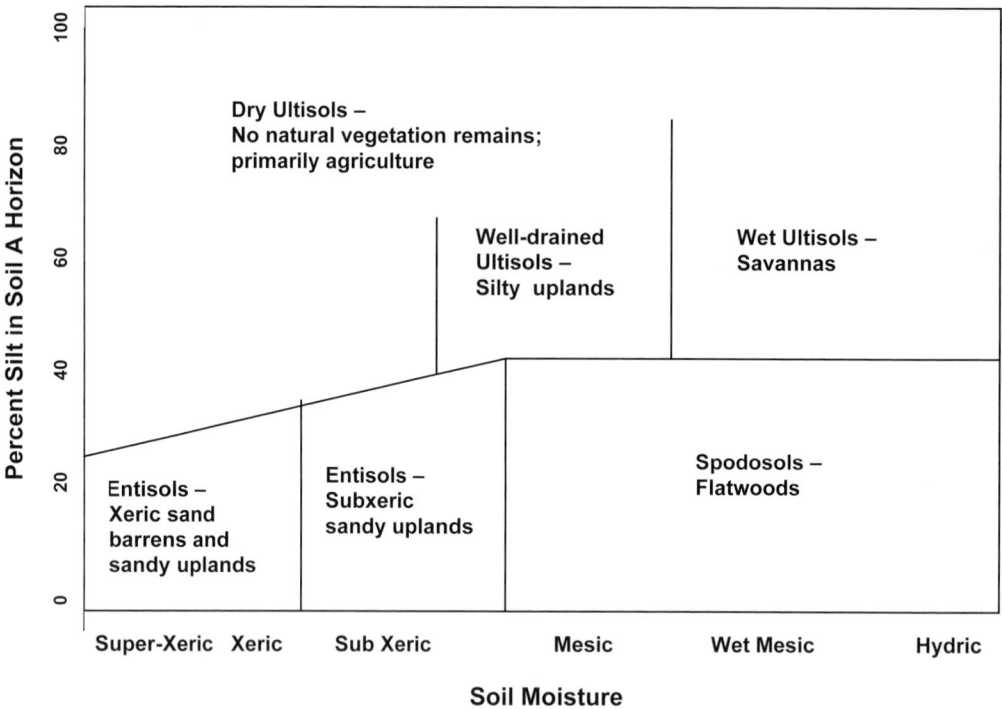

FIGURE 2. A model landscape of Coastal Plain longleaf pine vegetation showing dominant vegetation types in relation to soil silt content and soil moisture.

does this general, two-gradient model break down, being replaced by a model where soil clay and incident solar radiation are key variables and soil moisture is less significant.

Dominant soil orders vary in a manner consistent with the two-dimensional gradient model of variation in longleaf vegetation. Sandy, dry sites are dominated by Entisols. The most well-drained, coarse sands are extremely dry, intrinsically low in nutrients, and support open, sand-barren vegetation, whereas nearly all of the less extreme sites support a well-developed ground layer dominated by grasses. Soils of poorly drained, sandy sites are primarily Spodosols, and on the wettest sites Histisols can occur. The Spodosols, which are highly infertile, particularly with respect to nitrogen and phosphorus, support a vegetation type often called flatwoods. The portion of the gradient diagram with well-drained, silty soils is largely associated with Ultisols. Unfortunately, most such sites have been converted to agriculture. Poorly drained, silty soils are also generally Ultisols and support what I refer to here as pine savanna vegetation types.

Species diversity varies in a consistent pattern across the gradient model (Table 1). By all measures, the barrens of sandy, xeric sites are least diverse. At the scale of a 1000 m^2 plot, diversity increases with increasing soil silt content, but with no conspicuous difference between wet and well-drained sites. Examples of exceptionally species-rich plots with about 170 species per 1000 m^2 have been reported from well-drained soils of the Mississippi loess plains (W. Platt personal communication) and from silty seeps in the Tallahassee Red Hills (Peet, Carr and Gramling 2006). However, at scales between 10 m^2 and 0.01 m^2, the savanna vegetation types of wet, silty soils are the most diverse of the longleaf ecosystem. The average plot on moist, silty soils has on the order of 20 species per square meter, and where soils are particularly fine and fire occurs nearly annually, species richness can exceed 40 species per square meter.

TABLE 1. Species richness as a function of plot size for representative longleaf community types of southeast North Carolina and of Florida.[a]

	Area m²					
	0.01	0.1	1	10	100	1000
Xeric sand barrens & uplands						
North Carolina	0.5	1.3	3.2	6.6	12.9	22.5
Florida	1.1	3.2	9.5	20.9	42.6	74.1
Subxeric sandy uplands						
North Carolina	0.8	2.7	5.5	10.3	19.1	34.8
Florida	1.1	3.9	11.2	24.4	48.1	84.0
Silty & clayey uplands						
North Carolina	2.7	7.5	15.7	27.1	51.7	81.4
Florida	2.1	7.1	17.1	32.8	63.7	107.5
Flatwoods						
North Carolina	2.3	6.0	11.2	18.7	33.2	54.6
Florida	1.8	5.2	11.7	21.4	40.0	71.2
Savannas						
North Carolina	4.4	11.3	22.4	36.0	61.1	94.4
Florida	3.1	9.2	18.7	30.4	54.5	89.8

[a] (Values are based on 180 vegetation plots from southeastern North Carolina and 281 from Florida, each typically 1000 m² in area and containing eight subplots each of 0.01, 0.1, 1, and 10 m², and four subplots of 100 m². Averages were calculated for association types, and the associations within a community type were averaged within a major community type—types were averaged rather than plots to avoid weighting types by numbers of plots.

Major patterns of variation in vegetation composition also correspond well with the general gradient model. All longleaf vegetation except that of extremely xeric barrens sites has a well-developed grass-sedge layer. The best-known grass of the longleaf woodlands is wiregrass (*Aristida stricta* and *A. beyrichiana*[1]). From northern South Carolina north to roughly the Pamlico Sound, *Aristida stricta* dominates. On either side of the range of *Aristida stricta*, there is no wiregrass, but instead *Schizachyrium scoparium* dominates. South of the Santee River in South Carolina, wiregrass again occurs, here in the form of *Aristida beyrichiana*. Wiregrass extends southward through most of the Florida Peninsula and westward along the Gulf Coast into the southeastern corner of Mississippi. Wiregrass is generally the dominant grass type of Spodosols. Wiregrass also occurs on Ultisols, but the bluestems (*Andropogon* spp., *Schizachyrium* spp.) tend to be at least equally important. On wet, fine-textured soils numerous grasses and sedges share dominance with *Ctenium aromaticum, Muhlenbergia expansa, Dichanthelium* spp., and *Rhynchospora* spp. being nearly ubiquitous (though *Ctenium* is absent from the savannas of Texas, dropping out at the Sabine River).

Oaks (e.g., *Quercus laevis, Q. incana, Q. geminata*) frequently occur as co-dominants on xeric and subxeric sandy soils, with scrubby evergreen oak diversity (e.g., *Q. chapmanii, Q. myrtifolia*) increasing with proximity to peninsular Florida. The oaks decline in importance with increasing soil moisture or silt content, with two exceptions: live oak (*Q. virginiana*) occupies somewhat more mesic sites than most of its congeners, and runner oaks (*Q. minima, Q. pumila*) can be prominent on somewhat fine-textured Ultisols and on moderately drained Spodosols. Shrubs of the heath family are well developed in flatwoods and in subxeric sandhills types, but decrease significantly in importance with increasing silt content. In contrast, legumes are almost entirely confined to the fine-textured soils, where they can be exceptionally diverse and abundant (Gano 1917; Wells and Shunk 1931; Taggart 1994; Hainds et al. 1999; James 2000). The mechanism behind the distribution of legumes remains undocumented, though I have found experimentally in both North Carolina and Mississippi that legume abundance on wet silty soils can be significantly increased by application of phosphorus. Plants of the lily and orchid families are well known to be showy, diverse, and abundant on the wet longleaf sites over both Ultisols and Spodosols, but to drop off with decreasing soil moisture (Walker and Peet 1984; Peet and Allard 1993). Savannas and wet flatwoods are also the area of greatest concentration of the region's rich assemblage of insectivorous plants. Palms are largely plants of the flatwoods; from the southeastern corner of South Carolina southward, saw palmetto (*Serenoa repens*) can be an aspect dominant of the ground layer, and other palms appear in particular flatwoods types (e.g., *Sabal palmetto, Sabal etonia*).

Toward a Community Classification

The primary purpose of this chapter is to present and describe a comprehensive classification of longleaf pine vegetation based on a compilation and synthesis of prior work, both my own and that of others. Such a classification can serve as a framework for understanding ecological variation, planning and assessing restoration, and developing conservation strategies. I chose to define the focus of this study as fire-maintained longleaf pine vegetation, plus related vegetation from within the natural range of longleaf where other pines dominate, or pines are typically present but too sparse to generally be described as among the dominants.

The most mature work on classification of longleaf vegetation is represented by the longleaf vegetation types treated in the U.S. National Vegetation Classification. As a first step I compiled all qualifying vegetation types (associations) within the U.S. National Vegetation Classification as of 2004. This classification is maintained by NatureServe on behalf of and in compliance with the standards of the Vegetation Subcommittee of the U.S. Federal Geographic Data Committee (FGDC), and with advice on standards from the Vegetation Panel of the Ecological Society of America. Most of the current content of this classification is the result of efforts by NatureServe staff or the staff of state Natural Heritage Programs. Details can be examined at the NatureServe Explorer website (http://www.natureserve.org/explorer/; see also Anderson et al. 1998). The associations recognized in the National Classification are derived from multiple sources and range from types based on careful, quantitative analysis of multiple plots to types based only on old literature descriptions or author field notes.

I also compiled the results of a series of four, plot-based studies conducted by my own research group (Duncan et al. 1994; Peet et al. 1994; Kjellmark et al. 1998; Peet, Carr and Gramling 2006). These studies covered the range of variation my colleagues and I could identify in extant stands of longleaf and related vegetation types of North Carolina, South Carolina, Georgia, and Florida. The studies were based on approximately 900 inventory plots of typically 1000 m^2 each and collected following the protocol of Peet et al. (1998). The plot data are available for further analysis in VegBank (http://vegbank.org). Plots were selected only from stands that appeared to be in relatively natural condition with a sustained history of low-intensity fire. Sites severely fire suppressed, impacted by pine-straw ranking or military training, or otherwise degraded were avoided.

For each of the four studies we used agglomerative cluster analysis methods to identify groups of relatively similar plots and examined the results of each in an effort to identify groups that were both ecologically interpretable and consistent across multiple classification methods. For our final cluster analysis we generally used the flexible beta group linkage clustering method (McCune and Grace 2003) with Sorenson's metric to quantify ecological distances among plots. In each case we compared our results with the types in the then current version of the National Vegetation Classification. This led to proposals for changes in the classification documented below. I also reviewed available published articles and book chapters that have summarized longleaf vegetation for particular places or geographic regions. These are described in the relevant locations later in this section.

To create a single classification for this review, I compiled the recognized longleaf types into six ecological groups related to physiognomy and soil type: (1) xeric sand barrens and uplands, (2) subxeric sandy uplands, (3) silty uplands, (4) clayey and rocky uplands, (5) flatwoods, and (6) savannas and seeps. These largely conform to the types recognized in our general gradient model above (Fig. 2), with the exception that clayey and rocky uplands are recognized primarily from the Piedmont and Mountains. I then sorted the associations within a group into the six ecoregions described above (Fig. 1): (1) Atlantic Coastal Plain, (2) Fall-line Sandhills, (3) Southern Coastal Plain, (4) Eastern Gulf Coastal Plain, (5) Western Gulf Coastal Plain, and (6) Piedmont and Montane Uplands. Where our plot-based studies suggest a need for revision of the National Classification, these changes are implemented in the compilation of types.

In the following sections I refer to the 135 vegetation associations recognized in Table 2 by three-digit code where the first digit refers to the six ecological groups, the second refers to the six ecoregions, and the third is a sequence number. Relationships between types I recognize and types recognized in the 2004 National Classification are indicated in Table 2 using a standard set of relationship symbols. > and < indicate "includes" and "included in," respectively; no symbol indicates the community concepts are similar, and ( ) indicates the name has been revised. >< indicates that the two community concepts overlap, but each has unique components. Aff. indicates a weak or undefined affinity or relationship.

Xeric Sand Barrens and Uplands

Deep, coarse, well-drained sands can be found scattered across the coastal plain portions of the longleaf ecosystem. Vegetation on these sites consists of scattered overstory longleaf with an oak understory and often little else. The sand is typically bright white and is conspicuous for lack of ground cover. Grasses are often sparse and the open ground layer normally includes low shrubs such as *Gaylussacia dumosa* and often a discontinuous mat of *Selaginella* and fruticose lichens (e.g., *Cladonia* and *Cladina*) (Fig. 3).

Pine barren vegetation types generally have lower fire-frequency than the other longleaf types owing to insufficient fuel production to carry frequent fire. Nonetheless, the vegetation is prone to degradation from fire suppression in that in the absence of fire scrub oaks such as *Quercus laevis* assume a dense crown cover, shading out much of the sparse herb and grass layer and leaving a more fire resistant, desertlike *Quercus laevis* community (1.2.2 — see Table 2). On the Western Gulf Coast, beyond the range of *Quercus laevis*, the same phenomenon occurs but with *Quercus incana* and *Q. margarettiae* assuming dominance (1.5.2).

Atlantic Coastal Plain

Pine barrens and extreme xeric woodlands of the Atlantic Coastal Plain are best developed on eolian sands on the northeast sides of major rivers (e.g., Altamaha, Cape Fear, Peedee, Savannah; see Bozeman 1971) and northeast sides of Carolina bays (Wells and Shunk 1931), but also occur in other areas with extensive sand deposits. On these sites one encounters an understory typically dominated by *Quercus laevis*, a layer of low shrubs such as *Gaylussacia dumosa* and *Vaccinium stamineum*, and a very sparse herb layer of xerophytes such as *Stipulicida setacea, Cnidoscolus stimulosus, Rhynchospora megalocarpa, Minuartia caroliniana, Euphorbia ipecacuanhae, Polygonella polygama, Bulbostylis ciliatifolia,* and *Selaginella acanthonota* (1.1.1). Composition shifts with proximity to the coast where *Quercus geminata* and *Q. hemisphaerica* share dominance with *Q. laevis* (1.1.3, 1.1.4); inland on slightly silty soils, *Q. margarettiae* and *Q. incana* sometimes share dominance with *Q. laevis*. Grasses are typically unimportant and wiregrass (*Aristida stricta* north of

TABLE 2. Vegetation associations of fire-maintained longleaf pine and related ecosystems of the Southeastern United States.[a]

Community	States	Plots	NVC type
Xeric Sand Barrens and Uplands			
Atlantic Coastal Plain			
1.1.1 *Pinus palustris/Quercus laevis/Aristida purpurascens–Stipulicida setacea–(Rhynchospora megalocarpa, Selaginella acanthonota)* Woodland	GA NC SC	7	<(CEGL003590)
1.1.2 *Pinus palustris/Quercus laevis–Quercus incana/Gaylussacia dumosa–Gaylussacia (baccata, frondosa)* Woodland	NC VA		CEGL003592
1.1.3 *Pinus palustris/Quercus laevis–Quercus geminata/Vaccinium tenellum/Aristida stricta* Woodland	NC SC	6	CEGL003589
1.1.4 *Pinus palustris–Pinus taeda/Quercus geminata–Quercus hemisphaerica–Osmanthus americanus var. americanus/Aristida stricta* Woodland	NC SC	4	CEGL003577
1.1.5 *Pinus palustris/Quercus laevis–Quercus myrtifolia/Chrysoma pauciflosculosa/ Aristida beyrichiana* Woodland	GA	2	<(CEGL003590)
1.1.6 *Pinus palustris/Quercus incana–Quercus margarettiae/Licania michauxii/Aristida beyrichiana* Woodland	FL? GA SC	2	CEGL004492
Fall-line Sandhills			
1.2.1 *Pinus palustris/Quercus laevis/Aristida stricta–Baptisia cinerea–Stylisma patens* Woodland	NC,SC	17	
1.2.2 *Pinus palustris/Quercus laevis/Aristida stricta/Cladonia* spp. Woodland	NC SC	9	CEGL003584
1.2.3 *Pinus palustris/Quercus laevis/Gaylussacia dumosa/(Schizachyrium scoparium, Aristida beyrichiana)* Woodland	SC GA	11	
1.2.4 *Pinus palustris/Quercus laevis/Chrysoma pauciflosculosa* Woodland	NC SC GA	2	<(CEGL003590)
Southern Coastal Plain			
1.3.1 *Pinus palustris/Quercus laevis–Q. incana/Aristida beyrichiana–Sorghastrum secundum* Woodland	FL	20	>(CEGL004491)
1.3.2 *Pinus palustris/Quercus laevis/Aristida beyrichiana–Tephrosia chrysophylla–Bulbostylis warei* Woodland	FL GA	10	>(CEGL008569)
Eastern Gulf Coastal Plain			
1.4.1 *Pinus palustris/Quercus laevis/Licania michauxii/Euphorbia floridana* Woodland	AL FL GA MS	11	><(CEGL003583) >(CEGL003587)
1.4.2 *Pinus palustris/Quercus laevis/Serenoa repens/Aristida condensata* Woodland	MS		CEGL003588
Western Gulf Coastal Plain			
1.5.1 *Pinus palustris/Quercus incana–Quercus margarettiae/Vaccinium arboreum/ Cnidoscolus texanus–Stylisma pickeringii var. pattersonii* Woodland	LA TX		CEGL003602
1.5.2 *Pinus palustris–Pinus (echinata, taeda)–Quercus (incana, margarettiae)/Schizachyrium scoparium* Woodland	LA TX		CEGL007513
Subxeric Sandy Uplands			
Atlantic Coastal Plain			
2.1.1 *Pinus palustris/Quercus laevis–Quercus (incana, margarettiae)/Gaylussacia frondosa/ Aristida stricta* Woodland	NC SC	23	<CEGL003591
2.1.2 *Pinus palustris/Quercus laevis–Quercus incana/Gaylussacia dumosa/Schizachyrium scoparium* Woodland	SC	2	<(CEGL003586)

3. Ecological Classification of Longleaf Pine Woodlands

2.1.3	Pinus palustris/Quercus incana–Quercus stellata/Aristida beyrichiana–Sporobolus junceus–Nolina georgiana Woodland	SC, GA	4	CEGL004487
2.1.4	Pinus palustris/Gaylussacia dumosa/Vaccinium tenellum/Vaccinium crassifolium–Aristida stricta Woodland	NC	18	
2.1.5	Pinus palustris/Quercus hemisphaerica/Gaylussacia dumosa/Pteridium aquilinum var. pseudocaudatum–Schizachyrium scoparium–Tephrosia virginiana Woodland	SC	6	
2.1.6	Pinus palustris/Clethra alnifolia–Quercus pumila/Pteridium aquilinum var. pseudocaudatum–Schizachyrium scoparium–Symphyotrichum walteri Woodland	SC	6	
2.1.7	Pinus palustris/Quercus laevis/Serenoa repens–Vaccinium stamineum/Aristida beyrichiana Woodland	GA	4	CEGL004490
2.1.8	Pinus palustris/Quercus laevis/Gaylussacia dumosa/Aristida beyrichiana–Helianthus atrorubens Woodland	GA SC	4	CEGL004488
2.1.9	Pinus palustris/Quercus marilandica–Quercus laevis/Aristida beyrichiana–Nolina georgiana Woodland	GA	4	CEGL004489
Fall-line Sandhills				
2.2.1	Pinus palustris/Quercus laevis–Quercus incana/Aristida beyrichiana–Baptisia perfoliata Woodland	GA SC		CEGL007844
2.2.2	Pinus palustris/Quercus incana–Quercus marilandica/Aristida beyrichiana–Nolina georgiana Woodland	GA SC		CEGL007842
2.2.3	Pinus palustris/Quercus laevis–(Quercus incana)/Toxicodendron pubescens/Schizachyrium scoparium Woodland	SC,GA	13	>(CEGL003593)
2.2.4	Pinus palustris/Quercus laevis/Aristida stricta–Tephrosia virginiana Woodland	NC,SC	6	<CEGL003591
				<CEGL003586
2.2.5	Pinus palustris/Aristida stricta–Coreopsis major–Rhexia alifanus Woodland	NC,SC	4	
2.2.6	Pinus palustris/Quercus laevis–Quercus incana/Toxicodendron pubescens/Aristida stricta–Astragalus michauxii Woodland	NC	6	
Southern Coastal Plain				
2.3.1	Pinus palustris/Quercus (incana, margarettiae)/Aristida beyrichiana–Rhynchosia reniformis Woodland	FL	28	(CEGL008586)
2.3.2	Pinus palustris/Quercus minima–Serenoa repens/Aristida beyrichiana–Carphephorus odoratissimus Woodland	AL? FL GA? MS	14	
2.3.3	Pinus palustris/Quercus geminata/Serenoa repens–Polygonella gracilis Woodland	FL	4	aff CEGL003601
2.3.4	Pinus palustris/Quercus chapmanii/Serenoa repens/Aristida beyrichiana–Pterocaulon virgatum Woodland	FL	8	aff CEGL007714
2.3.5	Pinus palustris/Quercus geminata/Serenoa repens/Schizachyrium scoparium var. stoloniferum–Cyperus (croceus, retrorsus) Woodland	FL	7	<(CEGL007714)
2.3.6	Pinus palustris/Quercus geminata/Aristida condensata–Cyperus plukenetii Woodland	FL	6	
Eastern Gulf Coastal Plain				
2.4.1	Pinus palustris/Licania michauxii/Aristida beyrichiana–Schizachyrium scoparium var. stoloniferum–Ionactis linariifolius Woodland	FL	18	><(CEGL003583)
2.4.2	Pinus palustris/Quercus incana/Sporobolus clandestinus Woodland	LA MS?		CEGL004957
2.4.3	Pinus palustris/Quercus laevis/Serenoa repens–Clinopodium coccineum Woodland	AL, GA, MS		CEGL003601
2.4.4	Pinus palustris/Quercus laevis–Quercus margarettiae/Licania michauxii/Aristida beyrichiana–Dyschoriste oblongata Woodland	GA	3	
2.4.5	Pinus palustris–Pinus (echinata, taeda)/Quercus (marilandica, laevis)/ Schizachyrium scoparium Woodland	AL GA? MS?		CEGL008491

(cont.)

TABLE 2. (Continued)

Community	States	Plots	NVC type
Western Gulf Coastal Plain			
2.5.1 Pinus palustris/Quercus incana/Schizachyrium scoparium–Croton argyranthemus Woodland	LA? TX		CEGL008572
2.5.2 Pinus palustris/Quercus incana/Schizachyrium scoparium–Liatris elegans–Opuntia humifusa var. humifusa Woodland	LA TX		CEGL003580
2.5.3 Pinus palustris–Pinus (echinata, taeda)–Quercus falcata–Carya texana Woodland	LA TX		CEGL008571
Silty Uplands			
Atlantic Coastal Plain			
3.1.1 Pinus palustris/Quercus pumila–Vaccinium tenellum/Aristida beyrichiana–Schizachyrium scoparium Woodland	GA SC	16	
3.1.2 Pinus palustris/Quercus stellata/Vaccinium tenellum/Aristida beyrichiana–Pteridium aquilinum var. pseudocaudatum–Schizachyrium scoparium Woodland	GA SC	5	
3.1.3 Pinus palustris/Clethra alnifolia–Gaylussacia frondosa–Quercus pumila/Schizachyrium scoparium Woodland	SC	20	CEGL004496
3.1.4 Pinus palustris/Quercus marilandica/ Vaccinium tenellum–Quercus pumila/ Schizachyrium scoparium Woodland	SC	8	(CEGL007738)
3.1.5 Pinus palustris/Vaccinium tenellum/Aristida stricta–Desmodium tenuifolium Woodland	NC, SC?	8	<(CEGL003569)
3.1.6 Pinus palustris/Gaylussacia dumosa/Aristida stricta–Ionactis linariifolius–Liatris pilosa var. pilosa Woodland	NC, SC?	16	<(CEGL003569)
3.1.7 Pinus palustris/Quercus marilandica/Clethra alnifolia–Rhododendron atlanticum/ Aristida stricta–Pycnanthemum flexuosum Woodland	NC, SC?	9	><(CEGL003664)
Fall-line Sandhills			
3.2.1 Pinus palustris/Quercus laevis–Quercus marilandica/Toxicodendron pubescens/ Aristida stricta–Tephrosia virginiana Woodland	NC, SC	21	(CEGL003578)
3.2.2 Pinus palustris/Quercus marilandica/Aristida stricta–Parthenium integrifolium Woodland	NC,SC	14	(CEGL003595)
3.2.3 Pinus palustris/Aristida stricta–Panicum virgatum–Eupatorium rotundifolium Woodland	NC	7	<(CEGL003573)
3.2.4 Pinus palustris/Schizachyrium scoparium–(Aristida beyrichiana)–Tephrosia virginiana Woodland	SC,GA	17	
3.2.5 Pinus palustris–Pinus taeda/Carya pallida–Cornus florida/Aristida stricta Woodland	NC	9	
Eastern Gulf Coastal Plain			
3.4.1 Pinus palustris/Schizachyrium scoparium var. stoloniferum–Sorghastrum nutans–Dyschoriste oblongifolia Woodland	AL? FL GA	21	>(CEGL004485) >(CEGL007749)
3.4.2 Pinus palustris/Quercus falcata/Cornus florida/Aristida beyrichiana Woodland	AL, FL?, GA		CEGL004945
3.4.3 Pinus palustris/Vaccinium myrsinites/Aristida beyrichiana–Schizachyrium scoparium var. stoloniferum–Liatris gracilis Woodland	FL	9	
3.4.4 Pinus palustris/Schizachyrium scoparium/Verbesina aristata Loam–hill Woodland	AL GA		CEGL008452
3.4.5 Pinus palustris/Quercus falcata/Cornus florida/Schizachyrium scoparium Woodland	AL LA MS		CEGL003575

3. Ecological Classification of Longleaf Pine Woodlands

3.4.6	*Pinus palustris/Quercus marilandica/Schizachyrium scoparium–Schizachyrium tenerum–Rhexia alifanus* Woodland		AL? MS	CEGL003598
3.4.7	*Pinus palustris/Schizachyrium scoparium–Coreopsis tripteris–Baptisia bracteata* var. *leucophaea* Woodland		LA MS?	CEGL004955
3.4.8	*Pinus palustris–Pinus (echinata, taeda)/Schizachyrium tenerum–Vernonia angustifolia* Woodland		MS	CEGL004774
Western Gulf Coastal Plain				
3.5.1	*Pinus palustris/Schizachyrium scoparium–Liatris pycnostachya* Woodland		LA TX	CEGL003571
3.5.2	*Pinus palustris/Schizachyrium scoparium–Rudbeckia grandiflora* var. *alismifolia* Woodland		LA TX	CEGL003572
3.5.3	*Pinus palustris/Schizachyrium scoparium–Schizachyrium tenerum–Silphium gracile* Woodland		LA TX?	CEGL003581
3.5.4	*Pinus palustris/Quercus marilandica/Schizachyrium tenerum–Bigelowia nuttallii–Packera obovata* Woodland		LA	CEGL003597
Rocky and Clayey Uplands				
Atlantic Coastal Plain and Fall-line Sandhills				
4.1.1	*Pinus palustris/Kalmia latifolia–Vaccinium arboreum* Woodland	4	SC, NC?	>CEGL003599
4.1.2	*Pinus palustris/Quercus marilandica/Vaccinium crassifolium/Aristida stricta* Woodland	12	NC SC	>(CEGL007767)
4.1.3	*(Pinus palustris)/Bigelowia nuttallii–Talinum teretifolium–Allium cuthbertii–Penstemon dissectus* Altamaha Grit Herbaceous Vegetation		GA	CEGL004783
Eastern Gulf Coastal Plain				
4.4.1	Insufficient data to document; no extant stands			
Western Gulf Coastal Plain				
4.5.1	*Pinus palustris/Quercus marilandica/Schizachyrium scoparium–Silphium laciniatum–Ruellia humilis* Woodland		LA TX	CEGL003596
4.5.2	*Pinus palustris/Quercus marilandica/Ilex vomitoria/Schizachyrium scoparium* Woodland		LA TX	CEGL003579
4.5.3	*Pinus palustris/Quercus marilandica/Panicum virgatum* Woodland		TX	CEGL008580
4.5.4	*(Pinus palustris)/Schizachyrium scoparium–Bigelowia nuttallii/Cladonia* spp. Herbaceous Vegetation		LA	CEGL003600
Piedmont and Montane Uplands				
4.6.1	*Pinus palustris–Pinus echinata–(Pinus virginiana)/Quercus marilandica–(Quercus prinus)/Vaccinium pallidum* Woodland	1	AL GA NC SC?	CEGL008437
4.6.2	*Pinus palustris–(Pinus echinata)/Oxydendrum arboreum/Vaccinium pallidum–Vaccinium tenellum/Pteridium aquilinum* var. *pseudocaudatum* Woodland	5	NC	
4.6.3	*Quercus prinus–Quercus palustris* Forest		AL	CEGL004060
4.6.4	*Pinus palustris–Pinus echinata/Schizachyrium scoparium–Manfreda virginica* Serpentine Woodland	1	GA	CEGL003608
4.6.5	*Pinus palustris–Pinus echinata/Quercus coccinea–Quercus georgiana* Woodland		GA	CEGL004432

(*cont.*)

TABLE 2. (Continued)

Community	States	Plots	NVC type
Flatwoods			
Atlantic Coastal Plain			
5.1.1 Pinus palustris–Pinus serotina/Gaylussacia dumosa/Vaccinium crassifolium–Aristida stricta Woodland	NC SC	28	(CEGL003648)
5.1.2 Pinus palustris–Pinus serotina/Ctenium aromaticum–Muhlenbergia expansa–Carphephorus odoratissimus Woodland	GA? NC SC		CEGL003658
5.1.3 Pinus palustris–Pinus serotina/Pleea tenuifolia–Aristida stricta Woodland	NC	17	CEGL003661
5.1.4 Pinus palustris/Leiophyllum buxifolium/Aristida stricta Woodland	NC	12	CEGL003649
5.1.5 Pinus palustris/(Pinus serotina)/Ilex glabra–Gaylussacia frondosa–(Kalmia carolina) Woodland	NC VA		CEGL003647
Southern Coastal Plain			
5.3.1 Pinus palustris/Serenoa repens–Vaccinium myrsinites/Aristida beyrichiana–Sporobolus curtissii Woodland	FL GA	8	CEGL004486 ><(CEGL003653)
5.3.2 Pinus palustris–(Pinus elliottii var. elliottii)/Sporobolus pinetorum–Oclemena reticulata–(Sporobolus curtissii) Woodland	GA	17	>CEGL004967 >(CEGL004969)
5.3.3 Pinus (palustris, elliottii var. elliottii)/Gaylussacia nana–Serenoa repens–Vaccinium myrsinites/Sporobolus (floridanus, curtissii) Woodland	FL GA	7	<(CEGL004791) >(CEGL003797)
5.3.4 Pinus palustris/Serenoa repens–Quercus pumila/Aristida beyrichiana–Balduina uniflora Woodland	AL FL GA	11	>(CEGL003808) ><(CEGL003653)
5.3.5 Pinus elliottii var. elliottii–Pinus palustris/Lyonia ferruginea–Serenoa repens/ Aristida beyrichiana Woodland	FL	6	
5.3.6 Pinus elliottii var. elliottii/Lyonia lucida–Serenoa repens/Aristida beyrichiana–Befaria racemosa Woodland	FL	5	
5.3.7 Lyonia fruticosa–Serenoa repens–Quercus minima/Aristida beyrichiana–Elephantopus elatus Shrubland	FL	19	(CEGL004236)
5.3.8 Serenoa repens/Aristida beyrichiana–Scleria muehlenbergii–Eragrostis elliottii Herbaceous Vegetation	FL	9	aff CEGL003795
5.3.9 Pinus elliottii var. densa/Quercus minima/Panicum abscissum Woodland	FL		CEGL003650
5.3.10 Panicum abscissum Herbaceous Vegetation	FL	1	CEGL004113
5.3.11 Quercus (chapmanii, myrtifolia)–Serenoa repens/Aristida beyrichiana–Chapmannia floridana Shrubland	FL	8	(CEGL007750)
5.3.12 Pinus elliottii var. elliottii/Quercus chapmanii–Quercus geminata–Serenoa repens/ Rhynchospora megalocarpa Woodland	AL FL GA? LA MS SC	2	<(CEGL004658)
5.3.13 Pinus palustris–Pinus elliottii var. elliottii/Quercus chapmanii–Quercus myrtifolia–Quercus geminata–Lyonia ferruginea Woodland	FL GA		CEGL003662
Eastern Gulf Coastal Plain			
5.4.1 Pinus elliottii var. elliottii/Ilex vomitoria–Serenoa repens–Morella cerifera Woodland	AL FL LA MS		<CEGL004658
5.4.2 Pinus elliottii var. elliottii/Serenoa repens–Ilex glabra–Morella cerifera–Ilex vomitoria Woodland	AL FL		CEGL004680
5.4.3 Pinus palustris–(Pinus elliottii var. elliottii)/Ilex coriacea–Cyrilla racemiflora Woodland	MS		CEGL003656

3. Ecological Classification of Longleaf Pine Woodlands

Savannas and Seeps				
Atlantic Coastal Plain				
6.1.1	*Pinus palustris/Sporobolus pinetorum–Schizachyrium scoparium–Eryngium integrifolium* Woodland	NC SC	16	>(CEGL004501)
6.1.2	*Pinus palustris–Pinus serotina/Ctenium aromaticum–Muhlenbergia expansa–Rhynchospora latifolia* Woodland	GA NC SC	5	>(CEGL004086)
6.1.3	*Pinus palustris–Pinus (serotina, taeda)/Sporobolus curtissii*–Woodland	SC	2	CEGL003660
6.1.4	*Pinus palustris–Pinus serotina/Aristida palustris–Sarracenia flava* Woodland	SC	1	CEGL004085
6.1.5	*Pinus palustris–Pinus serotina/Sporobolus pinetorum–Ctenium aromaticum–Eriocaulon decangulare* Woodland	NC	7	CEGL004498
6.1.6	*Pinus palustris–Pinus serotina/Magnolia virginiana/Sporobolus teretifolius–Carex striata* Woodland	NC	2	CEGL004502
6.1.7	*Pinus palustris–Pinus serotina/Ctenium aromaticum–Scleria pauciflora–Sarracenia flava* Woodland	SC	4	CEGL004500
6.1.8	*Pinus palustris/Arundinaria gigantea* ssp. *tecta–Liquidambar styraciflua/Andropogon glomeratus–Sarracenia minor* Woodland	SC NC	5	CEGL004499
				CEGL004495
Fall-line Sandhills				
6.2.1	*Pinus serotina–Pinus palustris/Osmunda cinnamomea/Dichanthelium dichotomum* var. *ensifolium* Woodland	NC,SC	7	CEGL003659
6.2.2	*Pinus palustris–Pinus serotina/Ctenium aromaticum–Muhlenbergia expansa–Calamovilfa brevipilis* Woodland	NC SC		
Southern Coastal Plain				
6.3.1	*Pinus palustris–Pinus elliottii* var. *elliottii/Styrax americanus/Sporobolus floridanus* Woodland	GA SC	2	CEGL004497
6.3.2	*Pinus palustris–Pinus elliottii* var. *elliottii/Ctenium aromaticum–Aristida beyrichiana–(Sporobolus floridanus)*	GA SC		CEGL004790
6.3.3	*Pinus palustris/Ilex glabra/Aristida beyrichiana–Helianthus radula–Galactia erecta* Woodland	FL	6	<(CEGL003653)
6.3.4	*Aristida beyrichiana–Rhynchospora oligantha–Scleria muehlenbergii–Sabatia macrophylla* Herbaceous Vegetation		6	>(CEGL004154)
				>(CEGL004152)
6.3.5	*Pinus elliottii* var. *elliottii/Aristida beyrichiana–Eriocaulon decangulare–Lachnanthes caroliana* Woodland	FL	5	>(CEGL003673)
				>(CEGL004153)
6.3.6	*Pinus elliottii* var. *elliottii/Ilex glabra/Andropogon glaucopsis–Panicum rigidulum–Panicum hemitomon* Woodland	FL GA SC	3	
6.3.7	*Pinus elliottii* var. *elliottii–Sabal palmetto/Morella cerifera/Panicum virgatum* Woodland	FL	3	<(CEGL004958)
6.3.8	*Aristida beyrichiana–Amphicarpum muehlenbergianum–Scleria muehlenbergii–Oxypolis filiformis* Herbaceous Vegetation	FL	4	>(CEGL003795)
6.3.9	*Amphicarpum muehlenbergianum–Aristida palustris–Sporobolus floridanus* Herbaceous Vegetation	FL	1	aff CEGL004790
6.3.10	*Panicum tenerum–Aristida beyrichiana–Schizachyrium rhizomatum–Rhynchospora tracyi* Herbaceous Vegetation	FL	3	<(CEGL004105)
Eastern Gulf Coastal Plain				
6.4.1	*Aristida beyrichiana–Ctenium aromaticum–Andropogon arctatus–Rhynchospora chapmanii* Herbaceous Vegetation	AL FL	13	>(CEGL004155)
				>(CEGL003645)
6.4.2	*Rhynchospora oligantha–Sarracenia* (*alata, psittacina*)*–Carphephorus pseudoliatris* Herbaceous Vegetation	LA MS		CEGL004687
6.4.3	*Pinus palustris/Schizachyrium scoparium–Muhlenbergia expansa–Helianthus radula* Woodland	LA MS?		CEGL004956
6.4.4	*Quercus pumila–Rhododendron canescens/Ctenium aromaticum–Tephrosia spicata* Herbaceous vegetation	FL	5	
6.4.5	*Dichanthelium wrightianum–Panicum verrucosum–Dichanthelium erectifolium* Herbaceous Vegetation		2	<(CEGL004105)

(*cont.*)

TABLE 2. (Continued)

Community		States	Plots	NVC type
6.4.6	Pinus serotina/Arundinaria gigantea ssp. tecta/Ctenium aromaticum–Pycnanthemum flexuosum Woodland	FL GA?		
6.4.7	Pinus palustris–(Pinus elliottii var. elliottii)/Ilex vomitorial–Agalinis filicaulis Woodland	LA? MS	2	CEGL004792
Western Gulf Coastal Plain				
6.5.1	Pinus palustris/Rhynchospora elliottii–Lobelia flaccidifolia–Platanthera nivea–(Helenium drummondii) Woodland	LA TX		CEGL007802
6.5.2	Pinus palustris/Eryngium integrifolium–Rhynchospora spp.–(Ctenium aromaticum) Woodland	LA TX?		CEGL003646
6.5.3	Sarracenia alata–Rhynchospora gracilenta–Rudbeckia scabrifolia–Schoenolirion croceum Herbaceous Vegetation	LA TX		CEGL004175
6.5.4	Pinus palustris/Sporobolus silveanus–Muhlenbergia capillaris–Chaetopappa asteroides Woodland	LA TX?		CEGL003654
Piedmont and Montane Uplands				
6.6.1	Pinus palustris–Pinus taeda–Pinus serotina/Chasmanthium laxum–Panicum virgatum Piedmont Woodland	NC SC?	2	CEGL003663

[a] Associations shown were collected from (1) the U.S. National Vegetation Classification (USNVC) as maintained by NatureServe (see NatureServe 2005), (2) Duncan and Peet (1994), (3) Peet and Duncan (1994), (4) Kjellmark et al. (1998), and (5) Peet, Carr and Gramling (2006). Where the type was derived or validated in one of the last four references, the number of supporting vegetation plots is indicated. Synonymy relationships with the USNVC are indicated by NatureServe codes. Where the name used differs from the Name in the USNVC, the NatureServe code is shown in parentheses. Where the concept or circumscription of the association differs from the USNVC type, the relationship is indicated symbolically as < (included in), > (includes), >< (overlaps, but each type includes occurrences not in the other), () (roughly equivalent, though with a different name), aff. (weak or undefined affinity or similarity).

FIGURE 3. Xeric sand barrens and uplands, Atlantic Coastal Plain. *Pinus palustris* and *Quercus laevis* barren on the rim of a Carolina bay. The characteristic sparse ground layer consists primarily of small mats of *Selaginella acanthonota* and *Minuartia caroliniana*. Salters Lake, Bladen County, NC.

central South Carolina, *A. beyrichiana* to the south) is limited to the favorable microsites. There is a conspicuous latitudinal gradient in composition; the few examples remaining in northeastern North Carolina and southeastern Virginia (1.1.2) have a conspicuously low diversity of xerophytic forbs and woody plants compared with examples from the southern extreme of the ecoregion. In particular, *Chrysoma pauciflosculosa* and *Ceratiola ericoides* occur occasionally northward into South Carolina and *Quercus chapmanii* and *Q. myrtifolia* occupy extreme sites on dune systems associated with rivers in Georgia (1.1.5). *Licania michauxii* occurs on a few sites north of the Savannah River where it defines the northern boundary of a longleaf, mixed oak type of xeric sands with the xerophytes more typical of Georgia where additional southern taxa such as *Nolina georgiana* and *Serenoa repens* occur (1.1.6).

Fall-line Sandhills

The extreme xeric sites of the Fall-line Sandhills resemble those of the Atlantic Coastal Plain in the co-dominance of *Pinus palustris* and *Quercus laevis*, as well as the occurrence of extreme xerophytic herbs such as *Stipulicida setacea*, *Cnidoscolus stimulosus*, and *Euphorbia ipecacuanhae*. Grasses remain sparse but are somewhat more continuous than on the extreme Coastal Plain sites and with the same latitudinal gradient in dominance: *Aristida stricta* in the north (1.2.1), *Aristida beyrichiana* in the south (1.2.3), and *Schizachyrium scoparium* throughout but particularly conspicuous between the ranges of the two wiregrasses in South Carolina. As on the true Coastal Plain, some extreme sites in the Georgia and South Carolina sandhills support a shrub layer with specialist xerophytes such as *Chrysoma pauciflosculosa* and *Ceratiola ericoides* (1.2.4).

TABLE 3. Scientific names in text and associated common names.[a]

Scientific name	Common name
Agalinis filicaulis	Jackson false foxglove
Aletris	colicroot
Allium cuthbertii	striped garlic
Amphicarpum muehlenbergianum	Muhlenberg maidencane
Andropogon arctatus	pinewoods bluestem
Andropogon glaucopsis	purple bluestem
Andropogon glomeratus	bushy bluestem
Andropogon gyrans	Elliott's bluestem
Andropogon gyrans var. *stenophyllus*	Elliott's bluestem
Andropogon mohrii	Mohr's bluestem
Andropogon virginicus	broomsedge bluestem
Anthaenantia rufa	purple silkyscale
Aristida beyrichiana	Beyrich threeawn (Southern wiregrass)
Aristida condensata	piedmont threeawn
Aristida mohrii	Mohr's threeawn
Aristida palustris	longleaf threeawn
Aristida purpurascens	arrowfeather threeawn
Aristida stricta	pineland threeawn (Carolina wiregrass)
Arnoglossum floridanum	Florida cacalia
Arundinaria gigantea ssp. *tecta*	switchcane
Astragalus michauxii	sandhills milkvetch
Balduina angustifolia	coastalplain honeycombhead
Balduina uniflora	oneflower honeycombhead
Baptisia alba	white wild indigo
Baptisia bracteata var. *leucophaea*	longbract wild indigo
Baptisia cinerea	grayhairy wild indigo
Baptisia perfoliata	catbells
Befaria racemosa	tarflower
Berlandiera subacaulis	Florida greeneyes
Bigelowia nuttallii	Nuttall's rayless goldenrod
Bulbostylis ciliatifolia	capillary hairsedge
Bulbostylis warei	Ware's hairsedge
Calamovilfa brevipilis	pine barren sandreed
Calopogon	grasspink
Carex lutea	sulphur sedge
Carex striata	Walter's sedge
Carphephorus corymbosus	coastalplain chaffhead
Carphephorus odoratissimus	vanillaleaf
Carphephorus paniculatus	hairy chaffhead
Carphephorus pseudoliatris	bristleleaf chaffhead
Carya pallida	sand hickory
Carya texana	black hickory
Chaetopappa asteroides	Arkansas leastdaisy
Chapmannia floridana	Florida alicia
Chasmanthium laxum	slender woodoats
Chrysoma pauciflosculosa	woody goldenrod
Chrysopsis mariana	Maryland goldenaster
Cladina	reindeer lichen
Cladonia	cup lichen
Cleistes	rosebud orchid
Clethra alnifolia	coastal sweetpepperbush
Clinopodium coccineum	scarlet calamint
Clinopodium georgianum	Georgia calamint
Cnidoscolus stimulosus	finger rot (spurge-nettle)
Cnidoscolus texanus	Texas bullnettle

3. Ecological Classification of Longleaf Pine Woodlands

TABLE 3. *(Continued)*

Scientific name	Common name
Conradina canescens	false rosemary
Coreopsis major	greater tickseed
Coreopsis tripteris	tall tickseed
Cornus florida	flowering dogwood
Croton argyranthemus	healing croton
Ctenium aromaticum	toothache grass
Cyperus croceus	Baldwin's flatsedge
Cyperus plukenetii	Plukenet's flatsedge
Cyperus retrorsus	pine barren flatsedge
Cyrilla racemiflora	swamp titi
Desmodium tenuifolium	slimleaf ticktrefoil
Dichanthelium dichotomum var. *ensifolium*	cypress panicgrass
Dichanthelium erectifolium	erectleaf panicgrass
Dichanthelium leucothrix	rough panicgrass
Dichanthelium wrightianum	Wright's rosette grass
Dionaea muscipula	Venus flytrap
Diospyros virginiana	common persimmon
Drosera	sundew
Dyschoriste oblongifolia	oblongleaf snakeherb
Echinacea sanguinea	sanguin purple coneflower
Elephantopus elatus	tall elephantsfoot
Eragrostis elliottii	field lovegrass
Eriocaulon decangulare	tenangle pipewort
Eriocaulon decangulare var. *decangulare*	tenangle pipewort
Eriogonum tomentosum	dogtongue buckwheat
Eryngium integrifolium	blueflower eryngo
Eupatorium mohrii	Mohr's thoroughwort
Eupatorium rotundifolium	roundleaf thoroughwort
Euphorbia floridana	Greater Florida spurge
Euphorbia ipecacuanhae	American ipecac
Euthamia tenuifolia var. *tenuifolia*	slender goldentop
Galactia erecta	erect milkpea
Gaylussacia baccata	black huckleberry
Gaylussacia dumosa	dwarf huckleberry
Gaylussacia frondosa	blue huckleberry
Gaylussacia nana	Confederate huckleberry
Gaylussacia tomentosa	hairytwig huckleberry
Helenium drummondii	fringed sneezeweed
Helianthus atrorubens	purpledisk sunflower
Helianthus radula	rayless sunflower
Hypericum fasciculatum	peelbark St. Johnswort
Hypericum lloydii	sandhill St. Johnswort
Ilex coriacea	large gallberry
Ilex glabra	inkberry
Ilex vomitoria	yaupon
Ionactis linariifolius	flaxleaf whitetop aster
Kalmia carolina	Carolina laurel
Kalmia hirsuta	hairy laurel
Kalmia latifolia	mountain laurel
Lachnanthes caroliana	Carolina redroot
Leiophyllum buxifolium	sandmyrtle
Liatris elegans	pinkscale blazing star
Liatris gracilis	slender blazing star
Liatris pilosa var. *pilosa*	shaggy blazing star
Liatris pycnostachya	prairie blazing star
Licania michauxii	gopher apple

TABLE 3. (Continued)

Scientific name	Common name
Lilium	lily
Liquidambar styraciflua	sweetgum
Lobelia flaccidifolia	foldear lobelia
Ludwigia linifolia	southeastern primrose-willow
Lyonia ferruginea	rusty staggerbush
Lyonia fruticosa	coastalplain staggerbush
Lyonia lucida	fetterbush lyonia
Magnolia grandiflora	southern magnolia
Magnolia virginiana	sweetbay
Manfreda virginica	false aloe
Minuartia caroliniana	pinebarren stitchwort
Morella cerifera	wax myrtle
Muhlenbergia capillaris	hairawn muhly
Muhlenbergia expansa	cutover muhly
Nolina georgiana	Georgia beargrass
Nyssa sylvatica	blackgum
Oclemena reticulata	pinebarren whitetop aster
Opuntia humifusa var. *humifusa*	devil's-tongue (Eastern pricklypear)
Osmanthus americanus var. *americanus*	devilwood (wild olive)
Osmunda cinnamomea	cinnamon fern
Oxydendrum arboreum	sourwood
Oxypolis filiformis	water cowbane
Packera obovata	roundleaf ragwort
Panicum abscissum	cutthroat grass
Panicum hemitomon	maidencane
Panicum rigidulum	redtop panicgrass
Panicum tenerum	bluejoint panicgrass
Panicum verrucosum	warty panicgrass
Panicum virgatum	switchgrass
Parnassia caroliniana	Carolina grass of Parnassus
Parthenium integrifolium	wild quinine
Penstemon dissectus	dissected beardtongue
Pinguicula	butterwort
Pinus echinata	shortleaf pine
Pinus elliottii var. *densa*	Florida slash pine
Pinus elliottii var. *elliottii*	Honduras pine (slash pine)
Pinus palustris	longleaf pine
Pinus serotina	pond pine
Pinus taeda	loblolly pine
Pinus virginiana	Virginia pine
Pityopsis aspera	pineland silkgrass
Platanthera	fringed orchid
Platanthera nivea	snowy orchid
Pleea tenuifolia	rush featherling
Pogonia	pogonia
Polygonella gracilis	tall jointweed
Polygonella polygama	October flower
Pteridium aquilinum var. *pseudocaudatum*	western brackenfern (tailed bracken)
Pterocaulon virgatum	wand blackroot
Pycnanthemum flexuosum	Appalachian mountainmint
Pyxidanthera barbulata	flowering pixiemoss
Quercus chapmanii	Chapman oak
Quercus coccinea	scarlet oak
Quercus falcata	southern red oak
Quercus geminata	sand live oak
Quercus georgiana	Georgia oak

TABLE 3. *(Continued)*

Scientific name	Common name
Quercus hemisphaerica	Darlington oak (sand laurel oak)
Quercus incana	bluejack oak
Quercus inopina	sandhill oak
Quercus laevis	turkey oak
Quercus margarettiae	runner oak (sand post oak)
Quercus marilandica	blackjack oak
Quercus minima	dwarf live oak
Quercus myrtifolia	myrtle oak
Quercus palustris	pin oak
Quercus prinus	chestnut oak
Quercus pumila	running oak
Quercus stellata	post oak
Rhexia alifanus	savannah meadowbeauty
Rhododendron atlanticum	dwarf azalea
Rhododendron canescens	mountain azalea
Rhynchosia cytisoides	royal snoutbean
Rhynchosia reniformis	dollarleaf
Rhynchospora	beaksedge
Rhynchospora chapmanii	Chapman's beaksedge
Rhynchospora elliottii	Elliott's beaksedge
Rhynchospora gracilenta	slender beaksedge
Rhynchospora latifolia	sandswamp whitetop
Rhynchospora megalocarpa	sandyfield beaksedge
Rhynchospora oligantha	featherbristle beaksedge
Rhynchospora tracyi	Tracy's beaksedge
Rudbeckia grandiflora var. *alismifolia*	rough coneflower
Rudbeckia scabrifolia	roughleaf coneflower
Ruellia humilis	fringeleaf wild petunia
Sabal palmetto	cabbage palmetto
Sabatia macrophylla	largeleaf rose gentian
Sarracenia	pitcherplant
Sarracenia alata	yellow trumpets
Sarracenia flava	yellow pitcherplant
Sarracenia minor	hooded pitcherplant
Sarracenia psittacina	parrot pitcherplant
Schizachyrium rhizomatum	Florida little bluestem
Schizachyrium scoparium	little bluestem
Schizachyrium scoparium var. *stoloniferum*	creeping bluestem
Schizachyrium tenerum	slender little bluestem
Schoenolirion croceum	yellow sunnybell
Scleria muehlenbergii	Muehlenberg's nutrush
Scleria pauciflora	fewflower nutrush
Selaginella acanthonota	spiny spikemoss
Serenoa repens	saw palmetto
Silphium compositum	kidneyleaf rosinweed
Silphium gracile	slender rosinweed
Silphium laciniatum	compassplant
Sorghastrum nutans	Indiangrass
Sorghastrum secundum	lopsided Indiangrass
Spiranthes	ladies'-tresses
Sporobolus clandestinus	rough dropseed
Sporobolus curtissii	Curtis' dropseed
Sporobolus floridanus	Florida dropseed
Sporobolus junceus	pineywoods dropseed
Sporobolus pinetorum	Carolina dropseed
Sporobolus silveanus	Silveus' dropseed

TABLE 3. (Continued)

Scientific name	Common name
Sporobolus teretifolius	wireleaf dropseed
Stipulicida setacea	pineland scalypink
Stylisma patens	coastalplain dawnflower
Stylisma pickeringii var. *pattersonii*	Patterson's dawnflower
Styrax americanus var. *pulverulentus*	downy American snowbell
Styrax americanus	American snowbell
Symphyotrichum adnatum	scaleleaf aster
Symphyotrichum walteri	Walter's aster
Talinum teretifolium	quill fameflower
Tephrosia chrysophylla	scurf hoarypea
Tephrosia mohrii	pineland hoarypea
Tephrosia spicata	spiked hoarypea
Tephrosia virginiana	Virginia tephrosia
Thalictrum cooleyi	Cooley's meadow-rue
Tofieldia	tofieldia
Toxicodendron pubescens	Atlantic poison oak
Utricularia	bladderwort
Vaccinium arboreum	farkleberry
Vaccinium crassifolium	creeping blueberry
Vaccinium darrowii	Darrow's blueberry
Vaccinium fuscatum	black highbush blueberry
Vaccinium myrsinites	shiny blueberry
Vaccinium pallidum	Blue Ridge blueberry
Vaccinium stamineum	deerberry
Vaccinium tenellum	small black blueberry
Verbesina aristata	coastalplain crownbeard
Verbesina chapmanii	Chapman's crownbeard
Vernonia angustifolia	tall ironweed
Zigadenus	deathcamas

[a] Scientific and common names follow USDA, NRCS (2005), except for *Sporobolus*, which follows Peterson et al. (2003), *Muhlenbergia*, which follows Peterson (2003), and *Styrax*, which follows Gonsoulin (1974). Alternate common names in common usage are suggested in parentheses.

Southern Coastal Plain

Although *Quercus geminata* is confined to the maritime climate of the extreme coastal fringe in the Atlantic Coastal Plain Ecoregion, it occurs throughout peninsular Florida, probably because of the more moderate climate. Here extreme xeric pine barrens are co-dominated by *Quercus geminata*, *Q. laevis*, and the ubiquitous longleaf pine. *Quercus incana* can also be important, but decreases in importance south of northern Florida. The sparse grass layer is diverse with *Aristida beyrichiana* being joined by southern specialties such as *Schizachyrium scoparium* var. *stoloniferum*, and *Sorghastrum secundum*. Several species act as strong indicators of Florida barrens including *Arnoglossum floridanum*, *Desmodium floridanum*, and *Berlandiera subacaulis* in north-central peninsular Florida (1.3.1) and *Bulbostylis warei*, *Tephrosia chrysophylla*, *Balduina angustifolia*, and *Carphephorus corymbosus* on sites in the central highlands (1.3.2). Where fire has been suppressed for some years, *Ceratiola ericoides* tends to invade, reducing herb layer cover, and shifting the fire regime toward less frequent, more catastrophic events.

Eastern Gulf Coastal Plain

Vegetation of the Eastern Gulf Coastal Plain is reminiscent of that of the Atlantic Coastal Plain, particularly the coastal fringe of western

Florida where the subcanopy is primarily dominated by *Quercus laevis* and the understory has a significant component of *Aristida beyrichiana*. However, *Aristida beyrichiana* is largely absent from the northern half of the western Panhandle, the range limit (and edge of the ecoregion) crossing Eglin Air Force Base (Rodgers and Provencher 1999) and eventually reaching the coast in eastern Mississippi (Peet and Allard 1993; Peet 1993). As in the southern portion of the Atlantic Coastal Plain, there is a high frequency of *Eriogonum tomentosum, Licania michauxii, Andropogon virginicus*, and *Pityopsis aspera*, but also present are such distinctive taxa as *Rhynchosia cytisoides, Tephrosia mohrii*, and *Aristida mohrii* (1.4.1). Westward the impact of the Mississippi River drainage is expressed in the increasing content of silt in sediments relative to sand such that extreme xeric sites are much less common west of Florida. A relatively distinctive form does occur in the vicinity of Camp Shelby, Mississippi *Quercus laevis* dominates the understory and *Serenoa repens* is an important component, but, being beyond the range of wiregrass, the grass layer is dominated by *Aristida condensata* and to a lesser extent by *Andropogon ternarius* and *Sorghastrum secundum* (1.4.2).

Western Gulf Coastal Plain

Extreme xeric sites of the Western Gulf Coastal Plain are distinctive because they are beyond the ranges of both wiregrass and *Quercus laevis* (see Bridges and Orzell 1989; Harcombe et al. 1993). The dominant oaks here are *Quercus incana* and *Q. margarettiae*, and the herb layer contains a number of western specialties closely related to eastern taxa, two clear examples being *Cnidoscolus texanus* and *Stylisma pickeringii* var. *pattersonii* (1.5.1). With fire suppression, diversity declines and oak increases in importance (1.5.2).

Subxeric Sandy Uplands

Longleaf pine landscapes with topographic relief of several meters and deep, sandy soils are consistently droughty as precipitation rapidly dissipates via percolation or evaporation. Density and height of understory tree vegetation depends heavily on past fire and land-use history with fire suppression leading to increased oak density, though upland oak species are common throughout. These species typically include *Quercus laevis, Q. incana, Q. margarettiae* and in the more southern areas *Q. geminata*. Unlike in the xeric barrens, the ground layer is a nearly continuous sward of grass. Wiregrass (*Aristida stricta, A. beyrichiana*) is the dominant grass within its range, and elsewhere *Schizachyrium scoparium* dominates (Figs. 4, 5).

Atlantic Coastal Plain

Subxeric longleaf woodlands of the Atlantic Coast Plain are generally of two main types, one occupying low sandhills and the other on flatter, moister terrain transitional to flatwoods—here called dry flatwoods. The sandhill systems generally have an understory layer of *Quercus laevis* and *Q. incana*. In the north, *Aristida stricta* dominates the ground layer (2.1.1); in the wiregrass gap of central South Carolina *Schizachyrium scoparium* dominates (2.1.2); and south of the gap from southern South Carolina southward *Aristida beyrichiana* dominates with a scattering of *Sporobolus junceus* (2.1.3). The dry flatwoods types also differ latitudinally. In the north *Aristida stricta* is again dominant, and *Vaccinium crassifolium* can be very important as it is in true flatwoods of the region (2.1.4). Coastward in the wiregrass gap there is a maritime form with *Quercus hemisphaerica* (2.1.5) and a more inland form with *Quercus pumila* important (2.1.6). From southeasternmost South Carolina south along the eastern portion of the Atlantic Coastal Plain are dry flatwoods where *Aristida beyrichiana* is the dominant grass and *Serenoa repens* is conspicuous (2.1.7). Throughout the dry flatwoods, as in the true flatwoods, *Pteridium aquilinum* var. *pseudocaudatum* is abundant. Finally, two more types characterize the uplands of the inner Atlantic Coastal Plain, one with a dense sward of *Aristida beyrichiana* and scattered diagnostic herbs such as *Helianthus atrorubens* (2.1.8), and the other characteristic of the Altamaha Grit

FIGURE 4. Subxeric sandy uplands, Southern Coastal Plain. *Pinus palustris, Quercus laevis* woodland on a fluvial dune. The ground layer of *Aristida beyrichiana* and *Sporobolus junceus* is continuous but relatively sparse. Fort Stewart, Bryan County, GA.

region with *Nolina georgiana* as a common understory species (2.1.9).

Fall-line Sandhills

Subxeric longleaf woodlands of the Fall-line Sandhills typically have an understory layer dominated by *Quercus laevis*, which on slightly silty sites ("yellow sands") is joined by *Quercus incana*. Where clay layers approach the soil surface, *Q. marilandica* is often abundant. As in the Atlantic Coastal Plain Ecoregion, vegetation types can be arranged latitudinally by dominant grasses. The southern types are dominated by *Aristida beyrichiana*. In this zone two types are recognizable, one of relatively dry sites with *Baptisia perfoliata* as an indicator (2.2.1), and one of more mesic sites with *Nolina georgiana* as an indicator (2.2.2). Northward in the wiregrass gap region of South Carolina, *Schizachyrium scoparium* dominates and has as common associates *Vaccinium staminium* and *Toxicodendron pubescens* (2.2.3). From northern South Carolina northward *Aristida stricta* dominates. These sites, like those in the south, can be divided into a dry type with *Tephrosia virginiana* as an indicator species (2.2.4), and a mesic type with *Rhexia alifanus* and *Coreopsis major* as indicators (2.2.5). One other type restricted to the northern sandhills is that of dry depressions in the xeric upland sands in which silt has collected. The increased silt ameliorates the normally harsh soil chemistry and moisture status. A typical indicator of this type is *Astragalus michauxii* (2.2.6).

Southern Coastal Plain

Subxeric pinelands of northern peninsular Florida and the eastern Panhandle typically support an understory of *Quercus incana* and *Q. margarettiae* and a diverse ground layer

FIGURE 5. Subxeric sandy uplands, Eastern Gulf Coastal Plain. Old-growth longleaf pine over dry flatwood with *Serenoa repens* and *Quercus minima*. Patterson Natural Area. Eglin Air Force Base, Okaloosa County, FL.

dominated by *Aristida beyrichiana* mixed with *Schizachyrium scoparium*, *Andropogon gyrans*, *Sorghastrum secundum*, and *Sporobolus junceus* (2.3.1). Westward in the Apalachicola region this is replaced by subxeric woodlands with the understory dominated by *Quercus laevis* and with considerable *Q. minima* (2.3.2), likely owing to lower soil phosphorus content in the Panhandle than in peninsular Florida (see Peet, Carr and Gamling 2006). Further westward on the well-drained low terraces of the western Panhandle *Quercus geminata* dominates over *Serenoa repens* and *Aristida beyrichiana* (2.3.3). Throughout the Southern Coastal Plain are low sandhills somewhat intermediate to flatwoods with *Quercus geminata* as a subcanopy dominant. One phase in west central Florida and the adjacent Big Bend area is characterized by abundant *Quercus chapmanii* and other scrub oaks, *Sorghastrum secundum* and *Pterocaulon virgatum* (2.3.4), whereas a somewhat overlapping but more northern phase is characterized by fewer oaks and more *Schizachyrium scoparium* var. *stoloniferum* and *Sorghastrum nutans* (2.3.5). Small sand ridges isolated in relatively mesic habitats and surrounded by hammock vegetation can be very different with the sparse graminoid layer characterized by *Aristida condensata* and *Cyperus plukenetii* (2.3.6).

Eastern Gulf Coastal Plain

Subxeric sandy uplands are relatively uncommon on the Gulf Coastal Plain owing to the generally fine-textured sediments encountered with increasing proximity to the Mississippi River. The subxeric sandhill vegetation of the upper Panhandle (2.4.1) continues west sporadically into the eastern portion of the Gulf Coastal Plain to be replaced in Mississippi and eastern Louisiana by a somewhat depauperate version wherein *Aristida beyrichiana* is replaced by *Sporobolus clandestinus* as the dominant grass (2.4.2). A few exceptionally sandy areas in southern Mississippi support vegetation with a mixed scrub oak layer and a diversity of shrubs including *Clinopodium coccineum* (2.4.3). Occasional sandy areas also occur on the inner coastal plain. On the Tifton Upland and Dougherty Plain, *Aristida beyrichiana* dominates the groundlayer often with *Licania michauxii*, being joined by typical dry sandy site understory specialists such as *Quercus laevis*, and *Q. margarettiae* (2.4.4). On the inner Coastal Plain west of the range of wiregrass and near the range limit of longleaf, *Schizachyrium* dominates the grass layer and overall diversity tends to be low. Here *Pinus palustris* tends to share dominance with *Pinus echinata* and *P. taeda* (2.4.5).

Western Gulf Coastal Plain

On uplands and high terraces over deep sand on the Western Gulf Coast, the subxeric longleaf woodlands, being outside the range of *Q. laevis* and *Aristida beyrichiana*, have an understory generally dominated by *Quercus incana* and a ground layer of *Schizachyrium scoparium* with *Andropogon gerardii, A. ternarius, Panicum virgatum*, and *Sporobolus junceus*. These sites support dryland species such as *Croton argyranthemus, Tragia* spp., and *Pityopsis graminifolia* (2.5.1). More xeric, fine-sandy stream terraces are characterized by *Liatris elegans* and *Opuntia humifusa* var. *humifusa* (2.5.2). A final subxeric type is confined to sandy ridge tops in rolling landscapes where dominance is shared among *Pinus palustris, P. echinata*, and *P. taeda*, with hardwood species such as *Carya texana* and *Quercus falcata* scattered throughout (2.5.3).

Silty Uplands

Longleaf vegetation on silty upland soils is scarce in the modern landscape as most such sites were converted to agriculture long before 1900. Ultisol soils are abundant across the Coastal Plain, hinting at the one-time dominance of an ecosystem that has essentially vanished. Some authors such as Phillips (1994) have suggested that hardwoods might have been abundant on such sites, but inventory data in early surveys (e.g., Hale 1883; Ashe 1897) suggest longleaf woodlands to have been the predominant type. The few remnants remaining (e.g., Fig. 6) suggest a ground layer with high herb diversity, particularly with respect to legumes and composites. Throughout, longleaf pine is the canopy dominant, and subcanopy oaks are generally unimportant. *Pteridium aquilinum* var. *pseudocaudatum* and *Schizachyrium scoparium* are ubiquitous in the ground layer. Silty uplands are essentially absent from the Southern Coastal Plain ecoregion.

Atlantic Coastal Plain

Longleaf vegetation of the upland silty sites of the Atlantic slope can be conveniently sorted along two axes: a soil moisture gradient divided into mesic and subxeric, and a geographic axis. The four geographic regions include the range of southern wiregrass (*Aristida beyrichiana*) south of the Santee River system of South Carolina and across Georgia (3.1.1, 3.1.2), the wiregrass gap in central South Carolina dominated by *Schizachyrium scoparium* (3.1.3, 3.1.4), the outer coastal plain within the range of Carolina wiregrass (*Aristida stricta*) starting in northern South Carolina (3.1.5, 3.1.6), and the inner coastal plain within the range of Carolina wiregrass (3.1.7, cf. 3.2.2). Although longleaf vegetation on dry, silty uplands of the upper coastal plain is essentially gone, the few roadside scraps that persist suggest it to have been similar to the silty, dry uplands of the Fall-line Sandhills. The understory contains occasional oaks (*Quercus incana, Q. marilandica, Q. stellata*), and the shrub layer generally contains *Vaccinium tenellum* and *Quercus pumila*.

Fall-line Sandhills

Fine-textured soils are something of an anomaly in the Fall-line Sandhills. Some areas of silty soil are found associated with terraces of small streams where silts have been transported into the area by water and redistributed by wind. Perhaps the best-known examples

FIGURE 6. Silty uplands, Atlantic Coastal Plain. This rare remnant of fire-maintained silty soils supports a mixture of oaks (*Quercus marilandica, Q. stellata*) and longleaf pine over a grass layer of *Schizachyrium scoparium* and *Andropogon* spp. Jasper County, SC.

of silty soils occur in shallow depressions where fine-textured soils have either washed or blown in from the neighboring landscape and become trapped. Such depressions are relatively frequent within the range of *Aristida stricta* and are hypothesized to represent blow-outs similar to those that initially formed the Carolina bays of the Coastal Plain (James 2000). The vegetation of these sandhill depressions tends to be exceptionally species-rich and their abundance of legumes has earned them the colloquial name among botanists of "bean dips" (James 2000). Vegetation of silty sandhill soils can be arranged along a moisture gradient with the dry end characterized by species like *Toxicodendron pubescens, Tephrosia virginiana* and small trees like *Quercus margarettiae, Q. marilandica, Q. incana*, and *Diospyros virginiana* (3.2.1 in the Carolinas, 3.2.4 in Georgia). Intermediate sites have greater dominance by legumes, *Pteridium*, and forbs with prairie affinities such as *Parthenium integrifolium* and *Silphium compositum* (3.2.2). On mesic sites shrubs like *Ilex glabra*, grasses such as *Panicum virgatum*, and an abundance of taller forbs like *Eupatorium rotundifolium* give the vegetation a lush aspect (3.2.3). A different variant of silty vegetation can be found near the transition to Piedmont where hardwoods like *Carya pallida, Quercus stellata*, and *Cornus florida* give the aspect of a transition from longleaf woodland to open hardwood woodland (3.2.5).

Eastern Gulf Coastal Plain

The Eastern Gulf Coastal Plain is a region of predominantly silty soils. Unfortunately, quantitative studies of this vegetation have been confined to a few studies of specific localities (e.g., The Jones Center near Albany, GA, a few plantations in the Tallahassee Red Hills), and no comprehensive, regionwide study has

FIGURE 7. Silty uplands, Eastern Gulf Coastal Plain. Widely spaced old-growth longleaf pine over *Aristida beyrichiana* and *Schizachyrium* spp. Wade Tract. Thomas County, GA.

yet been undertaken. As a consequence, the current classification recognizes regional variants associated with different geomorphic surfaces, but little more. Doubtless more study would allow greater resolution with respect to moisture regime. Within its geographic range, *Aristida beyrichiana* typically co-dominates with *Schizachyrium scoparium* and various species of *Andropogon*. Variants of this vegetation have been recognized associated with the Tallahassee Red Hills, Mariana Lowlands, and Tifton Plain of southwestern Georgia and adjacent Florida (3.4.1; Fig. 7), the Dougherty Plain (3.4.2), and the coastal plain south and west of the Mariana Lowlands (3.4.3). All of these types support tall grasses and a highly diverse assemblage of forb species. North of the range of *Aristida beyrichiana* in western Georgia and the adjacent inner coastal plain of Alabama (e.g., southern Ft. Benning, Tuskegee and Talladega-Oakmulgee National Forests), *Schizachyrium scoparium* dominates along with *Andropogon gyrans, A. ternarius*, and *Danthonia sericea* (3.4.4). West of the range of wiregrass, distinctive silt-soil longleaf types have been described from the coastal flatlands and rolling hills (3.4.5), the Mississippi loam hills (3.4.6), and the loess soils immediately east of the Mississippi River (3.4.7, 3.4.8). The higher fertility of these sites leads to a greater abundance of hardwoods (e.g., *Quercus marilandica, Q. incana, Q. margarettiae, Q. stellata, Carya glabra*) and a particularly diverse forb layer. The longleaf vegetation of the loess soils of eastern Louisiana and adjacent Mississippi supports the highest known 1000 m^2 species richness for upland habitats in the United States with values reaching around 170 species (W. J. Platt personal communication).

Western Gulf Coastal Plain

Silty soils predominate on the Western Gulf Coastal Plain and support diverse longleaf woodlands somewhat intermediate between the longleaf woodlands of the Eastern Gulf Coastal Plain and the prairies of the eastern Great Plains. *Schizachyrium scoparium* is the ubiquitous dominant, in places mixed with *S. tenerum*. Longleaf vegetation of the high Pleistocene terraces (3.5.1) is extremely species rich, contains prairie taxa like *Liatris pycnostachya*, and alternates with the savanna vegetation of slightly lower sites (6.5.2). A somewhat wetter variant co-dominated by *Schizachyrium scoparium* and *S. tenerum* occupies the tops of pimple mounds on the lower Pleistocene terraces (3.5.2). More xeric silt-soil longleaf vegetation can be found on the dissected topography of ridge tops farther inland, similar floristically to the terrace type, but with such drier-site prairie taxa as *Rudbeckia grandiflora* var. *alismiflora* and *Echinacea sanguinea* (3.5.3). A rare vegetation type known only from Rapides Parish Louisiana is a glade type forming on silty soils over clay or siltstone with interdigitation of wet and dry microsites and a dominance of *Schizachyrium scoparium, S. tenerum, Muhlenbergia expansa,* and *Bigelowia nuttallii* (3.5.4).

Clayey and Rocky Uplands

Although stereotypic longleaf vegetation occurs on soft sediments of the southeastern Coastal Plain, longleaf pine occurs on other substrates such as ironstone and sandstone hills of the inner Coastal Plain and Fall-line Sandhills, as well as on clay soils derived from the bedrock of the Piedmont or southwesternmost Appalachian Mountains.

Atlantic Coastal Plain and Fall-line Sandhills

Scattered along the Fall-line Sandhills are patches of longleaf woodland with a relatively dense shrub layer of mountain laurel (*Kalmia latifolia*) and other woody species reminiscent of the Piedmont (e.g., *Vaccinium arboreum*), but with a sparse to absent herbaceous layer (4.1.1). These communities can occur on either eroded hills formed in marine kaolinite deposits or where rocky hills and scarps associated with ironstone caps punctuate the otherwise relatively gentle topography. Vegetation intermediate to the more typical subxeric longleaf woodlands occur where a thin veneer of sandy soil overlies the ironstones. These sites support an abundance of such classic longleaf species as *Vaccinium crassifolium* and *Aristida stricta*, as well as species more strongly indicative of the clayey site conditions, such as *Leiophyllum buxifolium*, *Pyxidanthera barbulata*, and *Hypericum lloydii*, (4.1.2; Fig. 8). In the Altamaha Grit region of Georgia, rocky glade vegetation can be found over similar isolated occurrences of indurated sandstones although with such distinctive rockland species as *Talinum teretifolium, Allium cuthbertii,* and *Penstemon dissectus* (4.1.3).

Eastern Gulf Coastal Plain

The innermost Coastal Plain of Alabama and adjacent Georgia contains significant areas of clay hills that in the original landscape supported longleaf-dominated vegetation. There is little if any vegetation of this type left to study, though brief descriptions can be found in early descriptive works (4.4.1; e.g., Mohr 1901; Harper 1943). This vegetation superficially appears to have resembled that of clay and rocky soils of the Atlantic Coastal Plain (4.1.1), particularly in the occurrence of *Kalmia latifolia*, xeric oaks like *Q. marilandica* and *Q. laevis*, and the scarcity of herbs. Southern Mississippi and Alabama contain scattered clay hills somewhat similar in character but lacking *Kalmia* and supporting a richer understory.

Western Gulf Coastal Plain

Outcroppings of calcareous sandstones (4.5.1) and high-calcium, shrink-swell clay soils (4.5.2, 4.5.3) on Tertiary terraces of the Western Gulf Coastal Plain support a typically stunted woodland that contrasts with surrounding *Pinus palustris* woodlands on deeper soils. *Quercus marilandica* and *Q. stellata* share dominance on the shrink-swell clays here as elsewhere on the Coastal Plain and Piedmont.

FIGURE 8. Rocky uplands, Fall-line Sandhills. Soils are shallow sands over impermeable ironstones formed over a thick clay layer. Diversity is low with a canopy of *Pinus palustris* over a subcanopy of *Quercus marilandica* and a ground layer of *Aristida stricta*. Fort Bragg, Hoke County, NC.

On drier sites *Andropogon gyrans, A. ternarius*, and *Schizachyrium scoparium* are the dominant grasses, whereas on moister clays the dominant grasses are *Panicum virgatum, Panicum anceps* var. *rhizomatum, Schizachyrium scoparium* var. *divergens*, and *Sporobolus junceus*. Also present are occasional herbaceous glades where sandstone is present at the surface (4.5.4). These sandstone glades support primarily *Schizachyrium scoparium* and *Bigelowia nuttallii*, though numerous forbs occur with less abundance.

Piedmont and Montane Uplands

Except for central Alabama, longleaf woodlands were probably always relatively uncommon in the Piedmont and mountain landscapes. On these upland sites little remains of the original longleaf vegetation owing to fire suppression combined with timber harvest and site conversion to agriculture. Consider that in 1860 Chatham County on the North Carolina Piedmont ranked fifth among North Carolina counties in amount of standing longleaf timber (Hale 1883), whereas today fewer than a dozen mature longleaf trees and perhaps only a single clump of wiregrass persist in the county. Some indication of the original composition and diversity of this vegetation can be found in Mohr's (1901) and Harper's (1943) summaries of the forests of Alabama, and the treatments of the Georgia highlands by Harper (1905) and Andrews (1917).

Remaining montane and Piedmont longleaf woodland vegetation types generally occur on exposed ridges and south-facing slopes. All these sites support relatively consistent, unremarkable vegetation as the flora is characteristic of hardwood-dominated xeric, acidic sites. *Pinus echinata* often occurs as a codominant along with such deciduous forest taxa as *Oxydendron arboreum*, *Quercus prinus*, and *Q. marilandica* (4.6.1, 4.6.2). On the few remaining sites from the Uwharrie National Forest, NC, to Fort McClennan, AL, one finds widespread taxa such as *Nyssa sylvatica*, *Pteridium aquilinum* var. *pseudocaudatum*, *Tephrosia virginiana*, *Pinus virginiana*, *Vaccinium tenellum*, and *Schizachyrium scoparium*. On slightly less exposed sites *Quercus prinus* and *Pinus palustris* can co-dominate (4.6.3) (Maceina et al. 2000).

A few peculiar, local longleaf types occur over unusual substrates. Burke Mountain on the Georgia Piedmont supports the only longleaf stand on serpentine soil (unusual taxa include *Clinopodium georgianum* and *Baptisia alba*; 4.6.4). On the quartzite ridges of Pine Mountain, GA, is another distinctive type where below the canopy of longleaf and shortleaf is a subcanopy dominated by *Quercus coccinea* and *Q. georgiana* (4.6.5).

Flatwoods

Although "flatwoods" vegetation has been described in all major treatments of the vegetation of the southeastern United States, the only long-term consistency in use of the term is its application to fire-maintained pine woodland over flat, Coastal Plain landscapes. Some of the confusion derives from the central place of flatwoods on the landscape. Flatwoods are found between the more extreme vegetation types of xeric sandhill, scrub, baygall, prairie, and savanna. When these physiognomically and topographically more extreme types are given narrow definitions, as in Abrahamson and Hartnett (1990) and Harper (1914), flatwoods become the broader concept referred to by Christensen (2000) as "intractable."

I reserve use of the term "flatwoods" for fire-maintained pinelands of low, flat terrain where marine sands were deposited during Pleistocene incursions and where the resultant soils are primarily poorly drained Spodosols, a convention also followed by Stout and Marion (1993). Vegetation consistent with this definition is found along the Atlantic coast north into southern North Carolina, and west along the Gulf coast to eastern Mississippi. North and west of these boundaries recent marine sediments are significantly more silty, and true Spodosols are uncommon to absent.

Atlantic Coastal Plain

Flatwoods are largely a phenomenon of the Southern Coastal Plain, which extends into southeastern South Carolina along the coastal fringe where it can be recognized by dominance of such characteristic species as *Serenoa repens*, *Vaccinium myrsinites*, and *Aristida beyrichiana*. True flatwoods are largely absent from the central coast of South Carolina where soils tend to be too silty for Spodosol development, but do occur on the outer coastal plain from the Santee River of South Carolina north to the Neuse River in North Carolina. In this region flatwoods are of relatively low diversity and are often co-dominated by *Pinus palustris* and *P. serotina* with a ground layer dominated by *Aristida stricta* and *Gaylussacia dumosa* (5.1.1; Fig. 9). Somewhat wetter sites tend to be dominated by *Ctenium aromaticum* and *Muhlenbergia expansa*, and often have *Pinus serotina* in the canopy (5.1.2). Among taxa that distinguish these northern flatwoods are *Dionaea muscipula* (Venus flytrap) and *Vaccinium crassifolium*. Aspect dominants of the groundlayer that distinguish unusual community variants include the lily *Pleea tenuifolia* (disjunct to the Apalachicola region of Florida; 5.1.3) and the low shrub *Leiophyllum buxifolium* (better known from the New Jersey pine barrens and Blue Ridge rock outcrops; 5.1.4). Although longleaf woodlands are now largely extirpated north of the Neuse River, a few persistent sites in northeastern North Carolina and southeastern Virginia have affinities with flatwoods vegetation and perhaps should be so classified (5.1.5), though their floristic composition is largely distinctive in the absence of species characteristic to the south, rather than species occurrences (see Frost and Musselman 1987).

Southern Coastal Plain

Flatwoods vegetation is best developed in the Southern Coastal Plain ecoregion. The canopy is generally dominated by *Pinus palustris*, but

FIGURE 9. Flatwoods. Northern variant of longleaf flatwoods missing *Serenoa repens*, and *Vaccinium myrsinites* of the Southern Coastal Plain, but with *Aristida stricta* and *Vaccinium crassifolium*. Atlantic Coastal Plain. Croatan National Forest, Carteret County, NC.

in much of Florida and southeast Georgia, *P. elliottii* var. *elliottii* replaces longleaf completely on the wettest sites (Clewell 1971; Gano 1917; Monk 1968), and in the deep south of the Florida peninsula *Pinus elliottii* var. *densa* can dominate (e.g., 5.3.9). The understory layer is dominated throughout by such widespread taxa as *Aristida beyrichiana*, *Vaccinium myrsinites*, *Serenoa repens*, *Pteridium aquilinum* var. *pseudocaudatum*, *Quercus pumila*, and *Q. minima*; their co-dominance makes flatwoods vegetation relatively easy to recognize. Other distinctive woody taxa of the southern flatwoods include *Kalmia hirsuta*, *Lyonia fruticosa* and *Lyonia ferruginea*, and *Gaylussacia tomentosa*.

Flatwoods vegetation shows substantial geographic variation along a longitudinal gradient from northeastern Florida and adjacent Georgia across to the Florida Panhandle. The eastern flatwoods are often co-dominated by such distinctive grasses as *Sporobolus curtissii* and *Ctenium floridanum*, in addition to *Aristida beyrichiana*. Forbs characteristics of the East include *Symphyotrichum walteri*, *Carphephorus paniculatus*, *Euthamia tenuifolia* var. *tenuifolia*, and *Eupatorium mohrii* (5.3.1). In addition to the typical *Sporobolus curtissii* flatwood type, there are types with *Sporobolus pinetorum* characteristic of the finer-texture soils of northeast Georgia (5.3.2) and with *Sporobolus floridanus* characteristic of the wettest sites (5.3.3). In contrast, flatwoods of the eastern Panhandle have the widespread *Vaccinium myrsinites* joined by its cousin *V. darrowii*, and commonly contain such herbaceous taxa as *Balduina uniflora*, *Carphephorus odoratissimus*, *Symphyotrichum adnatum*, and *Chrysopsis mariana* (5.3.4; Fig. 10), which elsewhere in the region are largely absent from flatwoods. Moister sites are dominated by *Pinus elliottii* var. *elliottii* and sort according to soil chemistry with near-coastal sites having high phosphorus and

FIGURE 10. Flatwoods. Southern Coastal Plain. Classic longleaf pine flatwoods with abundant *Serenoa repens*, *Pteridium aquilinum* var. *pseudocaudatum*, and *Vaccinium myrsinites* and *V. darrowii*. Apalachicola National Forest, Wakulla County, FL.

calcium characterized by presence of *Lyonia ferruginea* and *Kalmia hirsuta* (5.3.5), while the more sterile and acidic sites have abundant *Lyonia lucida* and *Befaria racemosa* (5.3.6).

The soils of the drier scrubby flatwoods and dry prairies of the central and more southern portions of peninsular Florida are somewhat sandier than those of the typical flatwoods and the vegetation tends to be more open. This vegetation supports only scattered stems of *Pinus palustris* and *P. elliottii* var. *elliottii*, though several characteristic flatwood species occur as dominants including *Serenoa repens*, *Vaccinium myrsinites*, and *Aristida beyrichiana*. *Quercus pumila* is largely absent. Two of these types of open vegetation are commonly recognized, one where *Lyonia fruticosa* and *Quercus minima* share dominance with *Serenoa* (5.3.7), and one where there is a near prairie of *Aristida beyrichiana* with *Serenoa* the only conspicuous woody species (5.3.8). An unusual vegetation type is that dominated by *Pinus elliottii* var. *densa* and *Panicum abscissum*, which appears intermediate in composition between the typical flatwoods and the dry prairies, but with an understory dominance of *P. abscissum* (5.3.9). In the extreme, we find sites lacking trees and appearing as a *P. abscissum* prairie (5.3.10).

Oaks, except for the running oaks (*Quercus pumila*, *Q. minima*), are generally absent from the true flatwoods. However, drier sites in central Florida and the Gulf Coast region of the Southern Coastal Plain (5.3.11) contain a mix of typical upland and "scrub oak" species as shrub and understory layer dominants, including *Quercus geminata*, *Q laevis*, *Q. chapmanii*, *Q. myrtifolia*, and *Q. inopina*. Other low shrubs with high constancy include *Ilex glabra*, *Quercus minima*, *Gaylussacia dumosa*, and *Morella pumila*. In addition, two types transitional to flatwoods, recognized from the barrier islands and extreme coastal fringe, contain abundant oak.

Along the barrier islands of the Gulf coast one finds typical flatwood vegetation but with an abundance of oaks like *Quercus chapmanii* and *Q. geminata* (5.3.12). On somewhat drier sites, especially in northeastern Florida and adjacent Georgia where the much reduced fire regime allows development of a nearly impenetrable understory thicket, evergreen shrubs like *Quercus chapmanii, Q. myrtifolia, Q. geminata*, and *Lyonia ferruginea* form a scrublike variant of flatwood (5.3.13).

Eastern Gulf Coastal Plain

West of Florida, sandy soils are largely replaced by silts, and longleaf pine flatwoods are generally confined to a narrow coastal fringe. One distinctive form of flatwoods of the Eastern Gulf Coastal Plain and adjacent coastal fringe of the Florida Panhandle is that of barrier island interdunal swales and flats strongly dominated by *Serenoa repens* and with an abundance of such maritime shrubs as *Morella cerifera* and *Ilex vomitoria* (5.4.1; cf. Huffman et al. 2004). Somewhat higher and drier barrier island and maritime fringe sites support a distinctive understory of oak including *Quercus virginiana, Q. geminata,* and *Q. hemisphaerica*, along with *Magnolia grandiflora* (5.4.2). The isolation of these islands keeps fire frequency low and allows shrub density to be high. On the mainland, flatwoods from the Apalachicola to the Pascagoula River tend to be attenuated versions of those of the eastern Panhandle (e.g., 5.3.5). West of the Pascagoula River soils are silty and flatwoods are mostly absent. The closest approximation in Mississippi is found in sandy wet longleaf areas of the De Soto National Forest (5.4.3), where this western extreme of flatwoods has few characteristic flatwoods taxa other than *Serenoa repens* and *Ilex glabra*.

Savannas, Seeps, and Prairies

The term "savanna" has many meanings, even within the context of the pinelands of the southeastern United States. However, in recent years usage of the term has generally converged on open pine woodlands of seasonally saturated, fine-textured soils (e.g., Peet and Allard 1993). The wet soil conditions of savannas often lead to the tree canopy being very open with composition dominated by any one of *Pinus palustris, P. serotina*, or *P. elliottii* var. *elliottii*, or a combination of these. This vegetation is generally rich in species across a broad range of spatial scales (10^{-3}–10^4 m^2) and characterized by an abundance of showy forbs. For example, such sites often contain a wealth of orchids (e.g., *Calopogon, Cleistes, Platanthera, Pogonia, Spiranthes*), insectivorous plants (e.g., *Drosera, Dionaea, Pinguicula, Sarracenia, Utricularia*), and lilies (e.g., *Aletris, Lilium, Tofieldia, Zigadenus*), to say nothing of numerous grasses, sedges, and composites. Legumes are conspicuously scarce in moist savannas, a phenomenon noted by Gano (1917), Wells and Shunk (1931), and Taggart (1990, 1994). The floristic novelty and diversity of pine savannas have led to this vegetation being perhaps the best known of the original longleaf community types (e.g., Kologiski 1977; Folkerts 1982; Walker and Peet 1983; Norquist 1984; Taggart 1994).

The pine savannas exhibit significant variation in composition driven by subtle changes in hydrologic regime, soil texture, and soil chemistry, more so than any other vegetation type of the fire-maintained southeastern pinelands (see Fig. 11). In addition to environmentally driven, local variation, geographic turnover is pronounced, and the insular distribution of wet pinelands has led to chance migration events generating significant stochastic variation in composition among sites. As a consequence, despite the diversity of savanna types, the range of variation in each recognized type can be significantly greater than in other vegetation types recognized in this chapter.

On the outer Coastal Plain where the landscape is extremely flat, individual savannas can cover considerable area. Moving inland, savanna areas are constrained to smaller areas, often narrow swales within the gently rolling topography. In the inner Coastal Plain, Fall-line Sandhill, and Piedmont regions where the landscape is far less flat and much

3. Ecological Classification of Longleaf Pine Woodlands

FIGURE 11. Subxeric sandy upland-to-savanna transition. Southern Coastal Plain. Sparse *Pinus palustris* over *Aristida beyrichiana* and *Schizachyrium scoparium* grading down slope into moist savanna. Apalachicola National Forest, Liberty County, FL.

more topographically complex, savannalike vegetation is confined to small seepage areas associated with impermeable clays or near-surface bedrock.

Atlantic Coastal Plain

Savannas are best developed on the outermost Coastal Plain where extensive areas of flat, fine-textured soils occur, representing terraces in regions of predominantly fine-textured sediments or one-time embayments behind now-vanished barrier island systems. Although these savannas are generally dominated by grasses, the dominant species shift significantly with soil moisture and texture.

In those southeastern North Carolina savannas on seasonally saturated soils that are sufficiently well-drained that surface water is short-lasting, silty sites are dominated by the grasses *Aristida stricta, Schizachyrium scoparium,* and the local endemic *Sporobolus pinetorum* (6.1.1; Fig. 12). On somewhat wetter sites *Aristida stricta* drops out to be replaced by greater dominance of *Ctenium aromaticum* and *Muhlenbergia expansa,* though with an admixture of various *Rhynchospora, Andropogon,* and other graminoids (6.1.2). Numerous orchids (often as many as a half dozen species in a 1000 m² plot) occur within these vegetation types, and insectivorous plants are especially well developed on the wetter sites (e.g., 6.1.2) where it is not uncommon for a single site to have two or three species of *Sarracenia,* two or three *Pinguicula,* two *Drosera,* and *Dionaea.*

Dominance shifts as one moves south toward the more loamy landscapes of the South Carolina low country. *Aristida stricta* drops out near the Pee Dee River system in northeast South Carolina to be replaced by stronger dominance of *Schizachyrium scoparium,* whereas *Sporobolus pinetorum* penetrates southward to

FIGURE 12. Savanna. Atlantic Coastal Plain. Species-rich longleaf savanna in an area with an average richness of 35 species/m². Green Swamp Preserve, Brunswick County, NC.

the latitude of Charleston, where it is limited to just a few sites (appearing again on a few sites in northeast Georgia). In contrast, *Sporobolus curtissii*, a grass primarily dominating flatwoods of the Southern Coastal Plain where it is abundant northward to Savannah, is absent from the Atlantic Coastal Plain region except for clayey savannas near Charleston (6.1.3). Very wet savannas of the loamy South Carolina low country sometimes have the grass layer dominated by *Aristida palustris* (6.1.4).

Subtle changes in soil chemistry can have a significant impact on savanna composition. Where soils are wet but somewhat influenced by a marl substrate, composition shifts toward stronger dominance by a combination of *Sporobolus pinetorum* and *Ctenium* (6.1.5). These sites have a suite of narrowly distributed savanna calciphiles such as *Carex lutea*, *Thalictrum cooleyi*, and *Parnassia caroliniana*. On those rare sites with extremely wet marl-clay soils, species richness drops and another regionally endemic *Sporobolus*, *S. teretifolius*, tends to dominate (6.1.6). Inland from the coastal flatlands, savannas are mostly confined to gentle seepage slopes influenced by groundwater (6.1.7). Where soils are more fertile, composition shifts toward greater abundance of *Arundinaria* and *Andropogon* with *Liquidambar* often present in the understory (6.1.8). On the more extreme examples in the original landscape, such sites supported canebrakes with nearly complete dominance by *Arundinaria*, though such vegetation has all been lost to either agriculture or fire suppression.

Fall-line Sandhills

The Fall-line Sandhills, like the inner coastal plain rolling hills, generally do not have the extensive flatlands with impeded drainage necessary to support true savanna. However, impermeable clay layers and indurated spodic horizons occur in these regions and, where

3. Ecological Classification of Longleaf Pine Woodlands

FIGURE 13. Seep. Fall-line Sandhills. Species-rich seep dominated by *Ctenium aromaticum, Muhlenbergia expansa*, and *Andropogon* spp. with abundant *Sarracenia lutea*. Fort Bragg, Hope County, NC.

they approach the surface, seepage sometimes occurs. These seeps are usually similar to true coastal plain savannas in their species composition (see Wells and Shunk 1931), but lower in diversity (6.2.1, 6.2.2; Fig. 13).

Southern Coastal Plain

The Southern Coastal Plain is largely a land of sandy soils with the fine-textured soils characteristic of savannas occurring on either side. In northeast Georgia and southeastern South Carolina the wet sites on fine-textured soils are dominated by *Sporobolus floridanus*, often with a scattering of *Morella cerifera* and the distinctive dwarf shrub *Styrax americanus* var. *pulverulentus* (6.3.1; Fig. 14), while on slightly drier sites *Aristida beyrichiana* shares dominance with *Ctenium aromaticum* (6.3.2). To reach the next area of extensive savanna it is necessary to jump over the Florida peninsula to treeless flats of silty soils in the Apalachicola region. Here can be found a diverse array of savannalike vegetation, again with compositional sorting by moisture levels. At the drier end of the gradient, mesic pineland taxa such as *Helianthus radula* and *Galactia erecta* occur (6.3.3). Where the soils become chronically wet *Rhynchospora* diversity greatly increases and *Verbesina chapmanii* is often important (6.3.4). On the wettest sites *Eriocaulon decangulare, Lachnanthes caroliana, Pleea tenuifolia, Sarracenia psittacina*, and *S. flava* are conspicuous (6.3.5). In peninsular Florida savannalike vegetation can occur despite sandy soils under special circumstances. Particularly wet soils can be dominated by grasses such as *Panicum rigidulum, Panicum hemitomon*, and *Andropogon glaucopsis* (6.3.6). Where the soil is particularly high in calcium so that the pH is about 6, a vegetation type sometimes referred to as "sweet flats" forms with *Sabal palmetto* and *Morella cerifera* sharing dominance with *Pinus elliottii* var. *elliottii* (6.3.7).

FIGURE 14. Savanna, Southern Coastal Plain. Savanna with saturated soils that limit the growth of trees to a few scattered longleaf and slash pines. *Sporobolus floridanus* dominates a ground layer punctuated by scattered *Morella cerifera* and *Styrax americanus* var. *pulverulentus*. Jasper County, SC.

Wet prairie vegetation, largely devoid of trees is relatively frequent on and largely unique to the Florida peninsula and adjacent eastern Panhandle. Soils of these wet grasslands are too sandy to match the definition of savanna I use here, consistently containing over 95% sand in both the A and B horizons. Considerable floristic variation occurs in these types with variation in soil moisture regime, soil chemistry, and geographic position. Nonetheless, *Aristida beyrichiana, A. palustris, Amphicarpum muehlenbergianum,* and *Sporobolus floridanus* are relatively widespread dominants suggesting close affinities with longleaf pine savannas (6.3.8–6.3.10).

Eastern Gulf Coastal Plain

True savanna vegetation is better developed on the Eastern Gulf Coastal Plain and adjacent Panhandle than the main body of the Southern Coastal Plain owing to the predominantly siltier soils of the ecoregion. Vegetation in these savannas bears considerable floristic similarity to the savannas of the Atlantic Coastal Plain as seen in southeastern North Carolina. Wet savannas similar in character to 6.1.2 occupy particularly wet sites. East of Mobile Bay these sites are still dominated by *Aristida beyrichiana* but with considerable *Ctenium* and *Muhlenbergia expansa,* and the hillside seepage areas of the more rolling higher terraces of the Florida Panhandle are largely dominated by *Aristida beyrichiana* with co-dominance of *Andropogon arctatus, Ctenium aromaticum,* and *Dichanthelium leucothrix* (6.4.1; Fig. 15), whereas west of Mobile Bay *Aristida beyrichiana* is spotty, dropping out completely at the Pascagoula River. In place of *Aristida beyrichiana* one see increased dominance of *Muhlenbergia expansa* and various *Andropogon* (e.g., *mohrii, gyrans* var. *stenophyllus*)

FIGURE 15. Savanna. Gulf Coastal Plain. Moist coastal savannas can support a wealth of orchids and insectivorous plants. Abundant pitcher plants (*Sarracenia alata*) and sundews (*Drosera tracyi*) are visible in the foreground. Sand Hill Crane National Wildlife Refuge, Jackson County, MS.

(6.4.2). Somewhat drier savannas (similar to 6.1.1) also occur in southwestern Mississippi and southeastern Louisiana, these dominated by *Schizachyrium scoparium* and *Muhlenbergia expansa* (6.4.3).

Higher on the Gulf Coastal Plain, above the coastal flatlands, true savannas are mostly absent but one finds similar species occurring on seepage slopes. The seeps of the silty Tallahassee red hills are particularly striking with the somewhat drier sites dominated by *Quercus pumila* and *Rhododendron canescens* with *Ctenium aromaticum* (6.4.4), and the very wet sites dominated by unusual *Dichantheliums* and *Panicum verrucosum* (6.4.5). These latter sites have species richness values that are among the highest found in longleaf communities with as many as 170 species per 1000 m^2. The overall silt dominance of this landscape can lead to somewhat more fertile soils, reflected in the increased abundance of *Arundinaria* (6.4.6). Seeps intermediate between those of the Florida Panhandle and the outer coastal plain of Mississippi can be found in the vicinity of the De Soto National Forest. These sites are dominated by grasses such as *Muhlenbergia expansa, Schizachyrium scoparium, S. tenerum,* and *Anthaenantia rufa* (6.4.7).

Western Gulf Coastal Plain

Wet savanna vegetation was at one time widely distributed on the Western Gulf Coastal Plain. In contrast to savannas east of the Mississippi bottomlands, these savannas occupy soils that are relatively calcareous and often contain a significant amount of shrink-swell clays. Nonetheless, there are strong floristic affinities between this savanna vegetation and the more eastern savannas of fine-textured soils as far away as the Atlantic Coastal Plain (e.g., 6.1.2, 6.1.7). These savannas can usefully be divided into the lower to middle terraces with soils predominantly Glossaqualfs (6.5.1) and the somewhat higher and better-drained savannas of the upper terraces where the soils are primarily Paleudults (6.5.2). Both savanna types are dominated by *Ctenium* (east of the Sabine River), *Muhlenbergia expansa, Schizachyrium scoparium*, and multiple species of *Rhynchospora*. Among the rolling hills of the older land surfaces of the West Gulf Coast, seepage bogs with *Sarracenia alata* and *Rhynchospora gracilenta* occur as small patches, though rather few are left for study (6.5.3; see Bridges and Orzell 1989). A rare savanna type is sometimes found on saline Pleistocene terraces where beneath the longleaf pine the dominant grasses are *Sporobolus silveanus* and *Muhlenbergia capillaris*. The flora is relatively rich, but the taxa have stronger affinities with the Midwestern prairies than the Eastern Gulf Coastal Plain (6.5.4).

Piedmont and Montane Uplands

Piedmont and montane longleaf sites are generally well drained and lacking in seepage vegetation. A few exceptions can be found in the Uwharrie National Forest of North Carolina where small depressions and flat, seepy small-stream flats occur. On these sites the mixed pine canopy has an groundlayer dominated by the grasses *Chasmanthium laxum* and *Panicum virgatum*. Shrubs are well developed in these sites, including such typically coastal plain taxa as *Lyonia mariana, Gaylussacia frondosa, Vaccinium fuscatum*, and *Ilex glabra* (6.6.1). Additional variants should be expected in eastern Alabama.

Concluding Remarks

Although the once extensive southeastern longleaf pine woodlands may appear to the casual observer as a homogeneous expanse of longleaf pine, grass, and scrub oak, this is a gross oversimplification. The Southeastern Coastal Plain is exceptionally rich in endemic species, and much of this endemism is manifest in the flora of the fire-maintained pinelands. Much fieldwork remains before we can claim to have carefully documented the compositional variation of the remaining longleaf pine vegetation. However, drawing on work embedded in the U.S. National Vegetation Classification, my own preliminary analysis of approximately 900 vegetation plots scattered over the eastern two-thirds of the range of longleaf, and the work of many other authors, I here tentatively accepted 135 longleaf vegetation associations. Although this may seem like a high number, I expect this number to increase substantially with increased collection and analysis of plot data. This represents what is likely the minimum number of units for classifying longleaf vegetation for conservation and for providing targets for ecological restoration.

The remarkable diversity of the greater longleaf ecosystem is being lost rapidly, both through active habitat destruction and through neglect. If even a fraction of the diversity of the longleaf ecosystem is to be preserved, action must be taken quickly to both protect and manage the best remaining examples of each of the longleaf community types. Of particular importance in any such endeavor is recognizing the considerable variation in longleaf vegetation that occurs with simple geographic distance and subtle environmental variation. Conservation and preservation of the longleaf ecosystem cannot simply focus on a small number of high-quality preserves, as these will inevitably capture only a modest fraction of the natural variation. We need to devise a reserve system that includes preserves and restored sites that span the geographic and environmental range of the original longleaf ecosystem.

Acknowledgments

Fieldwork on which this chapter was based was supported by grants from the U.S. Forest Service and the Florida Fish and Wildlife Conservation Commission. This work would not have been possible without the extensive work on the National Vegetation Classification provided by the employees of NatureServe, The

Nature Conservancy, and the various state Natural Heritage programs of the region. Portions of this chapter are based on collaborations with other individuals, including in particular Susan Carr, Richard Duncan, Eric Kjellmark, Patrick McMillan, Michael Schafale, Alan Weakley, and Tom Wentworth. To all of these people and programs, and many others, I am deeply indebted.

Endnote

1. Botanical nomenclature follows USDA, NRCS (2005); except for *Sporobolus and Muhlenbergia*, which follow Peterson et al. (2003) and Peterson (2003), and *Styrax*, which follows Gonsoulin (1974); common names are shown in Table 3.

References

Abrahamson, W.G., and Hartnett, D.C. 1990. Pine flatwoods and dry prairies. In *Ecosystems of Florida*, eds. R.L. Myers and J.J. Ewel, pp. 103–149. Orlando: The University of Central Florida Press.

Anderson, M., Bourgeron, P., Beyer, M.T., Crawford, R., Engelking, L., Faber-Langendoen, D., Gallyoun, M., Goodin, K., Grossman, D.H., Landaal, S., Metzler, K., Patterson, K.D., Pyne, M., Reid, M., Sneddon, L., and Weakley, A.S. 1998. *International Classification of Ecological Communities: Terrestrial Vegetation of the United States. Volume II. The National Vegetation Classification System: List of Types*. Arlington, VA: The Nature Conservancy.

Andrews, E.F. 1917. Agency of fire in propagation of longleaf pines. *Bot Gaz* 64:497–508.

Ashe, W.W. 1897. Forests of North Carolina. *NC Geol Surv Bull* 6:139–224.

Bailey, R.G. 1995. *Description of the Ecoregions of the United States: Second Edition*. Miscellaneous Publications No. 1391. Washington, DC: USDA Forest Service.

Bozeman, J.R. 1971. A sociologic and geographic study of the sand ridge vegetation of the coastal plain of Georgia. Ph.D. dissertation, University of North Carolina, Chapel Hill.

Bridges, E.L., and Orzell, S.L. 1989. Longleaf pine communities of the West Gulf Coastal Plain. *Nat Areas J* 9:246–263.

Brown, D., Reichenbacher, F., and Franson, S. 1998. *A Classification of North American Biotic Communities*. Salt Lake City: University of Utah Press.

Christensen, N.L. 2000. Vegetation of the Southeastern Coastal Plain. In *North American Terrestrial Vegetation, Second Edition*, eds. M.G. Barbour and W. D. Billings, pp. 397–448. London: Cambridge University Press.

Clewell, A.F. 1971. The vegetation of the Apalachicola National Forest: An ecological perspective. Contract 38–2249, Final Report. US Forest Service, Atlanta, GA.

Daniels, R.B., Kleiss, H.J., Buol, S.W., Byrd, H.J., and Phillips, J.A. 1984. Soil systems in North Carolina. North Carolina Agricultural Research Service, Bulletin 467, Raleigh.

DuBar, J.R., Johnson, H.S., Jr., Thom, B.G., and Hatchell, W.O. 1974. Neogene stratigraphy and morphology, south flank of the Cape Fear arch, North and South Carolina. In *Post-Miocene Stratigraphy, Central and Southern Atlantic Coastal Plain*, eds. R.Q. Oaks, Jr., and J.R. DuBar, pp. 139–173. Logan: Utah State University Press.

Duncan, R.P., Peet, R.K., Wentworth, T.R., Schafale, M.P., and Weakley, A.S. 1994. Vegetation of the fire-dependent pinelands of the Fall-line Sandhills of Southeastern North America. Final Report. USDA Forest Service, Southern Forest Experiment Station.

Folkerts, G.W. 1982. The Gulf Coast pitcher plant bogs. *Am Sci* 70:260–267.

Frost, C.C. 1993. Four centuries of changing landscape patterns in the longleaf pine ecosystem. *Proc Tall Timbers Fire Ecol Conf* 18:17–44.

Frost, C.C. 2000. Studies in landscape fire ecology and presettlement vegetation of the southeastern United States. Ph.D. dissertation, University of North Carolina, Chapel Hill.

Frost, C.C., and Musselman, L.J. 1987. History and vegetation of the Blackwater Ecological Preserve. *Castanea* 52:16–46.

Frost, C.C., Walker, J., and Peet, R.K. 1986. Fire-dependent savannas and prairies of the Southeast: Original extent, preservation status and management problems. In *Wilderness and Natural Areas in the Eastern United States*, eds. D.L. Kulhavy and R.N. Conner, pp. 348–357. Nacogdoches, TX: Center for Applied Studies.

Gano, L. 1917. A study of physiographic ecology in northern Florida. *Bot Gaz* 63:337–372.

Gonsoulin, G.J. 1974. A revision of *Styrax* (Styracaceae) in North America, Central America and the Caribbean. *Sida* 5:191–258.

Hale, P.M. 1883. The woods and timbers of North Carolina. P.M. Hale Publisher, Raleigh (including a reprint of M.A. Curtis. 1860. Geological and Natural History Survey of North Carolina. Part III—Botany).

Hainds, M.J., Mitchell, R.J., Palik, B.J., Boring, L.R., and Gjerstad, D.H. 1999. Distribution of native legumes (Leguminoseae) in frequently burned longleaf pine (Pinaceae)–wiregrass (Poaceae) ecosystems. *Am J Bot* 86:1606–1614.

Harcombe, P.A., Glitzenstein, J.S., Knox, R.G., Orzell, S.L., and Bridges, E.L. 1993. Vegetation of the longleaf pine region of the West Gulf Coast. *Proc Tall Timbers Fire Ecol Conf* 18:83–104.

Harper, R.M. 1905. Some noteworthy stations for *Pinus palustris*. *Torreya* 5:55–60.

Harper, R.M. 1906. A phytogeographical sketch of the Altamaha Grit Region of the coastal plain of Georgia. *NY Acad Sci* 7:1–415.

Harper, R.M. 1914. Geology and vegetation of North Florida. Florida Geological Survey. Sixth Annual Report 163–451.

Harper, R.M. 1930. The natural resources of Georgia. School of Commerce, Bureau of Business Research, Study No. 2. Bulletin of the University of Georgia 30(3):105.

Harper, R.M. 1943. Forests of Alabama. Alabama Geological Survey Monograph 10. Alabama Geological Survey, University of Alabama. University, AL.

Hodgkins, E.J. 1965. Southeastern forest habitat regions based on physiography. Agricultural Experiment Station, Auburn University, Forestry Department Series, No. 2. Auburn, AL.

Hodgkins, E.J., Golden, M.S., and Miller, W.F. 1979. Forest habitat regions and types on a photomorphic-physiographic basis: A guide to forest site classification in Alabama–Mississippi. Southern Coop Series 210, Alabama Agricultural Experiment Station, Auburn.

Huffman, J.M., Platt, W.J., Grissino-Mayer, H., and Boyce, C.J. 2004. Fire history of a barrier island slash pine (*Pinus elliottii*) savanna. *Nat Areas J* 24:258–268.

James, M.M. 2000. Legumes in loamy soil communities of the Carolina sandhills: Their natural distributions and performance of seeds and seedlings along complex ecological gradients. Master's thesis, Curriculum in Ecology, University of North Carolina, Chapel Hill.

Kjellmark, E.W., McMillan, P.D., and Peet, R.K. 1998. Longleaf pine vegetation of the South Carolina and Georgia Coastal Plain: A preliminary classification. Final report, USDA Forest Service, Southern Forest Experiment Station.

Kologiski, R.L. 1977. The phytosociology of the Green Swamp, North Carolina. North Carolina Agricultural Experiment Station, Technical Bulletin 250. Raleigh.

Maceina, E.C., Kush, J.S., and Meldahl, R.S. 2000. Vegetational survey of a montane longleaf pine community at Fort McClellan, Alabama. *Castanea* 65:147–154.

McCune, B., and Grace, J.B. 2003. *Analysis of Ecological Communities*. Gleneden Beach, OR: MjM Software Design.

Mohr, C. 1901. Plant life of Alabama. *Contrib US Natl Herb* 6.

Monk, C.D. 1968. Successional and environmental relationships of the forest vegetation of north-central Florida. *Am Midl Nat* 79:441–457.

Myers, N., Mittermeier, R.A., Mittermeier, C.G., Da Fosseca, G.A.B., and Kent, J. 2000. Biodiversity hotspots for conservation priorities. *Nature* 403:853–858.

Myers, R.K., Zahner, R., and Jones, S.M. 1986. Forest habitat regions of South Carolina from LANDSAT imagery. Clemson University, Department of Forestry Research Series No. 42.

NatureServe. 2005. NatureServe Explorer version 4.4. 7 April 2005. (http://www.natureserve.org/explorer/).

Norquist, H.C. 1984. A comparative study of the soils and vegetation of savannas in Mississippi. Masters thesis, Mississippi State University, Mississippi State.

Omernik, J.M. 1987. Ecoregions of the conterminous United States. Map (scale 1:7,500,000). *Ann Assoc Am Geogr* 77(1):118–125 (for updates consult http://www.epa.gov/wed/pages/ecoregions/ecoregions.htm).

Peet, R.K. 1993. A taxonomic study of *Aristida stricta* and *A. beyrichiana*. *Rhodora* 95:25–37.

Peet, R.K., and Allard, D.J. 1993. Longleaf pine vegetation of the Southern Atlantic and Eastern Gulf Coast regions: A preliminary classification. *Proc Tall Timbers Fire Ecol Conf* 18:45–81.

Peet, R.K., Carr, S. and Gramling, J. 2006. Fire-adapted pineland vegetation of northern and central Florida: A framework for inventory, management, and restoration. Florida Fish and Wildlife Commission. In press.

Peet, R.K., Duncan, R.P., Wentworth, T.R., Schafale, M.P., and Weakley, A.S. 1994. Vegetation of the fire-dependent pinelands of the North Carolina Coastal Plain. Final Report. USDA Forest Service, Southern Forest Experiment Station.

Peet, R.K., Wentworth, T.R., and White, P.S. 1998. A flexible, multipurpose method for recording vegetation composition and structure. *Castanea* 63:262–274.

Peterson, P.M. 2003. *Muhlenbergia* Schreb. In *Flora of North America Volume 25: Magnoliophyta:*

Commelinidae (in part): Poaceae, part 2, eds. M.E. Barkworth, K.M. Capels, S. Long, and M.B. Piep, pp. 145–200. New York: Oxford University Press.

Peterson, P.M., Hatch, S.L., and Weakley, A.S. 2003. *Sporobolus* R. Br. In *Flora of North America Volume 25: Magnoliophyta: Commelinidae (in part): Poaceae, Part 2*, eds. M.E. Barkworth, K.M. Capels, S. Long, and M.B. Piep, pp. 115–139. New York: Oxford University Press.

Phillips, J.D. 1994. Forgotten hardwood forests of the Coastal-plain. *Geogr Rev* 84:162–171.

Platt, W.J. 1999. Southeastern pine savannas. In *Savannas, Barrens, and Rock Outcrop Plant Communities of North America*, eds. R.C. Anderson, J.S. Fralish, and J.M. Baskin, pp. 21–51. London: Cambridge University Press.

Ricketts, T.H., Dinertstein, E., Loucks, C.J., Eichbuam, W., DellaSella, D., Kavanagh, K., Hedao, P., Hurley, P.T., Carney, K.M., Abell, R., and Walters, S. 1999. *Terrestrial Ecoregions of North America: A Conservation Assessment. World Wildlife Fund, United States and Canada*. Washington, DC: Island Press.

Rodgers, H.L., and Provencher, L. 1999. Analysis of longleaf pine sandhill vegetation in northwest Florida. *Castanea* 64:138–162.

Soller, D.R., and Mills, H.H. 1991. Surficial geology and geomorphology. In *The Geology of the Carolinas*, eds. J.W. Horton and V.A. Zullo, pp. 290–308. Knoxville: University of Tennessee Press.

Sorrie, B.A., and Weakley, A.S. 2001. Coastal Plain vascular plant endemics: Phytogeographic patterns. *Castanea* 66:50–82.

Sorrie, B.A., and Weakley, A.S. 2006. Developing a blueprint for conservation of the endangered longleaf pine ecosystem based on centers of Coastal Plain plant endemism. *Applied Vegetation Science* 9(1) (in press).

Stout, I.J., and Marion, W.R. 1993. Pine flatwoods and xeric pine forests of the Southern (lower) Coastal Plain. In *Biodiversity of the Southeastern United States: Lowland Terrestrial Communities*, eds. W.H. Martin, S.G. Boyce, and A.C. Echternacht, pp. 373–446. New York: John Wiley & Sons.

Taggart, J.B. 1990. Inventory, classification, and preservation of coastal plain savannas in the Carolinas. Ph.D. dissertation, University of North Carolina, Chapel Hill.

Taggart, J.B. 1994. Ordination as an aid in determining priorities for plant community protection. *Biol Conserv* 68:135–141.

USDA, NRCS. 2005. The PLANTS Database, Version 3.5 (http://plants.usda.gov). National Plant Data Center, Baton Rouge, LA 70874–4490 USA.

Walker, J., and Peet, R.K. 1984. Composition and species diversity of pine–wiregrass savannas of the Green Swamp, North Carolina. *Vegetatio* 55:163–179.

Ware, S., Frost, C., and Doerr, P.D. 1993. Southern mixed hardwood forest: The former longleaf pine forest. In *Biodiversity of the Southeastern United States: Lowland Terrestrial Communities*, eds. W.H. Martin, S.G. Boyce, and A.C. Echternacht, pp. 447–493. New York: John Wiley & Sons.

Wells, B.W., and Shunk, I.V. 1931. The vegetation and habitat factors of the coarser sands of the North Carolina coastal plain: An ecological study. *Ecol Monogr* 1:466–520.

Chapter 4

Longleaf Pine Regeneration Ecology and Methods

Dale G. Brockway, Kenneth W. Outcalt, and William D. Boyer

Introduction

Regenerating longleaf pine (*Pinus palustris*) is key to its long-term sustainable production of forest resources and its perpetuation as the dominant tree species in a variety of important ecosystems ranging from xeric to mesic to hydric site conditions. Early regeneration problems and the subsequent efforts to overcome these are significant features of the continuing longleaf pine saga. This chapter discusses recent restoration relevant to longleaf pine regeneration, disturbance dynamics including fire as an ecological process and describes the uniqueness of longleaf pine's regeneration environment. Fundamental information concerning reproductive biology (including genetics, flowering, pollination, fertilization, cone production, and seed dispersal) and seedling development (including germination, shoot growth, rooting, sprouting, competition, initiation of height growth, effects of fire, and seedling morality) is then presented. Various aspects of natural regeneration and artificial regeneration are discussed and the even-aged (i.e., clearcutting, seed-tree and shelterwood) and uneven-aged (i.e., group selection and single-tree selection) forest reproduction methods are introduced. We conclude by highlighting recent work that calls for application of silviculture techniques that more closely mimic natural disturbance regimes.

Ecological Relationships

Indispensable Nature of Regeneration

Successful reproduction is essential to perpetuate any population of organisms. Indeed, if an existing generation is unable to produce a succeeding generation, then the existing generation can appropriately be considered an ecological and evolutionary "dead end" for that population and perhaps the entire species. Despite the extended longevity of many tree species, some approaching 500 years, all individual organisms eventually die. If none of the offspring survive to the age of reproductive maturity, then the entire species will eventually perish. If the species is a dominant organism, then entire ecosystems will be degraded or lost, potentially threatening the survival of associated plant and animal species.

While species extinction and ecosystem loss may seem like rare events in the shorter term of human experience, from the longer-term

---

Dale G. Brockway and William D. Boyer • Southern Research Station, USDA Forest Service, Auburn, Alabama 36849. **Kenneth W. Outcalt** • Southern Research Station, USDA Forest Service, Athens, Georgia 30602.

FIGURE 1. Naturally regenerated even-aged second-growth longleaf pine forest on mesic uplands. Photo courtesy of the Forestry Images Organization.

perspective of geologic time, such events have not been uncommon. The survival of species and sustainability of ecosystems cannot always be assumed, even under the best circumstances. And when new species and/or cultures encounter native ecosystems, new pressures can stress the indigenous organisms and threaten ecological sustainability. Such has been the case in the southern United States, where the Age of Discovery and the Industrial Revolution brought substantial change to native longleaf pine forests (Frost this volume). Effective means of regenerating longleaf pine are important for continuation of this species and the long-term sustainability of longleaf pine forest ecosystems.

Declining Trend

Longleaf pine ecosystems were once among the most extensive in North America, occupying about 37 million ha prior to European settlement (Frost 1993, Landers et al. 1995). While the initial impact of immigrants during the eighteenth and nineteenth centuries was generally modest, as populations grew and logging activity expanded during the late nineteenth and early twentieth centuries, most native longleaf pine forests were harvested. The land was often converted to agricultural, residential and urban uses or planted with plantations of other easier-to-establish, faster growing trees such as slash pine (*Pinus elliottii*) and loblolly pine (*Pinus taeda*) (Croker 1987, Outcalt 2000). Although many second-growth longleaf pine forests naturally regenerated following this initial harvest (Fig. 1), recovery was impaired by irregular seed production, with good seed years occurring at intervals of five or more years (Boyer 1990a). Where longleaf pine seedlings did survive logging, they were often consumed by feral hogs (*Sus scrofa*), causing many areas of potential longleaf pine forest to be lost (Schwarz 1907, Croker 1987, Simberloff 1993, McGuire 2001). As the southern landscape became increasingly domesticated, the modified structure of expansive agricultural areas and linear transportation corridors fragmented previously contiguous habitat, thereby impeding the movement of natural surface fires across these lands (Walker 1999). During the twentieth century, organized programs of fire suppression and policies of fire exclusion from the forest further interrupted natural fire regimes (Croker 1987). Since the absence of frequent surface fires impedes the natural regeneration of longleaf pine and allows invasion of longleaf pine sites by hardwoods and more aggressive southern pines, interruption of natural fire regimes is believed to be the most ecologically significant cause for its continuing decline (Wright and Bailey 1982, Landers et al. 1990, Pyne 1997, Gilliam and Platt 1999). Longleaf pine forests have undergone a steady decrease to 8 million ha in 1935 (Wahlenberg 1946), 2 million ha by 1975, 1.5 million ha in 1985 (Kelly and Bechtold 1990) and less than 1.2 million ha currently (Outcalt and Sheffield 1996). Occupying less than 3% of their original range (Ware et al. 1993), longleaf pine ecosystems are now recognized as being at high risk (Noss et al. 1995, Kush 2002). Unfortunately, area reductions continue for stands in every diameter class below 41 cm (Kelly and Bechtold, 1990), an indication that most remaining longleaf pine forests are aging without replacement.

Ecological Restoration

Extending along the Gulf and Atlantic Coastal Plains from Texas to Virginia and inland to the Piedmont and mountains in Alabama and Georgia, longleaf pine forests, woodlands, and savannas may occupy a wide variety of sites, ranging from wet poorly drained flatwoods to mesic uplands, xeric sandhills, and rocky mountain ridges (Boyer 1990a; Stout and Marion 1993). Distinguished by a generally open, parklike stand structure (Schwarz 1907; Wahlenberg 1946), naturally regenerated longleaf pine forests are typically an uneven-aged mosaic of even-aged patches distributed across the landscape, which vary in size, structure, composition, and density (Platt and Rathbun 1993; Brockway and Outcalt 1998) and contain numerous embedded special habitats such as stream bottoms, wetlands, and seeps (Hilton 1999). The natural variability of these ecosystems makes them excellent habitat for a variety of game animals and numerous nongame and rare wildlife species (Kantola and Humphrey 1990; Engstrom 1993; Guyer and Bailey 1993; Crofton 2001; Engstrom et al. 2001; Brockway and Lewis 2003; Means this volume).

The complex natural patterns and processes unique to longleaf pine forests create extraordinarily high levels of biological diversity in these ecosystems, with the great number of plant species per unit area qualifying these as among the most species-rich terrestrial ecosystems outside the tropics. As many as 140 vascular plant species have been observed in a 1000 m^2 area and equally impressive counts of more than 40 species per m^2 have been recorded (Peet and Allard 1993), a large number of which are restricted to or found principally in longleaf pine habitats. Habitat reduction resulting from decline of longleaf pine ecosystems has caused the increased rarity of 191 vascular plant taxa (Hardin and White 1989, Walker 1993) and several vertebrate species. Concern over loss of this unique ecosystem (Means and Grow 1985; Noss et al. 1995) has led to many efforts focused on effectively restoring longleaf pine ecosystems (Walker and Boyer 1993; Walker 1995; Kush 1996, 1998, 1999, 2001; Johnson and Gjerstad 1998, 1999; Brockway et al. 1998; Seamon 1998; Outcalt et al. 1999; Brockway and Outcalt 2000; Provencher et al. 2001a,b; Mulligan et al. 2002). Since longleaf pine still occurs in isolated fragments over most of its natural range, it is reasonable to conclude that restoration of these ecosystems is possible (Landers et al. 1995). Effective methods for regenerating longleaf pine will no doubt play a key role in ecological restoration efforts.

Disturbance Dynamics and Fire as an Ecological Process

Longleaf pine ecosystems exist in an environment influenced by large-scale catastrophic disturbance, such as damaging tropical storms. Lightning is an important agent in individual tree morality and creation of small-scale disturbance in longleaf pine forests (Komarek 1968; Taylor 1974). The structure, pattern, and diversity of longleaf pine ecosystems are maintained by a combination of site factors and periodic disturbance events, including lightning strikes, tree mortality, and animal interactions at local scales and tropical storms, soils, and hydrologic regimes at broader scales. Disturbances across site gradients provide large living trees, snags, coarse woody debris, forest canopy gaps, and hardwood thickets that support numerous plant and animal species adapted to these disturbance-prone, yet largely stable ecosystems.

Longleaf pine is closely associated with wiregrass (*Aristida* spp.) in the eastern part and bluestem grasses (*Andropogon* spp. and *Schizachyrium* spp.) in the western portion of its range. The understories of longleaf pine forests are typically dominated by herbaceous plants because these bunchgrasses facilitate the ignition and spread of frequent surface fires (Landers 1991). In these ecosystems, longleaf pine and bunchgrasses function together as keystone species that facilitate but are resistant to fire (Platt et al. 1988; Noss 1989). They also exhibit substantial longevity and demonstrate nutrient and water retention to a degree that reinforces their site dominance and minimizes change in the plant community following disturbance (Landers et al. 1995). As a key ecological process and disturbance agent, the benefits of periodic fire include (1) maintaining the physiognomic character of longleaf pine ecosystems through excluding invasive plants that are ill-adapted to fire, (2) preparing a seedbed favorable for the establishment of longleaf pine seedlings, (3) reducing the density of understory vegetation thus providing microsites for a variety of herbaceous plants, (4) releasing nutrients immobilized in accumulated phytomass for recycling to the infertile soil and subsequently more rapid uptake by plants, (5) improving forage for grazing, (6) enhancing wildlife habitat, (7) controlling harmful insects and pathogens, and (8) reducing fuel levels and wildfire hazard (McKee 1982; Wade and Lewis 1987; Boyer 1990b; Wade and Lundsford 1990; Dickmann 1993; Brennan and Hermann 1994; Brockway and Lewis 1997).

Regeneration Environment

Difficulties encountered during early attempts to regenerate longleaf pine impeded its recovery and contributed to its historical decline (Croker 1987). Erratic seed production, poor seedling survival, and slow early growth of seedlings discouraged forestland managers from investing in longleaf pine. Management policies based on these initial observations further contributed to the decrease of longleaf pine forests, as harvested stands were deliberately converted to other southern pine species rather than being regenerated with longleaf pine. Fortunately, later research illuminated the ecological mechanisms and identified silvicultural methods for effectively regenerating longleaf pine by natural and artificial means (Boyer and White 1990; Barnett et al. 1990; Kush 2002) and the earlier policies of forest type conversion have now been largely reversed.

The unique structural and process dynamics characteristic of longleaf pine forests provide both challenges to and opportunities for applying science and adapting technology to efficiently obtain regeneration. Foremost, all longleaf pine forests are obligatorily pyrophytic ecosystems. Therefore, all regeneration techniques employed must be compatible with periodic surface fires. Longleaf pine forests are disturbance-prone and naturally regenerate in a variety of configurations ranging from relatively small circular or elliptical canopy gaps and attenuated strings to larger areas of partially blown down or almost completely blown down overstory trees (Croker and Boyer 1975; Palik and Pedersen 1996; Brockway and Outcalt 1998). As a tree species that is intolerant of competition, whether for light, moisture, or nutrients (Boyer 1990a), its seedlings become established and flourish as opportunists responding to resource availability. Although longleaf pine displays many traits consistent with an intolerant, early seral species, its seedlings often persist in the forest understory for prolonged periods similar to those of more tolerant, late-seral species. Suppressed longleaf pine seedlings will not respond with improved growth until released from competition with overstory trees and long-suppressed individuals are, through time, increasingly prone to mortality. Whether a specific longleaf pine seedling survives to eventually become a dominant member of the forest canopy may depend as much on stochastic events (i.e., drought severity and duration, fire intensity and season, disturbance that results in timely release from competition) as it does on the competitive vigor (i.e., genetic attributes) of the individual. Whether by artificial or natural methods, by even-aged or uneven-aged silvicultural techniques, not only is efficiently regenerating longleaf pine feasible, it is also imperative for achieving ecological restoration and ecosystem sustainability goals.

Reproductive Biology

Genetics

Longleaf pine is a tree species of considerable genetic diversity, with variation among individuals typically greater than that among stands or geographically different seed sources (Snyder et al. 1977; Lynch 1980). Although genetic variation among populations is thought to be a result of the diversity of environments in which longleaf pine occurs throughout its native range (Boyer 1990a), measures of genetic diversity appear unrelated to climate variables (Schmidtling and Hipkins 1998). While the pattern of genetic variation for longleaf pine is similar to that of other southern pines (Schmidtling 1999), its unique pattern of allozyme variation is indicative of a very different history during the recent Ice Age (Schmidtling et al. 2000). Unlike other pines, longleaf pine appears to have migrated eastward across the southeastern United States from a single refuge in southern Texas and/or northeastern Mexico after the Pleistocene (Schmidtling and Hipkins 1998). The progressive decrease in allozyme diversity from western to eastern longleaf pine populations represents a loss in genetic variability from stochastic events during migration.

Longleaf pine may form a natural hybrid with loblolly pine, referred to as "Sonderegger pine" (*Pinus sondereggeri*). Since flowering of these two species frequently overlaps, there is no phenological barrier to natural crossing (Boyer 1990a). Hybridization between longleaf pine and slash pine is far less likely because of differences between their dormancy and heat requirements for flowering (Boyer 1981); however, artificial crossing can be easily achieved. Hybrids between longleaf pine and shortleaf pine (*Pinus echinata*) have not been observed in nature but have been artificially produced (Snyder et al. 1977).

With a pattern of genetic variation similar to that in other southern pines, longleaf pine is suitable for genetic improvement. However, the effort expended on this species is easily dwarfed by the immense resources devoted to loblolly pine and slash pine improvement programs (Schmidtling 1999). Traditional tree improvement approaches, which select "plus" trees in the forest based on size and form, have not proven to be suitable for longleaf pine. Variation in the "grass" stage of longleaf pine makes it impossible to determine the true age of a tree and thus its true growth potential. Therefore, tree improvement programs for longleaf pine have shifted their emphasis to a progeny test approach, with the duration of the grass stage and resistance to brown-spot fungus (*Mycosphaerella dearnessii*) the most important inherited traits of interest. When focusing efforts on accelerating the early height growth of longleaf pine, one of the greatest dangers is the possibility of incorporating loblolly pine genes that will result in a hybrid that begins growing earlier but has poor form and increased susceptibility to fusiform rust fungus (*Cronartium quercuum* f. sp. *fusiforme*). There is no ecotypic differentiation in longleaf pine based on site conditions and no important difference in survival or growth between eastern and western populations, as occurs with loblolly pine (Schmidtling 1999).

Flowering, Pollination, and Fertilization

Longleaf pine is monoecious, with male strobili (catkins) predominating in the lower crown and female strobili (conelets) occurring most frequently in the upper crown of the same tree (Schopmeyer 1974). Development of catkins

FIGURE 2. Elongated catkins, subtending a terminal bud, shown after their pollen has been shed. Photo courtesy of the Forestry Images Organization.

(Fig. 2) and conelets (Fig. 3) is initiated during the growing season before buds emerge, with catkins beginning to form during July and conelets being formed during a short period in August (Boyer 1990a). First appearing at the base of vegetative buds between mid-November and early December, purple catkins remain dormant for several weeks before resuming development between late December and early February (Boyer 1981). Conelet buds appear in January or February and conelets, upon emerging from the bud, are red until pollinated, after which they fade to yellowish green. The development rates of both catkins and conelets are almost entirely dependent on ambient temperature (Boyer 1990a).

The number of flowers produced appears related to weather conditions during the year of initiation. Catkin production is favored by abundant rainfall throughout the growing season, while conelet production is promoted by a wet spring and early summer followed by a dry period in late summer (Shoulders 1967). Because of the differential conditions that favor each sex, large crops of male and female flowers do not necessarily coincide (Boyer 1990a). Heavy annual losses of longleaf pine conelets can usually be expected, with observed losses ranging from 65% to 100% (Boyer 1974a; McLemore 1977; White et al. 1977). Insects, weather extremes, and insufficient pollen appear to be the primary causes for these losses, which primarily occur during spring pollination or the following summer.

Peak pollen shed and conelet receptivity typically vary from late February in the southern portion of the native range to early April in more northern areas (Boyer 1990a). Although pollen shedding and receptivity coincide on individual trees, there appears to be very little synchrony among trees in a longleaf pine forest, with some trees being consistently early and others being consistently late and overall dates highly influenced by air temperatures before and during the flowering period. Pollen shedding typically occurs during a period of 5 to 21 days, with an average of 13 days (Boyer 1981). Year-to-year variation in the time of pollen shedding appears to be related to accumulation of degree-day heat sums, with all temperatures greater than $10°C$ after January 1 promoting development of the male strobili (Boyer and Woods 1973; Boyer 1973, 1978).

While pollination takes place in late winter or spring, fertilization does not occur until the following spring. Conelets grow rapidly after fertilization, increasing in length from 2.5 cm to 18 cm by May or June (Boyer 1990a). Cones reach maturity between mid-September and mid-October of their second year and range in length from 10 cm to 25 cm. Although cone color changes from green to brown as they ripen, cones may be ripe before changing color (Schopmeyer 1974).

Cone Production and Seed Dispersal

Longleaf pine cone crops are highly variable from year to year (Fig. 4), with 1860 cones/ha

FIGURE 3. Conelets, located peripherally to terminal buds, are most often observed in the upper crown. Photo courtesy of the Forestry Images Organization.

(e.g., 30 cones/tree and 62 seed trees/ha) normally required for successful natural regeneration (Boyer 1996). While cone production may be influenced by the density of airborne pollen during flowering, low cone crop frequencies appear to be more a result of flower losses rather than a failure to produce flowers (Boyer 1974a, 1987a). Since 1986, cone crops on coastal plain sites from Louisiana to North Carolina have increased to an average of 36 cones/tree from an earlier average of only 14 (Boyer 1998). This increase in cone production appears to be a result of both an increase in flower production and an increase in the fraction of flowers surviving to become mature cones. Cone production of individual trees is influenced foremost by genetics and secondarily by tree size, crown class, stand density, and site quality. The greatest cone production occurs on dominant, open-grown longleaf pines having large crowns (Croker and Boyer 1975). Trees of 38–48 cm diameter at breast height produce on average 65 cones/year compared with 15 cones/year from trees in the 25–33 cm diameter at breast height size class (Boyer 1990a). The number of viable seeds per cone varies with the seed crop for a specific year, ranging from 50 seeds/cone in good years to 35 seeds/cone during average years to 15 seeds/cone in poor years (Croker 1973).

Peak seed production is observed in longleaf pine forests having stand densities between 6.9 and 9.2 m²/ha, when principally comprised of dominant and codominant trees of cone-bearing size (Boyer 1979). Such stands produce seed crops adequate for natural regeneration once every 4 to 5 years on average (Croker and Boyer 1975). When forests of substantially greater density are thinned to this level, increased cone production resulting from decreased intraspecific competition does not occur for three growing seasons (Croker 1952). Release following conelet initiation

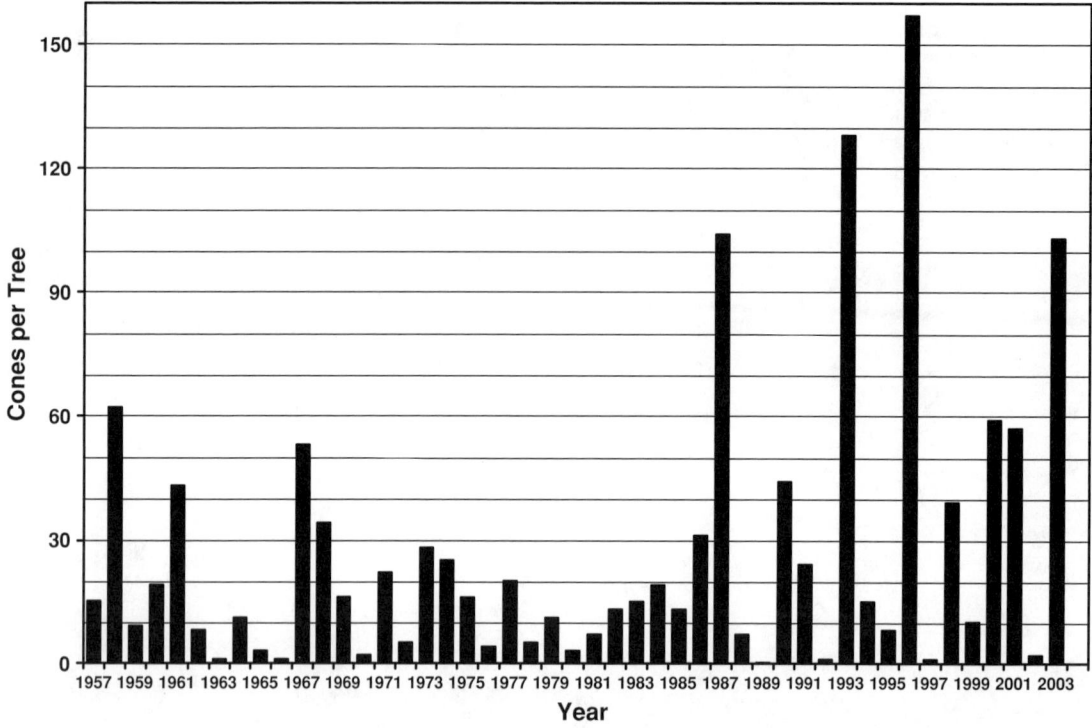

FIGURE 4. Example of high temporal variability for longleaf pine seed production: Cone crops at the Escambia Experimental Forest, southern Alabama, 1957–2003.

benefits the crop only by reducing moisture stress during dry periods (Boyer 1990a). Seeds are dispersed by wind during a 2- to 3-week period between late October and late November, depending on weather conditions. Longleaf pine seeds are the largest of the southern pines and their dispersal distance is limited, with 71% of sound seeds falling within 20 meters of the base of a parent tree (Croker and Boyer 1975).

Seedling Development

Germination and Shoot Growth

Longleaf pine seeds typically germinate (Fig. 5) within a week of contacting the ground (Boyer 1990a). While rapid germination may be an adaptation to reduce the risk of exposure to seed predators, newly germinated seedlings are also vulnerable to mortality from animals, pathogens, and adverse weather conditions (Croker and Boyer 1975). Although germination and establishment require that seeds contact mineral soil, the large seed and wing can impede penetration through dense grass or accumulated litter on the forest floor. Root systems of premature germinants fail to reach mineral soil and readily die from desiccation. The risk of regeneration failure can be significantly reduced by using mechanical treatment and/or prescribed fire to prepare a suitable seedbed before seedfall. Germination begins with emergence of the radicle and an almost simultaneous elongation of the cotyledons (Allen 1958). Hypocotyl elongation begins soon after radicle emergence but is limited. Growth of the cotyledons lifts the seedcoat from the ground. Limited hypocotyl growth causes the cotyledons of new germinants to remain at or near the ground line. Newly germinated seedlings are relatively inconspicuous with their small primary needles. Although secondary needles appear within 2 months, the epicotyl reaches a length of only 0.38 cm

4. Longleaf Pine Regeneration Ecology and Methods

FIGURE 5. Newly germinated longleaf pine seedlings emerging from thin forest litter (foreground) and among cones fallen from the parent trees. Photo courtesy of the Forestry Images Organization.

and, during the next 8 weeks, does not elongate as in other pines. This stemless condition is one of the unique characteristics of longleaf pine, commonly referred to as the "grass stage" (Fig. 6). Depending on ambient conditions, longleaf pine seedlings may remain in this stage for 2 to as long as 15 years, during which they are most susceptible to their major disease, brown-spot needle blight (Boyer 1990a).

Roots and Sprouts

While in the grass stage, longleaf pine seedlings devote much of their energy to root production. Following germination, the radicle forms a taproot that develops very rapidly. Growth rates may be as high as 50 cm in 15 days following germination in sandy soils under greenhouse conditions (Wahlenberg 1946). More typical rates are 0.8 cm/day during the first 60 days (Allen 1958). Taproot development is inversely related to moisture conditions, with greater root elongation occurring in drier soils. The general root structure of natural seedlings is unaffected by soil conditions, but these seedlings have slower root growth than those measured in greenhouse tests. The root system of a typical 1-year-old seedling consists of a taproot 60–70 cm long and a number of strong laterals 50–60 cm long in the upper soil layers with numerous attached feeder tips (Wahlenberg 1946). Laterals extend outward, and though they often change direction due to obstructions, they remain at a uniform depth from origin to tip (Heyward 1933). On wetter flatwoods sites, root systems have the same general architecture, but root distribution is restricted mostly to surface horizons, above the spodic layer that occurs in soils typical of these

FIGURE 6. Longleaf pine seedling in the grass stage. Photo courtesy of the Forestry Images Organization.

areas. Even in well-drained sands, 90% of all lateral roots are found in the upper 30 cm of the soil.

Longleaf pine seedlings can sprout from the root collar if the top is killed. Under natural conditions, sprouts often result following top kill from fire. This sprouting ability diminishes rapidly once seedlings emerge from the grass stage and begin stem elongation. As many as 40% of the grass-stage seedlings that were cut off at the groundline have been found to have living sprouts 1 year later (Farrar 1975). Only 14% of the seedlings, that had initiated height growth but were less than 1.37 m tall, produced sprouts when cut at the groundline. None of the trees more than 1.37 m tall produced sprouts after cutting. Three years after cutting, 30% of the grass-stage seedlings still had surviving sprouts, but these were growing very slowly. Although fewer of the seedlings that had begun height growth sprouted, those sprouts were much larger 3 years after treatment. Sprouts from smaller seedlings are rather weak and often die following subsequent fires.

Competition

Longleaf pine seedlings are intolerant of interspecific and intraspecific competition (Harrington, this volume). Seedling growth rates improve as the distance from adult pines increases, with the suppressive effect from stands of overstory trees adjacent to clearcut strips being greater than that from single overstory trees (Boyer 1963). The relationship of declining seedling growth rates with increasing amounts of overstory competition (i.e., overstory basal area) follows that of a general exponential decay curve. The root collar diameter for 4-year-old seedlings has been reported to sharply decrease from 1.2 cm to 0.7 cm as overstory basal area increased from 0 to 6.9 m^2/ha. More recent work has substantiated the suppressive effect of overstory adults upon longleaf pine seedlings on redhills sites (Grace and

Platt 1995) and flatwoods sites (Gagnon et al. 2003, 2004) and the exponential relationship between seedling growth and overstory basal area (Palik et al. 1997).

Overstory hardwood trees are even greater competition for longleaf pine seedlings. While 60% of the seedlings in the 0–3 m zone near adult longleaf pines reached sufficient root collar diameter by age 9 to begin height growth (Smith 1962), only 24% of the pine seedlings near oak trees had initiated height growth. Removal of overstory oaks significantly improved height growth of longleaf seedlings (Walker 1954). Understory vegetation, including other longleaf pine seedlings and herbaceous plants, also competes with longleaf pine seedlings. Single season growth increases of 18% were measured in seedling densities of 247,100/ha and increases of 64% were observed at densities of 2471/ha (Pessin 1938). When herbaceous competition was removed, growth increased by 25 and 456% for the same respective seedling densities. Thus, intraspecific competition in the understory appears much less important than competition from other species in this layer.

Ascertaining the underlying causes of seedling growth reduction has proven challenging. Root competition was believed to be much more important than light because growth depression extends well beyond the height of adjacent overstory trees (Walker and Davis 1956; Croker and Boyer 1975). This inhibitory effect is also greater on sites with poor soils, another indication of belowground competition. The importance of belowground competition was more recently substantiated by research on xeric sandhills sites in Florida, where total daily light influx did not differ significantly with distance from adult trees, but the fine root biomass of overstory trees did decline with distance (Brockway and Outcalt 1998). Others working on more fertile soils with higher densities of woody plants have shown that light does increase with distance from gap edge (Palik et al. 1997; McGuire et al. 2001; Gagnon et al. 2003). Palik et al. (1997) also documented a positive relationship between available N and seedling growth. Longleaf pine seedling growth was found to be significantly related to N, soil water, and their interaction in a greenhouse study (Jose et al. 2003). Light was important only if water was not limiting. Therefore, competition affecting longleaf pine seedling growth follows the fundamental principle of limiting factors, with growth most impaired by whichever essential factor is most limited in a specific environment.

Initiation of Height Growth

Although varying considerably, longleaf pine seedlings usually begin emerging from the grass stage when their root-collar diameter reaches about 2.5 cm (Boyer 1990a). Infection by brown-spot needle blight can substantially delay grass stage emergence. The height growth of seedlings also depends on the intensity of competition in the ambient environment (Ramsey et al. 2003). Therefore, treatments that reduce competition will increase the proportion of seedlings that emerge from the grass stage (Haywood 2000). Differences in levels of intraspecific and interspecific competition and individual genetic control of growth rates and resistance to brown-spot infection result in considerable variation in emergence of longleaf pine seedlings from the grass stage. This high within-stand variation is actually beneficial as it leads to the early establishment of dominant seedlings, preventing stand stagnation even when seedlings occur at very high densities. The key to height growth initiation is accumulation of carbohydrate reserves in the root sufficient to support rapid upward expansion of the stem. This interval is known as the bolting stage in which rapid growth of the needle-covered stem occurs with limited lateral branch development, producing the characteristic "bottlebrush" structure (Fig. 7). Under favorable conditions, longleaf pine seedlings can grow 30–90 cm in a single season (Wahlenberg 1946).

Effects of Fire

Longleaf pine evolved in ecosystems subject to frequent low-intensity surface fires and has developed many adaptations for enhancing survival and growth in this environment (Landers 1991). Although newly germinated

FIGURE 7. Longleaf pine seedlings in the bolting stage. Photo courtesy of the Forestry Images Organization.

FIGURE 8. This larger grass-stage longleaf pine survived a recent fire; note the thick root collar and partially consumed needles that adequately protected its terminal bud. Photo courtesy of the Forestry Images Organization.

and small grass-stage seedlings are susceptible to fire-caused mortality, once they attain a root-collar diameter of 1.3 cm, they become quite fire-resistant (Boyer 1974b). Such larger grass-stage seedlings have thicker bark and a large tuft of needles that protect the central meristem (Fig. 8) and, as noted previously, even if the top is killed, they will often sprout. However, since many natural seedlings do not attain this size until they are 2 or 3 years old, mortality from fires can be quite high. A single growing-season fire, at age 2, can result in as much as 80% of the total seedling mortality for the 3-year period following germination. Fire can also be beneficial to longleaf pine seedlings by consuming brown-spot-infected needles. Fire stimulates height growth (Grelen 1983) by reducing competition and releasing nutrients immobilized in organic matter.

When initiating height growth, seedlings once again become more susceptible to fire-caused mortality. During this stage, the terminal bud is situated directly in the flaming zone, where it may be exposed to maximum temperatures. These seedlings are especially vulnerable during early season bud break when the candles are rapidly expanding and there is very little protective needle mass. Once they have bolted above the flame zone (about 1 m), seedlings are again quite resistant to fires of moderate intensity. Although fire does not kill many longleaf pines once they reach this small sapling stage, burning can still cause reductions in growth (Boyer 1987b).

Seedling Mortality

Seedlings are susceptible to losses from frost heaving, flooding, disease, logging, fire,

FIGURE 9. Close-up view of longleaf pine needles infected with brown-spot fungus; note highly variegated color tones and distinctive dark spots on the needles. Photo courtesy of the Forestry Images Organization.

drought, and animals. As expected, losses are greatest during the first growing season. Frost heaving is a threat only near the northern limit of the natural range on finer-textured soils. Flooding can be a problem on wet coastal plain sites where prolonged standing water can kill newly germinated seedlings (Croker and Boyer 1975). Brown-spot needle blight (Fig. 9) may at times become so severe that it destroys all the foliage and directly kills the seedling. More often, the seedling becomes so weakened that it dies during the next fire. While depending on the harvest method, season, and volume of trees removed, losses from logging generally average about 50% (Boyer 1990a). To minimize logging losses, harvesting should be done when seedlings are in the grass stage. As noted above, fire is a significant cause of seedling mortality during the first 2 years, where losses up to 90% may often occur (Grace and Platt 1995; Provencher et al. 2001b). The probability of seedling mortality from fire is greatest in the zone around adult trees, because greater needle litter accumulations there often result in hotter fires. Higher intraspecific competition in this zone also results in slower seedling growth, thus predisposing them to higher rates of fire-related mortality. Before the first fire occurs, seedlings growing close to adult pines rarely attain the 1.3 cm root-collar diameter needed for survival.

Drought can cause significant losses of first-year seedlings, especially on sandhills sites where 50% mortality is common during the spring drought period. Many of the seedlings that perish during dry periods are growing on locations where the forest floor is somewhat thicker. Although bare mineral soil is ideal for longleaf pine germination and growth, sufficient moisture is all that is needed to initiate the germination process. Therefore, many of the seedlings that germinate on thicker forest floor material in the fall, fail to survive the spring drought because their root system has not penetrated sufficiently into the mineral soil to obtain adequate moisture. Litterfall is of course not distributed evenly, but rather is highest near adult trees. The greater forest floor thickness, increased root competition, and reduced rainfall from crown interception all combine to substantially lower the probability of longleaf pine seedling survival near adult overstory trees. This typical spring mortality significantly contributes to development of the often-observed zone of seedling exclusion near adult pines (Brockway and Outcalt 1998).

Both wild and domestic animals can decimate stands of longleaf pine seedlings. Locally high populations of rabbits can cause significant damage to newly germinated seedlings by clipping off the needles (Croker and Boyer 1975). Local losses can also result from pocket gophers feeding on the starch-laden roots of grass-stage seedlings. While cattle will also graze on longleaf pine seedlings, most damage and mortality results from trampling where sites are allowed to suffer from overstocking and excessive use. However, the most destructive of all animals is the wild hog, a nonnative species introduced from Europe. The roots of grass-stage seedlings, rich in starch, are a favorite food of wild hogs. A single hog can uproot and consume a great many seedlings in a single day. The destructive power of the vast herds of wild hogs was noted in the early 1900s (Schwarz 1907). An early fencing study demonstrated their effect on longleaf pine seedling densities, with 14,826 stems/ha occurring within the protective fence and a mere 20 stems/ha surviving outside (Mattoon 1922). In some locations, wild hogs still pose a problem for regeneration. However, removing hogs for only 2 years resulted in successful regeneration of longleaf pine in the South Carolina Coastal Plain (Lipscomb 1989).

Natural Regeneration

Regeneration Problems

Perhaps the greatest impediment to wider acceptance of longleaf pine, as a tree worthy of forest management investment, is its reputation as a difficult species to successfully regenerate, whether by natural or artificial means. Schwarz (1907) was among the first to identify poor regeneration as a major threat to the future of longleaf pine. Insufficient numbers of seed trees, infrequent seed crops, cone infestation by insects, limited dispersal distance of heavy seeds, seed predation, seed perching in litter above mineral soil, vulnerability of early germinating seed to temperature extremes, brown-spot fungus infection, slow early seedling growth, untimely fires, and fire exclusion favoring competing species are among the primary reasons for regeneration failure of longleaf pine (Wahlenberg 1946; Croker and Boyer 1975; Boyer 1979). Fortunately, the extensive research conducted on longleaf pine regeneration during the second half of the twentieth century has now reduced these problems to considerations that can be effectively addressed through appropriate management strategies and practices (Kush 2002).

Regeneration Requirements

When the fundamental requirements can be met, natural regeneration is a practical and inexpensive management option for existing longleaf pine forests (Boyer 1993a). The foremost requirements for successfully regenerating longleaf pine by natural means are an adequate seed source and a receptive seedbed (Boyer and White 1990). A sufficient number of sexually mature parent trees, having desirable characteristics and being well distributed throughout the area, typically serves to meet this need (Dennington and Farrar 1991). Since longleaf pine seed crops are infrequent, averaging one every 5 to 7 years, annual monitoring of conelet and cone production is necessary to appropriately forecast years when adequate seed will be available (Boyer 1996). For initiating even-aged stands, the size, number, and distribution of seed-bearing trees must be such that a minimum of 1860 and preferably 2470 or more cones/ha will be provided (Boyer and White 1990). Peak seed production may be encouraged in future years by reducing stand density to between 6.9 and 9.2 m^2/ha (Boyer 1979).

One of the most pervasive problems impeding the natural regeneration of longleaf pine is high-density understory and midstory layers dominated by woody plants. If preestablishment competition control is not achieved, longleaf pine regeneration success will be highly doubtful (Boyer and White 1990). Therefore, competing hardwood trees and shrubs must be effectively controlled before seed dispersal. The presence of competing woody plants is not only deleterious to seedlings as they are becoming established, but may also interfere with dispersed seeds as they attempt to reach mineral soil. Although prescribed fire conducted well in advance of seedfall may be sufficient to control these woody plants, herbicide application or mechanical treatment may also be necessary to obtain adequate control under certain circumstances (Croker and Boyer 1975).

Longleaf pine seeds need to be in contact with mineral soil for successful germination and establishment (Boyer and White 1990).

Since seed supplies are typically limited, a well-prepared mineral seedbed is necessary to provide optimum opportunity for regeneration success. A suitably prepared seedbed can be created by a prescribed fire conducted within 1 year prior to seedfall. Such a fire will remove enough forest litter to double the rate of longleaf pine seedling establishment compared to unburned sites. While only 14% of longleaf pine seeds became established as seedlings on unburned areas, 36% became established on winter-burned areas and 24% became established on fall-burned areas (Croker 1975). The lower establishment level on fall-burned areas is believed to result from lighter subsequent litter accumulation causing higher rates of seed predation by birds.

Successful regeneration requires an adequate number of established seedlings well-distributed throughout the area (Boyer and White 1990). Although exact criteria must be left up to individual land managers with specific management objectives, a reasonable goal when using the shelterwood method might be about 14,800 established 1-year-old or older seedlings per hectare. This standard allows for about 50% mortality among seedlings, when the stand is later logged, and anticipates high rates of mortality from other sources, especially brown-spot fungus, during the early years following establishment. It is reasonable to expect that at least 1235 well-distributed seedlings per hectare could result that are about 1 m in height and therefore free from brown-spot infection and relatively safe from damage by fire (Croker and Boyer 1975).

Since longleaf pine is highly sensitive to competition, established seedlings develop optimally in the absence of competitors. Although hardwood trees and shrubs comprise the most severe competition for young longleaf pine, these should be largely eliminated during preestablishment treatments. While eliminating all competitors is not practical, action to obtain postestablishment competition control is indicated when hardwood densities exceed 2.3 m^2/ha (Boyer and White 1990). Herbaceous understory plants and adult longleaf pine trees in the overstory typically remain as the major competitors for newly established seedlings. Seedling growth beneath adults remains inhibited until nearby overstory trees are removed (Boyer and White 1990). Periodic prescribed fire should adequately curtail understory plants, with herbicide application reserved for use under specific circumstances.

Control of the brown-spot fungus, the most serious pathogen of longleaf pine, is essential for successful natural regeneration (Boyer and White 1990; Boyer 1990a; Dennington and Farrar 1991). Timely application of prescribed fire, which consumes the infected needles, is the most effective remedy for this needle blight. In newly established seedlings, prescribed fire should be deferred until an adequate number of the seedlings reach a fire-resistant size. Unfortunately, seedlings beneath adult trees grow slowly and delaying fire could allow brown-spot needle blight to spread in a stand, further inhibiting seedling growth. This risk can be abated by maintaining relatively low basal areas where seedlings are less likely to be suppressed by overstory trees.

Young seedlings must also be protected from hogs, which can rapidly destroy seedlings by consuming them while in the grass stage (Boyer 1990a; Boyer and White 1990). Cattle grazing will also reduce the fine fuel needed to effectively carry fire in these ecosystems and may directly damage seedlings if conducted at high intensities (Dennington and Farrar 1991). Regenerating longleaf pine by natural means requires careful attention to details relevant to advanced planning, monitoring site conditions, prescription development, and timing of cultural treatments. Through identifying the area, assessing the seed source, monitoring cone crops, controlling competition, preparing a suitable seedbed, conducting regeneration surveys of seedlings, appropriately adjusting overstory density, and conducting postestablishment burns and other treatments, the likelihood of successfully regenerating longleaf pine forests will be substantially increased.

Unique Considerations

Largely as a result of ignorance concerning the unique life history of longleaf pine, mismanagement may be the rule rather than the exception (Kush 2002). Unlike other southern pines, longleaf pine is both resistant to and dependent on fire and its seedlings must pass through a stemless grass stage. If unwise management or the absence of management allows competing species to grow freely while longleaf pine is in the grass stage, then it can be lost from a site and will not become reestablished and regain dominance without intervention that provides the requisite disturbance regime. Although periodic surface fires are essential for sustaining longleaf pine forests, these must be prudently timed because longleaf pine seedlings remain highly vulnerable to fire-caused mortality until they reach a groundline diameter of 7.5 mm and a height of 1 m (Kush 2002). Because of their high susceptibility to competition for aboveground and belowground resources, their slower growth near adult trees, and the greater needle litter (i.e., fuel) accumulation under tree crowns, longleaf pine seedlings located beneath adult trees are more vulnerable to mortality from fire than are those in openings (Croker and Boyer 1975). Prescribed burning, with the objectives of adequately reducing fuels and competition while affording protection for seed-bearing trees and seedlings, should be conducted within 2 to 3 days following a saturating rain, at temperatures no greater than 27°C, with relative humidity higher than 40% and wind speeds steady at 8–13 kph (Croker and Boyer 1975).

Longleaf pine is a competition-intolerant species, with its regeneration largely confined to canopy gaps (Wahlenberg 1946). Abundance and growth of seedlings have long been reported to be negatively related to the presence of adult longleaf pine (Walker and Davis 1954; Davis 1955; Smith 1961), with the competitive influence of individual trees and forest edges on seedlings observed to extend up to 16 m (less on better quality sites) from adults into forest gaps (Smith 1955; Walker and Davis 1956; Boyer 1963). This competitive effect is so pronounced that the most frequently surviving and vigorous growing longleaf pine seedlings typically cluster near the center of canopy gaps (Brockway and Outcalt 1998). Although longleaf pine seedlings may persist for years beneath the crowns of adult trees, their growth is impeded by competition from adults and continuing long-term suppression makes them highly vulnerable to mortality from a variety of individual and cumulative stress factors (Boyer 1974b; Boyer and White 1990). Therefore, unless natural or anthropogenic disturbances remove nearby adults from the canopy, it is doubtful that seedlings present beneath mature pines will ever ascend to become dominant members of the forest canopy.

The impoverished (low available moisture and nutrients) and relatively simple (very few woody plants other than longleaf pine) ecosystem represented by the well-drained and somewhat excessively drained, coarse-textured soils in north central Florida provided a unique opportunity to assess the results of gap-phase regeneration in longleaf pine that are the cumulative product of resource competition and repeated fires over several decades. In examining these long-term patterns of natural longleaf pine regeneration in numerous naturally occurring forest gaps, Brockway and Outcalt (1998) observed that most surviving longleaf pine seedlings aggregated near the center of canopy gaps and were encircled by a zone approximately 12–16 m wide from which they were generally absent. This "seedling exclusionary zone" (SEZ) was found to spatially coincide with an area of greater fine root biomass and relatively intense intraspecific root competition between adult and juvenile longleaf pines. The center of the canopy gap, where most longleaf pine seedlings cluster, spatially coincides with a fine-root gap, where seedlings are under less competitive stress and thus enjoy higher rates of survival and growth. Although all three factors are no doubt of some degree of importance, in these relatively open longleaf pine stands (57% cover) growing on xeric sandhills, competition for soil moisture and possibly competition for nutrients appear to be proportionally more important than competition for light.

Others studying gap-phase regeneration in longleaf pine have typically examined short-term responses of planted seedlings on higher quality forest sites (Palik et al. 1997; McGuire et al. 2001; Gagnon et al. 2003). On more mesic sites, with richer soils resulting in higher stand density and a greater number of woody plants with roots innervating the surface soil layers, various root clusters are not readily distinguishable and a SEZ was not identified. On these better sites, competition for light and nutrients is reportedly more important than competition for soil water. Should the root geometry characteristic of xeric sites eventually develop on these mesic sites, it is likely that the resulting SEZ would be considerably narrower (i.e., shorter foraging distance for adult roots), perhaps only 5–10 m wide, making its effects difficult to distinguish from the influence of the 4–7 m radius directly beneath the crown of a mature longleaf pine tree.

Survival and growth of longleaf pine seedlings is negatively influenced by the density of adult trees in the canopy and the proximity of adults to the seedlings (Palik et al. 1997; Brockway and Outcalt 1998; Gagnon et al. 2004). Timely seedling release from intraspecific competition is essential to achieve successful natural regeneration. Although release can be affected through stand treatments that thin the forest to a uniform overstory density (i.e., even-aged management methods), recent appreciation for the ecological values protected by maintaining an uneven-aged structure has increased interest in achieving natural regeneration through selection silviculture techniques (Farrar 1996). Group selection, which mimics the natural stand replacement dynamics of gap-phase regeneration, may be one of the most useful approaches for regenerating longleaf pine forests. Although canopy gaps created on xeric sandhill sites should perhaps range from 0.13 to 0.8 ha (radius = 20–50 m) to sufficiently reduce the competitive influence of adults on seedlings (Brockway and Outcalt 1998), gaps as small as 0.1–0.14 ha (radius = 18–21 m) may be suitable for regenerating longleaf pine on mesic sites (Palik et al. 1997; McGuire et al. 2001; Gagnon et al. 2003).

Artificial Regeneration

Direct Seeding

Initiating a stand of longleaf pine from seed sown directly onsite can be a cost-effective regeneration method, especially for small landowners. However, this approach is potentially quite risky. Many longleaf pine stands have been successfully established by direct seeding, but there have also been many failures resulting from adverse weather, seed predation, or simply not following the required procedures. The first step in attempting to achieve successful regeneration by this method is proper seed selection. Longleaf pine has five seed zones and seed should not be transported for use more than one zone in any direction (Schmidtling 2001). Since there is no genetic evidence for east–west or ecotypic site variation, seed can be collected from anywhere within a zone for use in that zone. Seed from one zone warmer should grow faster and seed from one zone colder will typically grow slower than local seed.

Seedlots should be 95% pure and have a germination rate of at least 75% (Barnett et al. 1990). To obtain this germination rate, cones must not be collected until they are fully mature (Barnett and McGilvray 2002). Mature cones have a specific gravity of 0.89, which can be determined by the water displacement method (Barnett and McGilvray 2002). Guidelines specify that cones should not be collected until 19 of 20 cones have a specific gravity equal to 0.89, which means the average specific gravity will be about 0.81 (Wakeley 1954). Cones must be stored in a dry location having adequate ventilation. Cone storage for a short period will increase seed yields, but storage beyond 4 weeks may reduce seed quality. Cones will dry in a kiln within 24–48 hours depending on initial moisture and weather conditions. Kiln drying temperatures should not exceed 45°C (Barnett and McGilvray 2002). Once cones are dry and open, seed may be removed by using a tumbler.

Newly extracted seed will have a relatively high moisture content and, therefore, should be refrigerated and then dried to an 8–10% moisture level. Dewinging the seed requires appropriate equipment and processing techniques to prevent damage and a corresponding decline in quality (Barnett and McGilvray 2002). The wing is not really removed, but rather reduced to a stub. Debris is removed by screening and air, followed by removal of empty and partially filled seeds on a gravity table. Processed seed can then be stored in moisture-proof containers at no more than 2°C for up to 3 years. Storage for longer periods requires a temperature of -18°C.

Seed should be tested for its germination rate prior to use or sale. Since longleaf pine seed is nondormant, stratification is not necessary. However, a 10-minute drench with benomyl fungicide prior to sowing will improve germination. Since longleaf pine seed is quite large and nutritious, some type of repellent is needed to deter consumption by birds and rodents (Nolte and Barnett 2000). On most sites, seed is best sown during the fall, when moisture is adequate and the maximum daytime temperature is below 29°C (Dennington and Farrar 1991). On other sites, such as those in northern areas with finer-textured soils prone to frost heaving or those having high rabbit populations that may damage new seedlings, seed should be sown in late winter or early spring. Seed can be broadcast onsite by air or ground equipment. Since most sites are now rather small, aerial seeding is rarely employed, but ground application with a cyclone seeder is a commonly used method. Row and spot seeders operate more slowly, but use less seed per unit area and provide increased spacing control. Following dispersal, the seed can be covered with soil, which improves seedling stocking especially on dry sandy sites.

Direct seeding should not be attempted on poorly drained soils or sites having a high water table, because new germinants are highly susceptible to mortality from wet conditions (Dennington and Farrar 1991). Steep sites where seed is likely to be displaced down slope and deep sands during drought periods are also poor candidates for direct seeding. Preparing the site prior to sowing, by burning, chemical, or mechanical means to reduce plant competition and expose mineral soil, will increase the probability of regeneration success. If competition is mostly composed of herbaceous plants and low shrubs, a prescribed fire in the spring or summer prior to sowing may be useful (Mann 1970). On sites with substantial competition from woody plants, mechanical or herbicide treatments may be required, before fire can be effectively used to prepare the seedbed.

A scalper attachment may be used with the row seeder to remove competition from the area being seeded. Recommended minimum seeding rates are 32,000 repellent-treated viable seeds/ha for broadcast application. When row seeding, use 16,800 seeds/ha with one seed placed every 23–30 cm within a row. When spot seeding, 5–6 seeds should be deposited at each spot with spots arrayed at a 2×2 m spacing. Covering seeds lightly with about 6 mm of soil improves germination and establishment success (Barnett et al. 1990). In assessing seeding success, a minimum of 100 survey plots (each 4 m^2) per site is recommended (Mann 1970). If seedling stocking is greater than 50%, then a suitable stand of longleaf pine should develop.

Bare-Root Nursery Seedlings

The nursery management objective is to produce high-quality longleaf pine seedlings with a root-collar diameter of 10 mm, at least six primary lateral roots, a stout taproot, well-developed terminal bud, and many needle fascicles (Barnett et al. 1990). The first step in this process is to select good-quality seed with 95% purity and a germination rate of at least 75% (Cordell et al. 1990). A benomyl application to seeds will reduce fungal pathogens and improve seedling establishment in the nursery bed (Barnett et al. 1999). Seedbeds should consist of sandy or loamy sand soils with a pH of 5–6. Weeds, insects, and pathogens are reduced by seedbed fumigation prior to sowing. If mulch is used, it should also be fumigated to prevent introduction of pathogens. Seed may be sown in either fall or spring, although fall sowing will usually produce larger seedlings (Barnett et al. 1990). However, a cold winter can damage young seedlings so spring sowing may be favored at more northerly locations. Longleaf pine seedlings should be grown at a density of 108–160/m^2. Seedlings produced at higher densities have lower survival following outplanting, while growing them at lower densities increases the per unit cost of production. A precision sowing machine greatly aids in obtaining the desired planting density in the seedbed.

Irrigation and fertilization must both be carefully controlled to produce good-quality longleaf pine seedlings (Cordell et al. 1990). Too much water will promote disease, while too little will impede growth and development. The pH of irrigation water must also be controlled, maintaining it at 5–6. Substantial practical experience is needed to develop the skill necessary to determine when seedlings require watering. Fertilization decisions should be based on a chemical analysis of the soil and seedling nutrition requirements. Generally, numerous smaller applications rather than fewer larger inputs of nutrients are provided, with the highest rates applied during the phase of rapid growth. Although most large nursery operations use soluble fertilizers applied from tank sprays or as injections into the irrigation system, dry fertilizers can also be used. Standard practices should be followed for reducing fertilization during the seedling hardening phase.

Pruning of both vertical and horizontal roots improves longleaf pine seedling survival (Cordell et al. 1990). For production of 1-0 seedlings, two root-pruning treatments applied in August and October are recommended. Root pruning should be followed with irrigation to settle the soil and reduce desiccation. Needle clipping just before lifting can improve survival on dry sites (Barnett et al. 1990). The best lifting time is January and early February, but lifting should be coordinated with planting schedules to reduce time in storage. Care must be exercised during lifting to limit root system damage. Survival is improved by application of benomyl fungicide to the packing medium or as a root coating at the time of packing (Barnett et al. 1988). Grading should normally not be necessary if proper seedbed densities, fertilization, and irrigation have been used. If problems caused by nonuniform growth arise, then all seedlings with a root-collar diameter less than 7.5 mm should be discarded. Bare-root seedlings should be stored and transported in refrigerated containers maintained at a temperature of 1–3°C. Seedlings are best planted as soon as possible after lifting, with the duration in cold storage limited to 1 week. Seedling health declines rapidly when kept in cold storage longer than 2 weeks and survival becomes correspondingly low (White 1981).

Containerized Nursery Seedlings

Superior containerized longleaf pine seedlings are best grown outdoors in full sunlight. During initial germination, a 30% shade cloth is used to prevent seed wash from containers (Barnett and McGilvray 2000), but this material should be removed as soon as germination is complete. A 1:1 mixture of sphagnum peat moss and number 2 grade vermiculite is an optimum growing medium. Prior to mixing, the peat moss should be screened to remove large sticks and woody fragments. If necessary, medium pH should be adjusted to 4.5–5. Containers having a minimum depth of 11 cm, volume of 100 cm^3, and density of less than 535 seedlings/m^2 are recommended for growing longleaf pine (Barnett and McGilvray 1997). If possible, good-quality seed with a viability of at least 80% should be sown. Treatment with benomyl prior to sowing will reduce pathogens and improve germination, especially for lower quality seedlots. Seeds should be sown during spring, from March to May, putting one seed per cavity for good-quality seedlots or two per cavity if viability is 65–80% (Barnett and McGilvray 2000). The seed is then covered with a thin 3-mm layer of growing medium, grit, or vermiculite. Seedlings should be thinned to one per cavity when seedcoats are being shed.

Because longleaf normally germinates in the fall, it will do best if sown during cooler temperatures when the daily range is 15–27°C and the mean is 22°C. During the germination phase, seedlings need frequent but light watering. Following germination, water application is based on container weight. A weight of 80–85% of field capacity during rapid growth and 70–75% during the hardening phase is a good guideline for determining when water is needed. Sufficient water is typically added to wet the entire plug (Starkey 2002). While many nursery managers now incorporate slow-release fertilizers into the growing medium, water-soluble fertilizers can also be applied as needed. Starkey (2002) recommends starting with three fertilizer applications per week, adjusted as needed for rainfall amounts, using a balanced N-P-K fertilizer like 20-20-20 or 15-15-15. After germination is completed, use 50 ppm for 2 weeks during the establishment phase, followed by 200 ppm for 14–16 weeks and then 25–50 ppm during the hardening phase. If weeds become a problem, oxyfluorfen should be applied at about one-fourth the recommended rate to prevent seedling damage (Barnett and McGilvray 2000).

Watering to field capacity makes plug extraction easier. Poor-quality seedlings, where the root plug does not hold together or root-collar diameter is less than 7 mm, should be culled. Good-quality seedlings are then put into cardboard boxes and placed in cold storage at 1–3°C. Storage intervals in excess of 2 weeks should be avoided, as seedling quality will decline. Containerized seedlings can be planted in Florida during July and August because rainfall is normally plentiful. In other areas, seedlings should be planted in the fall as soon as soil moisture is adequate, to reduce the risk from freezing winter temperatures (Barnett and McGilvray 2000).

Planting Longleaf Pine Seedlings

In the past, longleaf pine was often not chosen for reforestation because high mortality showed it to be very difficult to establish using bare-root seedlings and, even if survival was acceptable, many seedlings remained in the grass stage for extended periods. More recently, however, longleaf pine has been shown to survive well and begin height growth quite rapidly if seedlings are planted properly. Successful regeneration with bare-root seedlings requires healthy, fresh planting stock and precision planting on a well-prepared site (Barnett 1992). The average survival and growth of containerized longleaf pine, however, remains better than those of bare-root seedlings. At five locations in Georgia, containerized seedling survival was 76%, while that for bare-root seedlings was only 51%, after 5 years (Boyer 1989). Although containerized seedlings are more costly, they are clearly the better choice especially on droughty sites prone to moisture stress and when planting outside the normal dormant-season period (Barnett 2002).

Since longleaf pine is sensitive to competition, some form of site preparation is required before seedlings are planted. Prescribed burning may be all that is required if the site has been regularly burned and is dominated by herbaceous species. On sites with heavy competition from woody plants, mechanical or herbicide treatments will be required. Intensive mechanical site preparation is often a poor choice for longleaf pine ecosystems because it destroys most native understory plants, can promote rapid growth of annual weeds, and may lead to soil erosion. Herbicide and fire in a combination sequence of "brown then burn" treatments may be used to prepare harvested sites for planting. A combination of herbicide and strip scalping works well on sites with native understory plants. The herbicide should target woody species, while the scalp reduces herbaceous competition within the planting zone. The native grasses will recover and reoccupy the scalped strip, providing the fuel necessary for future management with fire (Outcalt 1995). Herbicide and scalping is also a very effective site preparation technique for old-field sites (Hainds 2001).

Longleaf pine requires careful handling and seedlings must be kept in cold storage until they are planted. For small operations, bales or boxes of seedlings can be brought to the planting site a couple of times each day and kept in the shade until planted. For large operations, it is best to have a refrigerated truck present onsite. Both bare-root and containerized seedlings can be hand or machine planted, although machine planting is preferable for bare-root seedlings because of their large root system. Bare-root seedlings should be root pruned before lifting and not in the field. Obvious culls, which are too small or have poorly formed root systems or plugs, should be discarded. Planting depth is a critical factor in survival and growth. The bud of bare-root seedlings should be placed at ground level or slightly below to allow for soil settling (Barnett et al. 1990). Shallow planting, with the bud above the soil surface, is better for containerized seedlings that have smaller buds more prone to being buried (Hainds 2001).

Containerized seedlings can be planted anytime from July to March, when soil moisture is adequate. Winter planting should be avoided in northern portions of the range because cold temperatures may lead to mortality from freezing. Bare-root seedlings should be planted only during the dormant season from December to February, with January the best month in most areas. The planting objective is typically 740 well-distributed seedlings/ha, which means 1000–1200 seedlings/ha must be planted, assuming a survival rate of 60–75%. If the survival rate is lower, containerized seedlings can be used to fill in or, if survival is very low, the site can be burned and replanted the next season (Barnett et al. 1990). Seedling surveys, conducted 1 year after planting between October and February, are very important. Some sites may require post planting treatments to control competition and speed seedling emergence from the grass stage.

Forest Reproduction Methods

Concern for renewing the forest, through timely and successful regeneration, is the most significant difference between silviculture and exploitive logging. Reproduction methods describe the manner in which forests will be cut to ensure regeneration and are a component of silvicultural systems that define in a more comprehensive way the manner in which forest stands will be tended, harvested, and replaced (Daniel et al. 1979). Reproduction methods that are effective in achieving successful regeneration are not only important for the continuation of overstory tree species, but also have a profound impact on composition and structure of the understory plant community which provides habitat for numerous associated species and indicates to a substantial degree the ecological health of an ecosystem. Properly selecting and implementing forest reproduction methods that contribute to numerous conservation and utilitarian goals requires extensive knowledge of longleaf pine ecology, silviculture, and the variation in longleaf pine ecosystems across an extensive natural range. Although no single approach to obtaining longleaf pine regeneration can be appropriately prescribed everywhere, methods that seek to holistically sustain longleaf pine ecosystems need to be compatible with (1) frequent use of surface fires, (2) maintaining native understory vegetation, (3) retaining appropriate numbers of overstory trees onsite well beyond economic maturity, and (4) creating and maintaining small-scale canopy gaps interspersed among forest patches of varying ages within an uneven-aged landscape mosaic at larger scale (Landers et al. 1990).

Clearcutting Method

The clearcutting method consists of removing all overstory and midstory trees, resulting in an open site that is influenced little by the edge of adjacent forests. The principal objective of clearcutting is to establish a new forest stand that will have an even-aged structure (Daniel et al. 1979). Clearcutting may be prescribed for removing stands in poor condition owing to damage by insects, pathogens, or fire and may also be used to quickly remove an overstory composed of undesirable species prior to reforesting a site with longleaf pine. An area so cleared may more easily be treated with mechanical site preparation and subsequently regenerated by artificial means. Clearcutting may also be combined with natural regeneration, through application of alternative-strip or progressive-strip clearcutting. However, application of these techniques is somewhat challenging, considering the short dispersal distance of large longleaf pine seeds and the care that must be taken to protect the usually scarce advanced regeneration (Kush 2002).

Certain disadvantages of the clearcutting method suggest that it should not be applied to perpetuate existing longleaf pine forests in good condition. Clearcutting mature longleaf pine stands can destroy much of their advanced reproduction (Boyer and Peterson 1983), thus requiring reliance on more expensive planted seedlings rather than established natural regeneration. Managing longleaf pine in accordance with a "clearcut and plant" model is very expensive, because advanced regeneration is destroyed and all stand establishment costs must be carried through a long rotation period until harvest. Although longleaf pine is a competition-intolerant tree species, its does not require complete canopy removal to regenerate and its need for sunlight cannot be used as an appropriate rationale for implementing the clearcutting method (Brockway and Outcalt 1998). Also, complete canopy removal through clearcutting can diminish aesthetics and degrades habitat quality for certain at-risk species. If clearcut longleaf pine sites are somehow captured by other tree species, not only does this contribute to further attrition of these endangered ecosystems, but plant community succession on such areas will be driven along very different trajectories that can fundamentally alter important ecological processes and structures (Brockway and Lewis 2003).

Seed-Tree Method

The seed-tree method consists of removing most of the mature overstory while leaving enough good seed-producing trees scattered across a site to ensure acceptable future stocking (Daniel et al. 1979). The 20–25 trees/ha typically left as residuals afford some degree of site amelioration and ensure an even distribution of seed over the area (Kush 2002). The residual seed-trees can provide a more aesthetic appearance and improve composition of the future stand if good-quality phenotypes (and presumably genotypes) are selected. The seed-trees are typically harvested after regeneration becomes established; however, in a variation of this method known as "deferment" cutting, these residuals may be retained onsite along with the regeneration throughout the entire next rotation (Smith et al. 1989). While the objective of the seed-tree method is usually to establish a new forest with an even-aged stand structure, use of the deferment-cut variation will result in more than one age-class present in the future forest.

The seed-tree method has a number of disadvantages for longleaf pine. Very rarely will the low number of residual overstory trees be capable of producing sufficient quantities of longleaf pine seed to obtain satisfactory natural regeneration. Therefore, a longer time period will typically be needed to achieve regeneration success with the seed-tree method. The limited seed dispersal distance of longleaf pine seed requires that cleared areas be within about 30 m of residual seed-trees (Boyer and Peterson 1983). If a good seed crop does not occur for a number of years, open areas can become occupied by competing hardwoods and other southern pines. The sparse longleaf pine overstory will produce very little needle litter, resulting in surface fires that may burn with insufficient intensity to satisfactorily control this competition (Kush 2002). If these invaders escape to grow to fire-resistant size, then chemical and/or mechanical treatments may be necessary (Boyer and Peterson 1983). Also, seed-trees retained onsite represent a timber volume loss to managers interested in maximizing the output of forest products (Kush 2002). However, the greatest problem with application of the seed-tree method in existing longleaf pine forests is that most or all of the large overstory trees will be removed, changing ecosystem structure and degrading habitat that supports numerous at-risk species.

Shelterwood Method

The shelterwood method consists of removing a moderate portion of the overstory while retaining onsite sufficient numbers of seed-bearing trees so that regeneration becomes established under the protective partial shade of mature residual trees (Fig. 10). Compared with the relatively rigid conditions resulting from application of clearcutting and seed-tree methods, the shelterwood method is the most flexible method for establishing new forests of longleaf pine having an even-aged structure and is

4. Longleaf Pine Regeneration Ecology and Methods

FIGURE 10. Longleaf pine forest being regenerated by the shelterwood method; an overstory of 60 trees/ha having a basal area of 7 m^2/ha remains following the seed cut. Photo courtesy of the Forestry Images Organization.

capable of producing a variable degree of site amelioration (Daniel et al. 1979). When applied in longleaf pine forests, the 50–62 residual trees/ha typically produce an abundance of seed that is uniformly distributed across a site (Kush 2002). Selecting high-quality phenotypes as residual trees provides an opportunity to improve genetic composition of the future stand. With no large areas left barren, a substantial forest canopy deposits sufficient needle litter to support surface fires with intensities high enough to control competing woody plants (Kush 2002). The more substantial forest canopy of this method also enhances environmental aesthetics and improves habitat for numerous wildlife species. The principal disadvantages of most forms of the shelterwood method are related to the eventual removal of all overstory trees. During overwood harvest, excessive damage may occur to established longleaf pine seedlings and, following overstory removal, forest structure is substantially transformed and habitat quality for at-risk species may remain reduced for decades.

The shelterwood method can be applied only in existing stands with sufficient dominant and/or codominant trees of seed-bearing size (Boyer and Peterson 1983). The most commonly used approach is the uniform shelterwood method, which may be implemented with either a two-cut or three-cut technique. As the name suggests, each cut in the uniform shelterwood method is applied throughout the entire stand resulting in a uniform appearance and even-aged structure. If a forest is highly stocked and requires thinning or improvement cutting, the three-cut technique is necessary. The first cut in the three-cut technique is the preparatory cut, which is optional in regularly tended stands and normally performed 10 years before the planned overwood harvest. The preparatory cut should leave a well-distributed population of dominant and codominant trees totaling

13.8–16.1 m²/ha (Boyer and Peterson 1983). This will promote crown development and enhance cone production in the residual seed-bearing trees. If the density of competing woody plants is greater than 2.3 m²/ha, herbicide and/or fire treatments should be implemented (Kush 2002).

The second cut in the three-cut technique and the first cut in the two-cut technique is the seed cut (Croker and Boyer 1975). The seed cut is performed 5 years or more before the planned overwood harvest to reduce the stand density to 6.9 m²/ha, leaving well-distributed high-quality seed-bearing trees in the overstory (Kush 2002). The most desirable residuals are trees equal to or greater than 40 cm diameter at breast height that have evidence of past cone bearing ability. Large hardwood and other undesirable trees may also be removed during the preparatory cut, and smaller hardwoods can subsequently be controlled by prescribed fire conducted during the growing season (Boyer and Peterson 1983).

Seed crops are then monitored by annual springtime counts of flowers and conelets, watching carefully for crops that produce 1860–2470 cones/ha, the range typically required for adequate natural regeneration (Boyer and Peterson 1983). Once an acceptable cone crop is noted, a seedbed preparation burn can be conducted during the year prior to seedfall (Kush 2002). While prescribed fire during summer or autumn will control woody plant competition better than earlier cool-season burns, increased seed predation by birds has been associated with fire during the fall season. Annual seedling surveys are also conducted to assess the degree of regeneration success, the principal criterion being a minimum of 9880–14820 seedlings/ha that are at least 1 year old and well distributed across the site (Boyer 1993a). This abundance provides sufficient seedling numbers to compensate for future mortality during overwood removal and from other factors (Kush 2002). Once an adequate number and distribution of established longleaf pine seedlings are present, the residual seed-bearing trees in the overstory may be harvested by the removal cut (Boyer and Peterson 1983). Although longleaf pine seedling survival beneath overstory trees may be unaffected for 8 or more years, seedling damage during overwood harvest is minimized if the removal cut is conducted when seedlings are no more than 2 years old and in the stemless grass stage (Boyer 1975, 1993a). While overwood harvest may be scheduled with a reasonable amount of flexibility to meet economic and other management objectives, conducting the removal cut on overstories of higher density above seedlings of increased size will result in substantially greater seedling mortality rates (Maple 1977). When the residual overstory density is equal to or greater than 9.2 m²/ha, overwood harvest is best accomplished by two separate removal cuts. The principal postharvest need is application of prescribed fire to control competing plants and the brown-spot fungus; however, timing is crucial, since mistimed fires can harm vulnerable seedlings (Boyer and White 1990). Prescribed burning should be avoided during the first 2 years following the removal cut, because accumulated logging slash will produce fires too hot for newly released seedlings (Boyer and Peterson 1983). Periodic burning should generally not resume until dominant longleaf pine seedlings have attained a root-collar diameter of 7.5 mm (Boyer 1979). Prescribed fire is indicated when surveys reveal a brown-spot fungus infection rate of equal to or greater than 20% among dominant seedlings (Boyer 1993a).

Variations of the shelterwood method include its application in block, strip, and group configurations (Boyer and White 1990; Boyer 1993a). The block approach is associated with even-aged forests and identifies stands typically from 4 to 40 ha as management units delineated by various natural or artificial boundary features such as creeks and roads. While each unit may be regenerated individually, several adjacent blocks could also be treated at the same time, resulting in a relatively large area being in the same stages of regeneration. The progressive strip approach can convert a large even-aged stand into a number of smaller even-aged stands covering an entire range of age-classes from seedling to mature. Strips created by cutting are long and narrow, not exceeding 60 m in width, so that most or all of the strip will be within seed-dispersal range of the adjacent mature trees. Within each strip, the uniform shelterwood method is applied as described above, using either the three-step or two-step technique. When the seed cut is made in the first strip, a preparatory cut is made in the second strip. When seedlings are established in the first strip, a removal cut is made there, with a seed cut then made in the second strip and a preparatory cut made in an entirely new third strip. Thus, a "rolling wave" of regeneration may proceed indefinitely across the landscape. The group approach is applied to management units that are too small to be considered blocks, typically 0.2–2 ha. Using this approach, numerous patches are treated with either the three-step or two-step technique. When the first patches are treated with a seed cut, the second group of patches is treated with a preparatory cut. Once seedlings are established in the first patches, they are treated with a removal cut, the second patches are treated with a seed cut, and a third group of patches are treated with a preparatory cut. The substantial dispersion and smaller size of areas under treatment at any one time allow this shelterwood approach to more closely achieve management objectives related to aesthetic and ecological values. However, eventually removing all canopy trees still raises objections from those concerned about the rarity of older forests and welfare of associated species.

By contrast, the irregular shelterwood method retains a moderately light (less than 7 m^2/ha) canopy of mature trees onsite throughout an entire stand rotation. This method may be applied using any of the above techniques and variations, except the removal cut is mostly or entirely eliminated (Boyer 1999). Unlike the uniform method, the irregular shelterwood method can maintain a residual canopy of mature trees onsite over many rotations and potentially in perpetuity. Mature overstory residuals plus newly regenerated seedlings initially comprise a two-aged forest that may be maintained by thinning from below or allowed to eventually achieve an uneven-aged structure, as successive waves of regenerating seedlings establish and develop during ensuing decades. With this method, overstory residuals may range in age from newly mature (50–100 years) to very mature (300–400 years), depending on the health of individual trees and specific management objectives. Although the growth of longleaf pine seedlings will no doubt be slowed by competition from residual adult trees (Table 1), the continuously maintained forest canopy is beneficial in sustaining species dependent on such structural conditions (Boyer 1993b). The irregular shelterwood method is the most rapid way to convert an even-aged longleaf pine forest (or plantation) to a two-aged and eventually an uneven-aged stand structure (Boyer 1999).

Uneven-Aged Silviculture

In the most recent decade, ecosystem management policies have substantially increased interest in managing forests through uneven-aged silviculture (Guldin 1996). The chapter by Guldin (this volume) discusses uneven-aged silviculture of longleaf pine forests in greater detail. Protecting the native plant community, maintaining a continuous forest canopy, and facilitating development of large, old trees are among the desirable habitat features potentially resulting when uneven-aged silviculture is applied in an adaptive ecosystem management framework. Interest has also grown in methods for converting even-aged stands to uneven-aged forests, like uniform partial cutting and patch cutting combined with thinning (Nyland 2003). Whether by even-aged or uneven-aged methods, silviculture is fundamentally applied at varying spatial and temporal scales to mimic disturbance and guide succession in a forest. Considering the wide natural variation in frequency, intensity, and area impacted by disturbance, forests tended according to ecosystem management principles might well contain a mixture of even-aged and uneven-aged stands. Where stages of advanced succession are sought and perpetuating those stages is desirable (and ecologically feasible), uneven-aged silviculture can

TABLE 1. Inhibitory influence of residual overstory trees on stand growth during a 40-year period (1957–1996) for longleaf pine forests treated with the uniform shelterwood (initial basal area = 0) and irregular shelterwood (initial basal area > 0) methods (tree dbh > 8.9 cm).

Initial stand basal area (m²/ha)	Residual pine		Ingrowth pine		Total pine	
	Basal area (m²/ha)	Volume[a] (m³/ha)	Basal area (m²/ha)	Volume[a] (m³/ha)	Basal area (m²/ha)	Volume[a] (m³/ha)
Block A						
0	0	0	18.2	131.9	18.2	131.9
0	0	0	17.0	125.7	17.0	125.7
2.1	2.2	18.6	11.7	70.4	13.9	89.0
4.1	7.3	62.0	6.8	30.1	14.1	92.1
6.2	8.9	74.7	2.3	6.2	11.2	80.9
8.3	12.9	109.6	0.8	4.4	13.7	114.0
10.4	15.4	126.8	0.2	1.1	15.6	127.9
Block B						
0	0	0	20.1	139.9	20.1	139.9
2.1	2.5	21.4	11.8	68.7	14.3	90.1
4.2	6.3	53.7	4.6	17.6	10.9	71.3
6.2	9.3	78.8	1.8	5.2	11.1	84.0
8.3	12.7	106.9	0.4	1.5	13.1	108.4
10.4	16.1	135.8	0.1	0.2	16.2	136.0

[a] All volumes reported as inside bark and derived from local forest volume tables.

be used to achieve such conditions (Guldin 1996).

Uneven-aged silviculture affords a major advantage, in that natural regeneration is more or less continuous, as late successional forest dynamics are emulated (Guldin 1996). The Dauerwald (i.e., continuous forest) of Germany is a prime example of the uneven-aged, continuously regenerating forest that can result (Schabel and Palmer 1999). Loblolly pine and shortleaf pine, growing on gentle terrain, under aggressive or limited competition, with the worst trees cut and the best trees left to serve as seed-bearing residuals, have been successfully managed with uneven-aged methods (Guldin 1996). Although longleaf pine was formerly thought to be too competition-intolerant for uneven-aged silviculture, recent evidence suggests this to be a viable and desirable forest management alternative (Farrar and Boyer 1990; Farrar 1996; Palik et al. 1997; Brockway and Outcalt 1998; McGuire et al. 2001; Gagnon et al. 2003). The group selection method and the single-tree selection method are the principal means by which uneven-aged silviculture can be practiced.

Group Selection Method

As an uneven-aged mosaic of even-aged patches distributed across the landscape (Platt and Rathbun 1993), natural longleaf pine forests maintain a continuous overstory canopy while establishing naturally regenerated seedlings in canopy gaps created by lightning and other local disturbance agents. Of all available forest reproduction methods, group selection most closely mimics this natural gap-phase regeneration pattern and is perhaps best suited in the long term for sustaining the ecological character of longleaf pine ecosystems (Brockway and Outcalt 1998). The group selection method can be used to create circular, elliptical, and irregularly shaped gaps ranging from 0.1 to 0.8 ha distributed throughout the forest to effectively simulate the desired uneven-aged mosaic (Fig. 11). If volume control is preferred, the volume-guiding diameter limit (V-GDL) and basal area-maximum diameter at breast height-q (BDq) stand regulation procedures in a modified group selection method are suitable for managing longleaf pine forests, when used in combination with prescribed fire on a 3-year

FIGURE 11. Longleaf pine regeneration freely ascending within a forest canopy gap (background); note numerous seedlings that remain suppressed by intraspecific competition from nearby adults (foreground). Photo courtesy of the Forestry Images Organization.

cycle to control competing vegetation and maintain seedbeds in appropriate condition (Farrar and Boyer 1990; Farrar 1996). Alternatively, longleaf pine stands may be quite easily regulated using an area control procedure.

Group selection implementation begins during year 0, with an entry to create the first series of canopy gaps by cutting groups of overstory trees, placing emphasis on removing poor-quality trees and retaining good-quality seed-bearing trees around the periphery of each gap (Table 2). Ideally, gaps should be cut in areas where advanced longleaf pine regeneration is already present, thereby decreasing the likelihood that newly created gaps will become occupied by competing woody species. Gaps should be created of sufficient size (at least 0.1 ha) to minimize the suppressive effects of intraspecific competition between longleaf pine adults and seedlings. The forest will be burned several times during this decade to maintain good-quality seedbeds and control woody competitors. If a 10-year cutting cycle is used, the forest will be again entered during year 10 to create a second series of gaps. If longleaf pine seedlings are established in the first series of gaps, they can be protected by skipping one fire cycle, if necessary, and allowing dominant seedlings to reach a more fire-resistant size. Overstory trees impeding seedling development along the gap periphery may now be removed. During year 20, another entry is made to create a third series of gaps in the canopy. Periodic surface fires need to be conducted throughout this decade. The seedlings in the first series of gaps should be sapling size by now and young seedlings may be present in the second series of gaps. Competing adults may be removed here as needed. During the next entry in year 30, a fourth series of canopy gaps is initiated. Seedlings may now be present in the third series of gaps and competing adults along the periphery may be removed where necessary. Seedlings in the

TABLE 2. Schedule for implementing the Group Selection Method using a 10-year cutting cycle during the first 100 years of application.

Year	Create gaps; retain cone-bearing trees along edges	Seedlings established in openings; may remove adjacent size adults; skip one fire cycle	Seedlings grown to sapling size	Saplings grown to pole size	Poles grown to sawlog	Sawlogs continue to increase in size
			gap series			
0	1st					
10	2nd	1st				
20	3rd	2nd	1st			
30	4th	3rd	2nd	1st		
40	5th	4th	3rd	2nd	1st	
50	6th	5th	4th	3rd	2nd	1st
60	7th	6th	5th	4th	3rd	2nd
70	8th	7th	6th	5th	4th	3rd
80	9th	8th	7th	6th	5th	4th
90	10th	9th	8th	7th	6th	5th
100	11th	10th	9th	8th	7th	6th

second series of gaps should be sapling size by now while saplings in the first gap series should have grown to pole size. By year 40, a fifth series of gaps will be created, seedlings should become established in the fourth series of gaps, saplings will have developed in the third series, poles will have grown in the second series, and sawlogs will have emerged in the first series of gaps. If retained for a prolonged period, these sawlogs will eventually become very large trees. This continuous serial pattern of entry, cutting, and reproduction in the longleaf pine forest may be continued, along with application of periodic prescribed burning and other cultural treatments, to create a potentially infinite series of regenerating canopy gaps that indefinitely sustain an uneven-aged mosaic.

The principal advantages of the group selection method include (1) high-forest cover is constantly maintained with no degradation of habitat for at-risk species, (2) the diversity of tree size-classes results in an aesthetically pleasing environment, (3) regeneration is continuous and not confined to short high-risk periods, (4) regeneration develops under even-aged conditions within canopy gaps, (5) the larger openings permit establishment of intolerant species, (6) more concentrated harvest results in lower logging costs, less damage to seedlings and residual overstory trees, and easier conduct of inventories compared to the single-tree selection method, (7) full stand regulation is more easily achieved, with area control being an easier-to-apply procedure than volume control procedures (i.e., BDq, V-GDL), (8) even small areas can be economically managed for regular, even product flow, (9) larger trees may be grown by adjusting the cutting cycle and/or maximum diameter at breast height with no change in stand area, and (10) the stand is afforded a greater degree of protection from disturbance-induced losses, because regeneration is present to replace fallen overstory trees (Daniel et al. 1979; Farrar and Boyer 1990; Farrar 1996; Kush 2002). Among the disadvantages of group selection are (1) if healthy groundcover is not maintained through periodic prescribed burning, gaps without advanced regeneration may become occupied by woody competitors, necessitating costly mechanical or herbicide treatments, (2) need for periodic vegetation control treatments other than prescribed fire (e.g., herbicide application every 20 years) on better quality sites with more severe woody plant competition, (3) higher management costs, greater inventory information needs and lower product outputs relative to even-aged management methods, (4) higher logging costs and greater logging damage of residuals compared with those for even-aged methods, and (5) it can be difficult to apply on sites with severe understory competition such as saw palmetto

(*Serenoa repens*) dominated flatwoods (Farrar and Boyer 1990; Farrar 1996; Kush 2002).

Single-Tree Selection Method

In the single-tree selection method, individual trees are removed from the forest canopy and regenerating seedlings become established in their place. Since only small openings are created by this method, it is best applied to tree species that are tolerant of competition for site resources (Daniel et al. 1979). Individual trees are harvested in a series of partial cuts that typically occur in a stand at 10-year intervals (though the actual cutting cycle may be somewhat shorter or longer). Application of this method results in development and maintenance of an uneven-aged forest structure, having a reverse-J size-class distribution that can be quantified by the diminution quotient, q. The q factor is derived from de Liocourt's law, ranges from 1.2 to 1.5, and is applicable to all uneven-aged forests (de Liocourt 1898; Meyer 1952).

Selection was first applied in southern pines using annual harvests for the initial 15 years, followed by periodic cuts every 5–7 years thereafter (Reynolds 1969; Reynolds et al. 1984). Although this approach was neither group selection nor single-tree selection, volume control was successfully achieved in loblolly pine and shortleaf pine stands by using the V-GDL regulation procedure and following the rule of "cut the worst and leave the best" trees (Guldin 1996). The keys to successful regeneration with application of the single-tree selection method in this forest type are regulation of stocking and stand structure, careful logging, and control of competing vegetation (Shelton and Cain 2000). Uneven-aged stands are generally easier to create and maintain on poor-quality sites because of less competing vegetation and greater ease of obtaining natural pine regeneration. Implementing the single-tree selection method with the BDq stand regulation procedure calls for postcutting guidelines of 10–14 m^2/ha for basal area, 35–55 cm for maximum diameters, and a q factor of about 1.2 for 2.5 cm diameter at breast height size classes (Shelton and Cain 2000). However, since the q factor is the most difficult to control (and least important of the variables), most operational-scale applications result in pine stands consisting of multiple size classes, rather than strictly corresponding to a classically balanced reversed-J distribution. With seed-bearing trees greater than 40 cm diameter at breast height favored as residual overstory trees, basal area should not exceed 17 m^2/ha to limit the adverse influences of shading and root competition on regenerating seedlings (Shelton and Cain 2000). Application of the single-tree selection method results in a pine stand with an irregular canopy and many gaps of various sizes, up to 0.1 ha. Although seedlings become established throughout stands managed under single-tree selection, those in gaps created by harvest are of greatest interest since they have the greatest potential to eventually ascend to a dominant position in the forest canopy.

Application of the single-tree selection method in longleaf pine forests has received less study than other southern pines. The infrequency of good seed years and inclination of seedlings to be suppressed by competition from nearby adults, appears to make longleaf pine a less suitable candidate for application of the single-tree selection method. Guidance for implementing the single-tree selection method in longleaf pine forests emphasizes thinning removal of individual trees (especially those of poor quality) on a 10-year cutting cycle, establishment of seedlings in small gaps, and removal of adjacent overstory trees to progressively enlarge these gaps thereby releasing seedlings from competition with adult longleaf pine trees (Moore 2001). Although each tree is marked and harvested as an individual, application of this guidance eventually results in the removal of groups of overstory trees, creating canopy gaps that approach the dimensions of those created by the group selection method. The principal difference is that two or more stand entries are required to create sufficiently large gaps using the single-tree selection method, while only a single entry is typically needed using the group selection method. Such dependence on multiple stand entries for multiple-tree gap creation to achieve successful regeneration, raises concerns regarding suitability of the single-tree selection method for regenerating longleaf pine forests. Nonetheless, a management approach identified as "the single-tree selection method" has been applied for several decades on quail hunting plantations in southern Georgia (Nature Conservancy 2002). While regeneration has been obtained on these sites, the absence of an objective stand regulation procedure, such as V-GDL or BDq, impairs reliable replication of management results among different practitioners.

Advantages of the single-tree selection method include (1) forest cover and habitat for at-risk species is constantly maintained, (2) the variety of tree sizes creates high-quality aesthetics, (3) regeneration is continuous, (4) full stand regulation is easily and rapidly achieved, (5) small areas can be economically managed for regular, even product flow, (6) large trees can be produced by increasing the cutting cycle or maximum diameter at breast height without changing the stand area, and (7) the stand is afforded a higher degree of protection from windthrow, insect, pathogen, and fire losses, since regeneration is always present to replace damaged overstory trees (Daniel et al. 1979; Farrar and Boyer 1990; Farrar 1996; Shelton and Cain 2000; Kush 2002). The disadvantages of single-tree selection are (1) requirement of well-trained staff, highly skilled in applying more difficult stand management concepts and procedures, (2) large amounts of stand information from inventory examinations and growth and yield projections that are difficult and time-consuming, (3) crop trees are scattered throughout the stand resulting in higher logging costs, greater logging damage to residuals, and lower product outputs relative to other forest reproduction methods, (4) very small single-tree gaps provide conditions unfavorable for establishment and growth of intolerant longleaf pine, (5) the resulting forest environment may be less suitable for some wildlife species, such as bobwhite quail (*Colinus virginianus*) that prefer low vegetation and edge areas created by other methods, (6) fire damage of pine seedlings that are often present in the youngest and most vulnerable size-classes, (7) need for periodic vegetation control other than prescribed fire on better quality sites with more severe woody plant competition and the difficulty in efficiently applying area-wide mechanical and chemical treatments, and (8) it is very difficult to apply on sites where intense competition from native plants, such as saw palmetto, or exotic plants, especially vines like kudzu (*Pueraria lobata*) and Japanese honeysuckle (*Lonicera japonica*), severely limit pine regeneration success (Daniel et al. 1979; Farrar and Boyer 1990; Farrar 1996; Shelton and Cain 2000; Kush 2002).

Since little saleable product results, intermediate cuts in the younger-age classes necessary for maintaining an uneven-aged stand structure have a tendency to be neglected (Daniel et al. 1979). Also, among the most serious dangers inherent in the single-tree selection method is the always present temptation to violate the "cut the worst trees first and leave the best" rule. In not observing this discipline, errant practitioners will harvest the best trees first, thereby allowing the single-tree selection method to degenerate into exploitive timber "high-grading" (Farrar 1996). Finally, it is important to note that applications of the selection methods in longleaf pine forests are still in their infancy and flawless results cannot be expected without substantially more scientific study and management experience.

Silviculture that Mimics Natural Disturbance

The above forest reproduction methods were first developed by foresters in Europe during the eighteenth century, when the newly created profession of forestry was confronted with the challenge of repairing the extensive damage done to forests by centuries of exploitive logging (Daniel et al. 1979). During the ensuing 300 years, various clearcutting, seed-tree, shelterwood, and selection methods have been employed by foresters in Europe and on other continents to regenerate and rehabilitate many forest types. Although their principal focus was often limited to renewing the forest overstory, many early foresters can rightly be considered among the first practitioners of restoration. With ecologists, foresters, and wildlife biologists more recently broadening the focus of forest management to include understory flora and fauna, appropriate forest reproduction methods applied through the practice of silviculture remain the fundamental means by which forest ecosystems will be restored and sustained.

Despite longtime recognition of the inherent efficiency of working in concert with nature, silviculture has traditionally placed little emphasis on the importance of natural disturbance dynamics in the development and maintenance of forests. Indeed, forest management has often implemented silvicultural systems that prescribe stand homogeneity to optimize stand-level tree growth. However, as the primary emphasis on timber production has shifted to also include the stewardship goals of sustaining ecological functions, conserving biological diversity, and protecting at-risk species, natural disturbance has become recognized as a potential source of guidance for forest management (Coates and Burton 1997). The fundamental hypothesis for such an approach is that the species, structures, and processes that characterize natural ecosystems are more likely to persist in managed forests if anthropogenic disturbances like logging and prescribed burning mimic the dynamics of natural disturbances (Mitchell et al. 2002). The greater spatial and temporal heterogeneity of natural disturbance regimes cause higher variation in the size and age structure of living trees and distribution of dead woody debris than typically result from application of traditional silviculture (Franklin et al. 1997; Spies 1997). Silviculture might achieve a more natural basis by striving to mimic this heterogeneity and leaving onsite appropriate patterns and suitable numbers of residual trees, snags, logs, and other woody debris as biological legacies (Mitchell et al. 2002). These legacies of the past can substantially influence forest development following disturbance and contribute to future conservation of biological diversity (Palik et al. 2002).

The forest reproduction methods of traditional silviculture are in some ways analogous to various natural disturbance events. The clearcutting method eliminates or greatly reduces the forest canopy, with results similar to those caused by a powerful windstorm (e.g., hurricane) or other large-scale disturbance (Fig. 12). However, tree removal from the site is not normally associated with such natural events and the magnitude and pattern of accompanying soil disturbance are very different (e.g., skid trails from logging rather than tip-up mounds from windthrow). Results of the seed-tree method are also like those produced by major windstorms, where a few large trees remain scattered across a site. Once again, removal of trees from the site and the resulting soil disturbance pattern are unnatural consequences. The shelterwood method is thought to closely mimic stand replacement dynamics following a damaging tropical storm, leaving the forest canopy moderately intact and facilitating prolific regeneration (Croker and Boyer 1975). While this may be true for the irregular shelterwood method (where seed-bearing trees of the overwood are retained onsite as biological legacies), eventual harvest of the overstory trees makes the uniform shelterwood method appear less similar to natural disturbance. The two selection methods result in stand conditions not unlike those created by small-scale disturbances (e.g., lightning, localized fire, disease, or insect outbreaks), where individuals or groups of trees suffer mortality and fall from the forest canopy. The major difference between these two methods is related to gap size and dispersion pattern. Group selection normally results in canopy gaps of at least 0.1 ha and well dispersed, while single-tree selection produces smaller gaps less than 0.1 ha (and often as small as the area under an individual tree crown about 0.01 ha) requiring less dispersion. As additional overstory trees are removed through time, these smaller gaps may coalesce, forming larger gaps difficult to distinguish from those created by the group selection method. Although tree removal from the site and periodic soil disturbance are also unnatural aspects of both selection methods, the abundant biological legacies retained onsite following each harvest make both of these methods appear consistent with the characteristics of a more natural version of silviculture that mimics natural disturbance.

As in numerous tropical and other temperate forest types (Brokaw 1985; Denslow

FIGURE 12. Naturally regenerating even-aged longleaf pine stand following Hurricane Camille. Photo courtesy of the Forestry Images Organization.

1987; Spies and Franklin 1989; Veblen 1989; Liu and Hytteborn 1991; Lertzman 1992), the importance of canopy gaps as fundamental ecological structures in the natural regeneration of longleaf pine ecosystems has also been recognized (Palik et al. 1997; Brockway and Outcalt 1998; McGuire et al. 2001; Gagnon et al. 2003). Canopy gaps are essential in maintaining the fine-scale variability created by small-scale natural disturbances and are also generated by forest reproduction methods that remove overstory trees in a patch configuration (Coates and Burton 1997). Forests having a wide range of gap sizes afford a diversity of microenvironments for regeneration, with habitat conditions varying among gaps, within gaps (gap centers versus gap edges), and beneath the forest canopy. The resulting variation in available light, soil moisture, and nutrient resources will differentially affect regeneration success, as will the presence of advanced regeneration prior to gap formation. Gap sizes and distribution resulting from initial harvest and subsequent stand entries will have a substantial influence on the colonization of individuals and development of advanced regeneration, therefore the type and timing of silvicultural treatments will have important impacts on forest population dynamics (Coates and Burton 1997).

Canopy gaps in longleaf pine forests naturally result from a variety of disturbance agents that operate over a wide range of spatial and temporal scales (Palik and Pedersen 1996). Disturbances that remove single trees ultimately create canopy gaps of suitable size for unimpaired growth of regeneration (about 0.14 ha) as effectively as single events that remove groups of trees, a major difference being the longer time required for many episodes of single-tree removal to occur (Palik et al. 1997). Although different pathways are followed, the essential result is that similar outcomes for stand structure are eventually obtained (Palik et al. 2002). Thus, silviculturists may validly apply either the group selection method or single-tree selection method in achieving structures representative of natural forests as long as canopy gaps of sufficient size to facilitate regeneration are created.

Silviculture, because of social, economic, and other constraints, may be unable to perfectly emulate the forest ecosystem structures and processes sustained by a natural disturbance regime (Palik et al. 2002). However, improvements can be obtained by mitigating the negative effects of logging on understory plants, forest floor, and soil structure, and incorporating knowledge concerning ecosystem disequilibrium dynamics, the importance of biological legacies and the vital role of gap-phase replacement into appropriately implemented forest reproduction methods. The application of silviculture will then result in environmental conditions that more closely approximate those produced by natural disturbance and improve the likelihood that multiple constituencies (i.e., environmental, conservation, utilitarian, and industrial groups) will be served as longleaf pine forests are effectively regenerated and sustained in the future.

Acknowledgments

The authors express their appreciation to Becky Estes for assistance in searching the literature to identify numerous relevant publications. We are also grateful to Shibu Jose, Eric Jokela, Dennis Hardin, Larry Bishop, and one anonymous reviewer for comments helpful in improving this manuscript.

References

Allen, R.M. 1958. A study of the factors affecting height growth of longleaf pine seedlings. Ph.D. dissertation, Duke University, Durham, NC.

Barnett, J.P. 1992. The South's longleaf pine, it can rise again! *For People* 41(4):14–17.

Barnett, J.P. 2002. Longleaf pine: Why plant it? Why use containers? In *Proceedings of Workshops on Growing Longleaf Pine in Containers*, eds. J.P. Barnett, R.K. Dumroese, and D.J. Moorhead, pp. 5–7. USDA Forest Service, Southern Research Station, General Technical Report SRS–56, Asheville, NC.

Barnett, J.P., and McGilvray, J.M. 1997. Practical guidelines for producing longleaf pine seedlings in containers. USDA Forest Service, Southern Research Station, General Technical Report SRS–14, Asheville, NC.

Barnett, J.P., and McGilvray, J.M. 2000. Growing longleaf pine seedlings in containers. *Native Plants J.* 1(1):54–58.

Barnett, J.P., and McGilvray, J.M. 2002. Guidelines for producing quality longleaf pine seeds. USDA Forest Service, Southern Research Station, General Technical Report SRS–52, Asheville, NC.

Barnett, J.P., Brissette, J.C., Kais, A.G., and Jones, J.P. 1988. Improving field performance of southern pine seedlings by treating with fungicides before storage. *South J Appl For* 12:281–285.

Barnett, J.P., Lauer, D.K., and Brissette, J.C. 1990. Regenerating longleaf pine with artificial methods. In *Management of Longleaf Pine*, ed. R.M. Farrar, pp. 72–93. USDA Forest Service, Southern Forest Experiment Station, General Technical Report SO–75, New Orleans, LA.

Barnett, J.P., Pickens, B., and Karfalt, R. 1999. Improving longleaf pine seedling establishment in the nursery by reducing seedcoat microorganisms. In *Proceedings of 10th Biennial Southern Silvicultural Research Conference*, ed. J.D. Haywood, pp. 339–343. USDA Forest Service, Southern Research Station, General Technical Report SRS–30, Asheville, NC.

Boyer, W.D. 1963. Development of longleaf pine seedlings under parent trees. USDA Forest Service, Southern Forest Experiment Station, Research Paper SO–4, New Orleans, LA.

Boyer, W.D. 1973. Air temperature, heat sums and pollen shedding phenology of longleaf pine. *Ecology* 54(2):420–426.

Boyer, W.D. 1974a. Longleaf pine cone production related to pollen density. In *Seed Yield from Southern Pine Seed Orchards*, ed. J. Krause, pp. 8–14. Macon: Georgia Forest Research Council.

Boyer, W.D. 1974b. Impact of prescribed fire on mortality of released and unreleased longleaf pine seedlings. USDA Forest Service, Southern Forest Experiment Station, Research Note SO–182, New Orleans, LA.

Boyer, W.D. 1975. Timing overstory removal in longleaf pine. *J For* 73(9):578–580.

Boyer, W.D. 1978. Heat accumulation: An easy way to anticipate the flowering of southern pines. *J For* 76(1):20–23.

Boyer, W.D. 1979. Regenerating the natural longleaf pine forest. *J For* 77:572–575.

Boyer, W.D. 1981. Pollen production and dispersal as affected by seasonal temperature and rainfall patterns. In *Pollen Management Handbook*, ed. E.C. Franklin, pp. 2–9. USDA Agricultural Handbook 587, Washington, DC.

Boyer, W.D. 1987a. Annual and geographic variations in cone production by longleaf pine. In *Proceedings of the 4th Biennial Southern Silvicultural Research Conference*, comp. D.R. Phillips, pp. 73–76. USDA Forest Service, Southeastern Forest Experiment Station, General Technical Report SE–42, Asheville, NC.

Boyer, W.D. 1987b. Volume growth loss: A hidden cost of periodic prescribed burning in longleaf pine? *South J Appl For* 11:154–157.

Boyer, W.D. 1989. Response of planted longleaf pine bare–root and container stock to site preparation and release: Fifth–year results. In *Proceedings 5th Biennial Southern Silvicultural Research Conference*, ed. J.H. Miller, pp. 165–168. USDA Forest Service, Southern Forest Experiment Station, General Technical Report SO–47, New Orleans, LA.

Boyer, W.D. 1990a. *Pinus palustris*, Mill. longleaf pine. In *Silvics of North America*, technical coordinators R.M. Burns and B.H. Honkala, pp. 405–412. Vol. 1, Conifers Washington, DC: USDA Forest Service.

Boyer, W.D. 1990b. Growing-season burns for control of hardwoods in longleaf pine stands. USDA Forest Service, Southern Forest Experiment Station, Research Paper SO–256, New Orleans, LA.

Boyer, W.D. 1993a. Regenerating longleaf pine with natural seeding. In *Proceedings of the 18th Tall Timbers Fire Ecology Conference*, ed. S.M. Hermann, pp. 299–309. Tall Timbers Research Station, Tallahassee, FL.

Boyer, W.D. 1993b. Long–term development of regeneration under longleaf pine seedtree and shelterwood stands. *South J Appl For* 17(1):10–15.

Boyer, W.D. 1996. Anticipating good longleaf pine cone crops: The key to successful natural regeneration. *Alabama's Treasured Forests* 15(3):24–26.

Boyer, W.D. 1998. Long-term changes in flowering and cone production by longleaf pine. In *Proceedings of the 9th Biennial Southern Silvicultural Research Conference*, ed. T.A. Waldrop, pp. 92–98. USDA Forest Service, Southern Research Station, General Technical Report SRS–20, Asheville, NC.

Boyer, W.D. 1999. Longleaf pine: Natural regeneration and management. *Alabama's Treasured Forests* 18(3):7–9.

Boyer, W.D., and Peterson, D.W. 1983. Longleaf pine. In *Silvicultural Systems for the Major Forest Types of the United States*, tech. comp. R.M. Burns, pp. 153–156. USDA Forest Service, Agricultural Handbook No. 445, Washington, DC.

Boyer, W.D., and White, J.B. 1990. Natural regeneration of longleaf pine. In *Management of Longleaf Pine*, ed. R.M. Farrar, pp. 94–113. USDA Forest Service, Southern Forest Experiment Station, General Technical Report SO–75, New Orleans, LA.

Boyer, W.D., and Woods, F.W. 1973. Date of pollen shedding by longleaf pine advanced by increased temperatures at strobili. *For Sci* 19(4):315–318.

Brennan, L.A., and Hermann, S.M. 1994. Prescribed fire and forest pests: Solutions for today and tomorrow. *J For* 92(11):34–37.

Brockway, D.G., and Lewis, C.E. 1997. Long–term effects of dormant-season prescribed fire on plant community diversity, structure and productivity in a longleaf pine wiregrass ecosystem. *For Ecol Manage* 96(1,2):167–183.

Brockway, D.G., and Lewis, C.E. 2003. Influence of deer, cattle grazing and timber harvest on plant species diversity in a longleaf pine bluestem ecosystem. *For Ecol Manage* 175(1–3):49–69.

Brockway, D.G., and Outcalt, K.W. 1998. Gap-phase regeneration in longleaf pine wiregrass ecosystems. *For Ecol Manage* 106(2,3):125–139.

Brockway, D.G., and Outcalt, K.W. 2000. Restoring longleaf pine wiregrass ecosystems: Hexazinone application enhances effects of prescribed fire. *For Ecol Manage* 137(1–3):121–138.

Brockway, D.G., Outcalt, K.W., and Wilkins, R.N. 1998. Restoring longleaf pine wiregrass ecosystems: Plant cover, diversity and biomass following low-rate hexazinone application on Florida sandhills. *For Ecol Manage* 103(2/3):159–175.

Brokaw, N.V.L. 1985. Gap-phase regeneration in a tropical forest. *Ecology* 66(3):682–687.

Coates, K.D., and Burton, P.J. 1997. A gap-based approach for development of silvicultural systems to address ecosystem management objectives. *For Ecol Manage* 99:337–354.

Cordell, C.E., Hatchell, G.E., and Marx, D.H. 1990. Nursery culture of bare-root longleaf pine seedlings. In *Management of Longleaf Pine*, ed. R.M. Farrar, pp. 38–51. USDA Forest Service, Southern Forest Experiment Station, General Technical Report SO–75, New Orleans, LA.

Crofton, E.W. 2001. Flora and fauna of the longleaf pine–grassland ecosystem. In The Fire *Forest: Longleaf Pine Wiregrass Ecosystem*, ed. J.R. Wilson. *Georgia Wildlife* 8(2):69–77.

Croker, T.C. 1952. Early release stimulates cone production. USDA Forest Service, Southern Forest Experiment Station, Southern Forestry Note 79, New Orleans, LA.

Croker, T.C. 1973. Longleaf pine cone production in relation to site index, stand age and stand density. USDA Forest Service, Southern Forest Experiment Station, Research Note SO–156, New Orleans, LA.

Croker, T.C. 1975. Seedbed preparation aids natural regeneration of longleaf pine. USDA Forest Service, Southern Forest Experiment Station, Research Paper SO–112, New Orleans, LA.

Croker, T.C. 1987. Longleaf pine: A history of man and a forest. USDA Forest Service, Southern Region, Forestry Report R8–FR7, Atlanta, GA.

Croker, T.C., and Boyer, W.D. 1975. Regenerating longleaf pine naturally. USDA Forest Service, Southern Forest Experiment Station, Research Paper SO–105, New Orleans, LA.

Daniel, T.W., Helms, J.A., and Baker, F.S. 1979. *Principles of Silviculture*. New York: McGraw–Hill.

Davis, V.B. 1955. Don't keep longleaf pine seed trees too long! USDA Forest Service, Southern Forest Experiment Station, Research Note 98:3, New Orleans, LA.

de Liocourt, F. 1898. De l'amenagement des sapinieres. *Societe Forestiere de Franche-Comte et Belfort Bulletin* 6:396–405.

Dennington, R.W., and Farrar, R.M. 1991. Longleaf pine management. USDA Forest Service, Southern Region, Forestry Report R8–FR3, Atlanta, GA.

Denslow, J.S. 1987. Tropical forest gaps and tree species diversity. *Annu Rev Ecol Syst* 18:431–451.

Dickmann, D.I. 1993. Management of red pine for multiple benefits using prescribed fire. *North J Appl For* 10(2):53–62.

Engstrom, R.T. 1993. Characteristic mammals and birds of longleaf pine forests. In *Proceedings of the 18th Tall Timbers Fire Ecology Conference*, ed. S.M. Hermann, pp. 127–138. Tall Timbers Research Station, Tallahassee, FL.

Engstrom, R.T., Kirkman, L.K., and Mitchell, R.J. 2001. The natural history of the fire forest. In *The Fire Forest: Longleaf Pine Wiregrass Ecosystem*, ed. J.R. Wilson. *Georgia Wildlife* 8(2):5–11, 14–17.

Farrar, R.M. 1975. Sprouting ability of longleaf pine. *For Sci* 21:189–190.

Farrar, R.M. 1996. Fundamentals of uneven-aged management in southern pine. Tall Timbers Research Station, Miscellaneous Publication No. 9, Tallahassee, FL.

Farrar, R.M., and Boyer, W.D. 1990. Managing longleaf pine under the selection system: Promises and Problems. In *Proceedings of the 6th Biennial Southern Silvicultural Research Conference*, eds. S.S. Coleman and D.G. Neary, pp. 357–368. USDA Forest Service, Southeastern Forest Experiment Station, General Technical Report SE–70, Asheville, NC.

Franklin, J.F., Berg, D.R., Thornburgh, D.A., and Tappeiner, J.C. 1997. Alternative silvicultural approaches to timber harvesting: Variable retention harvest systems. In *Creating a Forestry for the 21st Century: The Science of Ecosystem Management*, eds. K.A. Kohm and J.F. Franklin, pp. 111–139. Washington, DC: Island Press.

Frost, C.C. 1993. Four centuries of changing landscape patterns in the longleaf pine ecosystem. In *Proceedings of the 18th Tall Timbers Fire Ecology Conference*, ed. S.M. Hermann, pp. 17–43. Tall Timbers Research Station, Tallahassee, FL.

Gagnon, J.L., Jokela, E.J., Moser, W.K., and Huber, D.A. 2003. Dynamics of artificial regeneration in gaps within a longleaf pine flatwoods ecosystem. *For Ecol Manage* 172:133–144.

Gagnon, J.L., Jokela, E.J., Moser, W.K., and Huber, D.A. 2004. Characteristics of gaps and natural regeneration in mature longleaf pine flatwoods ecosystems. *For Ecol Manage* 187:373–380.

Gilliam, F.S., and Platt, W.J. 1999. Effects of long-term fire exclusion on tree species composition and stand structure in an old-growth *Pinus palustris* (longleaf pine) forest. *Plant Ecol* 140:15–26.

Grace, S.L., and Platt, W.J. 1995. Effects of adult tree density and fire on the demography of pregrass stage juvenile longleaf pine (*Pinus palustris* Mill.). *J Ecol* 83:75–86.

Grelen, H.W. 1983. May burning favors survival and early height growth of longleaf pine seedlings. *South J Appl For* 7:16–20.

Guldin, J.M. 1996. The role of uneven–aged silviculture in the context of ecosystem management. *West J Appl For* 11(1):4–12.

Guyer, C., and Bailey, M.A. 1993. Amphibians and reptiles of longleaf pine communities. In *Proceedings of the 18th Tall Timbers Fire Ecology Conference*, ed. S.M. Hermann, pp. 139–158. Tall Timbers Research Station, Tallahassee, FL.

Hainds, M. 2001. Scalping aids survival of longleaf. *Alabama's Treasured Forests* 20(3):24–27.

Hardin, E.D., and White, D.L. 1989. Rare vascular plant taxa associated with wiregrass (*Aristida* stricta) in the southeastern United States. *Nat Areas J* 9:234–245.

Haywood, J.D. 2000. Mulch and hexazinone herbicide shorten the time longleaf pine seedlings are in the grass stage and increase height growth. *New For* 19:279–290.

Heyward, F. 1933. The root system of longleaf pine on the deep sands of western Florida. *Ecology* 14:136–148.

Hilton, J. 1999. Biological diversity in the longleaf pine ecosystem. *Alabama's Treasured Forests* 18(4):28–29.

Johnson, R., and Gjerstad, D. 1998. Landscape-scale restoration of the longleaf pine ecosystem. *Restor Manag Notes* 16(1):41–45.

Johnson, R., and Gjerstad, D. 1999. Restoring the longleaf pine forest ecosystem. *Alabama's Treasured Forests* 18(4):18–19.

Jose, S., Merritt, S., and Ramsey, C.L. 2003. Growth, nutrition, photosynthesis and transpiration responses of longleaf pine seedlings to light, water and nitrogen. *For Ecol Manage* 180:335–344.

Kantola, T.A., and Humphrey, S.R. 1990. Habitat use of Sherman's fox squirrel (*Sciurus niger shermanni*) in Florida. *J Mammal* 71:411–419.

Kelly, J.F., and Bechtold, W.A. 1990. The longleaf pine resource. In *Management of Longleaf Pine*, ed. R.M. Farrar, pp. 11–22. USDA Forest Service, Southern Forest Experiment Station, General Technical Report SO–75, New Orleans, LA.

Komarek, E.V. 1968. Lightning and lightning fires as ecological forces. In *Proceedings of the 9th Tall Timbers Fire Ecology Conference*, pp. 169–198. Tall Timbers Research Station, Tallahassee, FL.

Kush, J.S., comp. 1996. Longleaf pine: A regional perspective of challenges and opportunities. Proceedings of the 1st Longleaf Alliance Conference. Longleaf Alliance Report No. 1, Auburn, AL.

Kush, J.S., comp. 1998. Proceedings of the longleaf pine ecosystem restoration symposium. Longleaf Alliance Report No. 3, Auburn, AL.

Kush, J.S., comp. 1999. Longleaf pine: A forward look. Proceedings of the 2nd Longleaf Alliance Regional Conference. Longleaf Alliance Report No. 4, Auburn, AL.

Kush, J.S., comp. 2001. Restoration and management of longleaf pine ecosystems: Silvicultural, ecological, social, political and economic challenges. Proceedings of the 3rd Longleaf Alliance Regional Conference, Longleaf Alliance Report No. 5, Auburn, AL.

Kush, J.S. 2002. Natural regeneration of longleaf pine: Adaptations to site conditions and management systems. Ph.D. dissertation, School of Forestry and Wildlife Sciences, Auburn University, Auburn, AL.

Landers, J.L. 1991. Disturbance influences on pine traits in the southeastern United States. In *Proceedings of the 17th Tall Timbers Fire Ecology Conference*, pp. 61–98. Tall Timbers Research Station, Tallahassee, FL.

Landers, J.L., Byrd, N.A., and Komarek, R. 1990. A holistic approach to managing longleaf pine communities. In *Management of Longleaf Pine*, ed. R.M. Farrar, pp. 135–167. USDA Forest Service, Southern Forest Experiment Station, General Technical Report SO–75, New Orleans, LA.

Landers, J.L., Van Lear, D.H., and Boyer, W.D. 1995. The longleaf pine forests of the Southeast: Requiem or renaissance? *J For* 93(11):39–44.

Lertzman, K.P. 1992. Patterns of gap-phase replacement in a subalpine old-growth forest. *Ecology* 73(2):657–669.

Lipscomb, D.J. 1989. Impacts of feral hogs on longleaf pine regeneration. *South J Appl For* 13:177–181.

Liu, Q., and Hytteborn, H. 1991. Gap structure, disturbance and regeneration in a primeval *Picea abies* forest. *J Veg Sci* 2:391–402.

Lynch, K.D. 1980. A phenotypic study of selected variable in longleaf pine. Ph.D. dissertation, School of Forestry, Auburn University, Auburn, AL.

Mann, W.F. 1970. Direct seeding longleaf pine. USDA Forest Service, Southern Forest Experiment Station, Research Paper SO–57, New Orleans, LA.

Maple, W.R. 1977. Planning longleaf pine regeneration cuttings for best seedling survival and growth. *J For* 75:25–27.

Mattoon, W.R. 1922. Longleaf pine. U.S. Department of Agriculture, Bulletin No. 1061, Washington, DC.

McGuire, J.P. 2001. Living on longleaf: How humans shaped the piney woods ecosystem. In *The Fire Forest: Longleaf Pine Wiregrass Ecosystem*, ed. J.R. Wilson. *Georgia Wildlife* 8(2):42–53.

McGuire, J.P., Mitchell, R.J., Moser, E.B., Pecot, S.D., Gjerstad, D.H., and Hedman, C.H. 2001. Gaps in a gappy forest: Plant resources, longleaf pine regeneration and understory response to tree removal in longleaf pine savannas. *Can J For Res* 31:765–778.

McKee, W.H. 1982. Changes in soil fertility following prescribed burning on Coastal Plain pine sites. USDA Forest Service, Southeastern Forest Experiment Station, Research Paper SE–234, Asheville, NC.

McLemore, B.F. 1977. Strobili and conelet losses in four species of southern pines. USDA Forest Service, Southern Forest Experiment Station, Research Note SO–226, New Orleans, LA.

Means, D.B., and Grow, G. 1985. The endangered longleaf pine community. ENFO, Florida Conservation Foundation, Inc., Winter Park, FL 85(4):1–12.

Meyer, H.A. 1952. Structure, growth and drain in balanced uneven-aged forests. *J For* 50:85–92.

Mitchell, R.J., Palik, B.J., and Hunter, M.L. 2002. Natural disturbance as a guide to silviculture. *For Ecol Manage* 155:315–317.

Moore, J.H. 2001. Managing the forest and the trees: A private landowner's guide to conservation management of longleaf pine. The Nature Conservancy, East Gulf Coastal Plain Ecoregional Team and Southeast Conservation Science Team, Baton Rouge, LA, Theo Davis and Sons, Inc., Zebulon, NC.

Mulligan, M.K., Kirkman, L.K., and Mitchell, R.J. 2002. Aristida beyrichiana (wiregrass) establishment and recruitment: Implications for restoration. *Restor Ecol* 10(1):68–76.

Nature Conservancy. 2002. Conserving Greenwood Plantation. Eye on Nature, Georgia Chapter, The Nature Conservancy, Atlanta, GA. Summer Issue, p. 4.

Nolte, D.L., and Barnett, J.P. 2000. A repellent to reduce mouse damage to longleaf pine seed. *Int Biodeterior Biodegr* 45:169–174.

Noss, R.F. 1989. Longleaf pine and wiregrass: Keystone components of an endangered ecosystem. *Nat Areas J* 9:211–213.

Noss, R.F., LaRoe, E.T., and Scott, J.M. 1995. Endangered ecosystems of the United States: A preliminary assessment of loss and degradation. USDI National Biological Service, Biological Report 28, Washington, DC.

Nyland, R.D. 2003. Even- to uneven-aged: The challenges of conversion. *For Ecol Manage* 172: 291–300.

Outcalt, K.W. 1995. Maintaining the native plant community during longleaf pine establishment. In Forest Research Institute, Bulletin No. 192, pp. 283–285. Rotorua, New Zealand.

Outcalt, K.W. 2000. The longleaf pine ecosystem of the South. *Native Plants J* 1(1):42–44, 48–53.

Outcalt, K.W., and Sheffield, R.M. 1996. The longleaf pine forest: Trends and current conditions. USDA Forest Service, Southern Research Station, Resource Bulletin SRS-9, Asheville, NC.

Outcalt, K.W., Williams, M.E., and Onokpise, O. 1999. Restoring *Aristida stricta* to *Pinus palustris* ecosystems on the Atlantic Coastal Plain, USA. *Restor Ecol* 7:262–270.

Palik, B.J., and Pedersen, N. 1996. Overstory mortality and canopy disturbances in longleaf pine ecosystems. *Can J For Res* 26:2035–2047.

Palik, B.J., Mitchell, R.J., Houseal, G., and Pedersen, N. 1997. Effects of canopy structure on resource availability and seedling responses in a longleaf pine ecosystem. *Can J For Res* 27:1458–1464.

Palik, B.J., Mitchell, R.J., and Hiers, J.K. 2002. Modeling silviculture after natural disturbance to sustain biodiversity in the longleaf pine (*Pinus palustris*) ecosystem: Balancing complexity and implementation. *For Ecol Manage* 155:347–356.

Peet, R.K., and Allard, D.J. 1993. Longleaf pine–dominated vegetation of the southern Atlantic and eastern Gulf Coast region, USA. In *Proceedings of the 18th Tall Timbers Fire Ecology Conference*, ed. S.M. Hermann, pp. 45–81. Tall Timbers Research Station, Tallahassee, FL.

Pessin, L.J. 1938. The effect of vegetation on the growth of longleaf pine seedlings. *Ecol Monogr* 8:115–149.

Platt, W.J., Evans, G.W., and Rathbun, S.L. 1988. The population dynamics of a long-lived conifer (*Pinus palustris*). *Am Nat* 131(4):491–525.

Platt, W.J., and Rathbun, S.L. 1993. Dynamics of an old-growth longleaf pine population. In *Proceedings of the 18th Tall Timbers Fire Ecology Conference*, ed. S.M. Hermann, pp. 275–297. Tall Timbers Research Station, Tallahassee, FL.

Provencher, L., Herring, B.J., Gordon, D.R., Rodgers, H.L., Galley, K.E. M., Tanner, G.W., Hardesty, J.L., and Brennan, L.A. 2001a. Effects of hardwood reduction techniques on longleaf pine sandhill vegetation in northwest Florida. *Restor Ecol* 9:13–27.

Provencher, L., Litt, A.R., Gordon, D.R., Rodgers, H.L., Herring, B.J., Galley, K.E.M., McAdoo, J.P., McAdoo, S.J., Bobris, N.M., and Hardesty, J.L. 2001b. Restoration fire and hurricanes in longleaf pine sandhills. *Ecol Restor* 19(2):92–98.

Pyne, S.J. 1997. *Fire in America: A Cultural History of wildland and Rural Fire*. Seattle: University of Washington Press.

Ramsey, C.L., Jose, S., Brecke, B.J., and Merritt, S. 2003. Growth response of longleaf pine (*Pinus palustris* Mill.) seedlings to fertilization and herbaceous weed control in an old field in southern USA. *For Ecol Manage* 172:281–289.

Reynolds, R.R. 1969. Twenty-nine years of selection timber management on the Crossett Experimental Forest. USDA Forest Service, Southern Forest Experiment Station, Research Paper SO–40, New Orleans, LA.

Reynolds, R.R., Baker, J.B., and Ku, T.T. 1984. Four decades of selection management on the Crossett Farm Forestry Forties. Arkansas Agricultural Experiment Station Bulletin 872.

Schabel, H.G., and Palmer, S.L. 1999. The Dauerwald: Its role in the restoration of natural forests. *J For* 97(11):20–25.

Schmidtling, R.C. 1999. Longleaf pine genetics. In *Longleaf Pine: A Forward Look. Proceedings of the 2nd Longleaf Alliance Regional Conference*, Longleaf Alliance Report No. 4, comp. J.S. Kush, pp. 24–26. Auburn, AL.

Schmidtling, R.C. 2001. Southern pine seed sources. USDA Forest Service, Southern Research Station, General Technical Report SRS–44, Asheville, NC.

Schmidtling, R.C., and Hipkins, V. 1998. Genetic diversity in longleaf pine (*Pinus palustris*): Influence on historical and prehistorical events. *Can J For Res* 28:1135–1145.

Schmidtling, R.C., Hipkins, V., and Carroll, E. 2000. Pleistocene refugia for longleaf and loblolly pines. *J Sustain For* 10(3/4):349–354.

Schopmeyer, C.S., tech. coord. 1974. *Seeds of Woody Plants in the United States*. USDA Agricultural Handbook 450, Washington, DC.

Schwarz, G.F. 1907. *The Longleaf Pine Virgin Forest: A Silvical Study*. New York: John Wiley & Sons.

Seamon, G. 1998. A longleaf pine sandhill restoration in northwest Florida. *Restor and Manage Notes* 16:46–50.

Shelton, M.G., and Cain, M.D. 2000. Regenerating uneven-aged stands of loblolly and shortleaf pines: The current state of knowledge. *For Ecol Manage* 129:177–193.

Shoulders, E. 1967. Fertilizer application, inherent fruitfulness and rainfall affect flowering of longleaf pine. *For Sci* 13:376–383.

Simberloff, D. 1993. Species-area fragmentation effects on old-growth forests: Prospects for longleaf pine communities. In *Proceedings of the 18th Tall Timbers Fire Ecology Conference*, ed. S.M. Hermann, pp. 1–13. Tall Timbers Research Station, Tallahassee, FL.

Smith, H.C., Lamson, N.I., and Miller, G.W. 1989. An esthetic alternative to clearcutting. *J For* 87(3):14–18.

Smith, L.F. 1955. Development of longleaf pine seedlings near large trees. *J For* 53(4):289–290.

Smith, L.F. 1961. Growth of longleaf pine seedlings under large pines and oaks in Mississippi. USDA Forest Service, Southern Forest Experiment Station, Research Paper 189, New Orleans, LA.

Smith, L.F. 1962. Growth of longleaf pine seedlings under large pines and oaks in Mississippi. USDA Forest Service, Southern Forest Experiment Station, Paper 189, New Orleans, LA.

Snyder, E.B., Dinus, R.J., and Derr, H.J. 1977. Genetics of longleaf pine. USDA Forest Service, Research Paper WO-33, Washington, DC.

Spies, T. 1997. Forest stand structure, composition and function. In *Creating a Forestry for the 21st Century: The Science of Ecosystem Management*, eds. K.A. Kohm and J.F. Franklin, pp. 11–30. Washington, DC: Island Press.

Spies, T.A., and Franklin, J.F. 1989. Gap characteristics and vegetation response in coniferous forests of the Pacific Northwest. *Ecology* 70(3):543–545.

Starkey, T.E. 2002. Irrigation and fertilization type, rate and frequency of application. In *Proceedings of Workshops on Growing Longleaf Pine in Containers*, eds. J.P. Barnett, R.K. Dumroese, and D.J. Moorhead, pp. 30–34. USDA Forest Service, Southern Research Station, General Technical Report SRS–56, Asheville, NC.

Stout, I.J., and Marion, W.R. 1993. Pine flatwoods and xeric pine forests of the southern lower coastal plain. In *Biodiversity of the Southeastern United States: Lowland Terrestrial Communities*, eds. W.H. Martin, S.G. Boyce, and A.C. Echternacht, pp. 373–446. New York: John Wiley & Sons.

Taylor, A.R. 1974. Ecological aspects of lightning in forests. In *Proceedings of the 13th Tall Timbers Fire Ecology Conference*, pp. 455–482. Tall Timbers Research Station, Tallahassee, FL.

Veblen, T.T. 1989. Tree regeneration responses to gaps along a transandean gradient. *Ecology* 70(3):541–543.

Wade, D.D., and Lewis, C.E. 1987. Managing southern grazing ecosystems with fire. *Rangelands* 9(3):115–119.

Wade, D.D., and Lundsford, J. 1990. Fire as a forest management tool: Prescribed burning in the southern United States. *Unasylva* 162(41):28–38.

Wahlenberg, W.G. 1946. Longleaf pine: Its use, ecology, regeneration, protection, growth and management. C.L. Pack Forestry Foundation and USDA Forest Service, Washington, DC.

Wakeley, P.C. 1954. Planting the southern pines. U.S. Department of Agriculture, Agricultural Monograph 18, Washington, DC.

Walker, J.L. 1993. Rare vascular plant taxa associated with the longleaf pine ecosystem. In *Proceedings of the 18th Tall Timbers Fire Ecology Conference*, ed. S.M. Hermann, pp. 105–125. Tall Timbers Research Station, Tallahassee, FL.

Walker, J.L. 1995. Longleaf pine ecosystem restoration: Toward a regional strategy. USDA Forest Service, Southern Research Station, Asheville, NC and Southern Region, Atlanta, GA.

Walker, J.L. 1999. Longleaf pine forests and woodlands: Old growth under fire! In *The Value of Old Growth Forest Ecosystems of the Eastern United States*, ed. G.L. Miller, pp. 33–40. Asheville: University of North Carolina.

Walker, J.L., and Boyer, W.D. 1993. An ecological model and information needs assessment for longleaf pine ecosystem restoration. In *Silviculture from the Cradle of Forestry to Ecosystem Management*, comp. L.H. Foley, pp. 138–147. USDA Forest Service, Southeastern Forest Experiment Station, General Technical Report SE–88, Asheville, NC.

Walker, L.C. 1954. Early scrub-oak control helps longleaf pine seedlings. *J For* 52:939–940.

Walker, L.C., and Davis, V.B. 1954. Forest walls retard young longleaf pine. USDA Forest Service, Southern Forest Experiment Station, Research Note 93:3, New Orleans, LA.

Walker, L.C., and Davis, V.B. 1956. Seed trees retard longleaf pine seedlings. *J For* 54(4):269.

Ware, S., Frost, C.C., and Doerr, P.D. 1993. Southern mixed hardwood forest: The former longleaf pine forest. In *Biodiversity of the Southeastern United States: Lowland Terrestrial Communities*, eds. W.H. Martin, S.G. Boyce, and A.C. Echternacht, pp. 447–493. New York: John Wiley & Sons.

White, J.B. 1981. The influence of seedling size and length of storage on longleaf pine survival. *Tree Planters' Notes* 32(4):3–4.

White, T.L., Harris, H.G., and Kellison, R.C. 1977. Conelet abortion in longleaf pine. *Can J For Res* 7:378–382.

Wright, H.A., and Bailey, A.W. 1982. *Fire Ecology of the United States and Southern Canada*. New York: John Wiley & Sons.

Chapter 5

Plant Competition, Facilitation, and Other Overstory–Understory Interactions in Longleaf Pine Ecosystems

Timothy B. Harrington

Introduction

Many of the stand structural characteristics of longleaf pine (*Pinus palustris* Mill.) forests that existed prior to European colonization have been altered or lost from past disturbance histories (Frost this volume). For example, often missing are the widely spaced, large-diameter trees, the all-aged stand structure that included a vigorous cohort of grass-stage longleaf pine seedlings, and the understory community composed of numerous woody and herbaceous species of short stature embedded within the flashy fuels of a wiregrass (*Aristida beyrichiana* Trin. & Rupr.) or bluestem (*Andropogon* spp.) matrix. Some of these structural features, such as the understory community, can be restored through modern silvicultural methods, vegetation management, prescribed fire, pine thinning, and artificial regeneration (i.e., planting or seeding) (Johnson and Gjerstad this volume; Walker and Silletti this volume). Other structural features, such as an all-aged distribution of longleaf pines, must be allowed to develop over time given appropriate disturbance regimes and the presence of keystone species (i.e., longleaf pine and wiregrass or bluestem) to "jump-start" the system. A mechanistic understanding of overstory and understory interactions will provide a sound basis for prescribing treatments designed to restore and maintain longleaf pine communities.

Overstory trees in forest stands affect understory vegetation by modifying growing conditions, either directly or indirectly. These modifications are manifested in a variety of ways, including consumption of growth-limiting resources (i.e., light, soil water, and nutrients) and alteration of other physical characteristics that impact growing conditions (i.e., temperature, litterfall accumulation, and fire behavior). In a similar way, understory vegetation can influence the growing conditions of overstory trees, potentially affecting their survival, stem growth, and crown morphology.

This chapter will focus on two common interactions in forest communities, competition and facilitation, and their potential influences on overstory and understory responses in southern pine forests, with emphasis on longleaf pine. First I will discuss basic concepts of plant interactions, including types and associated responses, and attempt to classify overstory and understory interactions commonly observed in longleaf pine forests. Implications of plantation silviculture to these interactions will be considered. Next I will review previous

---

Timothy B. Harrington • USDA Forest Service, Pacific Northwest Research Station, Olympia, Washington 98512.

research on overstory and understory interactions with emphasis on southern pine communities. Finally I will discuss overstory and understory interactions observed in two case studies conducted in longleaf pine plantations at the Savannah River Site, a National Environmental Research Park near Aiken, SC. The discussion will conclude with implications of these interactions to restoration and maintenance of longleaf pine communities.

The plant interactions that are the primary focus of this chapter imply a stand structure of at least two canopy layers, the overstory and the understory. To simplify the discussion, I will define "overstory" as the pine trees that comprise the upper canopy of a forest stand, where the minimum height of a "tree" is equal to 6 m (Daniel et al. 1979). "Understory" will be defined as herbaceous species (forbs and grasses) plus woody species (vines, shrubs, hardwoods, and pines) that have a stem diameter at breast height (1.37 m) less than 2.5 cm growing under or in the proximity of overstory trees. Midstory layers also are possible strata in this hypothetical stand structure but are not the focus of this chapter.

Concepts of Plant Interactions

Plant interactions encompass a broad variety of positive and negative relationships that can exist when plants are grown in close proximity such that they influence each other's survival, growth, or reproduction (Harper 1977). In this section, I will rely on the conceptual framework proposed by Goldberg (1990) to characterize plant interactions, with emphasis on competition and facilitation. Most interactions between plants are indirect (i.e., they do not physically injure one another) and occur through an intermediary such as resources, natural enemies, or plant-produced toxins. The net result from a plant interaction (i.e., is it positive or negative and what is the magnitude of the effect?) is the combination of one plant's effect on abundance of the intermediary and the "target" plant's response to abundance of the intermediary (Goldberg 1990).

Interference includes those negative plant interactions that result either from competition for limited resources or allelopathy (production of chemical toxins by one plant that inhibit the functions of another). Because of differences in plant size or other traits, competition is often asymmetrical such that one plant is negatively affected while the other may show little or no signs of a response (Grace 1990). Resource gradients, such as spatial differences in soil nitrogen availability, can change competitive relationships that exist among plants to favor species or individuals that are most effective at resource capture (Goldberg and Miller 1990; Kalmbacher and Martin 1996). Such modifications in competitive relationships often result in declines in plant species diversity because of dominance by a few species.

Table 1 provides examples of plant interactions that have been observed in southern pine communities, classified according to the conceptual framework proposed by Goldberg (1990). The simple case of exploitation of a limiting resource has been defined as "uptake effects" of competition. "Nonuptake effects" of competition occur when an intermediary, such as litterfall from overstory trees, changes resource availability indirectly. In southern pine communities, both uptake and nonuptake effects of competition from overstory pines play a prominent role in limiting abundance and species diversity of the understory (Monk and Gabrielson 1985; Harrington and Edwards 1999; Harrington et al. 2003). These effects on the understory occur largely through direct exploitation of light, soil water, and nitrogen resources by the overstory, but also indirectly by accumulation of overstory needle litter that limits light availability to forest floor plants.

"Apparent competition" results when the intermediary is not a growth-limiting resource but rather it is a natural enemy, such as fire, disease, or herbivory, that is promoted by one plant so that the net effect of the enemy on the target plant is similar to that resulting from exploitation competition. For example,

TABLE 1. Examples of plant interactions observed in southern pine communities as classified according to the framework of Goldberg (1990).[a]

Interaction	Intermediary	Overstory → understory interactions				Understory → overstory interactions			
		E	R	N	References	E	R	N	References
Exploitation competition (uptake effects)[b]	light	−	+	−	Monk and Gabrielson 1985; Means 1997; Harrington and Edwards 1999; McGuire et al. 2001; Harrington et al. 2003				
	soil water	−	+	−	Monk and Gabrielson 1985; Harrington et al. 2003				Harrington and Edwards 1999
	nitrogen	−	+	−	Palik et al. 1997; McGuire et al. 2001; Harrington et al. 2003				
Exploitation competition (nonuptake effects)[c]	light	−	+	−	Monk and Gabrielson 1985; Shelton 1995; Harrington et al. 2003				
Apparent competition[d]	fire	+	−	−	(1) Grace and Platt 1995; Brockway and Outcalt 1998; Glitzenstein et al. 1995	−	+	−	(2) Richardson and Williamson 1988
Allelopathy[e]	toxin					+	−	−	Richardson and Williamson 1988
Positive facilitation[f]	soil water	+	+	+	(3) Ginter et al. 1979; Boyer and Miller 1994				
	nitrogen	+	+	+	(4) Harrington et al. 2003	+	+	+	(6) Hendricks and Boring 1992; Hendricks and Boring 1999; Wilson et al. 1999; Hiers et al. 2003
	fire	+	+	+	(5) Streng et al. 1993; Brewer and Platt 1994; Anderson and Menges 1997	+	+	+	(7) Platt et al. 1988a

[a] The influence of overstory pines on understory vegetation (overstory → understory interactions) are shown as the direction (+ or −) of effects (E) of the overstory on abundance of an intermediary, response (R) of the understory to abundance of the intermediary, and net effect (N) to the understory. Understory → overstory interactions follow in a similar manner. Numbers in parentheses are explained in the footnotes.
[b] Plant consumption of a limiting resource (light, soil water, or nitrogen) restricts its availability to "target" plants.
[c] Litterfall of overstory pines limits light availability to ground-layer vegetation.
[d] Apparent competition results when (1) pine litterfall increases fire intensity causing mortality of longleaf pine seedlings and injury of hardwoods, and (2) understory scrub vegetation decreases frequency of ground fires needed to regenerate longleaf pine.
[e] Above- or belowground exudates from understory scrub vegetation chemically suppress growth of pine seedlings.
[f] Positive facilitation results when pine litterfall (3) protects soil structure needed to maintain its water-holding capacity, (4) provides nitrogen inputs to the soil, and (5) promotes frequent ground fires that stimulate growth and flowering of herbaceous species, and when understory plants (6) provide nitrogen inputs to the soil through symbiotic nitrogen fixation and (7) promote frequent ground fires needed to regenerate longleaf pine.

accumulations of needle litter near overstory trees can support fire intensities capable of killing grass-stage seedlings of longleaf pine to create a vegetation-free zone similar in appearance to what would occur from intense resource competition (Grace and Platt 1995; Brockway and Outcalt 1998).

"Allelopathy" is a direct interaction in which plants produce a toxin in their foliage or roots that chemically alters the functions of the target plant. For example, in scrub vegetation communities of Florida, rosemary (*Ceratiola ericoides* Michx.) and false rosemary (*Conradina canescens* [Torr. & Gray] A. Gray) shrubs can limit seedling growth of bluestem, wiregrass, longleaf pine, and sand pine (*Pinus clausa* [Chapm. ex Engelm.] Vasey ex Sarg.) when grown together in noncompetitive environments (Richardson and Williamson 1988). This allelopathic relationship fosters the development of a fire-tolerant community that favors survival of shrubs at the expense of fire-dependent species, such as wiregrass and longleaf and sand pines.

"Facilitation" is a plant interaction in which one or both plants benefit from their relationship with each other. For example, needle litter from overstory pines is a source of nitrogen to understory plants that may aid their survival and growth (Harrington et al. 2003). Similarly, needle litter can act as a mulch to conserve soil water (Ginter et al. 1979) and it can protect soil structure needed to preserve water-holding capacity (Boyer and Miller 1994). Greater first-year survival of longleaf pine seedlings under uncut forest versus seedlings in experimentally created gaps suggests a beneficial effect of the overstory (McGuire et al. 2001). Note that, in contrast to the common plant interaction of facilitation, "mutualism" implies an obligatory relationship between plant species such that each benefits when grown in proximity to the other and each suffers when grown separately (Radosevich and Holt 1984).

As indicated by the majority of the plant interactions listed in Table 1, competition in southern pine communities is asymmetric with the overstory having the predominant influence in its relationship with understory vegetation. However, fire, as a common disturbance agent in these communities, has a dual role in regulating interactions between overstory and understory vegetation. It has a negative effect (apparent competition described above) in directly injuring or killing understory species incapable of tolerating its influence (Glitzenstein et al. 1995; Grace and Platt 1995; Brockway and Lewis 1997; Brockway and Outcalt 1998). But fire also has a positive effect (facilitation) in stimulating vegetative growth and flowering of specific understory species (Streng et al. 1993; Brewer and Platt 1994; Anderson and Menges 1997) and in maintaining wiregrass, a critical keystone species in longleaf pine communities (Platt et al. 1988a).

Overstory and understory interactions may operate differently in even-aged plantations than in natural stands because of differences in structural attributes (Oliver and Larson 1996). Even-aged plantations generally have uniform spacing and size of trees, similar crown morphologies and shapes largely dictated by spacing, and complete crown coverage when the stands are at full stocking. Stem size distributions are usually unimodal, and with time, they develop increasing positive skewness indicative of stand differentiation into different crown classes. Depending on spacing among trees, rates of crown closure and overall productivity of even-aged plantations often exceed those of natural stands.

Structure of understory vegetation in even-aged forest plantations is often symptomatic of overstory structure and site history, including previous land use and silvicultural treatments. Thus, an understory in an even-aged plantation is often uniform in plant size and species composition due to the homogeneity of the overstory and associated limitations in understory resource availability. At crown closure of the overstory, the stand enters the "stem exclusion stage" during which recruitment of new trees into the overstory ceases and understory abundance of shade-intolerant species and vigor of tolerant species decline (Oliver and Larson 1996). Invigoration of existing species and recruitment of new species in the understory may not occur until the "understory reinitiation stage" when canopy coverage of the overstory begins to decline.

The transition to the understory reinitiation stage may be delayed in even-aged plantations because their uniform spacing and size of trees prevents large canopy gaps from forming until late in stand development.

Given these differences in structural attributes and rates of development, it seems likely that some interactions between overstory trees and understory vegetation are likely to occur sooner, at greater intensity, and with increased duration in even-aged plantations versus natural stands. Exclusion or suppression of understory species as a result of overstory competition probably will occur more rapidly than in natural stands, depending on spacing among trees. However, understory reinitiation may be delayed substantially because of the prolonged uniformity of stand structure, even with the onset of density-dependent mortality of overstory trees. Although plantation silviculture can provide an effective means for reestablishing stands of a desired species composition and spacing, in its conventional usage it may impede restoration of understory species that rely on a heterogeneous stand structure. Thus, a complete understanding of overstory and understory interactions is needed to properly direct development of longleaf pine plantations toward the desired stand structure of the overstory and species composition of the understory.

Previous Research on Overstory and Understory Interactions

Early research on overstory and understory interactions in longleaf pine communities focused on factors influencing the rate at which longleaf pine seedlings exited the grass stage. Pessin (1938) compared 3-year growth of grass-stage longleaf pines growing at various densities (2470 to 247,000 seedlings ha^{-1}) with or without manual removal of herbaceous vegetation. Scrub oaks, primarily blackjack (*Quercus marilandica* Muenchh.) and post oaks (*Quercus stellata* Wangenh.), were removed in all but one plot. Height growth varied inversely with pine seedling density and also according to the presence (1–7 cm yr^{-1}) versus absence (2–31 cm yr^{-1}) of herbaceous vegetation. Emergence of pine seedlings from the grass stage clearly was limited by availability of belowground resources, because light availability (estimated by evaporation rates) was 60% of maximum intensity or greater in all but the scrub oak plot (34% of maximum). Not surprisingly, height growth of longleaf pine seedlings in the scrub oak plot (1 cm yr^{-1}) was among the lowest values observed in this study.

Research on production of grazing forage in longleaf pine rangelands has quantified the extent to which overstory trees limit abundance and biomass of understory herbaceous species. Abundance of understory vegetation varied inversely with increasing density of overstory longleaf pines (Wolters 1973, 1981). Pine thinning and prescribed burning combinations increased forage yields in natural stands of longleaf pine to about half that observed for treeless rangeland (Grelen and Enghardt 1973). Although these studies provide empirical evidence of the competitive effects of overstory longleaf pines on understory vegetation, they do not identify the primary mechanisms responsible for them.

Much of the research on overstory and understory interactions in forest communities has focused on the effects of three primary factors: shade, root competition, and litterfall. Overstory and understory competitive interactions have been studied effectively by trenching around experimental plots to eliminate belowground competition for soil water and nutrients while maintaining light availability in the understory at nominal levels. In an early study, Fricke (1904, cited in Spurr and Barnes 1992) cut the roots of overstory Scots pine (*Pinus sylvestris* L.) in a closed canopy forest and observed dramatic increases in growth of understory pines, indicating that soil moisture, and perhaps nutrients, were limiting their growth. Other trenching studies have demonstrated similar results with eastern white pine (*Pinus strobus* L.) in New Hampshire (Toumey and Kienholz 1931), loblolly pine (*Pinus taeda* L.) in North Carolina (Korstian and Coile 1938), grand fir (*Abies grandis* [Dougl. ex D. Don] Lindl.) in Montana (McCune 1986), and ponderosa pine (*Pinus ponderosa* Dougl. ex Laws) in eastern Oregon (Riegel et al. 1992). Using a different experimental approach, Riegel and Miller (1991) demonstrated for ponderosa pine that eliminating some of the overstory competition for soil water and nitrogen via irrigation and fertilization, respectively, stimulated a 36% increase in aboveground biomass of the understory relative to nontreated areas.

Of the three factors mentioned previously, litterfall is perhaps the one least studied in forest communities. Southern pines shed their needles throughout the year, and this accumulation forms a physical barrier that can prevent seeds from reaching mineral soil and seedlings from emerging into sunlight (Shelton 1995). In general, herbaceous species with erect, semiwoody growth habits can tolerate moderate levels of litterfall (Sydes and Grime 1981a,b). Litterfall can intercept, absorb, and facilitate evaporation of rainfall before it reaches mineral soil layers, but it also can act as a mulch to reduce evaporation from the soil and as a substrate for protecting soil structure for retention of its water-holding capacity (Ginter et al. 1979; Boyer and Miller 1994). Litterfall can reduce temperature fluctuations in the surface soil layers, and it can act as both a source and sink of nitrogen for plant nutrition. In addition, needle litter of longleaf pines is an important fuel component that can influence fire intensity and spread, especially where it accumulates around individual trees (Grace and Platt 1995; Brockway and Outcalt 1998). Monk and Gabrielson (1985) studied dynamics of species turnover in old-field herbaceous communities using treatments that combined artificial shade, presence or absence of litterfall, and trenching around isolated loblolly pines. The normal progression of changes in species composition during the 2-year study was an increase in density of perennials and a decrease in density of annuals. Presence of shade (4% of full sunlight) or litterfall accelerated the turnover or loss of annuals but only slowed the rate of succession to perennials resulting in fewer species and lower densities of individual plants. Combined effects of shade and litterfall stimulated the greatest reductions in plant density. Litterfall probably limited germination of annuals; however, the upright growth form of many of the perennial species enabled them to shed litter, especially when grown in full sun. In trenched plots, presence versus absence of litterfall strongly limited abundance of annuals but it had no detectable influence on perennials. However, in the absence of trenching, litterfall did not influence abundance of either annuals or perennials because apparently their populations were already being regulated by shade and root competition.

Recent research on longleaf pine communities has attempted to identify the critical factors limiting pine seedling development; however, as Harper (1977) points out, study design can strongly influence experimental outcomes. For example, in both natural and experimentally created gaps within longleaf pine forest, competition for light was identified as the primary factor limiting growth of longleaf pine seedlings (Palik et al. 1997; McGuire et al. 2001). Nitrogen availability increased with decreasing density of overstory trees, but the magnitude of its effect on seedling growth was secondary to light availability. Soil water availability did not vary in a systematic way with overstory density, perhaps because of broad fluctuations in rainfall. In these studies, understory vegetation either was not manipulated (McGuire et al. 2001) or it was eliminated up to 1.2 m around individual pine seedlings (Palik et al. 1997). Perhaps soil water availability did play a more prominent role in affecting seedling responses in these studies, but background effects of understory competition prevented its detection, either as edge effects in the study by Palik et al. (1997) or by varying inversely with overstory density in the study by McGuire et al. (2001).

In another example, abundance of longleaf pine natural regeneration in an uneven-aged stand did not increase substantially until the distance from overstory trees exceeded 12 m, and maximum seedling densities occurred at distances of 16 m or greater (Brockway and Outcalt 1998) (Fig. 1). Absence of strong differences in light availability and the limited size of the zone of higher fire intensity from needle

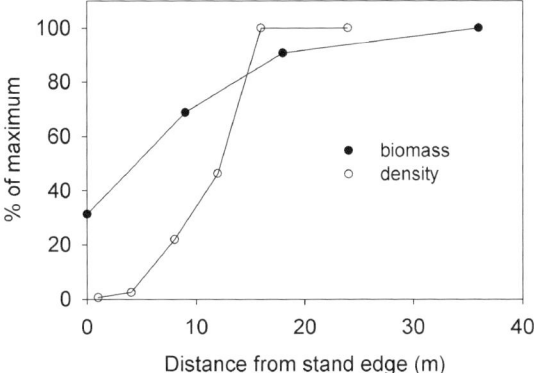

FIGURE 1. Relationships of average biomass (data from McGuire et al. 2001) and density (data from Brockway and Outcalt 1998) of longleaf pine seedlings growing within a gap versus distance from the edge of mature longleaf pine forest. Maximum values for these variables were approximately 9 g and 6700 seedlings ha^{-1} for biomass and density, respectively.

litter accumulations (up to 4 m away from individual trees) prompted the authors to conclude that competition for belowground resources was the primary factor limiting pine regeneration within gaps. In the study by McGuire et al. (2001), growth of planted longleaf pine seedlings was maximized when they occurred at least 18 m from overstory trees (Fig. 1). Results of these studies imply that gaps of radius 16 m or greater (0.08 ha or larger) are needed to eliminate overstory influences. Palik et al. (1997) advocated gap sizes of 0.14 ha or larger (i.e., a radius of 21 m or larger for circular gaps) to promote growth of pine seedlings free of overstory influences. However, in each of these examples, the experimental approaches did not adequately separate and quantify overstory and understory effects to determine if consumption of belowground resources by understory vegetation was a key factor limiting pine seedling development.

Results of the canopy gap research on longleaf pine indicate that group selection is an appropriate method of regeneration. For a given stand basal area, forest stand structures that have an aggregated distribution of overstory trees will provide a higher percentage of area in larger gaps than those that retain trees evenly dispersed across a given area of land (Palik et al. 1997). Such stand structures provide a higher percentage of area with sufficient availabilities of light and nitrogen to support regeneration of longleaf pine seedlings (Battaglia et al. 2002; Palik et al. 2003). The shape and orientation of individual gaps also will influence the duration of sunlight and spatial distribution of root competition from overstory trees.

Fire is an essential feature of longleaf pine forests because of their pyrogenic characteristics of needle drape, grass-stage pine seedlings, and uniformly distributed wiregrass and bluestem (*Andropogon* spp.) (Platt et al. 1988a). Numerous studies have affirmed the benefits of fire for reducing competition from hardwoods and shrubs and for reducing needle litter accumulations that can impede establishment of longleaf pine seedlings (Boyer 1990, 1993; Glitzenstein et al. 1995; Brockway and Lewis 1997). Where wiregrass forms a dense and continuous cover under longleaf pines, variability in fire intensity may play a role in generating gaps in ground-layer vegetation that promote expansion of surviving species and colonization of new species, such as golden aster (*Pityopsis graminifolia* [Michx.] Nutt.) (Brewer et al. 1996). Because they affect burn intensity, frequency and timing of prescribed fire influence the abundance, size, and composition of understory species (Boyer 1995) as well as the timing of their flowering and seed production (Platt et al. 1988b; Brewer and Platt 1994). Hardwood mortality also varies with frequency and timing of prescribed fire. Over an 18-year period, Boyer (1993) found that stand basal area of midstory hardwoods increased following winter biennial burns (from 0.8 to 2.2 m^2 ha^{-1}), while it decreased following spring biennial burns (from 0.9 to 0.3 m^2 ha^{-1}). Van Lear and Waldrop (1991) reported that over 80% of oak (*Quercus* spp.) and sweetgum (*Liquidambar styraciflua* L.) rootstocks were killed by 10 annual burns, while only 50% of rootstocks were killed by 10 biennial burns.

The research reviewed here indicates a complex suite of factors regulates overstory and understory interactions in longleaf pine forests. Overstory effects on understory vegetation can be direct, such as competition for limited resources or physical smothering from needle litter. These effects also can be indirect, such as spatial variation in fire intensity that results from variable rates of needle litter accumulation. Two case studies are discussed below to illustrate some of the overstory and understory interactions that occur in even-aged plantations of longleaf pine. As discussed previously, the uniform size and spatial distribution of overstory trees in even-aged plantations probably create a more homogeneous environment for studying effects of competition and facilitation than would be expected in a naturally regenerated, uneven-aged stand.

Case Studies on Overstory and Understory Interactions in Longleaf Pine Plantations

Two case studies are presented to illustrate responses to asymmetrical competition that occur between the overstory and understory of longleaf pine plantations. The studies provide a basis for prescribing silvicultural treatments aimed at restoring plant species native to longleaf pine communities. In the first study, community-level responses (abundance and diversity) of understory vegetation were investigated in response to pine thinning and hardwood and shrub control (Harrington and Edwards 1999). In the second study, a controlled experiment was established to separately quantify competition and needle litter effects of a longleaf pine overstory on fitness and fecundity of planted populations of several perennial herbaceous species native to longleaf pine forests (Harrington et al. 2003).

Study I: Understory Community Responses to Pine Thinning and Hardwood and Shrub Control

Initial research at the Savannah River Site on overstory and understory interactions in longleaf pine plantations focused on community responses (understory vegetation abundance and diversity) to thinning of overstory pines and control of hardwoods and shrubs with herbicides (Harrington and Edwards 1999). These silvicultural treatments were selected for study because they provided a wide range of light, soil water, and needle litter conditions in which to study understory responses. Six 8- to 11-year-old plantations of longleaf pine growing on sandhill sites were selected having average stand basal areas of 9.3 and 1.1 m^2 ha^{-1} of pines and hardwoods, respectively. Soils included loamy sands of the Blanton, Lakeland, or Troup series that were well drained to excessively well drained resulting in low to very low available water-holding capacities (Rogers 1990). A prescribed fire of moderate to high intensity was applied to each site in February 1994 and 1998. Each of the six sites was divided into four treatment areas of similar size (3 to 7 ha) and one of the following treatments was randomly assigned to each: (1) nontreated, (2) pine thinning in May 1994 to leave approximately half of the original stem density, (3) control of hardwoods and shrubs with herbicides in 1995 (grid application of hexazinone) and 1996 (spot treatments of triclopyr, imazapyr, and glyphosate), and (4) combined treatments of pine thinning and hardwood and shrub control. The experimental design is a randomized complete block with six replications (sites) of the four treatments arranged as a 2 × 2 factorial.

Within each of the 24 treatment areas, 10 sample points spaced on a 40-m grid were permanently marked for periodic vegetation measurements. At each sample point, cover was estimated for each understory species by the line intercept method (Mueller-Dombois and Ellenberg 1974) in August 1994–1996 and by visual estimation within 10 m^2 plots in August 1998. The data were pooled by categories of herbaceous (forbs and grasses) and woody species (hardwoods and shrubs). At the end of the 1994–1996, 1998, and 2002 growing seasons, stem diameter at breast height (millimeters at 1.37 m above ground) was measured on each pine and hardwood stem rooted within 6 m of a sample point. Height (centimeters) and crown width (centimeters in north–south and east–west directions) also were measured annually starting in 1995 on approximately 20% of measurement trees per sample point.

Periodic annual increments in individual-tree stem basal area, height, and crown width were calculated separately for pines and hardwoods. Understory vegetation and tree growth data were averaged first by sample point and then by treatment area. Data for each measurement year were subjected to analysis of variance to identify whether main effects (pine thinning or hardwood and shrub control) or their interaction were significant ($\alpha = 0.05$). Multiple comparisons of means were conducted with either Bonferroni adjusted probabilities (for significant interactions) or Tukey's test (if only main effects were significant) (Sokal and Rohlf 1981).

Understory Plant Abundance

Abundance of herbaceous vegetation responded dynamically to changes in stand structure, with initial decreases following the herbicide treatments, moderate increases following pine thinning, and, ultimately, large increases following the combination treatment (Fig. 2). In 1995, cover of both woody and herbaceous species varied significantly as a result of the interaction of pine thinning and hardwood and shrub control treatments. Of the four treatments, the combination treatment had the lowest overall abundance of vegetation probably because activity of hexazinone herbicide (a photosynthetic inhibitor) increased as a result of the greater light availability in the pine-thinning treatment. In 1996 and 1998, cover of understory hardwoods and shrubs was substantially less in the presence versus absence of the herbicide treatments (Fig. 2A). Cover of herbaceous species in 1996 and 1998 demonstrated strong increases in response to pine thinning. In addition, herbaceous cover in 1998 was greater in the presence versus absence of hardwood and shrub control. Pine thinning increased light availability throughout the study duration; however, it increased soil water availability only during May 1995 and 1996. In contrast, hardwood and shrub

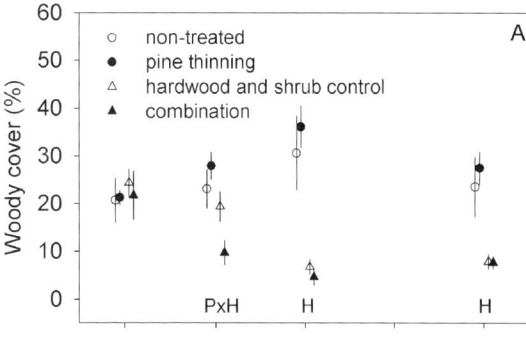

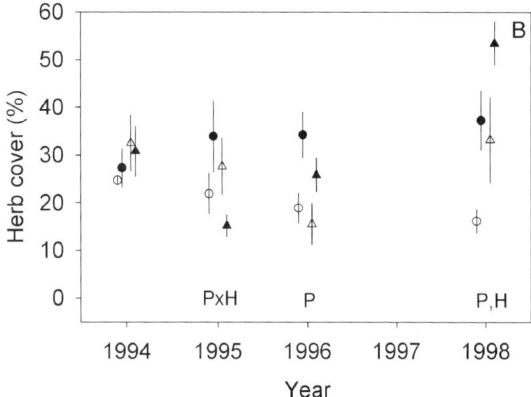

FIGURE 2. Average cover (± standard error) of (A) nonpine woody species (hardwoods and shrubs) and (B) herbaceous species during 4 to 5 years after combinations of thinning of overstory pines and control of hardwoods and shrubs with herbicides. Letters along the x-axis indicate factors and their interactions from the analysis of variance that were significant ($P \leq 0.05$) for a given year of the study: P = pine thinning and H = hardwood and shrub control.

control was associated with increases in soil water throughout the entire 1996 growing season. Relative differences in the magnitude of herbaceous cover responses to these treatments in 1998 indicated that light availability was the most influential factor limiting herbaceous cover (21% absolute cover reduction), although its effects were similar in magnitude to those resulting from differences in soil water availability (16% cover reduction) (Fig. 2B). Because needle litter accumulations were limited by the prescribed burns of 1994 and 1998, this factor did not appear to play a strong role in limiting development of herbaceous cover.

Herbaceous Species Diversity

Herbaceous species density (number of species per 40 m^2 sample area) in 1998 varied according to the interaction of pine thinning and hardwood and shrub control (Harrington and Edwards 1999). Pine thinning alone had a greater effect on species density (33 species) than either of the hardwood and shrub control or combination treatments (30 species), and each of these responses was greater than observed for nontreated areas (25 species). A comparison of relative differences in species density resulting from pine thinning versus hardwood and shrub control main effects indicated that only thinning (increased light availability) stimulated increased diversity of herbaceous species.

Tree Growth

Overstory and understory interactions were found to operate in both directions. That is, not only did the overstory influence understory vegetation abundance and species diversity, but the understory also influenced overstory tree growth. In each of the measurement years after 1994, pine thinning and/or hardwood and shrub control treatments were associated with growth increases in stem basal area and crown width of longleaf pine trees (Fig. 3). However, in 1995 and 1996, pine thinning was associated with reductions in height growth. Hardwood and shrub control was associated with marginal increases ($P = 0.07$) in height growth in 1996, a year noted for sustained increases in soil water from this treatment. However, 1998 height growth was less in the presence versus absence of hardwood and shrub control. Increased allocations of tree growth to stem diameter and crown width at the expense of growth in height also have been observed for loblolly pine soon after thinning an 8-year-old plantation (Ginn et al. 1991). Such shifts in growth allocation have been attributed to tree responses associated with the capture of newly available growing space rather than those associated with "thinning shock." In 1995, 1996, and 1998, pine thinning was associated with 67% to 91% in-

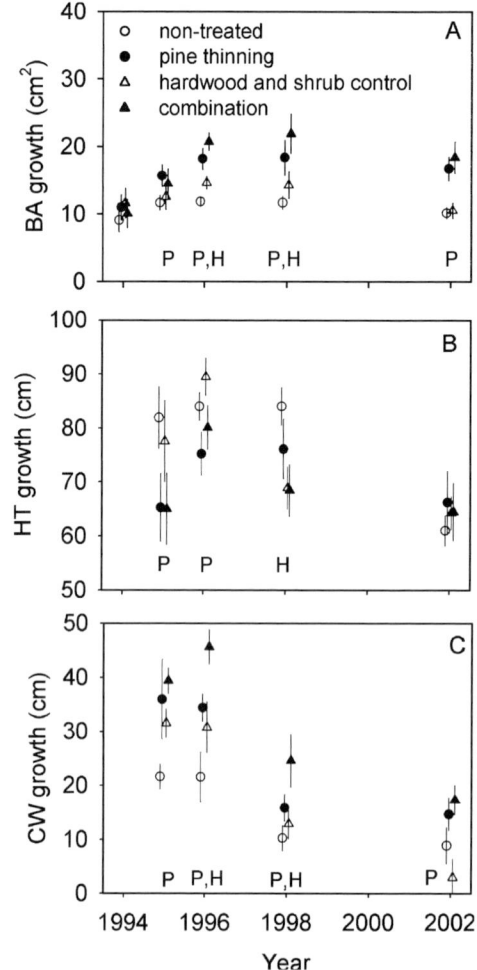

FIGURE 3. Average periodic annual growth (± standard error) in (A) stem basal area, (B) height, and (C) crown width of individual longleaf pine trees 7 to 8 years after combinations of thinning of overstory pines and control of hardwoods and shrubs with herbicides. Letters along the x-axis indicate factors from the analysis of variance that were significant ($P \leq 0.05$) for a given year of the study: P = pine thinning and H = hardwood and shrub control.

creases in average basal area growth of individual hardwood trees ($P \leq 0.08$; data not shown). In 2002, pine thinning was associated with a 117% increase in average height growth of hardwoods ($P = 0.05$); however, high levels of variability among plots prevented detection of significant growth increases for stem basal area ($P = 0.11$).

Conclusions

Results of Study I indicate that availabilities of both light (via pine thinning) and belowground resources (via hardwood and shrub control) play prominent roles in maintaining abundance of herbaceous species, while perhaps only availability of light influences diversity of herbaceous species. Similarly, availabilities of both light and belowground resources played similar roles in stimulating growth increases in stem basal area and crown width of longleaf pine trees. Height growth responses to treatment varied inversely with the stem and crown responses. Although this research may only be applicable to longleaf plantations of similar age, it demonstrates a useful approach for studying community dynamics in response to silvicultural treatments.

Study II: Effects of Above- and Belowground Competition and Needle Litter from Overstory Pines on Fitness and Fecundity of Reintroduced Herbaceous Species

A limitation of Study I was the confounding that existed between overstory competition (availability of light versus belowground resources) and needle litter because treatments were conducted at the stand level, making it impossible to separate the relative influence of each factor on understory vegetation. For example, understory conditions in nonthinned stands included growth limiting effects of competition for both light and belowground resources, making it impossible to distinguish which had the greatest influence on herbaceous vegetation. To isolate above- and belowground components of overstory competition from needle litter effects, a controlled experiment was established in 1998 according to methods described in Harrington et al. (2003). Nonsampled portions of three of the six sites from Study I were used in Study II (i.e., areas away from the sample points in the hardwood and shrub control and combination treatments). Four 0.09-ha plots (30 m × 30 m) were established in each of the 13- to 15-year-old plantations. Each plot was randomly assigned a thinning to 0, 25, 50, or 100% of the average basal area of nonthinned stands (20 m^2 ha^{-1}). In order to focus the research on overstory effects, plots were kept free of all nonpine vegetation with periodic applications of non-soil-active herbicides and hand weeding.

In the interior 20-m × 20-m area within each plot, four split-plot treatments were established to vary belowground resources and needle litter independently of pine stocking. Each of the following split-plot treatments was randomly assigned to one of four 1.2-m × 13.7-m strips in each plot: (a) trenching plus needle litter, (b) trenching minus needle litter, (c) absence of trenching plus needle litter, and (d) absence of trenching minus needle litter. Trenching was conducted in October 1998 with a Ditch Witch® (Perry, OK), aluminum flashing 0.51 m wide was installed along the vertical wall of the trench to an average depth of 0.43 m to prevent encroachment of pine roots, and the soil was replaced. In needle litter-present split plots, stand-produced needle litter was supplemented with monthly applications to result in a standardized rate (7893 kg ha^{-1} yr^{-1}) equal to twice that predicted for nonthinned stands of 20 m^2 ha^{-1} basal area using data from Study I. In needle litter-absent split plots, needle litter was removed manually each month. The experimental design of Study II was a randomized, complete block with three replications (sites) of the split-plot arrangement of treatments. Pine stocking was the whole-plot factor and split-plot factors were trenching and needle litter.

Fourteen perennial herbaceous species native to longleaf pine forests were selected for intensive study to represent different growth forms (e.g., upright versus prostrate) of forbs and grasses. One herbaceous species was randomly assigned to each of 10 1-m^2 quadrats per split plot. Containerized seedlings of the 14 species were grown in a greenhouse and planted at a fixed density (11 to 36 plants m^{-2}, depending on success of greenhouse propagation) in May 1999, 2000, or 2001. Containerized seedlings of longleaf pine were purchased from International Forest Company (Statesboro, GA) and planted at a density of 36 plants m^{-2} within a randomly assigned quadrat of each split plot.

Fitness of each herbaceous species was assessed as survival (%) and visually estimated cover (percentage) in October. Fecundity of each herbaceous species was assessed as inflorescence count (number of flower clusters per square meter), seed production (number of seeds produced per square meter; estimated from subsamples of 20 or more inflorescences), seed weight (grams per 1000 seeds), and seedling emergence (percentage) from standard germination tests (45 days of cold stratification followed by 3 weeks of germination in a greenhouse). In fall 2001, surviving longleaf pine seedlings growing in their assigned quadrat were severed, counted, and returned to the laboratory for biomass estimation. Fitness of longleaf pine was assessed as survival (percentage), average stem height (centimeters), average biomass (grams), and percentage of surviving trees that were exiting the grass stage, defined as those with stems 12 cm or greater in height (Haywood 2000).

To highlight the basic findings of Study II, this chapter covers responses of two perennial grasses (green silkyscale, *Anthaenantia villosa* [Michx.] Beauv., and pineywoods dropseed, *Sporobolus junceus* [Michx.] Kunth, planted May 1999 and assessed in October 2000), a perennial forb (beggar's ticks, *Desmodium ciliare* [Muhl. Ex Willd.] DC, planted May 2001 and assessed October 2001), and seedlings of longleaf pine (planted December 1998 and assessed October 2001). Analysis of variance was conducted to identify significant ($\alpha = 0.05$) treatment effects. Multiple comparisons of means were conducted with either Bonferroni adjusted probabilities (for significant interactions) or Tukey's test (if only main effects were significant) (Sokal and Rohlf 1981). Linear regression was used to quantify the relationship between average aboveground biomass and average height of longleaf pine seedlings.

Anthaenantia villosa Responses

Average survival of *Anthaenantia villosa* exceeded 90% in all treatments at the end of the second year after planting, although a slight reduction was observed in the presence of needle litter (Fig. 4A). In the absence of trenching (i.e., in the presence of root competition from overstory pines), cover, number of inflorescences, and number of seeds decreased with increasing pine stocking at a more rapid rate than in the presence of trenching (Fig. 4B–D). Cover in trenched plots averaged about 80% and it did not vary among pine stockings. Cover was greater in the presence of needle litter, an effect likely attributable to nitrogen inputs associated with this treatment (Harrington et al. 2003). Seed weight did not vary significantly among treatments (Fig. 4E). Seedling emergence was less at 0% pine stocking than at 25, 50, or 100% stockings (Fig. 4F). Although plant size, flowering, and seed production each were greatest in the absence of overstory pines (0% stocking), seed viability was not. Apparently some level of overstory competition stimulated increased seed viability; thus, total reproductive effort (i.e., the product of seed production and seedling emergence) was greatest at 25% pine stocking. *A. villosa* had an interactive response pattern in which its fitness and fecundity were maximized at specific combinations of above- and belowground competition. The species was able to survive in a wide range of competitive environments, it produced the most vegetation when belowground resources were not limiting, and it produced the most viable seed in a mildly competitive environment.

Sporobolus junceus Responses

In contrast to *Anthaenantia villosa*, *Sporobolus junceus* displayed a very different set of responses. Average survival of this species exceeded 60% for all treatments, but it decreased in the presence of either trenching or needle litter (Fig. 5A). Needle litter most likely caused shading and eventual death of suppressed individuals. Cover of *S. junceus* did not vary significantly among any of the treatments,

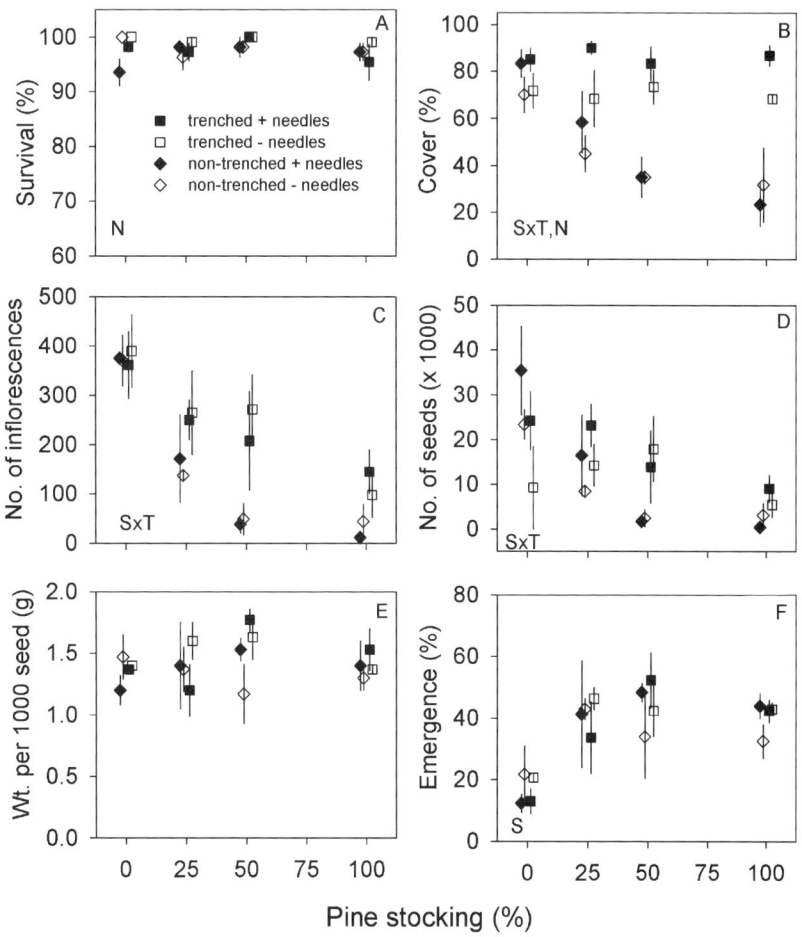

FIGURE 4. Average indices of fitness and fecundity (± standard error) of the perennial grass *Anthaenantia villosa* as influenced by stocking of overstory pines (basal area as a percentage of nonthinned stands) and presence or absence of trenching and needle litter. Letters in the lower left corner of each graph indicate factors and their interactions from the analysis of variance that were significant ($P \leq 0.05$): S = pine stocking, T = trenching, and N = needle litter.

suggesting that the species was relatively tolerant to the experimental range of resource availabilities and needle litter levels (Fig. 5B). For the trenching effects on survival, increases in belowground resources might have stimulated greater amounts of density-dependent mortality (i.e., "self-thinning"). This explanation was supported by the fact that, although survival declined slightly from trenching and needle litter effects, surviving individuals occupied the newly available growing spacing resulting in the absence of treatment effects on cover. These fitness responses contrasted sharply with the species' fecundity responses.

In the presence of needle litter, inflorescence number was less than a third of that observed in its absence (Fig. 5C). In the absence of trenching, inflorescence number at 100% pine stocking was less than at the other stocking levels; however, in the presence of trenching, inflorescence number did not vary significantly among stocking levels. Seed number at 100% stocking was less than at the other stocking levels (Fig. 5D). Presence of needle litter greatly limited seed number, seed weight (Fig. 5E), and seedling emergence (Fig. 5F). *S. junceus* demonstrated a fitness response that was relatively tolerant of

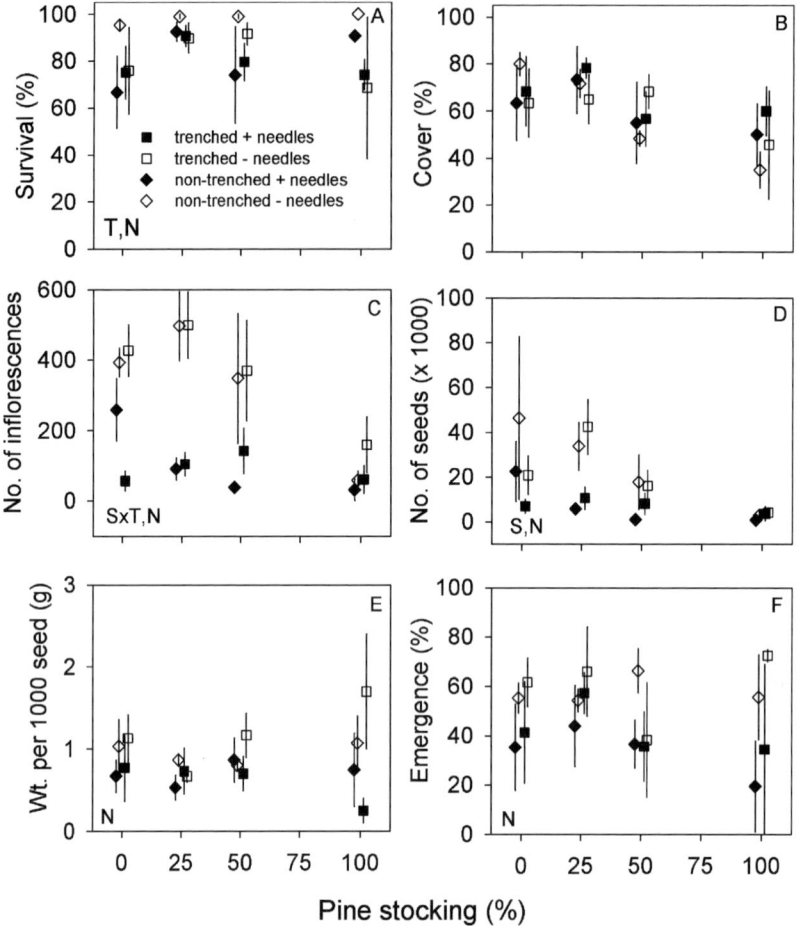

FIGURE 5. Average indices of fitness and fecundity (± standard error) of the perennial grass *Sporobolus junceus* as influenced by stocking of overstory pines (basal area as a percentage of nonthinned stands) and presence or absence of trenching and needle litter. Letters in the lower left corner of each graph indicate factors and their interactions from the analysis of variance that were significant ($P \leq 0.05$): S = pine stocking, T = trenching, and N = needle litter.

competition from overstory pines. The species was able to recover growing spacing through increases in plant size and fully occupy the site regardless of density-dependent mortality or needle litter. However, fecundity was greatly impacted primarily by needle litter (via reductions in flowering, seed production, seed size, and seed viability) and secondarily by stocking (via reductions in flowering and seed production). The mechanism underlying the strongly negative effects of needle litter on fecundity was not clear since similar effects on plant fitness were not apparent until October 2001 (Harrington et al. 2003). The species was able to thrive vegetatively in the wide range of growing conditions present in the study. Observed increases in nitrogen concentration of *S. junceus* foliage in the presence of needle litter (Harrington et al. 2003) suggested that improvements in plant nutrition might have stimulated vegetative growth at the expense of reproductive allocation. However, a more likely explanation for the fecundity responses could be the shading effects of needle litter. Fecundity responses to stocking indicated that a threshold in light availability existed below which flowering and seed production were strongly curtailed. This threshold

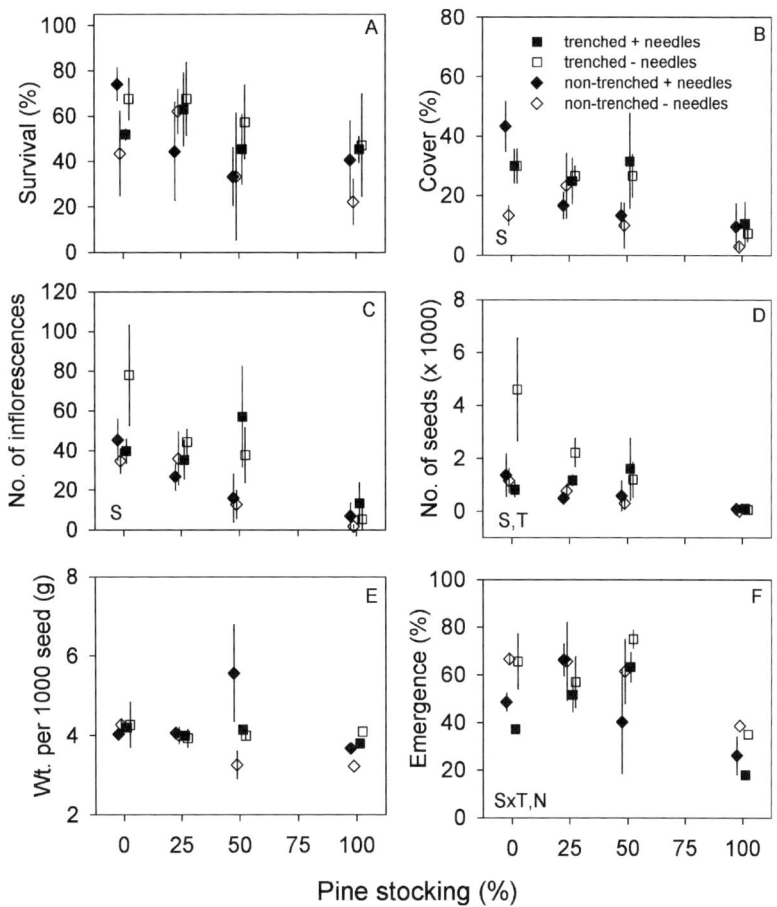

FIGURE 6. Average indices of fitness and fecundity (± standard error) of the perennial forb *Desmodium ciliare* as influenced by stocking of overstory pines (basal area as a percentage of nonthinned stands) and presence or absence of trenching and needle litter. Letters in the lower left corner of each graph indicate factors and their interactions from the analysis of variance that were significant ($P \leq 0.05$): S = pine stocking, T = trenching, and N = needle litter.

occurred between 50% and 100% stocking of overstory pines, because at lesser stockings, responses of these variables were relatively stable.

Desmodium ciliare Responses

Desmodium ciliare responses to overstory conditions were similar to those observed for other perennial forbs in that they indicated a strong degree of intolerance to both shade and root competition from overstory pines (Harrington et al. 2003). In general, these experimental effects were additive and not interactive,

suggesting that the species was able to respond to a variety of growing conditions. Although survival of *D. ciliare* declined somewhat with increasing pine stocking, these responses were not statistically significant (Fig. 6A). Cover decreased systematically with increasing pine stocking and it was especially limited at 100% pine stocking (Fig. 6B). Trenching effects on cover were marginally significant ($P = 0.06$), indicating a mild additive effect resulting from increased availability of belowground resources. Flowering and seed production responses were similar to those for cover (Fig. 6C,D): a decline with increasing stocking and a small to moderate increase from trenching. Seed weight did not vary significantly among treatments (Fig. 6E). The interaction of stocking and trenching was significant for seedling emergence because increases from trenching were observed only at 50% stocking; all other stockings demonstrated neutral effects from trenching (Fig. 6F). In addition, there was less seedling emergence in the presence versus absence of needle litter. These results suggested that seed viability of *D. ciliare* was regulated largely by light availability determined by either or both of overstory density and needle litter. In summary, *D. ciliare* had a strong requirement for light to survive, grow, and reproduce; its potential responsiveness to enhanced belowground resources was secondary to that resulting from increased light availability.

Pinus palustris Seedling Responses

Pinus palustris was able to survive but grew very little in response to shade and root competition from overstory pines. In the third year after planting (2001), seedling survival was greater in the presence versus absence of trenching, and it was reduced in the presence versus absence of needle litter (Fig. 7A). Stocking of overstory pines did not significantly influence either of these survival responses, suggesting that light availability was not a limiting factor for survival except when smothering from needle litter occurred. The fact that stocking had no detectable effect on survival of *P. palustris* seedlings probably was the result of adequate carbohydrate storage in the taproot of the nursery-grown seedlings that permitted the plant to tolerate severe growth-limiting conditions for several years. However, it remains uncertain how much longer these seedlings could have sustained themselves under these growing conditions. The moderate degree of survival (59% to 68%) observed in the absence of overstory competition and needle litter suggests that density-dependent mortality had occurred.

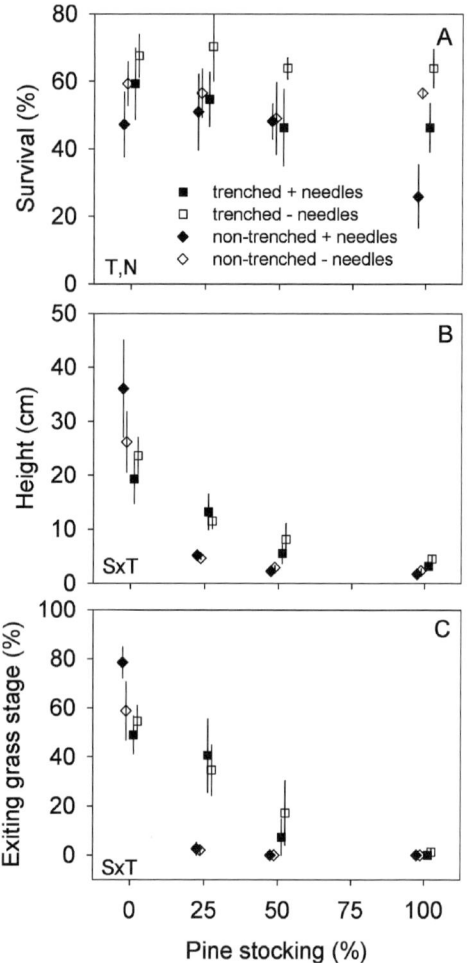

FIGURE 7. Average (A) survival, (B) stem height, and (C) percentage of survivors exiting the grass stage (± standard error) for seedlings of the tree *Pinus palustris* as influenced by stocking of overstory pines (basal area as a percentage of nonthinned stands) and presence or absence of trenching and needle litter. Letters in the lower left corner of each graph indicate factors and their interactions from the analysis of variance that were significant ($P \leq 0.05$): S = pine stocking, T = trenching, and N = needle litter.

About 97% of the variation in average aboveground biomass (grams) of longleaf pine seedlings was explained by its linear relationship with average height (centimeters) ($R^2 = 0.97; s_{y.x} = 12.2; n = 48$):

$$Y = -2.755 + 6.090(X)$$

Because of this high correlation, height and

aboveground biomass had essentially identical responses to the interaction of stocking and trenching (only height responses are shown in Fig. 7B). Both of these variables exhibited steep declines with increasing stocking but the steepness of the decline was much greater in the absence versus presence of trenching. This result indicated a strong threshold response to light availability because average height (and aboveground biomass) declined precipitously as stocking increased from 0% to 25% with subsequent increases in stocking causing only slight decreases in growth. The threshold growth response to light availability was moderated only slightly by increases in belowground resources from trenching. By the third year after planting, over half of longleaf pine seedlings growing in 0% stocking were exiting the grass stage, while less than 20% of seedlings were doing so in stockings of 50 and 100% (Fig. 7C). In contrast to the average height responses, trenching strongly increased the percentage of trees exiting the grass stage for stockings of 25 and 50%. This result is in agreement with previous research in which increased availability of belowground resources, such as resulting from control of competing herbaceous vegetation with herbicides or mulching (Haywood 2000), enabled a higher percentage of longleaf pine seedlings to exit the grass stage. However, Study II demonstrated that light availability had a much greater effect than availability of belowground resources since very few of the seedlings growing in trenched split plots at 100% stocking were exiting the grass stage. In terms of providing new saplings of *P. palustris* to ultimately replace overstory trees, only the lowest stockings (0 and 25%) averaged more than 20% of seedlings emerging from the grass stage to support their growth into the upper canopy.

Conclusions

Results from Study II confirm some of the findings from Study I: effects of overstory competition for light and for belowground resources on understory vegetation can be of similar magnitude. Study II further illustrates the great variability in species' responses to above- and belowground competition and needle litter. However, of all the species' responses, only one indicated a positive effect of overstory shade (facilitation): seed viability of *A. villosa* was greater at pine stockings of 25% to 100% than it was at 0%. Therefore, a primary finding from Study II is that the growing conditions in gaps (i.e., full sunlight) provided the resources needed to maximize fitness and fecundity of reintroduced herbaceous species. In addition, root competition had a critically negative effect on most of the species' responses, indicating that restoration plantings will be most successful in the absence of both overstory and understory vegetation. The vigorous responses of herbaceous species observed in the relatively small gaps (30 m × 30 m or 0.09 ha) of Study II indicated that the zone of overstory root competition from neighboring 13- to 15-year-old pines extended less than 10 m into the gap. Surprisingly, plant responses to needle litter varied from positive effects (facilitation via nitrogen addition) to negative effects (apparent competition via reduced light availability). Mulching effects from needle litter (i.e., increased retention of soil water) were not detected (Harrington et al. 2003).

Implications to Maintenance and Restoration of Longleaf Pine Communities

Previous research and the two case studies reviewed here illustrate the multiple interactions that occur between overstory pines and understory vegetation of longleaf pine forests. As indicated initially in this chapter, a mechanistic understanding of overstory and understory interactions will provide a sound basis for prescribing treatments designed to restore and maintain longleaf pine communities. In this concluding section I will discuss the implications of overstory and understory competition and facilitation to silvicultural regimes that are currently in use or being considered for longleaf pine forests. Three conditions will be considered: new longleaf pine communities; existing herbaceous communities within natural, uneven-aged longleaf pine stands; and existing herbaceous communities within even-aged, planted longleaf pine stands.

New Longleaf Pine Communities

Restoring all of the elements of a longleaf pine community requires coordinated establishment of longleaf pine seedlings and herbaceous species. Both vegetation components require an environment with an abundance of direct sunlight and soil water. Within existing forest, gaps 21 m in radius or larger provide such an environment. Recently harvested forest sites or old fields will have adequate light availability but not necessarily adequate soil water availability because of a potentially high abundance of competing vegetation. In order to increase soil water availability, competing vegetation abundance must be reduced significantly, at least in the individual spots where seedlings are to be established.

Spot (or banded) treatments provide a means of reducing competition without the widespread disruption in the ground layer community that can result from broadcast treatments (Brockway and Outcalt 2000). In addition, spot treatments enable existing desirable species growing between spots to be retained in an undisturbed condition. One potential technique is to create gaps in the ground layer vegetation 1 meter in radius or larger by chemical (e.g., nonsoil active herbicides, such as glyphosate or triclopyr) or mechanical methods (e.g., "Wombat" spot cultivator, Savannah Forestry Equipment, Savannah, GA).

By locating these ground-layer gaps at a spacing of 3 m or greater, the desired community components can be established over a broad area. Half of the gaps can be planted with a single longleaf pine seedling and no other vegetation, while perennial herbaceous species can be established in the remaining gaps by sowing seeds or planting containerized seedlings. The success of a spring planting of herbaceous will rely on absence of a severe frost and presence of adequate rainfall (Harrington et al. 2003); winter planting of dormant, containerized herbaceous seedlings is an untested approach that may avoid these pitfalls of spring weather.

Each ground-layer gap will have sufficient availabilities of light and soil water, at least during the first growing season, to enable establishment of the new seedlings. Alternatively, specific soil-active herbicides that are safe for pines and that provide residual control of competing vegetation (e.g., hexazinone or imazapyr) can be used specifically to create ground-layer gaps in which to plant longleaf pine seedlings. Herbaceous species susceptible to these herbicides can be established 1–2 years later after soil concentrations of such herbicides have declined sufficiently.

Nitrogen is probably the nutrient most limiting to establishment and growth of new herbaceous and pine seedlings. However, fertilizer should only be applied directly to the planting hole to avoid stimulating growth and potential dominance of other, more aggressive plant species, and the prescribed fertilizer rate should be safe for seedling roots.

Fire is a key ingredient for maintenance of new plantings, because it will suppress or kill nonadapted plant species and allow adapted species to flourish. Field observations from Study II indicated that, by the end of the first or second year after planting, most perennial herbaceous species native to longleaf pine communities can withstand dormant season fires.

Existing Herbaceous Communities within Natural, Uneven-Aged Longleaf Pine Stands

In this forest community, many of the stand structural components already exist, and only enrichment plantings are needed to reestablish the desired native herbaceous species. Understory light availability should be adequate, although not optimum, if the uneven-aged structure of the stand is composed of widely spaced individual trees (spacing of 11 m or greater; Platt et al. 1988a). If light availability is potentially limiting to longleaf pine seedlings and their associates, excess overstory trees should be harvested or girdled (to create wildlife habitat). As discussed previously, tree removals that result in aggregated, rather than dispersed, stand structures provide a higher percentage of growing space with high light availability. Additional canopy layers of hardwoods and shrubs cause further limitations to understory light availability, and their abundance should be reduced significantly by spot treatments that do not endanger desirable herbaceous species, such as injection with non-soil-active herbicides, girdling, or cutting.

Inherent limitations in availability of belowground resources, primarily water and nitrogen, are typical of longleaf pine communities because often they occur on sandy soils with low water-holding capacities and limited nitrogen retention. Because the pools of belowground resources can be in such short supply, full occupancy of belowground resources by overstory pines may occur at stem densities that do not provide complete crown coverage. Therefore, moderate densities of longleaf pine can be highly competitive for belowground resources. Soil water availability in this community also will be limited by understory vegetation. Although this recommendation conflicts with the findings of Palik et al. (1997) and McGuire et al. (2001) in which soil water availability did not vary systematically with overstory density, I hypothesize that if overstory and understory components were manipulated independently in natural stands of longleaf pine, their separate and significant competitive effects on soil water could be detected, as found in the longleaf pine plantations of Study II.

As discussed previously, spot (or banded) treatments can be used to create ground-layer gaps in areas at least 21 m from overstory pines. If advanced regeneration of longleaf pine is needed, mineral soil seedbeds can be created to foster germination and seedling development in years of adequate seed crops. Some of the ground-layer gaps can be located approximately 16 m from potential longleaf pine seed trees to allow pine seeds to germinate in areas of minimal competition and accumulation of needle litter from overstory trees. As discussed previously, root competition from overstory pines can create zones that exclude natural pine regeneration, and areas within 4 m of overstory pines have the potential for intense ground fires capable of killing young pine seedlings.

Reinstatement of prescribed fires in the dormant season or spring is essential for maintaining the vigor of the new plantings and to avoid overtopping by resprouting hardwoods and shrubs.

Existing Herbaceous Communities within Planted, Even-Aged Longleaf Pine Stands

Extensive plantations of longleaf pine have been established on public lands and, as part of the Conservation Reserve Program, on former agricultural land. Often these plantations were established at spacings typical for loblolly pine (e.g., up to 1800 trees ha^{-1}). As mentioned previously, belowground resources may not be adequate to support long-term development of longleaf pine established at close spacings, especially on sandhill sites. Under these conditions, mortality from bark beetles (*Dendroctonus* spp. and *Ips* spp.) is likely, as has been observed recently in nonthinned stands of Study I. To foster long-term health of the forest community, densely stocked plantations can be thinned to wide spacings (5 m or more) prior to engaging in understory restoration. Within the widely spaced stands, gaps 21 m in radius or larger can be established for understory restoration. Conversely, an aggregated stand structure can be created to maintain some areas of higher stocking for timber production while providing open areas for understory restoration.

Because of uniformity of spacing and size of trees, plantations will have regions of restricted light availability even after the preceding stocking guidelines have been followed. To provide adequate light and soil water availability, ground-layer gaps can be established away from overstory trees and on the north side of overstory gaps. Results from Study II suggest that the critical zone of root competition extends less than 10 m away from 13- to 15-year-old longleaf pines, unlike the 18-m zone observed for mature trees (Brockway and Outcalt 1998). Note that restrictions in light availability in longleaf pine plantations are likely to become more intense with decreasing space among trees and increasing site quality.

Based on Study II, reintroduced herbaceous and longleaf pine seedlings will have a wide range of fitness and fecundity responses in the understory of a longleaf pine plantation. Responses are likely to range from increases in vegetative growth at the expense of reproductive performance to gradual declines in response to needle litter accumulations. This diversity of plant responses indicates that a diversity of growing conditions would be needed to maintain a full gamut of species in the understory. Applications of prescribed fire, vegetation management, and pine thinning can be scheduled at appropriate times to create pockets of enhanced resource availability that foster flowering, seed production, and recruitment of new seedlings. In time and given appropriate maintenance treatments, reintroduced plant species are likely to form stable populations that are self-perpetuating.

USDA/Forest Service Disclaimers

The use of trade or firm names in this publication is for reader information and does not imply endorsement by the U.S. Department of Agriculture of any product or service. This publication reports research involving pesticides. It does not contain recommendations for their use, nor does it imply that the uses discussed here have been registered. All uses of pesticides must be registered by appropriate state or federal agencies, or both, before they can be recommended. CAUTION: Pesticides can be injurious to humans, domestic and wild animals, and desirable plants if they are not handled or applied properly. Use all pesticides selectively and carefully. Follow recommended practices for the disposal of surplus pesticides and pesticide containers.

References

Anderson, R.C., and Menges, E.S. 1997. Effects of fire on sandhill herbs: Nutrients, mycorrhizae, and biomass allocation. *Am J Bot* 84:938–948.

Battaglia, M.A., Mou, P., Palik, B., and Mitchell, R.J. 2002. The effect of spatially variable overstory on the understory light environment of an open-canopied longleaf pine forest. *Can J For Res* 32:1984–1991.

Boyer, W.D. 1990. Growing-season burns for control of hardwoods in longleaf pine stands. USDA Forest Service, Southern Forest Experiment Station, Research Paper SO-256.

Boyer, W.D. 1993. Season of burn and hardwood development in young longleaf pine stands. *Proceedings of the Seventh Biennial Southern Silviculture Research Conference* pp. 511–515.

Boyer, W.D. 1995. Responses of groundcover under longleaf pine to biennial seasonal burning and hardwood control. *Proceedings of the Eighth Biennial Southern Silviculture Research Conference* pp. 512–516.

Boyer, W.D., and Miller, J.H. 1994. Effect of burning and brush treatments on nutrient and soil physical properties in young longleaf pine stands. *For Ecol Manage* 70:311–318.

Brewer, J.S., and Platt, W.J. 1994. Effects of fire season and herbivory on reproductive success in a clonal forb, *Pityopsis graminifolia*. *J Ecol* 82:665–675.

Brewer, J.S., Platt, W.J., Glitzenstein, J.S., and Streng, D.R. 1996. Effects of fire-generated gaps on growth and reproduction of golden aster (*Pityopsis graminifolia*). *Bull Torrey Bot Club* 123:295–303.

Brockway, D.G., and Lewis, C.E. 1997. Long-term effects of dormant-season prescribed fire on plant community diversity, structure and productivity in a longleaf pine wiregrass ecosystem. *For Ecol Manage* 96:167–183.

Brockway, D.G., and Outcalt, K.W. 1998. Gap-phase regeneration in longleaf pine wiregrass ecosystems. *For Ecol Manage* 106:125–139.

Brockway, D.G., and Outcalt, K.W. 2000. Restoring longleaf pine wiregrass ecosystems: Hexazinone application enhances effects of prescribed fire. *For Ecol Manage* 137:121–138.

Daniel, T.W., Helms, J.A., and Baker, F.S. 1979. *Principles of Silviculture*, 2nd ed. New York: McGraw–Hill.

Ginn, S.E., Seiler, J.R., Cazell, B.H., and Kreh, R.E. 1991. Physiological and growth responses of eight-year-old loblolly pine stands to thinning. *For Sci* 37:1030–1040.

Ginter, D.L., McLeod, K.W., and Sherrod, C., Jr. 1979. Water stress in longleaf pine induced by litter removal. *For Ecol Manage* 2:13–20.

Glitzenstein, J.S., Platt, W.J., and Streng, D.R. 1995. Effects of fire regime and habitat on tree dynamics in north Florida longleaf pine savannas. *Ecol Monogr* 65:441–476.

Goldberg, D.E. 1990. Components of resource competition in plant communities. In *Perspectives on Plant Competition*, eds. J.B. Grace and D. Tilman, pp. 27–49. New York: Academic Press.

Goldberg, D.E., and Miller, T.E. 1990. Effects of different resource additions on species diversity in an annual plant community. *Ecology* 71:213–225.

Grace, J.B. 1990. On the relationship between plant traits and competitive ability. In *Perspectives on Plant Competition*, eds. J.B. Grace and D. Tilman, pp. 51–65. New York: Academic Press.

Grace, S.L., and Platt, W.J. 1995. Effects of adult tree density and fire on the demography of pregrass stage juvenile longleaf pine (*Pinus palustris* Mill.). *J Ecol* 83:75–86.

Grelen, H.E., and Enghardt, H.G. 1973. Burning and thinning maintain forage in a longleaf pine plantation. *J For* 71:419–425.

Harper, J.L. 1977. *Population Biology of Plants*. New York: Academic Press.

Harrington, T.B., and Edwards, M.B. 1999. Understory vegetation, resource availability, and litterfall responses to pine thinning and woody vegetation control in longleaf pine plantations. *Can J For Res* 29:1055–1064.

Harrington, T.B., Dagley, C.M., and Edwards, M.B. 2003. Above- and belowground competition from longleaf pine plantations limits performance of reintroduced herbaceous species. *For Sci* 49:681–695.

Haywood, J.D. 2000. Mulch and hexazinone herbicide shorten the time longleaf pine seedlings are in the grass stage and increase height growth. *New For* 19:279–290.

Hendricks, J.J., and Boring, L.R. 1992. Litter quality of native herbaceous legumes in a burned pine forest of the Georgia Piedmont. *Can J For Res* 22:2007–2010.

Hendricks, J.J., and Boring, L.R. 1999. N_2-fixation by native herbaceous legumes in burned pine ecosystems of the southeastern United States. *For Ecol Manage* 113:167–177.

Hiers, J.K., Mitchell, R.J., Boring, L.R., Hendricks, J.J., and Wyatt, R. 2003. Legumes native to longleaf pine savannas exhibit capacity for high N_2-fixation rates and negligible impacts due to timing of fire. *New Phytol* 157:327–338.

Kalmbacher, R., and Martin, F. 1996. Shifts in botanical composition of flatwoods range following fertilization. *J Range Manage* 49:530–534.

Korstian, C.F., and Coile, T.S. 1938. Plant competition in forest stands. Duke University School of Forestry Bulletin 3, Durham, NC.

McCune, B. 1986. Root competition in a low-elevation grand fir forest in Montana: A trenching experiment. *Northwest Sci* 60:52–54.

McGuire, J.P., Mitchell, R.J., Moser, E.B., Pecot, S.D., Gjerstad, D.H., and Hedman, C.W. 2001. Gaps in a gappy forest: Plant resources, longleaf pine regeneration, and understory response to tree removal in longleaf pine savannas. *Can J For Res* 31:765–778.

Means, D.B. 1997. Wiregrass restoration: Probable shading effects in a slash pine plantation. *Restor Manage Notes* 15:52–55.

Monk, C.D., and Gabrielson, F.C., Jr. 1985. Effects of shade, litter, and root competition on old-field vegetation in South Carolina. *Bull Torrey Bot Club* 112:383–392.

Mueller-Dombois, D., and Ellenberg, H. 1974. *Aims and Methods of Vegetation Ecology*. New York: John Wiley & Sons.

Oliver, C.D., and Larson, B.C. 1996. *Forest Stand Dynamics. Update edition.* New York: John Wiley & Sons.

Palik, B.J., Mitchell, R.J., Houseal, G., and Pederson, N. 1997. Effects of canopy structure on resource availability and seedling responses in a longleaf pine ecosystem. *Can J For Res* 27:1458–1464.

Palik, B., Mitchell, R.J., Pecot, S., Battaglia, M., and Mou, P. 2003. Spatial distribution of overstory retention influences resources and growth of longleaf pine seedlings. *Ecol Appl* 13:674–686.

Pessin, L.J. 1938. The effect of vegetation on the growth of longleaf pine seedlings. *Ecol Monogr* 8:115–149.

Platt, W.J., Evans, G.W., and Rathbun, S.L. 1988a. The population dynamics of a long-lived conifer (*Pinus palustris*). *Am Nat* 131:491–525.

Platt, W.J., Evans, G.W., and Davis, M.M. 1988b. Effects of fire season on flowering of forbs and shrubs in longleaf pine forests. *Oecologia* 76:353–363.

Radosevich, S.R., and Holt, J.S. 1984. *Weed Ecology*. New York: John Wiley & Sons.

Richardson, D.R., and Williamson, G.B. 1988. Allelopathic effects of shrubs of the sand pine scrub on pines and grasses of the Sandhills. *For Sci* 34:592–605.

Riegel, G.M., and Miller, R.F. 1991. Understory vegetation response to increasing water and nitrogen levels in a *Pinus ponderosa* forest in northeastern Oregon. *Northwest Sci* 65:10–15.

Riegel, G.M., Miller, R.F., and Krueger, W.C. 1992. Competition for resources between understory vegetation and overstory *Pinus ponderosa* in northeastern Oregon. *Ecol Appl* 2:71–85.

Rogers, V.A. 1990. Soil survey of Savannah River Plant area, parts of Aiken, Barnwell, and Allendale Counties, South Carolina. USDA Soil Conservation Service, Washington, DC.

Shelton, M.G. 1995. Effects of the amount and composition of the forest floor on emergence and early establishment of loblolly pine seedlings. *Can J For Res* 25:480–486.

Sokal, R.R., and Rohlf, F.J. 1981. *Biometry*, 2nd ed. New York: W.H. Freeman and Company.

Spurr, S.H., and Barnes, B.V. 1992. *Forest Ecology*, 3rd ed. Malabar, FL: Krieger Publishing Co.

Streng, D.R., Glitzenstein, J.S., and Platt, W.J. 1993. Evaluating effects of season of burn in longleaf pine forests: A critical literature review and some results from an ongoing long-term study. *Proceedings 18th Tall Timbers Fire Ecology Conference*, pp. 227–263.

Sydes, C., and Grime, J.P. 1981a. Effects of tree leaf litter on herbaceous vegetation in deciduous woodland. I. Field investigations. *J Ecol* 69:237–248.

Sydes, C., and Grime, J.P. 1981b. Effects of tree leaf litter on herbaceous vegetation in deciduous woodland. II. An experimental investigation. *J Ecol* 69:249–262.

Toumey, J.W., and Kienholz, R. 1931 (cited in Spurr and Barnes 1992). Trenched plots under forest canopies. Yale University School of Forestry Bulletin 30.

Van Lear, D.H., and Waldrop, T.A. 1991. Prescribed burning for regeneration. In *Forest Regeneration Manual*, eds. M. L. Duryea and P.M. Dougherty, pp. 235–250. Dordrecht, The Netherlands: Kluwer Academic Publishers.

Wilson, C.A., Mitchell, R.J., Hendricks, J.J., and Boring, L.R. 1999. Patterns and controls of ecosystem function in longleaf pine – wiregrass savannas. II. Nitrogen dynamics. *Can J For Res* 29:752–760.

Wolters, G.L. 1973. Southern pine overstories influence herbage quality. *J Range Manage* 26:423–426.

Wolters, G.L. 1981. Timber thinning and prescribed burning as methods to increase herbage on grazed and protected longleaf pine ranges. *J Range Manage* 34:494–497.

Chapter 6

Vertebrate Faunal Diversity of Longleaf Pine Ecosystems

D. Bruce Means

Introduction

In the southeastern U.S. Coastal Plain, all landscapes can be conceptually divided into aquatic, wetland, and upland habitats. Aquatic and wetland habitats account for a substantial percentage of the Coastal Plain, especially near the coast and in Louisiana and Florida, but overall from southeastern Virginia to east Texas, uplands constitute the largest proportion of the terrain. It has been estimated that, upon the arrival of Europeans and Africans in North America, upland ecosystems dominated by a single tree species, longleaf pine (*Pinus palustris*), accounted for about 60% of the Coastal Plain landscape (Ware et al. 1993). In other words, longleaf pine ecosystems were the principal ecosystems in a belt of land stretching about 2000 miles along the southeastern margin of the North American continent. Most of the range of longleaf pine was in the Coastal Plain, a gently undulating, low-elevation (0–200 m), sedimentary landform with soils developed from sandy clays (clayhills and some flatwoods) or pure sand (sandhills and flatwoods), sometimes underlain by limestone (Brown et al. 1990; Martin and Boyce 1993). Longleaf pine ecosystems and their vertebrate faunas are the focus of this chapter.

Before describing longleaf pine ecosystems, something must be said about how we know what we think we know about them. A good deal of knowledge is available about vertebrate species in longleaf ecosystems from numerous autecological studies. However, knowledge about longleaf pine communities as functioning ecosystems has been difficult to obtain because in the twentieth century, when plant ecology became a scientific discipline, most of the longleaf pine habitat had already disappeared, and that which remains had been dramatically impacted by man. As recently as 13 years ago, it was calculated that only 3% of the original extent of longleaf pine ecosystems remained (Ware et al. 1993). A few forestry statistics tell the story. In Florida, longleaf pine forests declined from 30,756 km^2 (7.6 million acres) in 1936 to only 3845 km^2 (0.95 million acres) in 1989, an 88% decrease (Cerulean 1991). In southeastern Georgia, the longleaf pine forest declined 36% between 1981 and 1988 (Johnson 1988). The first inventory of what was left of old-growth longleaf reported a meager 9975 acres out of an estimated 85 million acres, or about 0.01% (Means 1996). Only 5 years later, those valuable acres had dwindled further by 43% (Varner and Kush 2001). Our knowledge about plant species

D. Bruce Means • Coastal Plains Institute and Land Conservancy, Tallahassee, Florida 32303.

composition, structure, community function, and even geographical distribution of its components, therefore, has been pieced together from historical records and recent research in the few longleaf habitats that remain. Rather than possessing direct observational evidence for such information as fire frequency, stand density, groundcover species composition, and other properties, we have had to infer them from brief and incomplete historical accounts, anecdotal information, and experimental manipulations of small remaining tracts whose naturally functioning ecological processes have been interrupted for possibly more than several centuries. In discussions that follow about the ecology of longleaf pine ecosystems, I will sometimes use the past tense to refer to truly pristine, unaltered longleaf pine ecosystems, because none are left and probably haven't been since about the 1920s. What remains (less than a million acres) is a small amount of second-growth longleaf pine ecosystems mostly on publicly owned lands.

Longleaf pine ecosystems are not well described as forest because they more readily fit the definition of "savannas" (Vogl 1973), which are clumps of trees or sparsely distributed trees not forming continuous canopies, and groundcover dominated by abundant heliophilic herbaceous vegetation, particularly warm-season grasses (Werner 1991; McPherson 1997; Scholes and Archer 1997; Platt 1999). Historical accounts of early explorers, travelers, and botanists reveal that the rich grassland or prairie ground covers of pine savannas stretched across the Coastal Plain and Piedmont from the Atlantic to the Gulf of Mexico and continued beyond the limits of longleaf pine in east Texas into the tallgrass prairies of the Great Plains (reviews in Vogl 1973; Platt 1999; Frost this volume). To understand the vertebrate faunas of longleaf pine savannas, it is crucial to appreciate that longleaf pine ecosystems are forests to only a handful of species, but that for most vertebrates, they are, or were, grasslands.

This is important because terrestrial vertebrates (amphibians, reptiles, mammals, birds) respond to vegetation in specific ways. In closed-canopy forests where most of the primary productivity takes place high in the trees, vertebrates are volant, scansorial, and arboreal. Herbivores that feed far above ground in the branches are limited in body size. Larger, ground-dwelling mammalian herbivores are mainly browsers. Terrestrial amphibians and reptiles must feed on detrital invertebrates and are therefore specialized to burrow in litter and feed upon a detritivore-based food web. In open-canopied savannas and prairies, by contrast, a large percentage of ecosystem primary productivity takes place at ground level, supporting a diverse herbivore fauna characterized by large grazing mammals and even reptiles (tortoises). Insectivorous amphibians and other reptiles can feed on a contiguous supply of arthropods on the ground and in the low herbaceous vegetation. Vertebrates can be so specialized for the types of habitats in which they live that paleontologists regularly deduce the general types of vegetation that existed in the vicinity of a fossil site by whether the fossil fauna is dominated by animals whose bones and teeth give evidence that they were browsers, grazers, carnivores, walked, burrowed, or climbed. Supple-bodied, climbing snakes such as boas, for example, are indicative of forest habitats whereas elongate, fast-moving snakes such as racers and whipsnakes suggest open, grassland habitats. That longleaf pine savannas were open-canopied grasslands or prairies as opposed to closed-canopy hardwoods or closed-canopy coniferous forests, therefore, had a great deal of influence on both the kinds and diversity of vertebrates that lived in these ecosystems.

The ground cover of Coastal Plain longleaf pine savannas contained a diverse mixture of grasses, forbs, and shrubs (Walker and Peet 1983; Christensen 1988; Bridges and Orzell 1989; Harcombe et al. 1993; Peet and Allard 1993; Peet this volume). Platt (1999) listed the most dominant grasses, most of which belong to warm-season genera: *Andropogon, Aristida, Ctenium, Schizachyrium, Sorghastrum, Sporobolus*. Hundreds of forb species occurred in longleaf pine savannas, with Asteraceae, Lamiaceae, Fabaceae, Liliaceae, Scrophulariaceae, Ericaceae, Orchidaceae, and Xyridaceae occurring as dominants (Christensen 1988;

Platt et al. 1988; Hardin and White 1989; Walker 1993). Walker (1993) proposed that endemic species were distributed among four different regions within the overall range of longleaf pine: south Atlantic Coastal Plain, peninsular Florida, and the east and west Gulf Coastal Plains. About 200 of these species are rare (Hardin and White 1989; Walker 1993). Shrubs and subshrubs were common in the ground cover of pine savannas, including sprouts of trees, especially oaks (*Quercus laevis, Q. margaretta, Q. incana, Q. marilandica*), as well as runner oaks (*Q. minima, Q. pumila*), sumac (*Rhus*), ericaceous shrubs (*Vaccinium, Gaylussacia*), palms (*Serenoa, Sabal*), wax myrtle (*Myrica*), and hollies (*Ilex*) (Peet and Allard 1993; Olson and Platt 1995).

Peet and Allard (1993) found that the numbers of plant species present in individual samples in longleaf savannas across the Coastal Plain were among the highest reported for the temperate Western Hemisphere. Values for vascular plants ranged from 140/1000 m^2 and 90+/10 m^2 to counts of more than 40 species/m^2. Several hundred plant species may occur within an area smaller than a hectare (Platt et al. 1988; Peet and Allard 1993; Stout and Marion 1993). This species richness, the highest reported from North American savannas (Peet and Allard 1993), rivals that of tropical forests, but on a two-dimensional, ground-level, square-meter spatial scale. Moreover, diversity is also high on a regional scale. Peet and Allard (1993) recognized 23 different longleaf pine savanna communities across the range of longleaf pine east of the Mississippi River.

With such high plant species richness in such a large natural area as the Coastal Plain, one would expect vertebrate species richness to be commensurately high. And indeed, it is for amphibians and reptiles. The Coastal Plain of the southeastern United States has the highest species richness of turtles, frogs, and snakes in the United States and Canada (Kiester 1971), as well as a large salamander fauna (Means 2000). Only lizards, which flourish in arid environments, are not superabundant. On the other hand, bird species richness is not exceptionally great (Stout and Marion 1993), and the mammalian fauna is exceedingly impoverished (Layne 1974). Below, I review the scant literature on vertebrate faunas of longleaf pine savannas, then discuss the ecological roles of each vertebrate species that is a specialist adapted for living in longleaf pine savannas. At the end of this chapter, I discuss why the mammalian fauna is depauperate and speculate on what effect the impoverished mammalian fauna may have had on longleaf pine savannas.

Only a few overviews of the vertebrate faunas of longleaf pine savannas have been published. The amphibians and reptiles of Georgia and Florida longleaf pine sandhills communities were discussed by Landers and Speake (1980) and Campbell and Christman (1982), respectively. Means and Campbell (1982) addressed the effects of fire on amphibians and reptiles worldwide, but most of their examples came from longleaf pine savannas. The most thorough reviews of amphibians and reptiles were done by Stout and Marion (1993) and Guyer and Bailey (1993). A shorter review of amphibians and reptiles was done by Dodd (1995b).

Bird communities of Southeastern pinelands were reviewed by Jackson (1988), but he cautioned that no data were available on species composition or population abundances of birds in old-growth longleaf pine savannas. Stout and Marion (1993) discussed bird and mammal community assemblages of pine flatwoods and longleaf pine-turkey oak sandhills. Engstrom (1993) prepared the most comprehensive review of mammals and birds of longleaf pine savannas, recognizing those species that characteristically live but are not specialists in longleaf savannas and species that are largely sympatric with longleaf savannas and presumably specialists in it. Echternacht and Harris (1993) reviewed the vertebrate fauna of the entire southeastern United States but did not mention habitat associations.

The four vertebrate classes possess entirely different morphologies, physiologies, vagilities, life history characteristics, and ecologies, so a review of all the vertebrate species that live in longleaf pine savannas is best done by discussing each class separately. We must define, however, criteria for how species were chosen for discussion. Some species in each class may be widely ubiquitous across many different habitat types throughout the Coastal Plain, including living in longleaf pine savannas, or have geographic distributions that overlap that of longleaf pine savannas, although the species never lives in them. Other species may occupy longleaf pine savannas in those parts of their much wider geographic distributions in North America. And some species are strict endemics that live exclusively in longleaf pine savannas. Guyer and Bailey (1993) and Engstrom (1993) defined two groups of longleaf pine savanna vertebrates: (1) species that are characteristic, that is, they occur in longleaf savannas but whose distributional limits (both ecological and geographical) are not confined to the distribution of longleaf pine (i.e., residents) and (2) species whose limits are confined within and/or closely associated with those of longleaf pine (i.e., specialists). Below I adopt this classification scheme with slight modifications and then briefly discuss details of longleaf pine savanna specialist and resident species for each vertebrate class. Throughout the text, common names are used; corresponding scientific names are listed in Tables 1–5.

Amphibians

Most amphibians have a biphasic life cycle in which their natal stages (eggs, larvae) live in water, and they then take up permanent residence on land as adults after metamorphosis (Duellman and Trueb 1986). Of the 9 salamander and 26 frog species that live in longleaf pine savannas, 6 salamanders and 11 frogs have adults that are specialists in these ecosystems (Table 1). All require lentic situations—small ponds, often with short hydroperiods—in their natal stages (Moler and Franz 1988). Many others, stream-dwelling species, also occur in the Coastal Plain but live in floodplains and swamps as adults. Much is known about the breeding biology and larval life in ponds of longleaf pine savanna amphibians, because calling males of frogs are easy to locate, larvae can be seined or dipnetted, and metamorphosed individuals can be intercepted by drift fences as populations move in and out of natal ponds (Dodd 1992; Semlitsch et al. 1996). However, in every case among the 35 species of amphibians that live in longleaf pine savannas as adults, little is known about their terrestrial lives, because field studies of them are difficult to conduct. Both juveniles and adults are usually fossorial or semifossorial, do not vocalize away from water, and are small and cryptic morphologically and behaviorally.

Throughout the Coastal Plain, there are numerous kinds of ponds in the longleaf pine landscape, including 500,000 Carolina Bays (Prouty 1952; Savage 1982; Sharitz and Gibbons 1982), 7000 Citronelle ponds (Folkerts 1997), and probably a similar number of limestone dissolutional basins in flatwoods and sandhills. Ponds may be forested with cypress (*Taxodium*), tupelos (*Nyssa*), myrtle-leaved holly (*Ilex myrtifolia*), pop-ash (*Fraxinus carolina*), titi species (Cyrillaceae), or other shrubs, or ponds may be open and dominated by emergent grasses (*Panicum*) and sedges (*Cladium*). They may have very short hydroperiods of only a few months in winter or be nearly permanent. What seems most important to the 17 species of longleaf pine amphibian specialists is the absence of fish, so that the preferred ponds in which they lay their eggs have short hydroperiods (Moler and Franz 1988). To illustrate the importance of the fishless, temporary nature of ponds, some 26 species of anurans are known to characteristically breed in those ponds found in longleaf pine savannas of the Coastal Plain (Table 1) compared to only 8 species that breed in numbers (bullfrog, bronze frog, pig frog, southern leopard frog, carpenter frog, green treefrog, southern cricket frog, southern toad) in the extensive permanent lakes and marshes of the Coastal Plain.

Two breeding seasons are present in the Coastal Plain range of longleaf pine savannas, because this region is characterized by a distinctive precipitation regime of summer thundershowers (June–September) and rains brought on by the passage of winter cold fronts (December–March). The rainy seasons are typically separated by autumn

TABLE 1. Characteristic resident amphibians of longleaf pine savannas.[a,b]

Salamanders			
1. *Ambystoma cingulatum*	Flatwoods salamander	s	w
2. *Ambystoma mabeei*	Mabee's salamander	s	w
3. *Ambystoma talpoideum*	Mole salamander	b	w
4. *Ambystoma tigrinum*	Tiger salamander	s	w
5. *Notophthalmus perstriatus*	Striped newt	s	w
6. *Notophthalmus viridescens*	Eastern newt	b	w
7. *Eurycea quadridigitata*	Dwarf salamander	s	w
8. *Eurycea* n. sp.	Bog salamander	s	w
9. *Plethodon glutinosus* "complex"	Slimy salamander	c	m
Frogs			
1. *Bufo americanus*	American toad	a	w
2. *Bufo quercicus*	Oak toad	s	m
3. *Bufo terrestris*	Southern toad	b	m
4. *Bufo valliceps*	Gulf coast toad	a	m
5. *Bufo woodhousii*	Woodhouse's toad	a	m
6. *Acris crepitans*	Northern cricket frog	c	m
7. *Acris gryllus*	Southern cricket frog	d	m
8. *Hyla andersonii*	Pine barrens treefrog	c	m
9. *Hyla chrysoscelis*	Cope's gray treefrog	c	m
10. *Hyla cinerea*	Green treefrog	d	m
11. *Hyla femoralis*	Pinewoods treefrog	s	m
12. *Hyla gratiosa*	Barking treefrog	s	m
13. *Hyla squirella*	Squirrel treefrog	s	m
14. *Pseudacris brachyphona*	Mountain chorus frog	a	w
15. *Pseudacris brimleyi*	Brimley's chorus frog	s	w
16. *Pseudacris crucifer*	Spring peeper	c	w
17. *Pseudacris nigrita*	Southern chorus frog	s	w
18. *Pseudacris ocularis*	Little grass frog	s	m
19. *Pseudacris ornata*	Ornate chorus frog	s	w
20. *Pseudacris triseriata*	Western chorus frog	c	m
21. *Gastrophryne carolinensis*	Eastern narrowmouth toad	b	m
22. *Gastrophryne olivacea*	Great plains narrowmouth toad	a	m
23. *Rana areolata*	Crawfish frog	s	w
24. *Rana capito*	Gopher frog	s	w
25. *Rana sphenocephala*	Southern leopard frog	d	w
26. *Scaphiopus holbrookii*	Eastern spadefoot	s	w

[a] Adapted, with modifications, from Moler and Franz (1988), Stout and Marion (1993), Guyer and Bailey (1993), Dodd (1995b).
[b] a = small overlap with longleaf pine savanna range; b = ubiquitous among Coastal Plain habitats; c = visitor from other habitats; d = aquatic; s = endemic in longleaf pine savanna, a specialist; w = winter breeder; m = summer breeder.

(October–November) and spring (April–May) droughts (Wolfe et al. 1988; Martin and Boyce 1993).

Longleaf Pine Amphibian Specialists

All six salamander specialists in longleaf pine savannas are winter breeders. Flatwoods (Fig. 1a) and Mabee's salamanders breed in small, fishless ponds in longleaf pine savannas that usually have a canopy of pond cypress (*Taxodium ascendans*) and blackgum or black tupelo (*Nyssa sylvatica*). Adult flatwoods salamanders migrate to breeding ponds in October and November and then move out of ponds in late November and December (Palis and Means 2005). The flatwoods salamander is a true endemic, highly specialized for living in longleaf pine savanna, but only in flatwoods, not sandhills (Goin 1950; Means et al. 1996). Whether it is something about the quality of the terrestrial habitat of the adults or the natal ponds that confines the flatwoods salamander to

FIGURE 1. (a) Flatwoods salamander, *Ambystoma cingulatum*, a specialist in longleaf pine flatwoods. (b) Striped newt, *Notophthalmus perstriatus*, a specialist in longleaf pine sandhills, terrestrial stage called "eft."

flatwoods is unknown, but a similar, although reverse, proclivity exists in the striped newt (Fig. 1b), another strict longleaf pine savanna endemic species that is mainly found in sandhills rather than flatwoods (Franz and Smith 1999; Johnson 2002; Dodd et al. 2005).

Observations of migrating flatwoods salamander adults found in the field (Means et al. 1996), and a study of four individuals leaving breeding ponds that were monitored with cobalt-60 radioactive tags (Ashton and Ashton 2005), indicate that the species lives its adult life in longleaf pine savanna several hundred meters away from natal ponds. Over a 7- to 9-day period, Ashton and Ashton (2005) recorded moves of 1566, 1702, and 1708 m in three adults emigrating from the breeding pond, although not necessarily in a straight-line direction. Likewise, the smaller striped newt emigrates more than 500 m from its natal ponds into sandhill longleaf pine-turkey oak vegetation (Johnson 2003). Striped newts have lived 12 years in captivity (Snider and Bowler 1992; Linda LaClaire, personal communication). No studies of the upland habitat of Mabee's salamander have been published, although the kinds of ponds it breeds in (Mosimann and Rabb 1948; Hardy 1969; Mitchell and Hedges 1980), its geographic range (Petranka 1998), and morphological similarity to the flatwoods salamander (Hardy and Olmon 1974) suggest that it, too, is a longleaf pine specialist as an adult. Early accounts of the flatwoods salamander described the adult habitat as slash pine flatwoods (Goin 1950; Martof 1968; Conant and Collins 1998), but Means et al. (1996) and Palis (1996) demonstrated that the primary adult habitat is longleaf pine savannas.

During autumn breeding migrations, the flatwoods salamander has been observed moving to breeding ponds out of longleaf pine savanna (Means et al. 1996), but since few adults have been taken outside the breeding season, microhabitat utilization is virtually unknown except that the species has been reported using crayfish burrows when in the breeding habitat (Neill 1951b; Ashton 1992; Ashton and Ashton 2005). Observations of tagged adults during emigration from the breeding pond indicated tagged salamanders were never found anywhere inside a large slash pine plantation except in pine duff up to 2 cm deep even when rotting logs were present. Three post-breeding flatwoods salamanders completed emigration in 7–9 days and occupied underground burrows of undetermined origin in a mesic area at the edge of the pine plantation (Ashton and Ashton 2005). The mesic area was vegetated with Florida maple (*Acer barbatum*), sweet gum (*Liquidambar styraciflua*), black gum, gallberry (*Ilex glabra*), wax myrtle (*Myrica cerifera*), and grape (*Vitis rotundifolia*). No reports of microhabitat use are available during the terrestrial lives of the striped newt, common newt, or Mabee's salamander besides mention of an occasional individual found under logs (Petranka 1998).

Here, I classify the eastern tiger salamander as a longleaf pine savanna specialist. It has been documented living in old-growth longleaf pine clayhills on the Wade Longleaf Forest in Grady County, GA (Means and Campbell 1982). Little remains of clayhills longleaf pine savanna in first- or second-growth character (Means 1996), but the eastern tiger salamander sometimes is found in clayhills ruderal habitats such as farmland, pastures, and oldfield successional stands of mixed hardwoods and pines where old-growth longleaf pine savanna once grew (Means and Campbell 1982; Travis 1992). It is found in sandhills and xeric hammock habitats in peninsular Florida (Kevin Enge personal communication), but in the Coastal Plain portion of its large geographic distribution, it is not known from southern temperate (beech-magnolia) hardwood forests (as defined in Platt and Schwartz 1990) or sand pine (*Pinus clausa*) and coastal scrubs, the only other upland habitats in the Coastal Plain presettlement landscape. Like many ambystomatid salamanders, it is rarely found outside the breeding season, a time when individuals are seen crossing roads or scooped from swimming pools during winter migrations (November–January). This large salamander probably occupies the burrows or subterranean cavities of moles, rodents, and other animals, as well as excavates its own tunnels (Semlitsch 1983a,b). Studies of its upland habitat use and behavior have not been conducted, not even a determination of whether it comes to the ground surface at night to forage.

Larvae of the dwarf salamander (*Eurycea quadridigitata*) have been found in the field from late January to March following breeding migrations to ponds and oviposition in ponds in October and November (Brimley 1923; Harrison 1973; Semlitsch and McMillan 1980). Recently, populations in the Piedmont and upper Coastal Plain that previously were referred to this species have been found to be a distinct species, *E. chamberlaini*, that breeds in streams and whose adults probably live in hardwood forests (Harrison and Guttman 2003). The remaining populations that occur in flatwoods and sandhills of the lower Coastal Plain from east Texas to North Carolina consist of three distinct species (Richard Highton and Carla Hass personal communication). At least two of these species breed in temporary ponds, including *E. quadridigitata*, and another undescribed species breeds in seepage bogs and wet flats (Enge 2002; D. B. Means and J. Jensen unpublished). Adults of these three species live in, or spend time in, longleaf savannas, especially mesic or wet flatwoods. Drift fence studies of longleaf pine savanna amphibian faunas have reported captures of one or two of these species (Enge 1997, 2002; Means and Jensen unpublished; David Printiss personal communication). Very little has been published about the ecology of these species.

No less than 11 species of frogs are longleaf pine specialists (Table 1). Five of these are winter breeders—southern chorus frog, ornate chorus frog, crawfish frog, gopher frog (Fig. 2), eastern spadefoot—and the remaining

FIGURE 2. (a) Gopher frog, *Rana capito*, a specialist in longleaf pine sandhills. (b) Pine barrens treefrog, *Hyla andersonii*, dependent upon hillside seepage bogs maintained by fires sweeping downhill from longleaf pine savannas

six species breed in summer. Of the winter breeders, little is known of the habitats of the adult southern chorus frog, but drift fence studies of the ornate chorus frog clearly indicate that this species migrates long distances into all types of longleaf pine savannas, including those in flatwoods, clayhills, and sandhills (Means and Campbell 1982; Brown and Means 1984). In fact, the ornate chorus frog actively burrows forward using its robust arms and sometimes backward using its hind legs (Brown and Means 1984). Neill (1958) found a few ornate chorus frogs in sand while probing at the bases of wiregrass clumps. In a study of the use by both species of chorus frogs of Carolina Bay ponds in South Carolina over a 4-year period, Caldwell (1987) concluded that both were short-lived as adults, because juveniles mature early and disperse to nearby ponds when ponds are abundant in most years. She did not gather information about the habitat use of metamorphosed juveniles and adults, however.

Three other winter-breeding frogs (crawfish frog, gopher frog, eastern spadefoot) have adults that are long-lived (for the latter two species, 9 and 12 years in captivity, respectively, Snider and Bowler 1992) and breed only after exceptional rainfall bouts (Hansen 1958; Palis 1998; Bailey and Means 2004). Their larvae develop rapidly (Semlitsch et al. 1995; Palis 1998; Richter 1998; Means unpublished), and juvenile recruitment can be very high (Greenberg 2001; Means unpublished). Breeding events are less regular in these species and can occur over a longer span of time, mostly in winter from January through March but occasionally at any time of the year (Franz 1991; Semlitsch et al. 1995; Franz and Smith 1999; Means and Means 2005). Dispersal of juveniles into longleaf pine savannas and adult use of longleaf pine savanna have been documented for all three species. Richter et al. (2001) showed that adult crawfish frogs moved relatively short distances (less than 300 m) out of a pond surrounded by longleaf pine savanna in southern Mississippi. However, the land approximately 200 m north of the pond was a privately managed pine plantation, so further movements may have been constrained by a recent clearcut. After 2 days, frogs became stationary in underground retreats associated with stump holes, root mounds of fallen trees, or mammal burrows (Richter et al. 2001).

The use of gopher tortoise burrows by the gopher frog has been known for a long time, ergo the common name of the species (Test 1893). Carr (1940) reported that the gopher frog utilizes burrows of animals other than the gopher tortoise: cotton mouse (*Peromyscus polionotus*), stumpholes, tip-up mounds, and other subterranean cavities. Dispersal distances of up to 200 m from breeding ponds have been recorded for the gopher frog (Means and Means 2005). Gopher and crawfish frogs are found almost exclusively in sandhills and rarely in flatwoods (Wright 1932; Enge 1997). This would seem to be due to the dependency of adults on the friable soils of sandhills and the availability in sandhills of suitable subterranean refugia such as burrows of the gopher tortoise, which is itself restricted to well-drained soils vegetated with longleaf pine-oak, xeric hammocks, sand pine-oak ridges, and ruderal successional vegetations that have replaced these vegetations (Auffenberg and Franz 1982). The sandy ridges and slopes of longleaf pine clayhills vegetation support populations of the gopher tortoise but apparently not the gopher frog (Means and Campbell 1982; Means and Means 1998).

Summer-breeding frogs that are longleaf pine savanna specialists include little grass frog, oak toad, and pinewoods, barking, and squirrel treefrogs. All of these species can be heard calling from the same temporary ponds in March–August (Wright 1932; Wright and Wright 1949). Adults of each of these species have been reported in longleaf savannas from miscellaneous observations (Carr 1940; Wright and Wright 1949) and drift fence studies (Enge 1997), but few special investigations of feeding habits, nocturnal behavior, or microhabitat utilization have been conducted. The large barking treefrog has been found during daytime in terrestrial burrows in sand at the base of wiregrass clusters, some nestled in concavities and others in burrows 5–8 cm deep (Neill 1952). How much of its activity is restricted to the ground versus trees and shrubs is unknown, but at least some of its activity is arboreal (Neill 1958). The pinewoods treefrog is found throughout longleaf pine savannas and often occurs in crevices in the bark of longleaf pine trees (Means personal observation). Squirrel treefrogs are common in longleaf pine ground cover, especially when it contains a woody component of saw palmetto, turkey oak, or gallberry. The tiny oak toad forages on the ground and primarily eats ants (Punzo 1995). Oak toads were abundant in old-growth longleaf pine in clayhills but not in nearby shortleaf/loblolly pine oldfield successional forest on the same soil type (Means and Campbell 1982), suggesting that the oak toad is highly specialized for living in wiregrass ground cover or is unable to recolonize old longleaf pine sites that have been abandoned after agriculture. Very little is known about the adult habitat of the tiniest frog in the United States, the little grass frog. From March to August, it can be heard calling throughout pine flatwoods.

Other Longleaf Pine Amphibians

Three other salamanders and 15 frogs are residents that commonly occur but are not exclusively found in longleaf pine savannas. The slimy salamander is occasionally found in rotting logs in flatwoods and clayhills, especially where the longleaf pine savanna transitions into the southern temperate hardwood forest on slopes and moist bottomlands (listed by Guyer and Bailey 1993). It is uncommon in longleaf pine ecosystems and probably should not be considered a resident species. The mole salamander breeds in a wide diversity of pond habitats, including forested as well as open-canopied, grassy ponds in flatwoods, clayhills, and sandhills (Shoop 1960; Semlitsch et al. 1981). It breeds in farm ponds and ponds with or without fish and commonly breeds in the same ponds as most of the longleaf pine savanna specialists, both in sandhills and in flatwoods (Means and Means 2005). Adults have been captured in drift fence studies in bottomland and floodplain hardwood forests (Enge 1997), as well as in longleaf pine savannas, and they have been taken in drift fences in longleaf sandhills 200 m from breeding ponds (Means and Means 2005). The eastern newt breeds in an even greater diversity of aquatic habitats, including marshy streams, canals, lakes with fish, and temporary ponds (Petranka 1998). Its ubiquity probably is due to its noxious skin secretions, which deter predation. Metamorphs and adults of populations that utilize ponds in flatwoods and sandhills spend the terrestrial phase of their lives in longleaf pine savannas, presumably very similarly to the striped newt. Both species can be found using the same breeding ponds, as well as migrating into the same longleaf pine sandhills more than 200 m away from the pond (Means and Means 2005).

The 15 frogs listed as residents but not specialists can be grouped into four categories: (1) those whose large ranges just enter the distribution of longleaf pine (American toad, gulf coast toad, mountain chorus frog, Brimley's chorus frog); (2) ubiquitous over a range of habitats including longleaf pine savanna (southern toad, eastern narrowmouth toad); (3) visitors that are more common in other habitats (woodhouse's toad, northern cricket frog, pine barrens treefrog, Cope's gray treefrog, spring peeper, western chorus frog); and (4) more-or-less aquatic species (cricket frog, green treefrog, southern leopard frog). All of these species can be found in longleaf pine savannas as adults while crossing through these habitats to reach breeding ponds. Some (southern toad, eastern narrowmouth toad) are as abundant in longleaf pine savannas as the specialist frogs and might be classified as specialists were they not also found in other habitats. Others are clearly occasional visitors (all frogs in category 3, southern cricket frog, green treefrog). The southern leopard frog is an interesting case because it probably breeds in more ponds in the Coastal Plain landscape than any other frog, produces a huge annual crop of metamorphs, and is highly obvious during rainy nights dispersing through the uplands in long jumps. Strangely, the proportion of time the adult leopard frog spends away from the edges of ponds is unknown, but the evidence from drift fence studies indicates a substantial amount (Enge 1997).

Amphibians probably contribute more to the ecology of longleaf pine savannas than any other group of vertebrates. Huge numbers of metamorphosed salamanders and frogs have been recorded emerging from natal ponds and migrating into adjacent uplands. Over a 16-year period, 216,251 metamorphs of five species of salamanders and eight species of frogs emerged from Rainbow Bay, a 1-ha pond with maximum depth of about 1 m in Barnwell County, SC (Semlitsch et al. 1996). From a very small, 0.02-ha temporary pond in Leon County, FL, metamorphs of 5895 mole salamanders, 49,664 eastern spadefoots, 3154 southern leopard frogs, and 2235 gopher frogs emerged as a result of single breeding events during the study (Means 2001; Means unpublished). From a large 1.2-ha pond in Santa Rosa County, FL, about 150 female gopher frogs deposited approximately 322,660 eggs during one winter's breeding season (Palis 1998). The percentage that survived to metamorphosis was not studied, but if even 10% survived, the number of recruits into the adjacent longleaf pine sandhills would have been substantial. Over a 2-year period, Johnson (2002) captured 5731 recently transformed larvae (efts and pedomorphs) of the striped newt emigrating from a small pond in Putnam County, FL. Pearson (1955) calculated densities of between 1000 and 1250 eastern spadefoots per hectare that occurred on his longleaf pine sandhill study area in north-central Florida. And during a 5-year drought, Dodd (1992) captured a total of 5740 adult eastern narrowmouth toads moving in and out of a 0.16-ha temporary pond in Putnam County, FL, with only one successful reproduction yielding metamorphs. Interestingly, there was little variation in numbers trapped in each of the 5 years in spite of the low rainfall, which would indicate that a substantial population of adults can survive in longleaf pine sandhills even during long-term drought (Dodd 1995a). These data on only a few species indicate that amphibian biomass in longleaf pine savannas is on the order of magnitude of that reported for salamanders in a New Hampshire woodland (Burton and Likens 1975).

Amphibians make three major contributions to longleaf pine savanna ecology. First, nutrients acquired in natal ponds are transported in body tissues into the rather nutrient-poor uplands (Platt 1999) where, upon death or ingestion by predators, the nutrients eventually wind up in the savanna soils. Second, all juvenile and adult amphibians are carnivores that prey upon invertebrates and even small vertebrates (Wright 1932; Petranka 1998), so that by their sheer numbers, they are important predators in the longleaf pine savanna food webs. Third, the high numbers of amphibians that annually take up life in longleaf pine savannas provide a large prey base for all the carnivores (snakes, shrikes, hawks, owls, skunks, weasels, bobcats, opossums, raccoons, shrews, moles, etc.) that eat them.

Reptiles

With 13 specialist species—nine snakes, two lizards, one worm lizard, and one turtle—reptiles are the second most diverse group of vertebrate specialists in longleaf pine savannas after amphibians; altogether, 56 reptiles are residents or spend a significant portion of their lives in longleaf pine savannas (Table 2). Reptiles are very different from amphibians in significant ways. By laying a shelled egg or giving live birth, reptiles are freed from dependency upon an aquatic habitat during development. They hatch or are born as fully developed miniatures of the adults, do not require a biphasic life cycle, and take up residence in the same habitat that will be occupied when adult. Body size in reptiles is generally larger because many reptiles (especially snakes) feed upon amphibians. Insectivorous lizards such as the scincids, iguanids, and teiids are about the same size as salamanders and frogs, with the exception of the elongated glass lizards. Snakes that feed on earthworms and other invertebrates are also small in size. Some snakes that feed on mammals (pine snake, indigo snake, eastern diamondback rattlesnake) are as large, by weight, as some of the smaller mammalian carnivores.

TABLE 2. Characteristic reptiles resident in longleaf pine savannas.[a,b]

Snakes		
1. *Carphophis amoenus*	Eastern worm snake	c
2. *Cemophora coccinea*	Scarlet snake	s
3. *Coluber constrictor*	Black racer	p
4. *Drymarchon corais*	Eastern indigo snake	s
5. *Elaphe guttata*	Corn snake	b
6. *Elaphe obsoleta*	Rat snake	b
7. *Heterodon platirhinos*	Eastern hognose snake	p
8. *Heterodon simus*	Southern hognose snake	s
9. *Lampropeltis calligaster*	Mole kingsnake	p
10. *Lampropeltis getula*	Common kingsnake	b
11. *Lampropeltis triangulum*	Scarlet kingsnake	p
12. *Masticophis flagellum*	Coachwhip	p
13. *Pituophis melanoleucus*	Pine snake	s
14. *Rhadinaea flavilata*	Pine woods snake	s
15. *Stilosoma extenuatum*	Short-tailed snake	s
16. *Storeria dekayi*	Brown snake	b
17. *Storeria occipitomaculata*	Red-bellied snake	b
18. *Tantilla coronata*	Crowned snake	b
19. *Tantilla relicta*	Florida crowned snake	s
20. *Thamnophis sauritus*	Eastern ribbon snake	d
21. *Thamnophis sirtalis*	Common garter snake	d
22. *Virginia striatula*	Rough earth snake	b
23. *Virginia valeriae*	Smooth earth snake	b
24. *Micrurus fulvius*	Eastern coral snake	s
25. *Agkistrodon contortrix*	Copperhead	b
26. *Agkistrodon piscivorus*	Cottonmouth	d
27. *Crotalus adamanteus*	Eastern diamondback rattlesnake	s
28. *Crotalus horridus*	Timber rattlesnake	b
29. *Sistrurus miliarius*	Pygmy rattlesnake	p
Lizards		
1. *Ophisaurus attenuatus*	Slender glass lizard	b
2. *Ophisaurus compressus*	Island glass lizard	b
3. *Ophisaurus mimicus*	Mimic glass lizard	s
4. *Ophisaurus ventralis*	Eastern glass lizard	b
5. *Sceloporus undulatus*	Eastern fence lizard	b
6. *Sceloporus woodi*	Florida scrub lizard	c
7. *Anolis carolinensis*	Green anole	b
8. *Eumeces anthracinus*	Coal skink	d
9. *Eumeces egregius*	Mole skink	s
10. *Eumeces fasciatus*	Five-lined skink	d
11. *Eumeces inexpectatus*	Southeastern five-lined skink	b
12. *Eumeces laticeps*	Broadhead skink	c
13. *Scincella lateralis*	Ground skink	b
14. *Cnemidophorus sexlineatus*	Six-lined racerunner	b
Amphisbaenian		
1. *Rhineura floridana*		s
Turtles		
1. *Deirochelys reticularia*	Chicken turtle	d
2. *Pseudemys floridana*	Florida cooter	d
3. *Pseudemys nelsoni*	Florida redbelly turtle	d
4. *Trachemys scripta*	Yellowbelly slider	d
5. *Terrapene carolina*	Eastern box turtle	b
6. *Terrapene ornata*	Ornate box turtle	a
7. *Kinosternon baurii*	Striped mud turtle	d
8. *Kinosternon subrubrum*	Eastern mud turtle	p
9. *Sternotherus odoratus*	Stinkpot	d
10. *Gopherus polyphemus*	Gopher tortoise	s

[a] Modified from Guyer and Bailey (1993).
[b] a = just enters longleaf pine savanna range; b = ubiquitous among habitats; c = visitor from other habitats; d = aquatic or semiaquatic; s = endemic in longleaf pine savanna; p = prairie or grassland habitat preferences.

Some turtles are completely terrestrial (box turtles and tortoises) but most in the Coastal Plain are aquatic as adults. Nine aquatic species live in the same ponds as those discussed above for amphibians and also spend critical times in their lives in the uplands (Table 2). The eggs and hatchlings spend variable periods in the uplands (up to 9 months) during development and during migration to ponds (Bennett et al. 1978). Adult females, eggs, and hatchlings are vulnerable to predation and desiccation. And juveniles and adults of all species must either disperse from drying ponds in search of other water bodies during droughts or take refuge in the uplands until water levels are restored in ponds (Burke and Gibbons 1995). Because of these dependencies of many aquatic turtles on longleaf pine savannas, they must be included in any consideration of longleaf pine savanna faunas.

Snakes are the largest group of reptiles resident in longleaf pine savannas (Table 2). I have apportioned snakes into four groups based on different aspects of how they use longleaf pine savannas. Each snake group is discussed separately, and then the lizards and turtles are discussed.

Longleaf Pine Snake Specialists

Among the nine snake specialists are species that represent a wide diversity of feeding strategies ranging from the egg-eating scarlet snake and small burrowers such as the Florida crowned snake and pine woods snake to two large predators on warm-blooded prey, the pine snake and eastern diamondback rattlesnake, and culminating in the apex predator that feeds on both warm- and cold-blooded prey, the indigo snake. Five of these snakes' geographical distributions are entirely nested inside the range of longleaf pine savannas and the other four have extralimital distributions in the southwestern United States (Ernst and Ernst 2003).

The five snakes whose ranges are encompassed entirely inside the range of longleaf pine are the eastern diamondback rattlesnake, pine woods, short-tailed, Florida crowned, and southern hognose snakes. The short-tailed and Florida crowned snakes are Florida endemic species, both of which are highly fossorial in deep, well-drained sandy soils of longleaf pine-turkey oak sandhills, scrub oak, or xeric hammocks (Campbell and Christman 1982; Cambell and Moler 1992) and it is thought that the Florida crowned snake is the main prey of the fossorial short-tailed snake (Mushinsky 1984). The pine woods snake ranges from the Florida parishes of Louisiana to North Carolina but requires open-canopied grasslands maintained by periodic fire and cannot tolerate the invasion of these habitats by hardwoods (Ernst and Ernst 2003). Very little is known about the natural foods and ecology of any of these species, including especially what size and qualities an area must have to sustain minimal effective populations.

The southern hognose snake (Fig. 3b) is most commonly found in xeric sandy habitats such as longleaf pine sandhills (Martof et al. 1980; Palmer and Braswell 1995; Jensen 1996; Tennant 1997), sand pine–rosemary scrub, xeric hammock (Ashton and Ashton 1981), and associated ruderal habitats (Enge and Wood 2003). It is one of the most xeric-adapted snakes in the eastern United States, in spite of unsubstantiated comments that it is sometimes found in dry river floodplains and hardwood hammocks (Tuberville et al. 2000). It uses its upturned snout to dig up its principal food, toads (genus *Bufo*) and the eastern spadefoot (Carr 1940; Goin 1947; Wright and Wright 1957) and a few other longleaf pine savanna vertebrates (Deckert 1918; Neill 1958; Mount 1975; Beane et al. 1998). Data gathered by Tuberville et al. (2000) suggest that the southern hognose snake has disappeared or declined in a substantial portion of its historical range. A highly fossorial snake (Palmer and Braswell 1995), little information is available on its life history and ecology.

The principal native habitat of the eastern diamondback rattlesnake in presettlement times was longleaf pine savannas, including sandhills, clayhills, and flatwoods varieties, but since so much of what is left of the original longleaf pine habitat is now ruderal vegetation such as shortleaf/loblolly pine oldfield successional forest or pine plantations, the species

FIGURE 3. (a) Mole skink, *Eumeces egregius*, a specialist in longleaf pine sandhills. (b) Southern hognose snake, *Heterodon simus*, a specialist in longleaf pine savanna.

survives mainly in any open-canopied pine woodland (Martin and Means 2000). In sandhills vegetation, Timmerman (1995) found that two females had home ranges of 46.5 ha and four males averaged 84.3 ha, but he had no density data. In 80-year oldfield successional shortleaf/loblolly pine forest that had been burned annually in late winter, Means (1986) found a density of one adult per 20 acres, but on barrier islands along both Atlantic and Gulf coasts, high densities of up to one to six per hectare can be achieved (Allen 1968; Means 1986). The eastern diamondback has declined less dramatically than some other large longleaf pine savanna snakes probably because it can tolerate high densities, and dense populations can be maintained in small areas. However, data from rattlesnake roundups in

Georgia and Alabama show that the upper end of the size class distribution of adults has declined significantly over the past 20 years (Means, in preparation).

Four other longleaf pine savanna specialist snakes (scarlet, eastern indigo, pine, and eastern coral snakes) all have large portions of their distributions in the range of longleaf pine (Ernst and Ernst 2003). Three of these (scarlet, eastern indigo, eastern coral) have disjunct populations or close relatives in the arid southwestern United States, indicating that their ancestors may have come from arid habitats. And, in fact, all three species prefer the driest habitats in longleaf pine savannas (Ernst and Ernst 2003). The eastern indigo snake is here classified as a longleaf pine savanna specialist because, in the northern four-fifths of its range, the indigo snake is restricted to the vicinity of xeric longleaf pine–turkey oak sandhills inhabited by the gopher tortoise (Diemer and Speake 1983; Moler 1985a; Stevenson et al. 2003). This, the longest snake in the United States and Canada, is the top predator among snakes, eating any vertebrate it can overpower, especially other snakes, including venomous snakes (Keegan 1944; Belson 2000; Ernst and Ernst 2003). Summer home ranges have been calculated at about 20–200 ha in the Gulf Hammock region of northwest peninsular Florida (Moler 1985b, 1992), to nearly 400 ha in northern Georgia (Natalie Hyslop personal communication). Densities of the indigo snake have not been determined, but there is some indication that its ophiophagy and agonistic behavior might prevent many from living together in the same area (Means in preparation). Behavior-mediated low densities of indigo snake populations might be a major factor responsible for the federally threatened status of the indigo snake (Means in preparation).

Scarlet and eastern coral snakes are two small fossorial snakes that are rarely seen but not necessarily uncommon. The scarlet snake specializes on eating predominantly reptile eggs including those of the southeastern five-lined skink, six-lined racerunner, ringneck snake, corn snake, pine snake, turtles, and even eggs of its own kind (Ernst and Ernst 2003). It is mostly active at night (Palmer and Tregumbo 1970; Nelson and Gibbons 1972) and is capable of actively burrowing into the substrate (Wilson 1951). Reynolds (1980) estimated the mean home ranges of males and females to be about 1627 and 1395 m^2, respectively, in North Carolina sandhills. The scarlet snake is one of the most abundant snakes captured in drift fence studies in longleaf pine-dominated habitats in Florida (Enge 1997), but most captures occur during a limited activity period from April through August (Enge and Sullivan 2000).

The eastern coral snake is mostly active in early morning and late afternoon (Neill 1957; Jackson and Franz 1981), and rarely after dark. It is an inveterate snake-eater, with nearly every longleaf pine savanna snake recorded in its diet plus all the terrestrial lizards, the Florida worm lizard, and an occasional eastern coral snake (Ernst and Ernst 2003). The surface activity of the eastern coral snake has been studied, but almost nothing is known about its subterranean habits, home range, or the acreage of suitable longleaf pine savanna habitat that will sustain a minimal viable population (Jackson and Franz 1981).

The pine snake (*Pituophis melanoleucus* including the races *mugitus, lodingi,* and populations of *melanoleucus* in the Coastal Plain of the Carolinas) is a large-bodied, spectacular diurnal snake that makes its home in longleaf pine savannas where the southeastern pocket gopher lives. In Florida it lives in xeric habitats such as longleaf pine–turkey oak, sand pine scrub, pine flatwoods on well-drained soils, and oldfields on former sandhill sites (Franz 1992). Females occupy home ranges of about 12 ha and males between 24 and 96 ha, with greatest activity in May, June, July, and October (Franz 1992). It digs open southeastern pocket gopher mounds using its pointed nose and remains underground up to 85% of the time in the tunnels of the southeastern pocket gopher, and to a lesser extent, the burrows of the gopher tortoise (Franz 1992; Tennant 1997). The diet of the pine snake is ground-dwelling birds and their eggs, southeastern pocket gophers, mice, woodrats, and immature rabbits (Ernst and Ernst 2003).

Resident Snakes Common in Other Habitats

Eleven other snakes are found to varying degrees in longleaf pine savannas: rat snake, corn snake, common kingsnake, brown snake, red-bellied snake, southeastern crowned snake, rough earth snake, smooth earth snake, copperhead, timber rattlesnake, and pigmy rattlesnake. The wide-ranging rat snake comes in several color varieties in the Coastal Plain, ranging from the Texas rat snake in Texas and Louisiana; the gray rat snake in Mississippi, Alabama, and the Florida Panhandle; and the yellow rat snake in peninsular Florida to the black rat snake in the Carolinas (Conant and Collins 1998). It is the most arboreal snake in the United States and Canada (Jackson 1970, 1974, 1977, 1978; Jackson and Dakin 1982; Dodd and Franz 1995) and is the most common snake found entering homes (Dodd and Franz 1995). Its diet consists mainly of small mammals and birds (Brown 1979; Ernst and Ernst 2003). A number of studies have been published on the habitat and behavior outside the range of longleaf pine savannas, but Franz (1995) found that a north-central Florida population used winter and summer activity areas connected by extensive migratory corridors. The corn snake, also wide ranging but with no presently recognized subspecies, is a close relative that also has the ability to climb, but it is more fossorial in its spatial habits in longleaf pine savanna habitats (Dodd and Franz 1995). It also feeds on mammals, birds, and eggs, but takes a high percentage of lizards and frogs (Ernst and Ernst 2003). The corn snake moved long distances between summer and winter activity areas, much like the rat snake on the same study area in north-central Florida (Franz 1995). The common kingsnake, a large terrestrial constrictor, is fossorial in its habits (Krysko and Smith 2005). It is even more wide ranging than the two previous species (Pacific to Atlantic coasts and dipping into northern Mexico) and has four subspecies in the range of longleaf pine, all of which principally live in longleaf savannas (Mount 1975; Dundee and Rossman 1989; Palmer and Braswell 1995; Krysko 2001; Means and Krysko 2001). Little information is available on the spatial relationships (home range, movements) of this diurnal species, but Stickel and Cope (1947) reported that a recaptured individual had moved only 100 m in 1.8 years. It preys on a wide range of small vertebrates, especially snakes and lizards, but rats, mice, terrestrial birds, and the eggs of birds, lizards, and turtles are of secondary importance (Ernst and Ernst 2003).

Two other large snakes, the timber rattlesnake and copperhead, are found in longleaf pine savannas in the upper Coastal Plain, do not range very far into the Florida peninsula (Conant and Collins 1998), but do reach the Atlantic and Gulf coasts both northeast and west of Florida. Both of these species include more mesic hardwood forests in their habitat repertoires. The timber rattlesnake feeds mostly on mammals but occasionally takes birds (less than 15%), whereas the copperhead feeds on anything it can overpower with its venom, including vertebrates and invertebrates (Ernst and Ernst 2003). Studies of the spatial relationships and population biology in longleaf pine savannas are wanting for both of these species.

The Coastal Plain contains most of the distribution of the small pigmy rattlesnake, which lives in all the longleaf pine savannas, turkey oak sandhills, and scrubs but is never very far from water (Ernst and Ernst 2003). Pigmys eat a wide range of prey from small mammals, small snakes, and frogs to invertebrates (Ernst and Ernst 2003). Their numbers are often quite abundant locally (May et al. 1996, 1997).

Five longleaf pine savanna resident snakes are small species that have not been extensively studied (brown, red-bellied, southeastern crowned, rough earth, and smooth earth snakes). Because of their small size, the acreage required to maintain a breeding population is probably much smaller than for larger snakes. Earthworms seem to be the principal dietary item for all five species, and they all probably eat other invertebrates, but more complete diet studies are needed (Ernst and Ernst 2003). Population densities of these small snakes in longleaf pine savannas seem to be relatively sparse in the brown snake (Clark 1949; Ford et al. 1991; Dalrymple et al. 1991) and red-bellied snake (Semlitsch and Moran 1984; Ford et al. 1991); not especially dense in the smooth earth snake (Ernst and Ernst 2003); locally abundant in both the southeastern crowned snake (Neill 1951a; Brode and Allison 1958; Semlitsch et al. 1981) and Florida crowned snake (Campbell and Christman 1982; Mushinsky 1984); and unstudied and poorly known in the rough earth snake (Ernst and Ernst 2003).

Eight snakes that are common residents of longleaf pine savannas have large geographical distributions in the eastern United States with extensions of their range into the prairies or grasslands of the Midwest. These snakes—black racer and coachwhip (both active on ground surface), corn snake, eastern hognose snake, scarlet snake, mole kingsnake, scarlet kingsnake, and pine snake (all burrowers or fossorial)—highlight the fact that longleaf pine savannas are prairielike grasslands. The black racer and coachwhip are fast-moving hunters that actively hunt their prey. The remaining six species are fossorial animals that spend a lot of their time underground.

A number of species that are mostly associated with wetland margins are frequently found in longleaf pine savannas. The eastern ribbon snake and common garter snake may forage for small vertebrates (frogs) in uplands far from water, as deduced from their presence in many longleaf pine savanna drift fence studies (Enge 1997). Likewise, the banded water snake (*Nerodia fasciata*) is often encountered in longleaf pine savannas. Whether it forages there for frogs, which are part of its diet (Ernst and Ernst 2003), or is merely passing through en route to other water bodies, as may also be the case for garter and ribbon snakes, is unknown. The cottonmouth usually brings to mind swamps and the edge of aquatic habitats, but radiotelemetry studies have shown that some individuals also spend substantial amounts of their lives in uplands (Means 2005b).

Lizards

Lizards are an impoverished group in the southeastern United States because they flourish mostly in deserts and xeric conditions (Pianka 1973, 1975). Fifteen lizards and one amphisbaenian (a worm lizard, closely related to lizards) are common residents of longleaf pine savannas (Table 2). Of the three specialists, the mole skink (Fig. 3a) is a small, slender, fossorial lizard with a long tail. Its burrowing habits restrict it to friable, well-drained soils of longleaf pine–turkey oak sandhills, scrub, and xeric hammocks (Mount 1963). Because it is so fossorial, it is rarely seen but can be raked from bare soil exposed in mounds of the gopher tortoise, southeastern pocket gopher, and scarab beetles (Geotrupinae) or found occasionally under logs, boards, pieces of tin, and other objects (Mount 1963). Mount (1963) collected 326 of 422 (77%) specimens from push-up mounds of the southeastern pocket gopher and noted that the geographical distribution of the mole skink almost exactly overlaps that of the southeastern pocket gopher, both longleaf pine savanna specialists. The mimic glass lizard is found in longleaf pine flatwoods with wiregrass ground cover (Moler 1992) in a narrow strip of the Coastal Plain from North Carolina to Mississippi, excluding peninsular Florida (Palmer 1987). It seems to be most common in the heliophilic ground cover of seepage bogs (Paul Moler personal communication), but little else is known about it. The Florida worm lizard is found principally in longleaf pine–turkey oak sandhills and sometimes in xeric hammocks and sand pine scrub, where it burrows underground and feeds on earthworms, small insects, and possibly termites and ants (Ashton and Ashton 1985). Its relatively small geographic distribution in the Coastal Plain is restricted to the northern half of the Florida peninsula.

Eleven other lizards are common residents in longleaf pine savannas (Table 2) or live in adjacent habitats and are occasionally found in the interface between longleaf pine habitats and evergreen shrub wetlands (coal skink), hardwood forests (five-lined skink, broadhead skink), or sand pine scrub (Florida scrub lizard). The eastern fence lizard is abundant in longleaf pine sandhills where turkey oaks are small trees and at the edges of hardwood hammocks (Crenshaw 1955). A relative, the

Florida scrub lizard, is narrowly distributed in the Coastal Plain in the central Florida highlands and peninsular coastal strand, where it lives primarily in sand pine scrub (Jackson 1973) but occasionally gets into adjacent longleaf pine habitats (Carr 1940; Ashton and Ashton 1985). The green anole may be the commonest lizard in the Southeast, occurring in almost every habitat from cypress swamps to longleaf pine sandhills (Smith 1946). In longleaf pine savannas, this arboreal lizard is found on logs, dead trees, and where woody ground cover plants are abundant, such as saw palmetto, gallberry, scrub oaks, and others.

The six-lined racerunner, most abundant in xeric longleaf pine habitats where ground cover is sparse and bare patches of sandy ground are exposed, is a fast-running, diurnal species that is most active in the hottest part of the day (Smith 1946). It is completely terrestrial, preferring bare soil conditions that are best available in the year or two after ground fires in longleaf pine savanna (Mushinsky 1985) as well as in sand pine scrub (Greenberg et al. 1994). The little ground skink is another ecologically widespread lizard that is found wherever there is some ground cover, such as dead grass and leaf litter (Ashton and Ashton 1985). In longleaf savannas, it is commonest where leaf litter is built up under scrub oaks. Brooks (1967) estimated population density for the ground skink in a hardwood forest at about 649 lizards per hectare, a figure he thought was low. The southeastern five-lined skink is abundant in longleaf pine savannas and prefers down logs or snags where it hides under exfoliating dead bark and forages on dead down or standing tree trunks and in litter (Mushinsky 1992). Three glass lizards besides the mimic glass lizard are occasionally found in longleaf pine savannas (Ashton and Ashton 1985).

Many species of snakes, small mammals, and birds of prey feed on lizards, all of which are insectivores. Although probably not contributing as much as amphibians to the overall biomass of the vertebrate predator–prey web in longleaf pine savannas, lizards may be more abundant, and important, than is realized. The spatial, temporal, and dietary relationships compared among all the lizards in longleaf pine savannas would make a worthwhile study.

Turtles

The eastern box turtle ranges widely throughout the eastern United States and is found in many different habitat types from river swamps and other wetlands and hardwood forests to offshore barrier islands of the Atlantic and Gulf coasts (Dodd 2001). It is abundant in longleaf pine savannas of all types but probably is more common in flatwoods than sandhills. The fact that the eastern box turtle is one of the few vertebrates in longleaf pine savannas that suffer from considerable injury and mortality from fire (Means and Campbell 1982; Ernst et al. 1995) is probably an indication that the eastern box turtle is really only a visitor in longleaf pine ecosystems, foraging there commonly out of the many stringers of swampy hardwoods along sluggish flatwoods streams, and out of the southern temperate hardwood forests (Platt and Schwartz 1990) that are interspersed among clayhills and sandhills. Its great longevity and the abundance of wetland and hardwood forest habitats enable the survival of those segments of its Coastal Plain populations that experience heavy mortality from fire.

On the other hand, the gopher tortoise is a strict longleaf pine savanna specialist that avoids fire, summer heat, winter cold, and predators by excavating burrows up to 14.5 m long and 3 m deep (Hallinan 1923; Hansen 1953; Auffenberg 1969). The gopher tortoise (Fig. 4) is one of the important keystone vertebrates (Eisenberg 1983) in longleaf pine savannas, because its burrows are long-lasting microcaverns that are utilized by more than 300 species of other vertebrates and invertebrates (Jackson and Milstrey 1989). It is the primary grazer in longleaf pine savannas (Landers and Speake 1980), eating a variety of broad-leaved grasses, legumes, fleshy fruits, mushrooms, and wiregrass (Garner and Landers 1981). Auffenberg (1969) speculated that the gopher tortoise is important in the dispersal of seeds of plants that it eats and that pathways radiating away from its burrows are fertilized strips that

FIGURE 4. Gopher tortoise, *Gopherus polyphemus*, at mouth of its burrow. The burrows of this keystone species, averaging about 8 m long and 2 m deep, are refugia for more than 300 species of other animals. A specialist in longleaf pine sandhills and clayhills.

encourage the growth of these plants. Tortoise burrows disrupt the ground cover, producing local microhabitats that may be germination sites for plants (Kaczor and Hartnett 1990; Hermann 1993; Platt 1999). Densities range to 20 gopher tortoises per hectare in longleaf pine–turkey oak habitat but average 7.75 (Franz and Auffenberg 1978; Cox et al. 1987).

Seven species of aquatic turtles commonly live in the same small, temporary ponds utilized by amphibians in the Coastal Plain and nest in adjacent longleaf pine savannas (chicken turtle, Florida cooter, Florida redbelly turtle, yellowbelly slider, striped mud turtle, eastern mud turtle, stinkpot). Most of these same turtles can be found in other water bodies, such as rivers and streams and swampy lakes, so they are not specialists that live only in a longleaf pine savanna landscape. However, they are highly mobile animals when out of water and commonly migrate overland for three reasons. First, individuals disperse overland seeking water when ponds dry up; second, females of all species move into longleaf pine uplands to lay eggs in nests dug in friable upland soils; third, individuals of both sexes also move into the uplands during winter or following dry-down of ponds, where they hibernate or aestivate until weather and water levels are conducive to returning to their ponds (Bennett et al. 1978; Burke and Gibbons 1995). Another species that is a casual visitor in longleaf pine savannas for the purpose of nesting is the Florida softshell, *Apalone ferox*. Fossil tortoises that might have played an important role in longleaf pine savanna ecology are discussed below.

Birds

Birds are the most vagile terrestrial vertebrates, many of which engage in long-distance migrations from tropical winter habitats to habitats far north of Coastal Plain longleaf pine savannas to breed and raise young in summer (Stevenson and Anderson 1994). Migrants pass through longleaf pine savannas on both trips, feeding as they progress. Because vegetative structural diversity has been implicated in bird species richness (MacArthur 1958; James 1971), one would not expect the open-canopied, longleaf pine savannas to be especially rich in resident birds, but the opposite is true. Engstrom (1993) found that bird species richness of coniferous forests in Florida was not lower than that of deciduous forests.

Longleaf Pine Specialists

Only five species of birds can be considered as specialists in longleaf pine savannas (Table 3). The northern bobwhite had few other open-canopied savanna habitats available to it in the presettlement Coastal Plain, so it was primarily found in longleaf pine savannas in spite of its common occurrence in present-day, open, grassy ruderal habitats such as pastures, agricultural fields, and annually burned shortleaf/loblolly pine oldfields (Stoddard 1931). It is a terrestrial species that feeds on seeds of all kinds, leafy vegetation, fruits, grasses, acorns, and insects (Ehrlich et al. 1988). In optimal longleaf pine habitat, the northern bobwhite can reach densities of up to five birds per hectare (Kellogg et al. 1972), but densities are usually lower than this. A variety of food plants, insects for young birds, and adequate cover in which to hide are the most important habitat requirements, both of which are eliminated during plant succession leading to hardwoods in the absence of fire (Stoddard 1931).

The red-cockaded woodpecker is undoubtedly the bird species most specialized for living

TABLE 3. Characteristic birds of longleaf pine savannas.[a,b]

Birds—residents		
1. *Aix sponsa*	Wood duck	
2. *Cathartes aura*	Turkey vulture	
3. *Buteo jamaicensis*	Red-tailed hawk	
4. *Buteo lineatus*	Red-shouldered hawk	
5. *Falco sparvarius*	American kestrel	
6. *Colinus virginianus*	Northern bobwhite	s
7. *Zenaida macroura*	Mourning dove	
8. *Bubo virginianus*	Great horned owl	
9. *Melanerpes carolinus*	Red-bellied woodpecker	
10. *Melanerpes erythrocephalus*	Red-headed woodpecker	
11. *Picoides pubescens*	Downy woodpecker	
12. *Picoides villosus*	Hairy woodpecker	
13. *Picoides borealis*	Red-cockaded woodpecker	s
14. *Colaptes auratus*	Northern flicker	
15. *Dryocopus pileatus*	Pileated woodpecker	
16. *Campephilus principalis*	Ivory-billed woodpecker*	
17. *Cyanocitta cristata*	Blue jay	
18. *Corvus brachyrhynchos*	American crow	
19. *Parus carolinensis*	Carolina chickadee	
20. *Parus bicolor*	Tufted titmouse	
21. *Sitta carolinensis*	White-breasted nuthatch	s
22. *Sitta pusilla*	Brown-headed nuthatch	s
23. *Thryothorus ludovicianus*	Carolina wren	
24. *Sialia sialis*	Eastern bluebird	
25. *Dumetella carolinensis*	Gray catbird	
26. *Mimus polyglottos*	Northern mockingbird	
27. *Toxostoma rufum*	Brown thrasher	
28. *Lanius ludovicianus*	Loggerhead shrike	
29. *Dendroica pinus*	Pine warbler	
30. *Geothlypis trichas*	Common yellowthroat	
31. *Cardinalis cardinalis*	Northern cardinal	
32. *Pipilo erythrophthalmus*	Rufus-sided towhee	
33. *Aimophila aestivalis*	Bachman's sparrow	s
34. *Sturnella magna*	Eastern meadowlark	
35. *Agelaius phoeniceus*	Red-winged blackbird	
36. *Quiscalus quiscula*	Common grackle	

TABLE 3. (Continued)

Birds—breeders
1. *Coragyps atratus* — Black vulture
2. *Meleagris gallopavo* — Wild turkey
3. *Columbina passerina* — Common ground-dove
4. *Coccyzus americanus* — Yellow-billed cuckoo
5. *Strix varia* — Barred owl
6. *Chaetura pealagica* — Chimney swift
7. *Chordeiles minor* — Common nighthawk
8. *Caprimulgus carolinensis* — Chuck-will's-widow
9. *Contopus virens* — Eastern wood-pewee
10. *Myiarchus crinitus* — Great crested flycatcher
11. *Tyrannus tyrannus* — Eastern kingbird
12. *Progne subis* — Purple martin
13. *Corvus ossifragus* — Fish crow
14. *Polioptila caerulea* — Blue-gray gnatcatcher
15. *Hylocichla mustelina* — Wood thrush
16. *Vireo griseus* — White-eyed vireo
17. *Vireo flavifrons* — Yellow-throated vireo
18. *Vireo olivaceus* — Red-eyed vireo
19. *Parula americana* — Northern parula
20. *Dendroica discolor* — Prairie warbler
21. *Dendroica dominica* — Yellow-throated warbler
22. *Wilsonia citrina* — Hooded warbler
23. *Icteria virens* — Yellow-breasted chat
24. *Piranga rubra* — Summer tanager
25. *Guiraca caerulea* — Blue grosbeak
26. *Passerina cyanea* — Indigo bunting
27. *Spizella pusilla* — Field sparrow
28. *Molothrus ater* — Brown-headed cowbird
29. *Icterus spurius* — Orchard oriole

Birds—Winter Visitors
1. *Sphyrapicus varius* — Yellow-bellied sapsucker
2. *Ectopistes migratorius* — Passenger pigeon*
3. *Sayornis phoebe* — Eastern phoebe
4. *Tachycineta bicolor* — Tree swallow
5. *Certhia americana* — Brown creeper
6. *Sitta canadensis* — Red-breasted nuthatch
7. *Troglodytes aedon* — House wren
8. *Regulus satrapa* — Golden-crowned kinglet
9. *Regulus calendula* — Ruby-crowned kinglet
10. *Catharus guttatus* — Hermit thrush
11. *Turdus migratorius* — American robin
12. *Bombycilla cedrorum* — Cedar waxwing
13. *Vireo solitarius* — Solitary vireo
14. *Vermivora celata* — Orange-crowned warbler
15. *Dendroica coronata* — Yellow-rumped warbler
16. *Dendroica palmarum* — Palm warbler
17. *Spizella passerina* — Chipping sparrow
18. *Melospiza melodia* — Song sparrow
19. *Melospiza georgiana* — Swamp sparrow
20. *Junco hyemalis* — Dark-eyed junco
21. *Zonotrichia albicollis* — White-throated sparrow
22. *Carduelis pinus* — Pine siskin
23. *Carduelis tristis* — American goldfinch

[a] Modified from Engstrom (1993).
[b] s = endemic longleaf pine Savanna specialists; * = recently extinct.

FIGURE 5. Red-cockaded woodpecker, *Picoides borealis*, is the only North American bird that excavates its nest cavity exclusively in living trees. The long-persisting cavities of this keystone species are homes to dozens of other animals in the long life of the longleaf pine tree in which the cavity is made. A specialist in longleaf pine savanna.

in longleaf pine savannas (Conner et al. 2001) (Fig. 5). The only woodpecker that excavates nesting and roosting cavities in a living tree, it selects longleaf pine trees 90 years or older (Costa 1995), presumably because longleaf pines above 90 years of age are more likely to have heart rot disease which makes the excavation of nest cavities easier (Hooper et al. 1991). It forages mainly on the trunks of the longleaf pine, where its main food appears to be arboreal ants, especially *Crematogaster ashmeadi* (Hess and James 1998). Studies have shown that the red-cockaded woodpecker survives better with little or no understory (Van Balen and Doerr 1978), which was highly significantly related to the ground cover composition and the extent of natural pine regeneration, both of which are indirect indicators of local fire history (James et al. 1997). The red-cockaded woodpecker is another important keystone species in longleaf pine savannas, because it makes cavities in living trees that may persist for several hundred years and are used by many other animals over the long life of the tree. Forest fragmentation, fire suppression resulting in midstory development, and the elimination of old trees (over 100 years of age) have been implicated in the decline of the red-cockaded woodpecker (Baker 1981; Anonymous 1990; U.S. Fish and Wildlife Service 2003).

Bachman's sparrow is a characteristic species and permanent resident of longleaf pine savannas, "dependent through the ages on the frequent grass fires that kept the 'flatwoods' open and parklike, with the characteristic prairie-type flora" (Stoddard 1978). It has survived reduction of its longleaf pine habitat by accommodating to ruderal situations in short herbaceous vegetation of open woodlands, borders of cultivated fields, or even fence rows with other sparrows. Bachman's sparrow nests and forages in dense ground cover of open pine forests. It will breed in clearcuts but prefers open mature stands of timber with low, thick ground cover (Dunning and Watts 1990; Engstrom 1993).

In the middle of the geographic range of longleaf pine in south Georgia, the white-breasted nuthatch prefers mixed pine and hardwoods and is most numerous in big timber, whereas the brown-headed nuthatch is most abundant in older stands of longleaf and annually burned shortleaf/loblolly pine stands (Stoddard 1978). Both nuthatches nest in cavities they excavate in dead snags, and the brown-headed nuthatch, especially, forages on the trunks and among the limbs of pine trees.

Other Characteristic Bird Species

Birds commonly found in longleaf pine savannas have been grouped as residents, breeders, and winter visitors (Table 3) (Engstrom 1993). All five bird specialists are residents that also breed in longleaf pine savannas, but 32 other species live in these habitats year-round. Of the 29 breeding species, 21 migrate mostly to the neotropics, and 8 either shift to other habitats or migrate to south Florida in the winter (Robertson and Kushlan 1974; Engstrom 1993). Some, if not most, of the south-central Florida prairie fauna that are dependent on dry prairie also appear to be dependent on recently burned sites (i.e., Florida grasshopper sparrow) or prefer recently burned sites over fire-suppressed prairies (Walsh et al. 1995).

Tucker and Robinson (2003) studied the influence and frequency of prescribed burning on Henslow's sparrows (*Ammodramus henslowii*) wintering on Gulf Coast pitcher plant bogs, communities that grade downslope out of longleaf pine savannas and depend on fires sweeping into them from the uplands. They found that bogs burned during winter typically hosted Henslow's sparrows for only one winter, but bogs burned during the growing season hosted the species for at least three winters.

Birds perform many functions in longleaf pine ecosystems. Many are seed and fruit eaters that disperse seeds that pass through the gut unscathed. The rank growth of black cherry (*Prunus serotina*), sassafras (*Sassafras albidum*), or persimmon (*Diospyros virginiana*) along fence rows confirms this. Other birds are insectivores that consume large quantities of arthropods by means of foliage- and bark-gleaning, sallying forth for flying insects, ground-litter searching, and pecking into chambers made by wood-boring beetle larvae. And some are consummate aerial predators that prey on mice, rats, and other small vertebrates. Extinct bird species of the late Pleistocene that may have been characteristic of longleaf pine savannas are discussed below.

Mammals

The role of mammals in grassland ecosystems is unique. No other living group of vertebrates produces species with such large body mass. Amphibians, reptiles, and birds generally have body masses in the range of about 10 g to about 10 kg, but mammals push the upper size limits to between 300 and 6500 kg (Nowak 1999) in the largest herbivores. Continentwide since at least the Middle Eocene, grasslands have attracted large grazing animals (Webb 1977). The largest herbivore in longleaf pine savannas today is the white-tailed deer (*Odocoileus virginianus*), second to the American bison (*Bison bison*) that became extinct in the Coastal Plain in the early 1800s (Rostlund 1960; Humphrey 1992b). While many small and medium-sized mammals are commonly found in longleaf pine savannas, a strikingly obvious dearth of megafaunal mammals is evident (Layne 1974; Stout and Marion 1993). I discuss the extant mammals in this section and the fossil record of mammals and other vertebrates afterward.

Longleaf Pine Specialists

Thirty-six mammals are found commonly in longleaf pine savannas, but only three are specialists in them (Table 4). The fox squirrel (*Sciurus niger*) is an animal of mature open longleaf pine flatwoods, sandhills, and clayhills (Lowery 1974; Brown 1997). It forages extensively on the ground and does not do well in brushy, thick ground cover. Periodic fires that kill back encroaching hardwoods or keep scrub oaks suppressed are absolutely necessary for fox squirrel populations to thrive (Brown 1997). Weigl et al. (1989) hypothesized that there may be a coevolved interdependence between

TABLE 4. Characteristic mammals of longleaf pine savannas.[a,b]

Mammals		
1. *Didelphis virginiana*	Virginia opossum	
2. *Sorex longirostris*	Southeastern shrew	
3. *Blarina carolinensis*	Southern short-tailed shrew	
4. *Cryptotis parva*	Least shrew	
5. *Scalopus aquaticus*	Eastern mole	
6. *Pipistrellus subflavus*	Eastern pipistrelle	
7. *Myotis austroriparius*	Southeastern myotis	
8. *Lasiurus borealis*	Red bat	
9. *Lasiurus seminolis*	Seminole bat	
10. *Lasiurus intermedius*	Yellow bat	
11. *Nycticeius humeralis*	Evening bat	
12. *Dasypus novemcinctus*	Nine-banded armadillo	
13. *Sylvilagus floridanus*	Eastern cottontail	
14. *Sciurus carolinensis*	Gray squirrel	
15. *Sciurus niger*	Fox squirrel	s
16. *Glaucomys volans*	Southern flying squirrel	
17. *Geomys pinetus*	Southeastern pocket gopher	s
18. *Geomys bursarius*	Plains gopher	
19. *Sigmodon hispidus*	Hispid cotton rat	
20. *Reithrodontomys humulis*	Eastern harvest mouse	
21. *Reithrodontomys fulvescens*	Fulvous harvest mouse	
22. *Oryzomys palustris*	Marsh rice rat	
23. *Podomys floridanus*	Florida mouse	s
24. *Peromycus polionotus*	Oldfield mouse	
25. *Peromycus gossypinus*	Cotton mouse	
26. *Ochrotomys nutallii*	Golden mouse	
27. *Baiomys taylori*	Northern pygmy mouse	
28. *Microtus pinetorum*	Pine vole	
29. *Procyon lotor*	Raccoon	
30. *Mustela frenata*	Long-tailed weasel	
31. *Mephitis mephitis*	Striped skunk	
32. *Spilogale putorius*	Spotted skunk	
33. *Canis latrans*	Coyote	
34. *Vulpes vulpes*	Red fox	
35. *Urocyon cinereoargenteus*	Gray fox	
36. *Felis concolor*	Cougar	
37. *Lynx rufus*	Bobcat	
38. *Ursus americanus*	Black bear	
39. *Sus scrofa*	Feral pig	
40. *Odocoileus virginianus*	White-tailed deer	

[a] Adapted from Engstrom (1993).
[b] s= endemic longleaf pine savanna specialists.

longleaf pine, the fox squirrel, and hypogeous fungi. In the longleaf pine ecosystem, food appears to be the most important factor influencing populations of the large fox squirrel, especially the seeds of longleaf pine and at least eight genera of hypogeous fungi. The fungi form mutualistic mycorrhizal associations with longleaf pine and probably other species of plants in the longleaf ecosystem, increasing the surface area and nutrient absorption of the roots of the plants while the fungus receives carbohydrates and a substrate on which to live. Most of these fungi depend upon animals for dispersal of their spores, which Weigl et al. (1989) found in the gut of every fox squirrel they examined. The fitness of all three partners seems to be enhanced by the relationship, but the degradation of the longleaf pine ecosystem by plant succession, logging, and fragmentation has had a deleterious effect upon the fox squirrel (Weigl et al. 1989), which in turn could feed back negatively on the fungi and those plants with which the fungi form mycorrhizal associations.

The Florida mouse is narrowly restricted to fire-maintained, xeric, upland vegetation on deep, well-drained sandy soils covered with longleaf pine–turkey oak, sand pine scrub, south Florida slash pine–turkey oak, coastal scrub, and drier pine flatwoods vegetations (Layne 1992). It has a small range confined to the upper one-half of the Florida peninsula, but populations follow the sandy ridges paralleling the coastlines to Sarasota County on the west and Dade County on the east (Layne 1992; Brown 1997). The Florida mouse inhabits side passages it excavates in the burrows of the gopher tortoise, and it forages at night on acorns, pine seeds, palmetto berries, mushrooms, insects, and arthropods found in gopher tortoise burrows (Layne 1992; Layne and Jackson 1994). On the Ordway Preserve in Putnam County, FL, where the species is most common in high pine communities with the gopher tortoise, home ranges were unaltered after prescribed burns, and no fire mortality was noted (Jones 1995).

The southeastern pocket gopher (Fig. 6) is found exclusively in sandy soils of longleaf pine savannas in sandhills and xeric flatwoods from central Alabama and Georgia south to central Florida (Humphrey 1992a; Brown 1997). It has been able to accommodate to man-made vegetation on sandy soils, such as golf courses, lawns, pastures, roadside rights-of-way, and other open, grassy areas. Because of its subterranean habits, little is known about the key behaviors of the southeastern pocket gopher, but southeastern pocket gophers retard eluviation by returning leached materials to ground surface; in one study, casting up 81,600 kg per hectare of burrow soil per year (Kalisz and Stone 1984). Their sandy push-up mounds create microsites that may promote colonization and secondary succession of a rich herbaceous flora in longleaf pine savannas (Platt 1975, Hermann 1993, Simkin and Michener 2005). Pocket gophers serve as the principal prey of the pine snake, which spends about half of its life in the rodent's tunnels (Franz 2005). The southeastern pocket gopher is so specialized for living in longleaf pine savannas that Gates and Tanner (1988) predicted that if succession toward a hardwood hammock caused by lack of fire continued unimpeded, southeastern pocket gopher populations would decline. Apparently, declines have been underway for some time. Entomologists Peter Kovarik and Paul Skelley, studying arthropods associated with the southeastern pocket gopher, were unable to locate populations at many historical localities (Bailey 2001).

Other Characteristic Mammal Species

Although only three mammals are specialists in longleaf pine ecosystems, some 37 other species are commonly found in them (Table 4). Six of these are insectivorous bats that take their insect prey on the wing. Four (three shrews and a mole) are tiny fossorial or semifossorial carnivores that feed on terrestrial invertebrates. The gray squirrel is a visitor from hardwood forests, but the southern flying squirrel reaches highest densities in mature pine-oak (longleaf pine/turkey oak) woodlands and xeric oak hammocks, where it commonly utilizes snags and woodpecker cavities (Brown 1997). The eastern cottontail lives in virtually all upland communities except dense forest, feeding preferentially on legumes, grasses, and various broad-leaved forbs (Brown 1997). Ten species of rats and mice are common in longleaf pine savannas, especially white-footed mice (*Peromyscus* spp.) and the cotton rat. The cotton rat can reach high densities (25–35/ha) at times and feeds on leaves, stems, roots, and seeds of grasses, sedges, legumes, and other herbaceous plants; it can be quite destructive to agricultural crops but it is also a staple food item for most of the Coastal Plain's mammal, reptile, and avian predators (Brown 1997).

The largest mammal in longleaf pine savannas is the black bear, an omnivore that eats a large percentage of fruits, nuts, berries, and other plant material (Lowery 1974; Brown 1997), but will take vertebrates

FIGURE 6. The characteristic push-up mounds of the southeastern pocket gopher, *Geomys pinetus*, a specialist in longleaf pine savanna, especially sandhills and clayhills. Its burrows provide refuge for more than 60 species of arthropods.

opportunistically (Rogers 1999). The largest strict carnivore is the cougar, but it has nearly been extirpated from the entire eastern United States and only a small population (fewer than 80 individuals) remains in southern Florida (Wood 2001). All the rest of the longleaf pine savanna mammals are small, generalized carnivores (opossum, raccoon, long-tailed weasel, striped skunk, spotted skunk, red fox, gray fox, bobcat), and three are recent arrivals (feral pig, nine-banded armadillo, coyote).

The white-tailed deer is the only large, herbivorous, native, ungulate mammal that has the potential to impact vegetation. Mainly a browser that feeds on twigs, leaves, and tender shoots of many trees and shrubs, it also is a grassland feeder spending time in hardwood forests as well as longleaf pine savannas (Barlow and Jones 1965). That there are so few large, megafaunal mammals utilizing the vegetative largess of longleaf pine savannas is highly unusual (Layne 1974; Stout and Marion 1993) and warrants an explanation (see the following section).

What Can the Vertebrate Fossil Record Tell Us about Longleaf Pine Savannas?

The fossil record indicates that grazing mammals have been evolving in midcontinental North American woodland savannas (as opposed to treeless prairies or steppes) since at least the middle Eocene (Webb 1977). By mid-Miocene, a close faunal continuity was established between the Gulf Coast and the Great Plains, extending far south into Mexico by the late Miocene (Graham 1965; Webb 1977). Onset of drier conditions in the Pliocene brought about prairie or steppe conditions in the Great Plains, but the effects of summer rains, presumably, maintained a broad belt of savannas around the Gulf of Mexico (Webb 1977). Savannas, having scattered trees and riparian gallery forests along the margins of streams and swampy wetlands, support a more diverse herbivore fauna including browsing horses, tapirs, most of the peccaries, most of the proboscideans, and many browsing ruminants (Webb 1977) in addition to the grazing mammals. All of the modern extant vertebrates that inhabit longleaf pine savannas were present by the end of the Pleistocene (Kurten and Anderson 1980; Holman 1995; Emslie 1998). In the past 11,000 years, however, something happened to the large mammals and large birds, but not to most of the amphibians and reptiles.

At the end of the North American Pleistocene, 8 families, 46 genera, and 191 species of mammals became extinct (Holman 1991). Two families and 19 genera of North American birds became extinct, with a very large number of extinct species reported (Holman 1991). In sharp contrast, no families and no genera of either amphibians or reptiles became extinct in the Pleistocene of North America, and only 12 taxa out of a total of 229 identified Pleistocene taxa were unquestioned extinct species (Holman 1995). A list of Pleistocene megafaunal vertebrates is given in Table 5 and those that may have been characteristic species of longleaf pine savannas are noted.

Fossil vertebrates can give us information about two important aspects of longleaf pine savannas: the possible antiquity of savannas and how the savanna vegetation may have been affected by the megafauna. Numerous studies of vertebrate fossils, mostly from Florida localities, show that longleaf pine savanna specialists and residents were present at least as far back as the early Pleistocene, about 2 million years ago. The Inglis IA fauna from the earliest Irvingtonian in the earliest Pleistocene consisted of an essentially modern herpetofauna with 21 of 26 snake species, 2 of 4 lizard species, and the Florida worm lizard surviving in Florida today (Meylan 1982). The combined evidence of the herpetofauna, other vertebrates, and the sedimentary context suggest a mixed habitat of mature longleaf pine with xeric hammock interspersed (Meylan 1982). Xeric hammock would be expected in the depressions characteristic of a karst topography, similar to that which exists locally today. Slightly later in the early Pleistocene, Leisey IA must have been adjacent to an upland community, most likely high pine or xeric hammock, to account for the large number of highly terrestrial turtles, including *Gopherus, Terrapene,* and two species of *Hesperotestudo* (Meylan 1995).

The bed of Rock Springs Run in Orange County, FL, produced one of the richest avian fossil deposits known from North America (Wolfenden 1959). Of 1025 bird bones, about

TABLE 5. Late Pleistocene (Rancholabrean) vertebrates that may have been characteristic in longleaf pine savannas.[a,b]

Turtles		
1. *Geochelone crassiscutata*	Eastern large tortoise	s
2. *Geochelone incisa*	Eastern small tortoise	s
3. *Terrapene carolina putnami*	Giant box turtle	?
Birds		
1. *Phoenicopterus*	Flamingo	d
2. *Ciconia*	Stork	?
3. *Anabernicula*	Shell duck	d
4. *Breagyps*	Condor	s
5. *Teratornis*	Teratorn	s
6. *Cathartornis*	Teratorn	s
7. *Spizaetus*	Hawk-eagle	s
8. *Amplibuteo*	Eagle	s
9. *Wetmoregyps*	Walking-eagle	s
10. *Neophrontops*	Old World vulture	s
11. *Neogyps*	Old World vulture	s
12. *Milvago*	Caracara	s
13. *Dorypaltus*	Lapwing	?
14. *Burhinus*	Thick-knee	?
15. *Protocitta*	Jay	?
16. *Henocitta*	Jay	?
17. *Cremaster*	Hangnest	?
18. *Pandanaris*	Cowbird	?
19. *Pyelorhamphus*	Cowbird	?
Mammals		
1. *Dasypus bellus*	Beautiful armadillo	s
2. *Kraiglievichia floridanus*	Florida pampathere	s
3. *Holmesina septentrionalis*	Northern pampathere	s
4. *Glyptotherium floridanum*	Simpson's glyptodont	s
5. *Megalonyx jeffersoni*	Jefferson's ground sloth	b
6. *Megalonyx wheatleyi*	Wheatley's ground sloth	b
7. *Eremotherium rusconii*	Rusconi's ground sloth	b
8. *Glossotherium harlani*	Harlan's ground sloth	b
9. *Homo sapiens*	Human	b
10. *Borophagus diversidens*	Bone-eating dog	b
11. *Canis lepophagus*	Johnson's coyote	b
12. *Canis lupus*	Gray wolf	b
13. *Canis dirus*	Dire wolf	b
14. *Canis rufus*	Red wolf	b
15. *Ursus americanus*	Black bear	b
16. *Tremarctos floridanus*	Florida cave bear	b
17. *Arcodus pristinus*	Lesser short-faced bear	b
18. *Felis onca*	Jaguar	b
19. *Felis concolor*	Puma, cougar (Florida panther)	b
20. *Felis atrox*	Lion	b
21. *Felis yagouaroundi*	Jaguarundi	b
22. *Felis rufus*	Bobcat	b
23. *Felis amnicola*	River cat	d
24. *Smilodon fatalis*	Sabretooth	b
25. *Smilodon gracilis*	Gracile sabretooth	b
26. *Homotherium serum*	Scimitar cat	b
27. *Chasmaporthetes ossifragus*	American hunting Hyena	b
28. *Castoroides ohioensis*	Giant beaver	d
29. *Hydrochoerus holmesi*	Holmes's capybara	d
30. *Neocherus pinckneyi*	Pinckney's capybara	d

TABLE 5. (Continued)

31. *Cuvieronius tropicus*	Cuvier's gomphothere	b
32. *Mammuthus columbi*	Columbian mammoth	s
33. *Mammut americanum*	American mastodon	b
34. *Tapirus veroensis*	Vero tapir	b
35. *Nannippus phlegon*	Gazelle-horse	s
36. *Equus simplicidens*	American zebra	s
37. *Equus fraternus*	Brother horse	s
38. *Equus giganteus*	Giant horse	s
39. *Platygonus bicalcaratus*	Cope's peccary	b
40. *Platygonus compressus*	Flat-headed peccary	b
41. *Mylohyus floridanus*	Kinsey's peccary	b
42. *Hemiauchenia blancoensis*	Blanco llama	s
43. *Hemiauchenia macrocephala*	Large-headed llama	s
44. *Paleolama mirifica*	Stout-legged llama	s
45. *Odocoileus virginianus*	White-tailed deer	b
46. *Capromeryx arizonensis*	Skinner's pronghorn	s
47. *Ovibos* sp.	Muskox	b
48. *Bison latifrons*	Giant bison	s
49. *Bison bison*	American bison	s

[a] Modified from Brodkorb (1964), Webb (1974), Steadman and Martin (1984), Young and Laerm (1993).
[b] b = prairie grasslands inhabitant; d = aquatic; s = endemic longleaf pine savanna specialists.

half were identifiable to species and included 35 species; some, including bones of the passenger pigeon, cardinal, and red-cockaded woodpecker, indicated the nearby presence of longleaf pine forest. The spring lies immediately east of the Pamlico shoreline (MacNeill 1950), indicating that the bones were deposited during Pamlico or post-Pamlico times about 180,000 to 120,000 years ago (Webb and Wilkins 1984).

By the late Pleistocene Sangamon Interglacial age (about 100,000 years ago), all families and genera of the Williston IIIA herpetofauna represent extant taxa, and only 1 of 36 taxa is extinct at the species level (Holman 1996). A low-energy aquatic environment is suggested by some of the species, but the bulk of the species were characteristic of a distinctively dry and sandy longleaf pine savanna including eastern spadefoot, eastern small tortoise (the extinct form), gopher tortoise, Florida worm lizard, eastern indigo snake, southern hognose snake, eastern hognose snake, coachwhip, pine snake, Florida crowned snake, and eastern diamondback rattlesnake (Holman 1996).

The red-cockaded woodpecker is so specialized that it is difficult to imagine any kind of pine forest other than longleaf pine savanna in which fossil red-cockaded woodpeckers could have lived. The red-cockaded woodpecker requires longleaf pines at least 90 years old for the construction of cavities (Costa 1995), the presence of red heart disease (Jackson 1977), a specialized diet of ants that only live on longleaf pine trunks (Hess and James 1998), and intolerance for closed canopy or brushy conditions (Van Balen and Doerr 1978). The same can be said for another longleaf pine savanna specialist, the gopher tortoise, which is common in earliest Pleistocene sites. In addition, the blind cave arthropods that use the burrows of the gopher tortoise, and the tortoise tick that is endemic on the gopher tortoise, required some time to evolve, helping infer the antiquity of longleaf pine savannas. Some semblance of a fire-maintained, longleaf pine savanna, therefore, must have been present in the Coastal Plain for at least 2 million years and had antecedents far back into the Miocene.

If we can be reasonably certain that longleaf pine savannas existed throughout the Pleistocene, then we can explain why present-day longleaf pine savannas are wanting in megafaunal vertebrates that are to be expected in savannas having an abundance of grasses and forbs for grazing animals and woody plants for browsing animals. Longleaf pine savannas

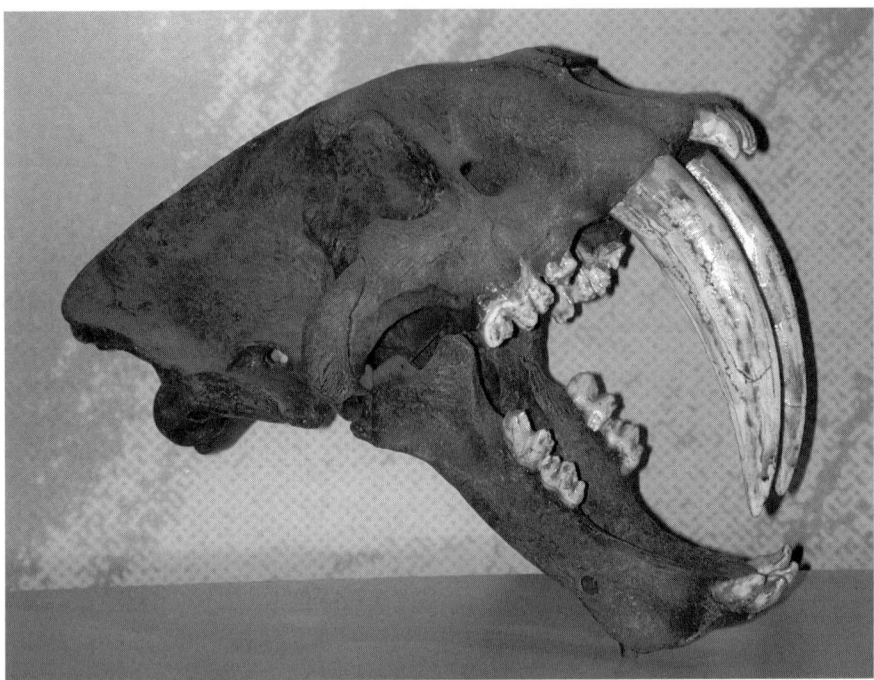

FIGURE 7. Pleistocene megafauna. Possibly 49 extinct species of large ungulates and carnivores, 19 species of extinct birds, and 3 giant turtles were inhabitants of longleaf pine savanna habitats prior to about 11,000 years ago. Here illustrated is the skull of one of the extinct large carnivores, the sabertooth (*Smilodon fatalis*). How longleaf pine savanna habitats were structured in the presence of these animals, and how the longleaf pine savanna vegetation has responded to their extinction, may never be known.

did, in fact, have a large megafauna that the fossil record tells us went extinct rather rapidly in an approximately 2000-year period from about 13,000 to 11,000 years ago (Webb 1990) (Fig. 7). The present-day biota of longleaf pine savannas evolved in concert with the megafauna, and we should wonder how many of their adaptations that we see, or do not see, are ghosts of evolution past (Janzen and Martin 1982; Barlow 2000). Not only should we ask how different must have been the vegetative characteristics of longleaf pine savannas during the late Pleistocene, but also how different must have been presettlement longleaf pine savannas from their naturally co-evolved conditions with the megafauna that were present only a few thousand years earlier?

The vegetative characteristics of Pleistocene longleaf pine savannas must have been greatly affected by the megafauna. Many small herb seeds survive the trip through cattle and other livestock digestive systems (Janzen 1984). Some of the herbaceous plants of longleaf pine savannas may well have depended on large mammals for seed dispersal to certain kinds of high-quality germination sites (e.g., edges of game trails, nitrogen-rich soils below legume trees or dung piles, heavily insolated sites, watercourse edges, gopher tortoise burrow aprons, tree tip-up mounds), or for seed removal from sites of easy harvest by small predators that concentrate on harvesting seeds from maturing fruits near ground level (Janzen 1984).

Mastodon dung found preserved under water in the Aucilla River, FL, consisted of cypress (*Taxodium*), wild grape (*Vitis*), buttonbush (*Cephalanthus occidentalis*), willow (*Salix*), pine (*Pinus*), pokeweed (*Phytolacca americana*), Mexican poppy (*Argemone*), and wild gourd seeds (*Cucurbit pepo*) (Webb et al. 1992). Most of these plants indicate a browsing diet, which corresponds well with the bunodont dentition of the mastodon. The site of deposition of the dung was the mouth of a cave that probably served as a watering hole. Some of the seeds recovered are upland species that could not exist near the watering hole. Mastodons probably went on feeding forays, ranging far from the water source as the food supply was diminished in an ever-growing ring centered around the watering hole during the driest months of the year.

There were no less than three proboscideans in the southeastern United States Coastal Plain in the late Pleistocene, the Columbian mammoth, American mastodon, and Cuvier's gomphothere, whereas of the two living elephants, only one species occurs today in either the savannas of Africa or India. Like living elephants, mammoths, mastodons, and gomphotheres could have debarked, stripped, and pushed over trees, thus creating a substantial impact on the environment. Owen-Smith (1987, 1988) pointed out that the feeding activities of elephants quickly change wooded savanna to open short-grass prairies dominated by rapidly regenerating plants that provide food for smaller herbivores. Elimination of elephants leads to massive environmental changes, and the extinction of megaherbivores at the end of the Pleistocene may have caused changes from grasslands to forests (Holman 1995). Pine- and oak-dominated savannas were present in the Southeast during the Wisconsin glaciation (Delcourt and Delcourt 1987). Pine and oak pollen alternated in dominance in lake sediments over the past 40,000 years in north Florida (Watts 1969, 1971, 1975, 1980; Watts et al. 1992), with pine pollen becoming dominant about 8000 years ago (Watts et al. 1992). Some have ascribed the ascendancy of pine pollen in lakes to climate change, but the largest shift to pines occurs shortly after the extinction of the megafauna.

In the Pleistocene, there were ground sloths, peccaries, and proboscideans that could have dug, rooted, or pulled up the large grass-stage longleaf pine seedlings and saplings to get at the cortex of the main taproot. A modern example verifies the palatability of longleaf pine taproots and the likelihood that this scenario is correct. Following their introduction into the New World by the Spaniards, hogs have been observed to kill more than 17,000 two-year-old, longleaf pine seedlings per hectare at rates of more than 200 seedlings per day, so one hog can obliterate a hectare of planted pines in 2 days (Simberloff 1993). On the De Soto National Forest, hogs greatly damaged seedlings 5–120 cm and saplings 1.5 – 4.5 m high. Chapman (1943) and Bruce (1947) described equally devastating damage by hogs in experimental plots in Louisiana. It may well be the case that longleaf pine ecosystems in the late Pleistocene were typically very sparsely treed, and that longleaf pine was only able to become established when migratory herds missed finding regeneration for a few years. One could even argue that many of the adaptations of longleaf pine such as masting, growing a deep, strong taproot, living to 500 years of age, and having dense, strong wood were evolved to avoid predation by megaherbivores.

One longleaf pine savanna resident vertebrate that had a strong impact on vegetation was humans, both directly and indirectly. Probably immediately after arriving in the New World, humans directly impacted vegetation by burning it to stampede game, remove hiding places for small game, make walking and hunting easier, and attract game to nutritious new growth following fire (Pyne 1982). Humans may have had an even greater impact on vegetation indirectly through predation on the megafauna. When Europeans first arrived, Amerindians were still using stone implements. After the earliest and unique Clovis, Suwannee, and Simpson stone technology of Paleoindians (about 13,000 to 10,000 years before present) that is thought to have been used to bring down the megafauna in the southeastern United States (Goodyear 1999), spear points went through an evolution of types. These ranged from Bolen Age (10,000 to 9000 years before present) projectile points that were corner and side-notched to Archaic Period (about 9000 to 3200 years before present) projectile points that were stemmed. Then, woodland technology (about 3200 to 1000 years before present) was characterized by basally notched stone spear points. Finally, the Mississippian Culture (1000 to

600 years before present) was characterized by temple mound complexes, farming, and small stone points that were true arrowheads. The evolution of stone spear points from very well crafted large, lanceolate forms to small points suitable as arrow tips, probably reflected the change in human diet from megafaunal animals to very small animals and fishing.

Some of the Coastal Plain megafauna survived to about 11,000 years ago (Grayson 2001). Paleoindians, arriving in North America about 13,000 years ago, lived contemporaneously for about 2000 years with the Coastal Plain megafauna and could have been the main reason for its extinction (Martin 1967). One way to understand how this might be possible is to calculate how rapidly a small band of, say, 20 humans (10 reproductive males and females) could have multiplied in the presence of the huge bounty of large, easy-to-kill, megafaunal animals that were completely naive about the danger presented by this newly arrived predator (Flannery 2001). Assume a simple doubling of the human population over a generational period of 20 years. After 20 generations in 400 years, the human population could theoretically multiply to more than 20 million people. There are five periods of 400 years in the 2000 years during which the megafauna went extinct, so there was ample time for human populations to become large enough to have had a significant hand in their demise, if not having been the sole reason. The first to go would have been the easiest vertebrates to overpower, such as the two giant tortoises and one giant box turtle (Table 5). These animals had no defense against a predator that could turn a tortoise on its back, fashion stone and wooden tools to kill it, break open its protective shell, and cut out the edible parts (Clausen et al. 1979). Likewise, other large, slow-moving animals such as the giant beaver, Florida pampathere, northern pampathere, and Simpson's glyptodont would have been similarly easy targets. Many of the late Pleistocene birds that went extinct were large carrion eaters or birds of prey (Table 5) that required open savannas and probably scavenged or hunted the ancient longleaf pine savannas. The demise of most of them probably followed the extinction of their megafaunal prey.

In summary, longleaf pine savannas are grasslands whose antecedents may go back at least into the early Miocene in Florida and probably earlier in other parts of the Coastal Plain. Savanna vegetation has been present in some parts of the Coastal Plain (especially Florida where the fossil evidence is most abundant) probably for much of the Cenozoic Era. During climate changes in the Pleistocene, the principal upland plant communities probably expanded or contracted their local distributions such as by moving farther upslope in the case of the southern temperate hardwood forest or shrinking to ridge tops and deeper sands in the case of longleaf pine savannas. Plant species composition may also have changed somewhat with the arrival or departure of northern species in or out of either of the two above-mentioned ecosystems, but it is difficult to believe that the fire-controlling relationships between fire-sensitive hardwood communities and fire-tolerant longleaf pine-grasslands ever disappeared. A large vertebrate fauna of at least 212 extant species plus an extinct component of up to possibly 49 megafaunal mammals, 3 giant tortoises and turtle, and between 9 and 17 birds evolved with the vegetation of longleaf pine savannas. Not only did the extinct fauna influence the characteristics and distribution of the longleaf pine savanna plants in the late Pleistocene, but an even greater effect might have been imposed on these plants by means of ecological release following the demise of these animals. What those effects might have been, we can only guess, but they would have been substantial. Even our conception of what longleaf pine savanna was like 500 years ago may be dramatically different from its late Pleistocene aspect. Interglacial periods, such as the Holocene in which we presently live, last only about 10,000 years whereas glacial periods with epicontinental ice sheets are about 100,000 years long. The evolutionary setting for vertebrates of the longleaf pine savanna was most likely more typical during the long glacial periods than the warmer interglacials.

Management Considerations from the Perspective of Vertebrates

Value of Upland Habitat Surrounding Temporary Ponds

As we have seen above, 35 amphibians that are characteristic species in longleaf pine savannas breed in temporary ponds. Semlitsch (1998) calculated that most of the salamanders spend 86–99% of their lives in the uplands and less than 15% in breeding ponds. A similar analysis has not been conducted for anurans, but because of the long captive life spans for many species such as the eastern spadefoot (12 years), gopher frog (9 years), narrowmouth toad (6 years, 9 months), barking treefrog (10 years), and others (Snider and Bowler 1992), the percentage of their lives spent in longleaf pine savannas is likely to be similarly high. Ten turtles that live in these same ponds as adults all require upland habitat for nesting, hibernation, and as a refuge when the pond goes dry (Burke and Gibbons 1995). The ecological quality of the upland habitat surrounding small isolated water bodies, therefore, is as important for vertebrates that utilize ponds as is the quality of the aquatic habitat to their larvae and aquatic stages (Burke and Gibbons 1995; Marsh and Trenham 2001; Semlitsch and Bodie 2003; Means and Means 2005). Successful management of temporary pond vertebrates requires restoring or maintaining the natural vegetative quality of their upland habitats as well as the ecological integrity of their breeding ponds (Means and Means 2005). Unfortunately, for most species, few data exist on how much adjacent upland habitat is necessary and sufficient for species to maintain minimum viable populations.

For only six salamanders for which he could find reliable data (including the mole and tiger salamanders of longleaf pine savannas), Semlitsch (1998) calculated that the terrestrial habitat that would likely encompass 95% of the population would extend 164.3 m (534 feet) from the wetland margins of ponds. Johnson (2003) estimated that at least 16% of the striped newts breeding in his study pond emigrated in excess of 500 m from the pond. He believed that a core of protected upland with a radius of approximately 800 m would be required to protect the vast majority of individuals that bred in the pond. Burke and Gibbons (1995) studied the distance from a South Carolina pond that three species of turtles (mud turtle, Florida cooter, pond slider) migrated into the uplands to lay eggs or hibernate. Their data indicated that the turtles required a 275-m zone of uplands around the pond to protect 100% of the nest and hibernation sites. Data on the amount of upland habitat needed for the local survival of populations of species that utilize temporary ponds are urgently needed for most longleaf pine savanna species. The upland habitats are not buffer or riparian "zones" but full-fledged "core" habitats that are as important to the vertebrates as the ponds themselves (Semlitsch and Jensen 2001; Semlitsch and Bodie 2003).

Observations of long-distance dispersal in a few species—2 km for the gopher frog (Moler and Franz 1988) and 1.7 km for the flatwoods salamander (Ashton and Ashton 2005)—does not necessarily mean that the appropriate upland habitat should be a zone of 2 km around a breeding pond, but it does illustrate another need of longleaf pine savanna amphibians and of all animals that utilize such ponds: connected corridors for dispersal among ponds (Fig. 8). Metapopulation dynamics are especially important for pond breeders because at a crucial juncture in their life cycles, and periodically, adults must make long journeys outside of their normal home ranges to breed in small habitats that are discontinuously distributed in the landscape. Moreover, the critical requirement of these pond environments—presence of water—is highly ephemeral, making amphibian and turtle populations much more vulnerable to local extinction than if they lived their entire lives only in a contiguous longleaf pine savanna like mammals, other reptiles, and birds. The ability to disperse long distances over land, therefore, is a mandatory requirement for pond-breeding amphibians and pond-inhabiting turtles, which allows them the possibility of finding other breeding

FIGURE 8. Temporary pond in longleaf pine sandhills in Leon County, FL. Throughout the range of longleaf pine, small ponds devoid of fish are the breeding grounds or homes of about 35 species of salamanders, frogs, turtles, and snakes that depend upon these ponds almost exclusively for their existence.

ponds or water to live in and for restocking locally extinct populations.

Unfortunately, small isolated wetlands that were once regulated by the U.S. Army Corps of Engineers, were removed in 2001 from federal jurisdiction by the U.S. Supreme Court decision in the case *Solid Waste Agency of Northern Cook County v. U.S. Army Corps of Engineers*. This bodes very badly for these extremely important habitats that contribute much to the biodiversity of the southeastern United States. In South Carolina, for instance, the elimination or alteration of more than 90% of Coastal Plain Carolina bay wetlands (Bennett and Nelson 1991) has reduced essential habitat for the black swamp snake (*Seminatrix pygaea*), Florida green water snake (*Nerodia floridana*), and chicken turtle (*Deirochelys reticularia*), all of whose distribution patterns are restricted primarily to seasonal wetlands (Buhlmann 1995; Dorcas et al. 1998; Gibbons et al. 2000). Not only is it imperative to manage the longleaf pine savanna uplands surrounding these small water bodies, but the water bodies themselves are in peril (Semlitsch and Bodie 1998; Semlitsch 2000).

Value of Dead Trees (Snags and Down Logs)

Dead standing trees (snags) and down logs are highly valuable to vertebrates in forest ecosystems and particularly so in the sparsely treed longleaf pine savannas (Baker 1974; Maser et al. 1979; Thomas et al. 1979; Miller and Marion 1995). Numerous studies of terrestrial vertebrates in longleaf pine savannas mention the use of snags by lizards (Goin and Goin 1951), frogs (Dodd 1996; Boughton et al. 2000), snakes (Brode and Allison 1958; Franz 1995; Tennant and Bartlett 2000 for the scarlet kingsnake), birds (Miller and Marion 1995), and mammals. For instance, nursery and roosting colonies of up to 20 evening bats were found under loose bark of a lightning-struck pine (Baker 1974). Bark only stays in a loose condition for a short time, making it a very temporary habitat for the bats. If roosting sites are a limiting factor for the evening bat, loose pine bark from lightning-struck trees may play a very important role in the local presence of the species, even though it might be usable as a roost for one season or less (Baker 1974).

In an experimental study of the use of snags by cavity-nesting birds, numbers and species of birds were consistently higher on longleaf pine sites than in slash pine plantations, even though longleaf sites had fewer snags (Miller and Marion 1995). They demonstrated clearly that, despite higher snag densities, pine plantations provided marginal habitat for cavity-nesting birds compared with longleaf pine areas. They concluded that retention of longleaf pine will support more species and numbers of birds than pine plantations.

Unfortunately, the remaining longleaf pine stands are almost all second-growth forests that germinated after the decimation of old-growth longleaf pine from the 1880s through the 1930s (Frost 1993). These forests, for the most part, are uniformly aged and contain few or no snags because they consist of relatively youthful trees (longleaf pine can live to more than 500 years; Wahlenberg 1946). A very important management consideration in longleaf pine savannas is to make sure that lightning-killed trees and trees that die from other causes are left for wildlife use and not cut by salvage operations (Miller and Marion 1995).

Value of Stumpholes or Tree Bases

Subterranean cavities (burrows, rootholes, stumpholes) may be important to more vertebrates than any other physical characteristic of longleaf pine savannas. Burrows of the gopher tortoise, for instance, are known to provide temporary or even permanent refuge to over 300 species of other animals (Jackson and Milstrey 1989) and burrows of the southeastern pocket gopher to more than 60 other species (Hubbell and Goff 1939). Not all vertebrates can create their own burrows, however, and yet many of the vertebrates that live in longleaf pine savannas are fossorial, living all or part of their time underground. This is especially true of the salamanders, whose moist skin would desiccate in the heat of midsummer sun. All of them live underground and very little is known about their adult lives. Many frogs hide in burrows or cavities during the day and emerge to forage at night. The eastern spadefoot, southern toad, and gopher frog all do this as well as do some treefrogs such as the barking treefrog and ornate chorus frog. Except for a few lizards that are arboreal, all of the reptiles are either active burrowers (gopher tortoise, pine snake, hognose snakes, red-tailed skink) or live in the burrows of other animals and other underground cavities. Many small mammals are consummate burrowers (oldfield mouse, cotton rat, southeastern pocket gopher, eastern mole, nine-banded armadillo, pine vole, and others) and make extensive tunnel systems in longleaf pine savannas that are utilized by fossorial animals which are not themselves burrowers. All of these animals periodically need to retreat underground from temperature extremes, predators, and fire. Probably the most overlooked refuge for longleaf pine savanna vertebrates is the subterranean base (butt, stump) of dead longleaf pine trees and their associated rotting roots (Means 2005b).

Longleaf pine is the only tree in the southeastern United States that grows a shaftlike, massive taproot up to 5 m deep. Its lateral roots, growing only 25 cm below ground surface, can reach out to 22 m from the base of the tree (Heyward 1933). As long as the tree is alive, its roots are tough and woody, consisting, like the bole of the tree, of a central core of heartwood that is heavily impregnated with oleoresin and surrounded by sapwood with much less oleoresin. When the tree dies, the dense, oleoresin-impregnated heartwood resists decomposition for many years, but the sapwood surrounding it rots quickly, creating a soft, moist substrate into which animals can easily dig or force their way (Means 2005b) (Fig. 9). Swelling as they grow, the roots leave behind cavities filled with soft and decaying wood after they die, but the

FIGURE 9. The long-persisting, rotting taproots of longleaf pine are a vitally important microhabitat for terrestrial vertebrates, most of which are fossorial.

hard, oleoresin-impregnated heartwood remains to keep the cavities partially filled and protected. During the frequent fires in longleaf pine savannas, the resinous heartwood may burn and the punky sapwood may smoulder underground for days after the ground fire has passed. The heartwood is so dense that it rarely burns completely, but it may ignite repeatedly in subsequent fires, keeping the underground cavity open. Numerous dead tree bases were scattered throughout the presettlement longleaf pine savannas, offering hideaways for fossorial animals around their rotting bases and root passageways (Means 2005b).

Kauffeld (1957, 1969) recorded the subterranean use of tree bases by 11 species of longleaf pine savanna vertebrates on a large game plantation in South Carolina (eastern box turtle, corn snake, yellow rat snake, common kingsnake, eastern hognose snake, copperhead, timber rattlesnake, eastern diamondback rattlesnake, eastern cottontail, cotton rat, Savannah sparrow). In the pulpy wood of decaying pine stumps or in tunnels formed by decaying pine roots in Richmond County, GA, Neill (1948) found the slimy salamander, southern toad, mud snake, black racer, corn snake, black rat snake, common kingsnake, and copperhead. During a radiotelemetry study of the eastern diamondback rattlesnake and cottonmouth in northern Florida, Means (1986, 2005b) compared the use by rattlesnakes of gopher tortoise burrows versus stumpholes (tree bases) and found that these snakes occupied stumpholes much more frequently than gopher tortoise burrows. Incidentally, he observed nine other species using stumpholes including the common kingsnake, cotton rat, coachwhip, common garter snake, black racer, northern bobwhite quail, eastern box turtle, gray rat snake, and opossum. Richter et al. (2001) radiotracked 12 crawfish frogs emigrating from a breeding pond in southern Mississippi for 24 to 88 days and noted that most of the frogs used stumpholes or root mounds during the period and all were in stumpholes at the end of the tracking study.

One might think that resinous stumpholes would be abundant after the cutting of the virgin longleaf pine forests at the turn of the twentieth century. Unfortunately for fossorial vertebrates, most of the old stumpholes have slowly disappeared after the naval stores industry discovered, in the early 1900s, that the resinous stumps could be pulled or dozed from the ground and cooked to extract turpentine, rosin, and pine oil (Wahlenberg 1946; Dyer and Sicilia 1990). Today, one has to be careful while walking in longleaf pine savannas not to step into stump extraction holes (Frost 1993). The harvest of resinous stumps may have been an important contributing factor in the decline of many pineland vertebrates in the Southeast (Moler 1992; Means 2005b).

In addition to underground refugia created by decaying tree bases and roots, burrows are made by many animals besides the gopher tortoise, southeastern pocket gopher, and, recently, armadillo. Some are excavated by earthworms (*Diplocardia* spp.), scarab beetles (*Geotrupes* spp.), cicada larvae (*Magicicada* spp.), mole crickets (Gryllotalpidae), ants (e.g., *Pogonomyrmex* spp., *Dolichoderus* spp., *Solenopsis* spp.), wolf spiders (Lycosidae), eastern mole (*Scalopus aquaticus*), pine vole (*Microtus pinetorum*), and oldfield mouse (*Peromyscus polionotus*), to name a few. No doubt all the burrows provide refuge for many of the small vertebrates that do not dig, such as the flatwoods salamander, common and striped newts, oak toad, eastern narrowmouth toad, scarlet snake, pine woods snake, short-tailed snake, crowned snakes, coral snake, glass lizards, and mole skink. A few studies have examined the ground surface disturbances of burrowing animals in longleaf savanna (Kalisz and Stone 1984, Hermann 1993, Simkin and Michener 2005), but no studies have examined the subterranean extent of burrows nor how important burrows are to the fossorial fauna. Moreover, if subterranean burrows and other cavities are limiting to the population survival of many longleaf pine savanna fossorial vertebrates, then disturbances to the soil and subterranean cavities from mechanical activities such as plowing, harrowing, bedding, roller chopping, stump extraction, or even conversion to pasture may be significantly harmful. Many of the recently reported declines in fossorial amphibians (Means et al. 1996, Franz and Smith 1999) and reptiles (Gibbons et al. 2000; Tuberville et al. 2000; Krysko and Smith 2005) might be at least partly explained by range-wide depletion of undisturbed longleaf pine savanna soil and its microcavities. Studies of the importance of subterranean cavities to longleaf pine savanna vertebrates are urgently needed.

The Importance of Fire to Wildleaf Pine Savannas and Adjacent Ecosystems

Fire is responsible worldwide, if not for the origin, then for the perpetuation of pine associations known as savannas (Mirov 1967). Pines may have evolved their special relationships with fire early in the evolutionary history of the group going back into the Jurassic, 150 million years ago (Mirov 1967). Longleaf pine, as a species, probably does not go nearly that far back in time, but it may have evolved from ancestors that were part of grassland/fire habitats throughout the later half of the Cretaceous and all of the Cenozoic. Longleaf pine savannas are pyrogenic communities that require periodic fires in order to exist through time (Means 1996; Platt 1999). In the absence of fire, longleaf pine savannas succeed to southern mixed hardwood forest (Platt and Schwartz 1990; Ware et al. 1993; Platt 1999), a completely different ecosystem, but fires associated with summer lightning probably have been characteristic of the climate of the Coastal Plain of the southeastern United States at least since the Miocene, 25 million years ago. Longleaf pine savanna plants are so completely adapted to the periodic effects of fire, and fire has been so continuously present in the Coastal Plain, that longleaf pine savanna is considered a climax vegetative type (Platt 1999).

Few studies of animal response to fire have been conducted, but Means and Campbell (1982) reviewed the effects of prescribed burning on amphibians and reptiles and concluded that for longleaf pine savanna vertebrates, few individuals are killed during fires, but some species experience limited mortality because they are casual visitors from hardwood communities, such as the eastern box turtle (Ernst et al. 1995), or are aggregated under unusually dense, flammable litter, such as the eastern glass lizard (Babbitt and Babbitt 1951). Under certain conditions of heavy fuel loads, strip burning, or when a snake is in its shed cycle and cannot see, a few individuals can be killed, but prescribed burning is beneficial to most herpetofauna by perpetuating their habitat (Means and Campbell 1982).

Prescribed burning in a Florida sandhill habitat increased diversity and abundance of amphibians and reptiles over control plots that had not been burned for 17 years, and some fire periodicities were better than others for maintaining high diversity (Mushinsky 1985). The six-lined racerunner, which likes open herbaceous vegetation and bare ground, was more abundant on sites burned annually (Mushinsky 1985), but the southeastern five-lined skink, which uses thick litter for foraging and shelter, was more abundant on 5- to 7-year burn cycles (Mushinsky 1992). On the other hand, fire periodicity had no effect on the peninsula crowned snake (*Tantilla r. relicta*), a longleaf pine savanna specialist (Mushinsky and Witz 1993). As expected, vegetation structure is affected by periodicity of fire (Mushinsky 1986, Mushinsky and Gibson 1991), so arboreal and ground-dwelling lizards will be affected differently. Both are adapted to living in longleaf pine savannas, so during frequent fires, the six-lined racerunner experiences a population increase while the southeastern five-lined skink declines, and vice versa when the fire periodicity is lengthened. So long as the habitat does not succeed into a hardwood forest, both species persist. Prescribed burning was recommended as a much more beneficial management tool than other techniques in maintaining optimal gopher tortoise habitat (Cox et al. 1987). Density of gopher tortoise populations is closely related to biomass of herbaceous food plants, which are promoted by frequent fires (Landers and Speake 1980).

Changes in the vertebrate community and vegetation can be dramatic on annually burned pinelands that are managed to resemble longleaf pine savannas. Shortleaf and loblolly pines grow up on abandoned agricultural land in rich clayhills soils of the Coastal Plain—soils that previously supported longleaf pine savanna. If these oldfield successional vegetations are prescribe burned annually after the pines reach about 10 years of age and are not so vulnerable to fire damage, a pine forest with a herbaceous ground cover mixed with fire-pruned hardwood sprouts (Ware et al. 1993) can be maintained. Following 15 years of fire exclusion on an 8.6-ha plot of one such oldfield pine forest that had been annually burned for over 70 years in northern Florida, only 11 of 43 bird species were encountered every year of the study. Most finches and brush-nesting species that were common at the beginning of the study no longer occurred on the study area, whereas several species associated with mesic hardwood forest conditions increased in abundance (Engstrom et al. 1984). The hardwood response in such habitats is vigorous, because hardwoods establish underground root systems that continue to grow and store nutrients in spite of having their annual stem growth killed. The same hardwood species seem to have a more difficult time becoming established in native wiregrass ground cover, possibly from root competition with the native herbs.

Spatial variation in fire temperatures appears to have much less effect on vegetation patterns and dynamics of vegetation in frequently burned savannas than does season of burn (Platt et al. 1988). Open-canopied longleaf pine savannas characterized by a ground cover of grasses and forbs appear to be a result primarily of lighting-initiated fires that occur repeatedly and frequently during the spring (May–June) of the year (Platt et al. 1988). Unfortunately, no experimental research has been done on the effects on vertebrates of the season of burn. On the other hand, many studies have shown that vertebrates respond positively to frequency of burn. For example, for the endangered red-cockaded woodpecker, a longleaf pine specialist whose ecology is best understood (Jackson 1994; Conner et al. 2001), experimental research has shown that social groups living with high percentages of wiregrass and low percentages of gallberry in the ground cover of their territories, and with larger areas of natural pine regeneration, have more adults and more neighboring groups, and produce more young than do other groups (James et al. 1997).

Longleaf pine savannas are vitally important to adjacent communities and their vertebrates. Fires in the pre-European Coastal Plain landscape probably were ignited by lightning mostly in the uplands of longleaf pine savannas and swept downslope into adjacent vegetations (Means 1996). Many of these vegetations, such as hillside seepage bogs, wet flats, and Atlantic white cedar (*Chamaecyparis thyoides*) stands, are wetland communities in which fires may not be easily ignited, but they nevertheless depend upon fires for their own characteristic nature (Folkerts 1982; Cerulean and Engstrom 1995; Peet this volume). This is especially true of seepage bogs that can be invaded by evergreen shrubs from downslope swamps in creek and stream bottoms (Folkerts 1982; Drewa et al. 2002a,b).

Means and Longden (1976) noticed that breeding choruses of male pine barrens treefrogs were more robust in landscapes in which herbaceous seepage bogs were free from woody shrubs that grew farther downslope in wetter soils. Means and Moler (1979) hypothesized that the shallow puddles of larval breeding sites of the pine barrens treefrog were dried up by the increased evapotranspiration resulting from the invasion of woody vegetation into seepage bogs. These same shallow seepage rills and puddles have recently been discovered as the breeding habitat of a new species of the dwarf salamander complex (Means and Jensen unpublished). In the long-term (5-plus years) absence of fire, seepage bogs dry up and become shaded by dense shrub growth. Heliophilic vegetation gives way to peaty soils covered with deep hardwood leaf litter and no exposed seepage water, all of which are unsuitable for the larvae of the pine barrens treefrog and the new species of the dwarf salamander complex.

Two other examples of forest vegetations that respond to fires coming into them from longleaf pine savannas are Atlantic white cedar forests (Ward and Clewell 1989; Frost 1995) and eastern and southern redcedar (*Juniperus virginiana, J. silicicola*). These species are sensitive to fire, but they are also sensitive to shade from hardwoods. They most often occur in wetland ecotones or steep upper slopes between longleaf pine savannas and southern temperate hardwood forests where fires periodically burn back the competing hardwoods and provide a zone for even-aged stands to grow (Monk 1968; Kucera 1981; Wright and Bailey 1982).

Kirkman (1995) reviewed the impacts of fire and hydrological regimes on vegetation in depression marshes (temporary ponds). Fire during drought promotes species richness of grasses and sedges. Winter fire followed by inundation can lead to shifts in the dominance of emergent vegetation, and long-term fire suppression can result in the replacement of herbaceous wetland vegetation with hardwood species (Kirkman 1995). Hardwood litter can significantly affect the occurrence of fires in wetlands and, from the point of view of vertebrates, the mix of amphibian species that utilize those wetlands (Pechmann et al. 2001).

Recently, Schurbon and Fauth (2003) reported a decrease of frogs and salamanders in plots burned on shorter versus longer rotations. They argued that amphibians would benefit from longer prescribed burn frequencies of 5 to 7 years. Means et al. (2004) pointed out many weaknesses in the Schurbon and Fauth (2003) study and maintained that no one has yet demonstrated unequivocally any explicit amphibian response to fire; therefore, managers should not refrain from burning longleaf pine savannas every 1 to 3 years, otherwise hardwood encroachment will eliminate the habitat altogether in time.

Until we know more about the details of the effects of fire on longleaf pine savannas and their vertebrates, the best fire management strategy probably should be to use prescribed fires frequently (1 to 3 years) and in the season in which lightning-ignited fires burned naturally (May–June). Frequent burning is required on most longleaf pine sites because of their recent long history (more than 80 years) of fire suppression. Early growing season fires are also important because they suppress the naturally woody components of longleaf pine savannas better than winter fires (Platt et al. 1988), and they stimulate flowering of the grasses and some forbs (Platt et al. 1988, 1991; Streng et al. 1993) that otherwise do not readily reproduce sexually. On the other hand, if for some reason (severe drought that forces regionwide burning bans) it is not possible to burn in May or June, prescribed fire at other seasons is better than no fire at all (Cox et al. 1987).

Problems Associated with Pine Plantation Silviculture

The largest known breeding migration of the flatwoods salamander (*Ambystoma cingulatum*) was monitored over a 22-year period following its discovery in 1970 in Liberty County, FL. Nightly migrations of 200–300 adults across a 4.3-km stretch of paved highway in 1970–1972 had dwindled to less than one individual per night in 1990–1992. Bedding and planting slash pine on the wet flat into which the flatwoods salamanders were migrating to breed may have interfered with migration, successful hatching, larval life, and feeding and finding suitable cover postmetamorphosis (Means et al. 1996).

Gopher tortoises abandoned their burrows in mature slash pine plantations at an average rate of 22%/year (Aresco and Guyer 1999a) and grew more slowly on slash pine plantations than they did in any other published study and were estimated to require at least 20 years to attain sexual maturity (Aresco and Guyer 1999b). Intensive soil disturbance associated with site preparation and conversion to pine plantations in the 1970s destroyed much of the native ground cover. Slow growth, which resulted in delayed maturity, was attributed to poor forage quality of sparse ground cover vegetation, especially legume and nonlegume forbs (Aresco and Guyer 1999b).

Breeding bird density, species richness, diversity (H'), and biomass were highest in the longleaf pine forest and differed ($P < 0.05$) from those found in all age-classes (1, 10, 24, 40 years) of slash pine plantations (Repenning and Labisky 1985). Although slash pine plantations in northern Florida do not provide habitat that will maintain the breeding bird community of the natural longleaf pine forest, older plantations (more than 40 years) do provide habitat for a wintering bird community that is reasonably similar to that of the natural longleaf pine forest (Repenning and Labisky 1985).

White et al. (1975) compared habitat differences 9 years after the planting of slash pine among three replicated levels of site preparation ranging from clearcut, burn, and plant (low intensity) to clearcut, burn, KG, single harrow, bed, and plant (high intensity). They found that where grasses and forbs were least productive (high site preparation intensity), ground arthropods, small mammals, herbivores, birds, and insectivores were least abundant. They concluded that growth and development of slash pine overstories was favored by intensive site preparation at the expense of understory wildlife habitat.

It is well documented that the wildlife habitat values of pine plantations are inferior to those of natural pine stands (Harris et al. 1974; Umber and Harris 1974; Kautz 1984; Repenning and Labisky 1985; McComb et al. 1986). Hence, the net result is that a large percentage of remaining pine forests in the Coastal Plain now provide poor-quality habitat for many formerly abundant species of wildlife (Kautz 1993). Throughout the Coastal Plain, there is no natural ecological analog of the modern pine plantation (Means 2005a). Because the main objective in pine plantations is to maximize cellulose production, planted pines that can tolerate close stocking and canopy closure (slash, loblolly, sand) have largely replaced native longleaf pine savannas. After 10–15 years of growth, plantations have very little or no ground cover and are not savannas, but instead, densely stocked forests where most of the photosynthesis takes place in the canopy and virtually none at ground level. One should not expect, therefore, that vertebrates adapted to open-canopied savannas with a rich, herbaceous ground cover would be suited for life in deep shade and pine detritus.

The Problem of Habitat Fragmentation

Habitat reduction poses obvious problems for vertebrate populations, but habitat fragmentation is sometimes quite subtle and often just as dangerous. It is not so obvious that pine plantations, for instance, create large holes in the population landscape of many longleaf pine savanna vertebrates. The red-cockaded woodpecker neither colonizes nor survives in planted pine stands of any commercial age, even though both habitats have an abundance of pines (Costa 1995). Likewise, small populations occupying a large area with a few patches of unconnected longleaf pine savanna scattered about are likely to have inbreeding problems, restricted genetic diversity, and lack of gene flow between patches, depending upon the landscape matrix in which the patches are embedded. Habitat fragmentation produces the same result as habitat reduction: breaking up into smaller areas. The species–area relationship, a well-documented principle of ecology, says that each fragment, at equilibrium, will have fewer species than the original (Simberloff 1993).

The metapopulation dynamics of vertebrate specialists in longleaf pine savannas differ for different groups. Those vertebrates, mostly amphibians, whose life cycles depend on breeding ponds must live in a landscape that not only provides a breeding pond for a local population and a sufficient upland habitat for the adults, but also alternative ponds in which to breed if one pond is destroyed, or when the hydroperiod fails or is altered. More than for many of the reptiles, mammals, and birds that live in the same longleaf pine savanna habitats during all life stages, the vertebrates that depend upon ponds need proximity of other ponds from which to recruit new propagules if the local population goes extinct. If ponds are too far apart or isolated by roads, agricultural fields, older silviculture stands, clearcuts, or urbanization, populations cannot be restored.

And yet, for the intact remnants of longleaf pine savannas, almost no data exist on the effects of roads on the smaller species of amphibians, reptiles, and mammals that may be loathe to cross such a hot, coverless, and hostile habitat—even dirt roads. In a Maine study of road effects on amphibian movements, forest roads served as a partial filter to the movements of some amphibian species, notably salamanders, including a longleaf pine savanna species, the eastern newt (DeMaynadier and Hunter 2000). Fossorial vertebrates such as the eastern mole, shrews, eastern pocket gopher, ground skink, Florida worm lizard, mole skink, southern hognose snake, mole kingsnake, and others might be inhibited from crawling on the surface of bare hard ground, pavement, or exposing themselves to birds of prey and other predators.

Means (1999) found fewer than expected mole salamanders immigrating into a pond next to a heavily trafficked federal highway through longleaf pine sandhills in north Florida and Smith et al. (2005) discussed the direct and indirect impacts highways have on amphibians and reptiles, generally, using the case study of U.S. Highway 441 through Payne's Prairie in north central Florida as an example. In the only study of its kind, Rudolph et al. (1999) trapped large snakes at incremental distances away from roads through longleaf pine savannas on the Angelina National Forest in east Texas. Their data strongly suggest that populations of large snakes (e.g., pine snake and timber rattlesnake) are reduced by 50% or more to a distance of 450 m from roads of moderate use (2400 vehicles per day).

What were thought to have been good populations of keystone species in longleaf pine savannas were found to be declining in spite of being on "large" reserves. McCoy and Mushinsky (1992) found that a population of the gopher tortoise had a 33% decline in active burrows over a decade on the J.N. "Ding" Darling National Wildlife Refuge in Florida. Similarly, in the largest remaining population of the red-cockaded woodpecker on Florida's Apalachicola National Forest, which is split into two roughly equivalent ranger districts, James (1991) discovered that the birds on the eastern district were declining and that the river separating the two districts may be a barrier to gene flow in the species. Moreover, the fact that artificially made cavities are readily occupied on the western district, which was deemed to have a healthy and stable red-cockaded woodpecker population of about 500 colonies, shows that cavity limitation was severe even there (James et al. 1997).

Declining Species

Between 1513, when Florida was first sighted by Ponce de Leon, and the mid-1900s, six indigenous vertebrates disappeared from the Coastal Plain at the hands of European man. The American bison (*Bison bison bison*) was extirpated in the late 1700s or early 1800s as a result of wanton slaughter by early settlers (Humphrey 1992b). The Florida red wolf (*Canis rufus floridanus*) disappeared in the early 1900s (Robson 1992). The Carolina parakeet (*Conuropsis carolinensis*) was virtually extinct by 1900, exterminated by man as agricultural pests, for food, and for sport (Hardy 1978a). The passenger pigeon was driven to extinction by 1914, the victim of mass slaughter for food and sport (Hardy 1978b). And the ivory-billed woodpecker, last sighted in Georgia in 1955 (Means 2004b) and in Florida in 1969 (Kautz 1993), met its demise due to the logging of mature stands of lowland hardwoods and, some think, to the removal of old-growth longleaf pines in the early part of the twentieth century (Tanner 1942). Not the least of the extinctions was indigenous man, who was completely gone from Florida between 1700 and 1800 (Tebeau 1971) and largely removed from the southeastern United States by 1839 (Perdue and Green 1995).

Among the 17 species of amphibians that are specialists in longleaf pine savannas, only the flatwoods salamander is federally threatened, but the striped newt is under review (Linda LaClaire personal communication) and the gopher and crawfish frogs are threatened (Godley 1992) or imperiled on various state lists (Bailey and Means 2004). All 17 species have declined when one takes into account that their primary habitat, longleaf pine savannas, have been reduced to less than 3% of their original extent (Ware et al. 1993). The best review of declining amphibians and threats to their populations in the southeastern United States is by Dodd (1997), and additional information is in Lannoo (2005).

Most of the longleaf pine savanna specialist snakes are in decline and in need of studies of their basic terrestrial biology. The eastern Indigo snake is a federally threatened species (U.S. Fish and Wildlife Service 1978). The short-tailed snake is listed as threatened in Florida (Campbell and Moler 1992). The Florida pine snake is a Species of Special Concern in Florida (Franz 1992), imperiled in Alabama (Means 2004a), and should rank at least as a species of special concern in Georgia and South Carolina. The southern hognose snake has declined severely throughout its range (Tuberville et al. 2000), and the historic range of the eastern diamondback rattlesnake has shrunk, with populations at the extremes of its range extinct in Louisiana and threatened in North Carolina (Martin and Means 2000), and experiencing severe age-class truncation in Florida, Georgia, and Alabama (Means in preparation). Gibbons et al. (2000) likened the global decline of reptiles to that of amphibians, citing several examples in the southeastern United States from small ponds surrounded by longleaf pine savannas (black swamp snake, Florida green water snake, chicken turtle) to residents of longleaf pine ecosystems (gopher tortoise, indigo snake, eastern diamondback rattlesnake, eastern box turtle, timber rattlesnake, southern hognose snake). Except for very limited data for a few species, very little is known about how much acreage of good habitat is required to support minimal breeding populations of any of them.

Fire ants have been proposed as a possible threat to ground-nesting and terrestrial egg-laying vertebrates (Mount 1981). The fire ant *Solenopsis geminata* was once reported from a wide variety of habitats but seems to have been partly displaced by another species of fire ant, *S. invicta*, which was accidentally introduced into Mobile, AL, from Brazil about 1940 (Wilson and Brown 1958; Buren 1972). In a study of the use of flatwoods versus sandhill savannas on the Apalachicola National Forest, FL, Tschinkel (1987) found that both species preferred sites whose soil had been mechanically disturbed, such as paved road shoulders, graded roadsides, and cleared and replanted flatwoods (*S. invicta*) or sandhills (*S. geminata*). The presence of both species in native ground cover of flatwoods and sandhills was greatly reduced in comparison to sites having mechanical disturbance, except that *S. invicta* was abundant in the littoral zones around the margins of ponds, which are naturally disturbed by rising and falling pond water levels. If the tendency for fire ants to stay out of undisturbed native ground cover is widespread, then fire-ant predation on vertebrate eggs, if it were in fact a problem, would not have much of an impact in remnant patches of the native longleaf pine savannas having undisturbed soils. This would give land managers all the more reason to protect native herbaceous vegetation from mechanical disturbance and to maintain it with appropriate prescribed fires.

One of the longleaf pine specialist birds, the red-cockaded woodpecker, has been federally threatened since 1970. In spite of receiving more research and management attention than any other longleaf pine savanna vertebrate, it remains on the decline in part of its largest surviving enclave (James 1991), and current management there for sawtimber is not sustainable (James et al. 2003). The northern bobwhite quail, whose 1920s population decline in the Southeast was responsible for stimulating the research that discovered the importance of fire in longleaf pine savannas (Stoddard 1931, 1962), has endured a decline of more than 65% over the last 20 years throughout its range (Brennan 1991). Among the factors of the decline is degradation and shrinking of habitat due to hardwood invasion of unburned pinelands, the very same cause of the decline in the 1920s. The white-breasted and brown-headed nuthatches and Bachman's sparrow are not yet listed species, but their longleaf pine savanna habitat continues to decrease.

All three species of mammals that are longleaf pine specialists have declined. The Florida mouse, having the smallest range and being associated with the gopher tortoise, whose populations are dwindling, is considered a threatened species in Florida, the state in which it is endemic (Layne 1992). The principal race of the fox squirrel in Florida is threatened (Kantola 1992), and the species has declined and become extirpated widely over its entire range (Weigl et al. 1989). Goff's gopher, a race of the southeastern pocket gopher that was endemic on a sandy ridge in east-central Florida, is thought to be extinct (Humphrey 1992a; Brown 1997). All over the range of the southeastern pocket gopher, however, populations that were abundant and obvious in the 1960s and 1970s have vanished (D. B. Means personal observation).

The drastic loss of longleaf pine savannas has had an even more severe impact on plants. Hardin and White (1989) listed 191 rare plant taxa that occur in ecosystems in the southeastern United States (Alabama, Florida, Georgia, Mississippi, North Carolina, South Carolina) where wiregrass (*Aristida stricta*) is an important component. Using the Nature Conservancy's Natural Heritage Program methodology, 122 taxa were considered endangered or threatened throughout their ranges. Seven taxa have been proposed or listed as endangered by the U.S. Fish and Wildlife Service, and 61 taxa are listed as endangered or threatened by rare plant laws in three states. Hardin and White (1989) estimated that 66 rare wiregrass associates are local endemics, one of the higher numbers reported for a regional ecosystem type in the United States.

Conclusions

The total number of resident vertebrates in longleaf pine savannas, 212 species including 38 species that are specialists occurring exclusively or primarily in longleaf pine savannas, is greater than for any other habitat type in the Coastal Plain of the southeastern United States and one of the largest vertebrate faunas in temperate North America. Such high species richness should be expected, given the antiquity of this type of ecosystem and that longleaf pine savannas once accounted for more than 60% of a landscape stretching from southeastern Virginia to east Texas. Unfortunately, the native longleaf pine savannas in which resident and specialist vertebrates evolved and to which, at least the specialists, are best adapted, have shrunk to less than 3% of their former expansive range, with the remnants highly fragmented and isolated.

Because rights of the private landowner are so strong, it is difficult to imagine how biodiversity might be governmentally regulated on private lands, so the main hope for conservation of longleaf pine savanna biodiversity is on publicly owned lands. Unfortunately, publicly owned lands are not distributed very well in the Coastal Plain. Some states have very few. The federal lands offer the largest opportunity, but these, too, are inequitably distributed throughout the Coastal Plain. Ensconcement of longleaf pine savannas on publicly owned lands is not enough to ensure their survival, however. Longleaf pine savannas need to be managed properly, but changing politics with respect to resource utilization (logging, recreation), air pollution (smoke from prescribed burning), and encroaching development (homes, municipalities, in-holdings) can present big challenges to management, especially prescribed burning.

Single-species management for a target species such as the red-cockaded woodpecker is probably not a good overall management strategy in any ecosystem. Considering how many longleaf pine savanna species are rare or threatened, the best management should be a multiple species program that focuses mainly on restoring and maintaining longleaf pine savannas to the most natural conditions possible. Prescribed fires on short rotations (1 to 3 years) in the early lightning season (May and June) are most likely to produce an ecosystem mosaic suitable for all the native species. As research produces new information about how the presettlement longleaf pine savannas really were structured, it should be integrated into the overall management program. A prime example comes from recent red-cockaded woodpecker research on the Apalachicola National Forest (ANF), the largest national forest in the eastern United States. James et al. (2001, 2003) found that present stocking of longleaf pines may be too dense and that not enough longleaf pine reproduction is taking place. In other words, current management for sawtimber on the ANF is not sustainable in this, the largest population of this endangered species. They recommended more vigorous burning and monitoring of the condition of the ground cover. When it is less than 30% herbaceous, they recommend a form of single tree selection that emphasizes thinning from below in combination with minigroup selection (harvest of patches of trees up to a radius equal to the height of canopy trees).

If we are to keep any semblance of the faunal diversity that the original longleaf pine savannas bequeathed us, we must recognize how valuable are the few remaining tracts and insure that they are properly studied and managed to prevent further losses of the ecosystems and the vertebrate species they contain. Whatever other values are gained from the preservation of the small percentage of remaining longleaf pine savannas and their faunal diversity, these remnants are, at the very least, exceedingly valuable repositories where most of the knowledge of the evolution and adaptation of the constituent species is stored. Learning why the red-cockaded woodpecker is a threatened species, for instance, would not be possible in ruderal habitats in which it did not evolve. To lose this storehouse of valuable information would be a loss to humanity far greater than loss of its simple parts.

Acknowledgments

I owe many thanks to the late Edwin V. Komarek, Sr. for empowering me to study longleaf pine savanna vertebrates and fire ecology. William J. Platt III provided intellectual stimulation and W. Wilson Baker abundant field knowledge and the love of fieldwork. Over the years some of my most helpful companions in the field were James B. Atkinson, Robert L. Crawford, Guy H. Means, and Ryan C. Means. I am especially grateful to my colleagues, W. Wilson Baker, Ken Dodd, Kevin Enge, Harley Means, and Ryan Means, who allocated their valuable time to review this manuscript.

References

Allen, E.R. 1968. The increase of rattlesnakes, *Crotalus adamanteus*, in Florida. American Association of Zoological Parks and Aquariums, 15 May 1968. Pittsburgh, PA.

Anonymous. 1990. Summary report, Scientific summit on the Red-cockaded Woodpecker. Unpublished report, Southeast Negotiation Network, Georgia Institute of Technology, Atlanta.

Aresco, M.J., and Guyer, C. 1999a. Burrow abandonment by gopher tortoises in slash pine plantations of the Conecuh National Forest. *J Wildl Manage* 63(1):26–35.

Aresco, M.J., and Guyer, C. 1999b. Growth of the tortoise *Gopherus polyphemus* in slash pine plantations of southcentral Alabama. *Herpetologica* 55(4):499–506.

Ashton, R.E., Jr. 1992. Flatwoods salamander, *Ambystoma cingulatum* (Cope). In *Rare and Endangered Biota of Florida, Volume III, Amphibians and Reptiles*, ed. P.E. Moler, pp. 39–43. Gainesville: University Press of Florida.

Ashton, R.E., Jr., and Ashton, P.S. 1981. *Handbook of Reptiles and Amphibians of Florida. Part One: The Snakes*. Miami, FL: Windward Publishing Co.

Ashton, R.E., Jr., and Ashton, P.S. 1985. *Handbook of Reptiles and Amphibians of Florida. Part Two: Lizards, Turtles and Crocodilians*. Miami, FL: Windward Publishing Co.

Ashton, R.E., Jr., and Ashton, P.S. 2005. Natural history and status of the flatwoods salamander, *Ambystoma cingulatum* (Cope) in Florida. In *Status and Conservation of Florida Amphibians and Reptiles*, eds. W. Meshaka and K. Babbitt, pp 62–73. Malabar, FL: Krieger Press.

Auffenberg, W. 1969. *Tortoise Behavior and Survival*. Chicago: Rand McNally.

Auffenberg, W., and Franz, R. 1982. The status and distribution of the gopher tortoise (*Gopherus polyphemus*). In *North American Tortoises: Conservation and Ecology*, ed. R.B. Bury, pp. 95–126. U.S. Fish and Wildlife Service Research Report 12.

Babbitt, L.H., and Babbitt, C.H. 1951. A herpetological study of burned-over areas in Dade County, Florida. *Copeia* 1951:79.

Bailey, M.A. 2001. The pocket gopher project. The tortoise burrow. *Newsl Gopher Tortoise Counc* 21(1):7.

Bailey, M.A., and Means, D.B. 2004. Gopher frog, *Rana capito*, and Mississippi gopher frog, *Rana sevosa*. In *Vertebrate and Selected Invertebrate Wildlife of Alabama*, ed. R.E. Mirarchi, pp. 15–17. Auburn, AL: Proceedings of the Second Nongame Wildlife Conference.

Baker, W.W. 1974. Longevity of lightning-struck trees and notes on wildlife use. *Proc Tall Timbers Fire Ecol Conf* 13:497–504.

Baker, W.W. 1981. The distribution, status, and future of the red-cockaded woodpecker in Georgia.

In *Proceedings of the Nongame and Endangered Wildlife Symposium*, eds. R.R. Odum and J.W. Guthrie, pp. 82–87. Georgia Department of Natural Resources, Game and Fish Division, Technical Bulletin WL5.

Barlow, C. 2000. *The Ghosts of Evolution: Nonsensical Fruit, Missing Partners, and Other Ecological Anachronisms*. New York: Basic Books.

Barlow, R.F., and Jones, F.K., Jr. 1965. The white-tailed deer in Florida. Florida Game and Fresh Water Fish Commission Technical Bulletin 9:1–239.

Beane, J.C., Thorp, T.J., and Jackson, D.A. 1998. *Heterodon simus* (southern hognose snake) diet. *Herpetol Rev* 29:44–45.

Belson, M.S. 2000. *Drymarchon corais couperi* (eastern indigo snake) and *Micrurus fulvius fulvius* (eastern coral snake). Predator-prey. *Herpetol Rev* 31:105.

Bennett, D.H., Gibbons, J.W., and Franson, J.C. 1978. Terrestrial activity in aquatic turtles. *Ecology* 51:738–740.

Bennett, S.H., and Nelson, J.B. 1991. Distribution and status of Carolina bays in South Carolina. Nongame and Heritage Trust Publication No. 1, SC Wildlife and Marine Resources Department, Columbia.

Boughton, R.G., Staiger, J., and Franz, R. 2000. Use of PVC pipe refugia as a sampling technique for hylid treefrogs. *Am Midl Nat* 144:168–177.

Brennan, L.A. 1991. How can we reverse the northern bobwhite population decline? *wildl Soc Bull* 19:544–555.

Bridges, E.L., and Orzell, S.L. 1989. Longleaf pine communities of the west Gulf Coastal Plain. *Nat Areas J* 9(4):246–263.

Brimley, C.S. 1923. The dwarf salamander at Raleigh, NC. *Copeia* 120:81–83.

Brode, W.E., and Allison, P. 1958. Burrowing snakes of the panhandle counties of Mississippi. *Herpetologica* 14(1):37–40.

Brodkorb, P. 1964. Catalogue of fossil birds: Part 2 (Anseriformes through Galliformes). *Bull Fla State Mus Biol Sci* 8(3):195–335.

Brooks, G.R., Jr. 1967. Population ecology of the ground skink, *Lygosoma laterale* (Say). *Ecol Monogr* 37:71–87.

Brown, E.E. 1979. Some snake food records from the Carolinas. *Brimleyana* 1:113–124.

Brown, L.E., and Means, D.B. 1984. Fossorial behavior and ecology of the chorus frog *Pseudacris ornata*. *Amphibia-Reptilia* 5:261–273.

Brown, L.N. 1997. *Mammals of Florida*. Miami, FL: Windward Publishing Co.

Brown, R.B., Stone, E.L., and Carlisle, V.W. 1990. Soils. In *Ecosystems of Florida*, eds. R. Myers and J. Ewel, pp. 35–69. Orlando: University of Central Florida Press.

Bruce, D. 1947. Thirty-two years of annual burning in longleaf pine. *J For* 45:809–814.

Buhlmann, K.A. 1995. Habitat use, terrestrial movements, and conservation of the turtle, *Dierochelys reticularia* in Virginia. *J Herpetol* 29:173–181.

Buren, W.F. 1972. Revisionary studies on the taxonomy of the imported fire ants. *J Ga Entomol Soc* 7:1–27.

Burke, V.J., and Gibbons, J.W. 1995. Terrestrial buffer zones and wetland conservation: A case study of freshwater turtles in a Carolina Bay. *Conserv Biol* 9(6):1365–1369.

Burton, T.M., and Likens, G.E. 1975. Salamander populations and biomass in the Hubbard Brook Experimental Forest, New Hampshire. *Copeia* 1975:541–546.

Caldwell, J.P. 1987. Demography and life history of two species of chorus frogs (Anura: Hylidae) in South Carolina. *Copeia* 1987:114–127.

Campbell, H.W., and Christman, S.P. 1982. The herpetological components of Florida sandhill and sand pine scrub associations. In *Herpetological Communities*, ed. N.J. Scott, pp. 163–171. USDI Wildlife Research Report 13.

Campbell, H.W., and Moler, P.E. 1992. Threatened short-tailed snake. In *Rare and Endangered Biota of Florida, Volume III, Amphibians and Reptiles*, ed. P. Moler, pp. 150–153. Gainesville: University Press of Florida.

Carr, A.F., Jr. 1940. A contribution to the herpetology of Florida. *Univ Fla Publ Biol Sci Ser* 3(1):1–118.

Cerulean, S.I. 1991. The Preservation 2000 report. Florida's natural areas—what have we got to lose? The Nature Conservancy, Winter Park, FL.

Cerulean, S.I., and Engstrom, R.T. 1995. Fire in wetlands: A management perspective. *Proc Tall Timbers Fire Ecol Conf* 19.

Chapman, H.H. 1943. A 27-year record of annual burning versus protection of longleaf pine reproduction. *J For* 41:71–72.

Christensen, N.L. 1988. Vegetation of the southeastern Coastal Plain. In *North American Terrestrial Vegetation*, eds. M. G. Barbour and W.D. Billings, pp. 317–363. London: Cambridge University Press.

Christman, S.P., and Means, D.B. 1992. Rare striped newt. In *Rare and Endangered Biota of Florida, Volume III, Amphibians and Reptiles*, ed. P.E. Moler, pp. 62–65. Gainesville: University Press of Florida.

Clark, R.F. 1949. Snakes of the hill parishes of Louisiana. *J Tenn Acad Sci* 24:244–261.
Clausen, C.J., Cohen, A.D., Emiliani, C., Holman, J.A., and Stipp, J.J. 1979. Little Salt Spring, Florida: A unique underwater site. *Science* 203: 609–614.
Conant, R., and Collins, J.T. 1998. *A Field Guide to Reptiles and Amphibians: Eastern and Central North America*, 3rd edition, expanded. Boston: Houghton Mifflin.
Conner, R.N., Rudolph, D.C., and Walters, J.R. 2001. *The Red-cockaded Woodpecker: Surviving in a Fire-Maintained Ecosystem*. Austin: University of Texas Press.
Costa, R. 1995. Biological opinion on U. S. Forest Service environmental impact statement for the management of the red-cockaded woodpecker and its habitat on national forests in the southern region. In *USDA Forest Service, Final Environmental Impact Statement*, Vol. II, pp. 1–192. Management Bulletin R8-MB–73.
Cox, J., Inkley, D., and Kautz, R. 1987. Ecology and habitat protection needs of gopher tortoise (*Gopherus polyphemus*) populations found on lands slated for large-scale development in Florida. Florida Game and Fresh Water Fish Commission, Nongame Wildlife Program Technical Report No. 4.
Crenshaw, J.W., Jr. 1955. The life history of the southern spiny lizard, *Sceloporus undulatus undulatus* Latreille. *Am Midl Nat* 54(2):257–298.
Dalrymple, G.H., Bernardino, F.S., Jr., Steiner, T.M., and Nodell, R.J. 1991. Patterns of species diversity of snake community assemblages, with data on two Everglades snake assemblages. *Copeia* 1991:517–521.
Deckert, R.F. 1918. A list of reptiles from Jacksonville, Florida. *Copeia* 1918(54):30–33.
Delcourt, P.A., and Delcourt, H.R. 1987. *Long-Term Forest Dynamics of the Temperate Zone: A Case Study of Late-Quaternary Forests in Eastern North America*. New York: Springer-Verlag.
DeMaynadier, P.G., and Hunter, M.L., Jr. 2000. Road effects on amphibian movements in a forested landscape. *Nat Areas J* 20(1):56–65.
Diemer, J.E., and Speake, D.W. 1983. The distribution of the eastern indigo snake, *Drymarchon corais couperi*, in Georgia. *J Herpetol* 17:256–264.
Dodd, C.K., Jr. 1992. Biological diversity of a temporary pond herpetofauna in north Florida sandhills. *Biodivers Conserv* 1:125–142.
Dodd, C.K., Jr. 1995a. The ecology of a sandhills population of the eastern narrow-mouthed toad, *Gastrophryne carolinensis*, during a drought. *Bull Fla Mus Nat Hist Biol Sci* 38, Pt. I(1):11–41.
Dodd C.K., Jr. 1995b. Reptiles and amphibians in the endangered longleaf pine ecosystem. In *Our Living Resources: A Report to the Nation on Distribution, Abundance, and Health of U.S. Plants, Animals, and Ecosystems*, eds. E.T. LaRoe, G.S. Farris, C.E. Puckett, P.D. Doran, and M.J. Mac, pp. 129–131. Washington, DC: National Biological Service.
Dodd, C.K., Jr. 1996. Use of terrestrial habitats by amphibians in the sandhill uplands of north-central Florida. *Alytes* 14:42–52.
Dodd, C.K., Jr. 1997. Imperiled amphibians: A historical perspective. In *Aquatic Fauna in Peril: The Southeastern Perspective*, Special Publication 1, eds. G.W. Benz and D.E. Collins, pp. 165–200. Southeast Aquatic Research Institute. Decatur, GA: Lenz Design and Communications.
Dodd, C.K., Jr. 2001. *North American Box Turtles: A Natural History*. Norman: University of Oklahoma Press.
Dodd, C.K., Jr., and Franz, R. 1995. Seasonal abundance and habitat use of selected snakes trapped in xeric and mesic communities of north-central Florida. *Bull Fla Mus Nat Hist* 38, Part I(2):43–67.
Dodd, C.K., Jr., Johnson, S.A., and Means, D.B. 2005. *Notophthalmus perstriatus*. In *Amphibian declines, the conservation status of United States species*, ed. M. Lannoo. Berkeley: University of California Press.
Dorcas, M.E., Gibbons, J.W., and Dowling, H.G. 1998. *Seminatrix* Cope, black swamp snake. Catalogue of American Amphibians and Reptiles 679.1–679.5.
Drewa, P.B., Platt, W.J., and Moser, E.B. 2002a. Community structure along elevation gradients in headwater regions of longleaf pine savannas. *Plant Ecol* 160:61–78.
Drewa, P.B., Platt, W.J., and Moser, E.B. 2002b. Fire effects on resprouting of shrubs in headwaters of southeastern longleaf pine savannas. *Ecology* 83(3):755–767.
Duellman, W.E., and Trueb, L. 1986. *Biology of Amphibians*. New York: McGraw–Hill Book Co.
Dundee, H.A., and Rossman, D.A. 1989. *The Amphibians and Reptiles of Louisiana*. Baton Rouge: Louisiana State University Press.
Dunning, J.B., and Watts, B.D. 1990. Regional differences in habitat occupancy by Bachman's sparrows. *Auk* 107:463–472.
Dyer, D., and Sicilia, D.B. 1990. *Labors of a Modern Hercules: The Evolution of a Chemical Company*. Boston: Harvard Business School Press.
Echternacht, A.C., and Harris, L.D. 1993. The fauna and wildlife of the southeastern United States. In *Biodiversity of the Southeastern United States: Lowland Terrestrial Communities*, eds. W.H. Martin, S.G.

Boyce, and A.C. Echternacht, pp. 81–116. New York: John Wiley & Sons.

Ehrlich, P.R., Dobkin, D.S., and Wheye, D. 1988. *The Birder's Handbook: A Field Guide to the Natural History of North American Birds*. New York: Simon & Schuster.

Eisenberg, J. 1983. The gopher tortoise as a keystone species. *Proceedings of the Annual Meeting of the Gopher Tortoise Council* 4:1–4.

Emslie, S.D. 1998. Avian community, climate, and sea-level changes in the Plio-Pleistocene of the Florida Peninsula. *Ornithol Monogra* 50:1–113.

Enge, K.M. 1997. A standardized protocol for drift-fence surveys. Florida Game and Fresh Water Fish Commission Technical Report No. 14.

Enge, K.M. 2002. Herpetofaunal drift-fence survey of two seepage bogs in Okaloosa County, Florida. *Fla Sci* 65:189–203.

Enge, K.M., and Sullivan, J.O. 2000. Seasonal activity of the scarlet snake, *Cemophora coccinea*, in Florida. *Herpetol Rev* 31:82–84.

Enge, K.M., and Wood, K.N. 2002. A pedestrian road survey of an upland snake community in Florida. *Southeast Nat* 1:365–380.

Enge, K.M., and Wood, K.N. 2003. A pedestrian road survey of the southern hognose snake (*Heterodon simus*) in Hernando County, Florida. *Fla Sci* 66:189–203.

Engstrom, R.T. 1993. Characteristic mammals and birds of longleaf pine forests. *Proc Tall Timbers Fire Ecol Conf* 18:127–138.

Engstrom, R.T., Crawford, R.L., and Baker, W.W. 1984. Breeding bird populations in relation to changing forest structure following fire exclusion: A 15-year study. *Wilson Bull* 96:437–450.

Ernst, C.H., and Ernst, E.M. 2003. *Snakes of the United States and Canada*. Washington, DC: Smithsonian Books.

Ernst, C.H., Boucher, T.P., Sekscienski, S.W., and Wilgenbusch, J.C. 1995. Fire ecology of the Florida box turtle, *Terrapene carolina bauri*. *Herpetol Rev* 26(4):185–187.

Flannery, T. 2001. *The Eternal Frontier, an Ecological History of North America and Its Peoples*. Melbourne, Australia: Text Publishing.

Folkerts, G.W. 1982. The Gulf coast pitcher plant bogs. *Am Sci* 70:260–267.

Folkerts, G.W. 1997. Citronelle ponds: Little-known wetlands of the Central Gulf Coastal Plain, USA. *Nat Areas J* 17(1):6–16.

Ford, N.B., Cobb, V., and Stout, J. 1991. Species diversity and seasonal abundance of snakes in a mixed pine–hardwood forest in eastern Texas. *Southwest Nat* 36:171–177.

Franz, R. 1986. Florida pine snakes and gopher frogs as commensals of gopher tortoise burrows. In *The Gopher Tortoise and Its Community*, eds. D.R. Jackson and R.J. Bryant, pp. 16–20. Proceedings 5th Annual Meeting of the Gopher Tortoise Council.

Franz, R. 1991. Remember the drought? *Fla Wildl* 45(6):10–12.

Franz, R. 1992. Florida pine snake. In *Rare and Endangered Biota of Florida, Volume III, Amphibians and Reptiles*, ed. P.E. Moler, pp. 254–258. Gainesville: University Press of Florida.

Franz, R. 1995. Habitat use, movements, and home range in two species of ratsnakes (genus *Elaphe*) in a north Florida sand hill. Florida Game and Fresh Water Fish Commission, Nongame Wildlife Program Project Report, Tallahassee, FL.

Franz, R. 2005. Up close and personal: A glimpse into the life of pine snakes in a north Florida sandhill. In *Status and Conservation of Florida Amphibians and Reptiles*, eds. W. Meshaka and K. Babbitt, pp. 120–131. Melbourne, FL: Krieger Press.

Franz, R., and Auffenberg, W. 1978. The gopher tortoise: A declining species. In *Proceedings of the Rare and Endangered Wildlife Symposium*, eds. R. Odom and L. Landers, Technical Bulletin WL 4, pp. 61–63. Georgia Department of Natural Resources, Game and Fish Division, Social Circle.

Franz, R., and Smith, L.L. 1999. Distribution and status of the striped newt and Florida gopher frog in peninsular Florida. Final Report, Florida Fish and Wildlife Conservation Commission, Tallahassee.

Frost, C.C. 1993. Four centuries of changing landscape patterns in the longleaf pine ecosystem. *Proc Tall Timbers Fire Ecol Conf* 18:17–43.

Frost, C.C. 1995. Presettlement fire regimes in Southeastern marshes, peatlands, and swamps. *Proc Tall Timbers Fire Ecol Conf* 19:39–60.

Garner, J.A., and Landers, J.L. 1981. Food and habitat of the gopher tortoise in southwestern Georgia. *Proc Annu Conf Southeast Assoc Fish Wildl Agenc* 35:120–134.

Gates, C.A., and Tanner, G.W. 1988. Effects of prescribed burning on herbaceous vegetation and pocket gophers (*Geomys pinetus*) in a sandhill community. *Fla Sci* 51(3/4):129–139.

Gibbons, J.W., Scott, D.E., Ryan, T.J., Buhlmann, K.A., Tuberville, T.D., Metts, B.S., Greene, J.L., Mills, T., Leiden, Y., Poppy, S., and Wynne, C.T. 2000. The global decline of reptiles, déjà vu amphibians. *Bioscience* 50(8):653–666.

Godley, J.S. 1992. Threatened gopher frog. In *Rare and Endangered Biota of Florida, Volume III,*

Amphibians and Reptiles, ed. P.E. Moler, pp. 15–19. Gainesville: University Press of Florida.

Goin, C.J. 1947. A note on the food of *Heterodon simus*. *Copeia* 1947:275.

Goin, C.J. 1950. A study of the salamander *Ambystoma cingulatum*, with the description of a new subspecies. *Ann Carnegie Mus* 31:299–321.

Goin, O.B., and Goin, C.J. 1951. Notes on the natural history of the lizard *Eumeces laticeps*, in Florida. *Q J Fla Acad Sci* 14:29–33.

Goodyear, A.C. 1999. The early Holocene occupation of the southeastern United States: A geoarchaeological summary. In *Ice Age Peoples of North America: Environments, Origins, and Adaptations*, eds. R. Bonnichsen and K.L. Turnmire, pp. 432–481. Corvallis: Oregon State University Press.

Graham, A. 1965. Origin and evolution of the biota of southeastern North America: Evidence from the fossil plant record. *Evolution* 18:571–585.

Grayson, D.K. 2001. The archaeological record of human impacts on animal populations. *J World Prehist* 15:1–68.

Greenberg, C.H. 2001. Spatio-temporal dynamics of pond use and recruitment in Florida gopher frogs (*Rana capito aesopus*). *J Herpetol* 35:74–85.

Greenberg, C.H., Neary, D.G., and Harris, L.D. 1994. Effect of high-intensity wildfire and silvicultural treatments on reptile communities in sand-pine scrub. *Conserv Biol* 8:1047–1057.

Guyer, C., and Bailey, M.A. 1993. Amphibians and reptiles of longleaf pine communities. *Proc Tall Timbers Fire Ecol Conf* 18:139–158.

Hallinan, T. 1923. Observations made in Duval County, northern Florida, on the gopher tortoise (*Gopherus polyphemus*). *Copeia* 1923:11–20.

Hansen, K.L. 1953. The burrow of the gopher tortoise. *Q J Fla Acad Sci* 26:353–360.

Hansen, K.L. 1958. Breeding pattern of the eastern spadefoot toad. *Herpetologica* 14:57–67.

Harcombe, P.A., Glitzenstein, J.S., Knox, R.G., Orzell, S.L., and Bridges, E.L. 1993. Vegetation of the longleaf pine region of the west Gulf Coastal Plain. *Proc Tall Timbers Fire Ecol Conf* 18:83–104.

Hardin, E.D., and White, D.L. 1989. Rare vascular plant taxa associated with wiregrass (*Aristida stricta*) in the southeastern United States. *Nat Areas J* 9(4):234–245.

Hardy, J.D., Jr. 1969. Reproductive activity, growth, and movements of *Ambystoma mabeei* Bishop in North Carolina. *Bull Md Herpetol Soc* 5(2):65–76.

Hardy, J.D., and Olmon, J. 1974. Restriction of the range of the frosted salamander, *Ambystoma cingulatum*, based on a comparison of the larvae of *Ambystoma cingulatum* and *Ambystoma mabeei*. *Herpetologica* 30:156–160.

Hardy, J.W. 1978a. Recently extinct Carolina parakeet. In *Rare and Endangered Biota of Florida*, ed. H.W. Kale, p. 120. Gainesville: University Press of Florida.

Hardy, J.W. 1978b. Extinct passenger pigeon. In *Rare and Endangered Biota of Florida. Vol. 2. Birds*, ed. H.W. Kale, pp. 120–121. Gainesville: University Press of Florida.

Harris, L.D., White, L.D., Johnston, J.E., and Milchunas, D.G. 1974. Impact of forest plantations on north Florida wildlife and habitat. *Proc Annu Conf Southeast Assoc Game Fish Comm* 28: 659–667.

Harrison, J.R. 1973. Observations on the life history and ecology of *Eurycea quadridigitata* (Holbrook). *HISS News J* 1:57–58.

Harrison, J.R., and Guttman, S.I. 2003. A new species of *Eurycea* (Caudata: Plethodontidae) from North and South Carolina. *Southeast Nat* 2(2):159–178.

Hermann, S.M. 1993. Small-scale disturbances in longleaf pine forest. In *The Longleaf Pine Ecosystem: Ecology, Restoration, and Management*, ed. S.M. Hermann, pp. 265–274. Proceedings of the Tall Timbers Fire Ecology Conference, Tallahassee, FL.

Hess, C.A., and James, F.C. 1998. Diet of the red-cockaded woodpecker in the Apalachicola National Forest. *J Wildl Manage* 62:509–517.

Heyward, F. 1933. The root system of longleaf pine on the deep sands of western Florida. *Ecology* 14(2):136–148.

Holman, J.A. 1991. North American Pleistocene herpetological stability and its impact on the interpretation of modern herpetofaunas: An overview. *Ill State Mus Sci Pap* 23:227–235.

Holman, J.A. 1995. *Pleistocene Amphibians and Reptiles in North America*. New York: Oxford University Press.

Holman, J.A. 1996. The large Pleistocene (Sangamonian) herpetofauna of the Williston IIIA site, north-central Florida. *Herpetol Nat Hist* 4(1):35–47.

Hooper, R.G., Lennartz, M.R., and Muse, H.D. 1991. Heart rot and cavity tree selection by red-cockaded woodpeckers. *J Wildl Manage* 55:323–327.

Hubbell, T.H., and Goff, C.C. 1939. Florida pocket gopher burrows and their arthropod inhabitants. *Proc Fla Acad Sci* 4:127–177.

Humphrey, S.R. 1992a. Recently extinct Goff's pocket gopher, *Geomys pinetis goffi*. In *Rare and Endangered Biota of Florida, Vol. 1, Mammals*, ed.

S.R. Humphrey, pp. 11–18. Gainesville: University Press of Florida.

Humphrey, S.R. 1992b. Recently extirpated Plains bison, *Bison bison bison*. In *Rare and Endangered Biota of Florida, Vol. 1, Mammals*, ed. S.R. Humphrey, pp. 47–53. Gainesville: University Press of Florida.

Jackson, D.L., and Franz, R. 1981. Ecology of the eastern coral snake (*Micrurus fulvius*) in northern peninsular Florida. *Herpetologica* 37:213–228.

Jackson, D.L., and Milstrey, E.G. 1989. The fauna of gopher tortoise burrows. In *Proceedings of the Gopher Tortoise Relocation Symposium*, eds. J.E. Diemer, D.R. Jackson, J.L. Landers, J.N. Layne, and D.A. Woods, pp. 86–98. Florida Game and Fresh Water Fish Commission Nongame Wildlife Program Technical Report 5. Tallahassee, FL.

Jackson, J.A. 1970. Predation of a black rat snake on yellow-shafted flicker nestlings. *Wilson Bull* 82:329–330.

Jackson, J.A. 1974. Gray rat snakes versus red-cockaded woodpeckers: Predator–prey adaptations. *Auk* 9(2):342–347.

Jackson, J.A. 1977. Notes on the behavior of the gray rat snake (*Elaphe obsoleta spiloides*). *J Miss Acad Sci* 22:94–96.

Jackson, J.A. 1978. Predation by a gray rat snake on red-cockaded woodpecker nestlings. *Bird-Banding* 49:187–188.

Jackson, J.A. 1988. The southeastern pine forest ecosystem and its birds: Past, present, and future. In *Bird Conservation, Vol. 3*, ed. J.A. Jackson, pp. 119–159. Madison: University of Wisconsin.

Jackson, J.A. 1994. Red-cockaded woodpecker (Picoides borealis). In *The Birds of North America*, No. 85, eds. A. Poole and F. Gill. Philadelphia: The Academy of Natural Sciences, Washington, DC: The American Ornithologist's Union.

Jackson, J.A., and Dakin, O.H. 1982. An encounter between a nesting barn owl and a gray rat snake. *Raptor Res* 16:60–61.

Jackson, J.F. 1973. Distribution and population phenetics of the Florida scrub lizard, *Sceloporus woodi*. *Copeia* 1973:746–761.

James, F.C. 1971. Ordinations of habitat relationships among breeding birds. *Wilson Bull* 83:215–236.

James, F.C. 1991. Signs of trouble in the largest remaining population of red-cockaded woodpeckers. *Auk* 108:419–423.

James, F.C., Hess, C.A., and Kufrin, D. 1997. Species-centered environmental analysis: Indirect effects of fire history on red-cockaded woodpeckers. *Ecol Appl* 7(1):118–129.

James, F.C., Hess, C.A., Kicklighter, B.C., and Thum, R.A. 2001. Ecosystem management and the niche gestalt of the red-cockaded woodpecker in longleaf pine forests. *Ecol Appl* 11:854–870.

James, F.C., Richards, P.M., Hess, C.A., McCluney, K.E., Walters, E.L., and Schrader, M.S. 2003. Sustainable forestry for the red-cockaded woodpecker's ecosystem. In *Red-Cockaded Woodpecker: Road to Recovery*, eds. R. Costa and S.J. Daniels. Blain, WA: Hancock House Publishers.

Janzen, D.H. 1984. Dispersal of small seeds by big herbivores: Foliage is the fruit. *Am Nat* 123:338–353.

Janzen, D.H., and Martin, P.S. 1982. Neotropical anachronisms: The fruits the gomphotheres ate. *Science* 215:19–27.

Jensen, J.B. 1996. *Heterodon simus* (southern hognose snake). Hatchling size. *Herpetol Rev* 27:25.

Johnson, S.A. 2002. Life history of the striped newt at a north-central Florida breeding pond. *Southeast Nat* 1(4):381–402.

Johnson, S.A. 2003. Orientation and migration distances of a pond-breeding salamander (*Notophthalmus perstriatus*, Salamandridae). *Alytes* 21:28–47.

Johnson, T.G. 1988. Forest statistics for southeast Georgia, 1988. USDA Forest Service Resource Bulletin SE-104.

Jones, C.A. 1995. Habitat use and home ranges of *Podomys floridanus* on the Ordway Preserve, Putnam County, Florida. *Bull Fla Mus Nat Hist* 38, Pt. II(7):195–209.

Kaczor, S.A., and Hartnett, D.C. 1990. Gopher tortoise (*Gopherus polyphemus*) effects on soils and vegetation in a Florida sandhill community. *Am Midl Nat* 123:100–111.

Kalisz, P.J., and Stone, E.L. 1984. Soil mixing by scarab beetles and pocket gophers in north-central Florida. *Soil Sci Soc Am J* 48:169–172.

Kantola, A.T. 1992. Threatened Sherman's fox squirrel. In *Rare and Endangered Biota of Florida. Volume 1. Mammals*, ed. S.R. Humphrey, pp. 234–241. Gainesville: University Press of Florida.

Kauffeld, C. 1957. *Snakes and Snake Hunting*. Garden City, NY: Hanover House.

Kauffeld, C. 1969. *Snakes, the Keeper and the Kept*. Garden City, NY: Doubleday & Co.

Kautz, R. 1984. Criteria for evaluaZing impacts of development on wildlife habitats. *Proc Annu Conf Southeast Assoc Fish Wildl agenc* 38:121–136.

Kautz, R. 1993. Trends in Florida wildlife habitat 1936–1987. *Fla Scie* 56(1):7–24.

Keegan, H.L. 1944. Indigo snakes feeding upon poisonous snakes. *Copeia* 1944:59.

Kellogg, F.E., Doster, G.L., and Komarek, E.V. 1972. The one quail per acre myth. *Natl Quail Symp* 1:15–20.

Kiester, A.R. 1971. Species density of North American amphibians and reptiles. *Syst Zool* 20:127–137.

Kirkman, L.K. 1995. Impacts of fire and hydrological regimes on vegetation in depression wetlands of southeastern USA. *Proc Tall Timbers Fire Ecol Conf* 19:10–20.

Krysko, K.L. 2001. Ecology, conservation, and morphological and molecular systematics of the kingsnake, *Lampropeltis getula* (Serpentes: Colubridae). Dissertation, University of Florida, Gainesville.

Krysko, K.L., and Smith, D.J. 2005. The decline and extirpation of the kingsnake (*Lampropeltis getula*) in Florida. In *Status and Conservation of Florida Amphibians and Reptiles*, eds. W. Meshaka and K. Babbitt, pp. 132–141. Gainesville: University Press of Florida.

Kucera, C.L. 1981. Grasslands and fire. In *Fire Regimes and Ecosystem Properties: Proceedings of the Conference; 1978 December 11–15*, eds. H.A. Mooney, T.M. Bonnicksen, and N.L. Christensen, pp. 90–111. Technical Report WO-26. Washington, DC: USDA Forest Service.

Kurten, B., and Anderson, E. 1980. *Pleistocene Mammals of North America*. New York: Columbia University Press.

Landers, J.L., and Speake, D.W. 1980. Management needs of sandhill reptiles in southern Georgia. Proceedings of the Annual Conference. *Annu Conf Southeast Assoc Fish Wildl Agenc* 34:515–529.

Lannoo, M. 2005. *Amphibian Declines: The Conservation Status of United States Species*. Berkeley: University Press of California.

Layne, J.N. 1974. The land mammals of south Florida. In *Environments of South Florida: Present and Past, Memoir 2*, ed. P.J. Gleason, pp. 386–413. Miami, FL: Miami Geological Society.

Layne, J.N. 1992. Threatened Florida mouse. In *Rare and Endangered Biota of Florida, Vol. 1, Mammals*, ed. S.R. Humphrey, pp. 250–264. Gainesville: University Press of Florida.

Layne, J.N., and Jackson, R.J. 1994. Burrow use by the Florida mouse (*Podomys floridanus*) in south-central Florida. *Am Midl Nat* 131:17–23.

Lowery G.H., Jr. 1974. *The Mammals of Louisiana and Its Adjacent Waters*. Baton Rouge: Louisiana State University Press.

MacArthur, R.H. 1958. Population ecology of some warblers of northeastern coniferous forest. *Ecology* 39:599–619.

MacNeill, F.S. 1950. Pleistocene shorelines in Florida and Georgia. *US Geol Surv Prof Pap* 221-F, pp. 95–107.

Marsh, D.M., and Trenham, P.C. 2001. Metapopulation dynamics and amphibian conservation. *Conserv Biol* 15:40–49.

Martin, P.S. 1967. Prehistoric overkill. In *Pleistocene Extinctions*, eds. P.S. Martin and H.E. Wright, Jr., pp. 75–120. New Haven, CT: Yale University Press.

Martin, W.H., and Boyce, S.G. 1993. Introduction: The Southeastern setting. In *Biodiversity of the Southeastern United States*, eds. W.H. Martin, S.G. Boyce, and A.C. Echternacht, pp. 1–46. New York: John Wiley & Sons.

Martin, W.H., and Means, D.B. 2000. Geographic distribution and habitat relationships of the eastern diamondback rattlesnake, *Crotalus adamanteus*. *Herpetol Nat Hist* 7(1):9–35.

Martof, B.S. 1968. *Ambystoma cingulatum* (Cope). Catalogue of American Amphibians and Reptiles 57.1–57.2.

Martof, B.S., Palmer, W.M., Bailey, J.R., Harrison, J.R., III, and Dermid, J. 1980. *Amphibians and Reptiles of the Carolinas and Virginia*. Chapel Hill: University of North Carolina Press.

Maser, C., Anderson, R.G., Cromack, K., Jr., Williams, J.T., and Martin, R.E. 1979. Dead and down woody material. In *Wildlife Habitats in Managed Forests, the Blue Mountains of Oregon and Washington*, ed. J.W. Thomas, pp. 78–79. USDA Forest Service, Agriculture Handbook 553, Washington, DC.

May, P.G., Farrell, T.M., Heulett, S.T., Pilgrim, M.A., Bishop, L.A., Spence, D.J., Rabatsky, A.M., Campbell, M.G., Aycrigg, A.D., and Richardson, W.E., II 1996. Seasonal abundance and activity of a rattlesnake (*Sistrurus miliarius barbouri*) in central Florida. *Copeia* 1996:389–401.

May, P.G., Heulett, S.T., Farrell, T.M., and Pilgrim, M.A. 1997. Live fast, love hard, and die young: The ecology of pigmy rattlesnakes. *Reptile and Amphibian Magazine* Jan./Feb. 1997:36–49.

McComb, W.C., Bonney, S.A., Sheffield, R.M., and Cost, N.D. 1986. Den tree characteristics and abundance in Florida and South Carolina. *J Wildl Manage* 50(4):584–591.

McCoy, E.D., and Mushinsky, H.R. 1992. Studying a species in decline: Changes in populations of the gopher tortoise on federal lands in Florida. *Fla Sci* 55:116–124.

McPherson, G. 1997. *Ecology and Management of North American Savannas*. Tuscon: University of Arizona Press.

Means, D.B. 1986. Life history and ecology of the eastern diamondback rattlesnake (*Crotalus adamanteus*). Final Report of Project No. GFC-84-013. Florida Game and Fresh Water Fish Commission, Tallahassee.

Means, D.B. 1996. Longleaf pine forest, going, going.... In *Eastern Old-Growth Forest: Prospects for Rediscovery and Recovery*, ed. M.B. Davis, pp. 210–229. Washington, DC: Island Press.

Means, D.B. 1999. The effects of highway mortality on four species of amphibians at a small, temporary pond in northern Florida. In *Proceedings of the Third International Conference on Wildlife Ecology and Transportation, September 13–16, 1999*, eds. G. Evink, P. Garrett, and D. Zeigler, pp. 125–128. Missoula, MT.

Means, D.B. 2000. Southeastern U. S. Coastal Plain habitats of the Plethodontidae: The importance of relief, ravines, and seepage. In *The Biology of Plethodontid Salamanders*, eds. R.C. Bruce, R.J. Jaeger, and L.D. Houck, pp. 287–302. New York: Plenum press.

Means, D.B. 2001. Reducing impacts on rare vertebrates that require small isolated water bodies along U.S. Highway 319. Final Report to the Florida Department of Transportation.

Means, D.B. 2004a. Florida pine snake, *Pituophis melanoleucus mugitus* (Barbour). In *Alabama Wildlife, Volume 3, Imperiled Amphibians, Reptiles, Birds, and Mammals*, eds. R.E. Mirarchi, M.A. Bailey, T.M. Haggerty, and T.L. Best, pp. 69–70. Tuscaloosa: University of Alabama Press.

Means, D.B. 2004b. Ghosts of the Red Hills. In *Between Two Rivers: Stories from the Red Hills to the Gulf*, eds. S. Cerulean, J. Ray, and L. Newton, pp. 89–99. Tallahassee, FL: Heart of the Earth and Red Hills Writer's Project.

Means, D.B. 2005a. Pine silviculture. In *Amphibian Declines: The Conservation Status of United States Species*, ed. M. Lanoo, pp. 139–145. Berkeley: University Press of California.

Means, D.B. 2005b. The value of dead tree bases and stumpholes as habitat for wildlife. In *Amphibians and Reptiles: Status and Conservation in Florida*, eds. W. Meshaka and K. Babbitt, pp. 74–78. Melbourne, FL: Krieger Press.

Means, D.B. In preparation. *Herpetophilia*. (Submitted).

Means, D.B., and Campbell, H.W. 1982. Effects of prescribed burning on amphibians and reptiles. In *Prescribed Fire and Wildlife in Southern Forests. Proceedings of a Symposium*, ed. G.W. Wood, pp. 89–97. Belle W. Baruch Forest Science Institute, Clemson University, Clemson, SC.

Means, D.B., and Krysko, K.L. 2001. Biogeography and pattern variation in the kingsnake, *Lampropeltis getula*, across the Apalachicola Region of Florida. Contemporary Herpetology 2001(5). http://www.calacademy.org/research/herpetology/ch/ch/2001/5/index.htm

Means, D.B., and Longden, C.J. 1976. Aspects of the biology and zoogeography of the pine barrens treefrog (*Hyla andersonii*) in northern Florida. Herpetologica 32(2):117–130.

Means, D.B., and Means, R.C. 1998. Red Hills survey for breeding pond habitat of the flatwoods salamander (*Ambystoma cingulatum*), gopher frog (*Rana capito*), and striped newt (*Notophthalmus perstriatus*) in the Tallahassee Red Hills of Leon, Gadsden, and Jefferson counties, Florida, and the Tifton Uplands of Thomas and Grady counties, Georgia. Final report to the U.S. Fish and Wildlife Service, Jackson, MS, for Order No. 43910-5-0091.

Means, D.B., and Means, R.C. 2005. Effects of sand pine silviculture on pond-breeding amphibians in the Woodville Karst Plain of north Florida. In *Amphibians and Reptiles: Status and Conservation in Florida*, eds. W. Meshaka and K. Babbitt, pp. 56–61. Malabar, FL: Krieger Press.

Means, D.B., and Moler, P.E. 1979. The pine barrens treefrog: Fire, seepage bogs, and management implications. In *Proceedings of the Rare and Endangered Wildlife Symposium, August 3–4, Athens, GA*, eds. R.R. Odum and L. Landers, pp. 77–83. Georgia Department of Natural Resources, Game and Fish Division Technical Bulletin WL-4.

Means, D.B., Palis, J.G., and Baggett, M. 1996. Effects of slash pine silviculture on a Florida population of flatwoods salamander. Conserv Biol 10(2):426–437.

Means, D.B., Dodd, C.K., Jr., Johnson, S.A., and Palis, J.G. 2004. Amphibians and fire in longleaf pine ecosystems. Conserv Biol 18(3):1149–1153.

Meylan, P.A. 1982. The squamate reptiles of the Inglis 1A fauna (Irvingtonian: Citrus Co., Florida). Bull Fla State Mus Biol Sci 27(3):1–85.

Meylan, P.A. 1995. Pleistocene amphibians and reptiles from the Leisey Shell Pit, Hillsborough County, Florida. Bull Fla State Mus Nat Hist 37, Pt. I(9):273–297.

Miller, S.H., and Marion, W.R. 1995. Natural and created snags and cavity-nesting birds in north Florida pine forests. Report for Project GFC-84-020, Florida Game and Fresh Water Fish Commission, Tallahassee.

Mirov, N.T. 1967. The genus *Pinus*. New York: Ronald Press Co.

Mitchell, J.C., and Hedges, S.B. 1980. *Ambystoma mabeei* Bishop (Caudata: Ambystomatidae): An addition to the fauna of Virginia. *Brimleyana* 3:119–121.

Moler, P.E. 1985a. Distribution of the eastern indigo snake, *Drymarchon corais couperi*, in Florida. *Herpetol Rev* 16(2):37–38.

Moler, P.E. 1985b. Home range and seasonal activity of the eastern indigo snake, *Drymarchon corais couperi*, in northern Florida. Final Performance Report, Study No. E-1-06, III-A-5, Florida Game and Fresh Water Fish Commission, Tallahassee.

Moler, P.E. 1992. Eastern indigo snake *Drymarchon corais couperi*. In *Rare and Endangered Biota of Florida, Volume III, Amphibians and Reptiles*, ed. P.E. Moler, pp. 181–186. Gainesville: University Press of Florida.

Moler, P.E., and Franz, R. 1988. Wildlife values of small, isolated wetlands in the Southeastern coastal plain. In *Proceedings of the Third Nongame and Endangered Wildlife Symposium*, eds. R.R. Odum, K.A. Riddleberger, and J.C. Ozier, pp. 234–241. Georgia Department of Natural Resources, Social Circle.

Monk, C.D. 1968. Successional and environmental relationships of the forest vegetation of north central Florida. *Am Midl Nat* 79(2):441–457.

Mosimann, J.E., and Rabb, G.B. 1948. The salamander *Ambystoma mabeei* in South Carolina. *Copeia* 1948:304.

Mount, R.H. 1963. The natural history of the red-tailed skink, *Eumeces egregius* Baird. *Am Midl Nat* 70:356–385.

Mount, R.H. 1975. *Reptiles and Amphibians of Alabama*. Alabama Agricultural Experiment Station, Auburn University, Auburn.

Mount, R.H. 1981. The red imported fire ant, *Solenopsis invicta* (Hymenoptera: Formicidae), as a possible serious predator on some native southeastern vertebrates: Direct observations and subjective impressions. *J Ala Acad Sci* 52(2):71–78.

Mushinsky, H.R. 1984. Observations of the feeding habits of the short-tailed snake, *Stilosoma extenuatum*, in captivity. *Herpetol Rev* 15:67–68.

Mushinsky, H.R. 1985. Fire and the Florida sandhill herpetofaunal community: with special attention to responses of *Cnemidophorus sexlineatus*. *Herpetologica* 41:333–342.

Mushinsky, H.R. 1986. Fire, vegetation structure and herpetological communities. In *Studies in Herpetology*, ed. Z. Rocek, pp. 383–387. Prague, Czechoslovakia: Charles University.

Mushinsky, H.R. 1992. Natural history and abundance of southeastern five-lined skinks, *Eumeces inexpectatus*, on a periodically burned sandhill in Florida. *Herpetologica* 48:307–312.

Mushinsky, H.R., and Gibson, D.J. 1991. The influence of fire periodicity on habitat structure. In *Habitat Complexity: The Physical Arrangement of Objects in Space*, eds. S.S. Bell, E.D. McCoy, and H.R. Mushinsky, pp. 237–259. New York: Chapman & Hall.

Mushinsky, H.R., and Witz, B.W. 1993. Notes on the peninsula crowned snake, *Tantilla relicta*, in periodically burned habitat. *J Herpetol* 27:468–470.

Neill, W.T. 1948. Hibernation of amphibians and reptiles in Richmond County, Georgia. *Herpetologica* 4:107–114.

Neill, W.T. 1951a. The eyes of the worm lizard, and notes on the habits of the species. *Copeia* 1951:177–178.

Neill, W.T. 1951b. Notes of the role of crawfishes in the ecology of reptiles, amphibians, and fishes. *Ecology* 32:764–766.

Neill, W.T. 1952. Burrowing habits of *Hyla gratiosa*. *Copeia* 1952:196.

Neill, W.T. 1957. Some misconceptions regarding the eastern coral snake, *Micrurus fulvius*. *Herpetologica* 13:111–118.

Neill, W.T. 1958. The varied calls of the barking treefrog, *Hyla gratiosa* LeConte. *Copeia* 1958:44–46.

Nelson, D.H., and Gibbons, J.W. 1972. Ecology, abundance, and seasonal activity of the scarlet snake, *Cemophora coccinea*. *Copeia* 1972:582–584.

Nowak, R.M. 1999. *Walker's Mammals of the World*, 6th ed. Baltimore: Johns Hopkins University Press.

Olson, M.S., and Platt, W.J. 1995. Effects of habitats and growing season fires on resprouting of shrubs in longleaf pine savannas. *Vegetatio* 119:101–118.

Owen-Smith, N. 1987. Pleistocene extinctions: The pivotal role of megaherbivores. *Paleobiology* 13:351–362.

Owen-Smith, R.N. 1988. *Megaherbivores: The Influence of Very Large Body Size on Ecology*. London: Cambridge University Press.

Palis, J.G. 1996. Element stewardship abstract: Flatwoods salamander (*Ambystoma cingulatum* Cope). *Nat Areas J* 16:49–54.

Palis, J.G. 1998. Breeding biology of the gopher frog, *Rana capito*, in western Florida. *J Herpetol* 32:217–223.

Palis, J.G., and Means, D.B. 2005. *Ambystoma cingulatum*. In *Amphibian declines, the conservation status of United States species*, ed. M. Lannoo. Berkeley: University of California Press.

Palmer, W.M. 1987. A new species of glass lizard (Anguidae: *Ophisaurus*) from the southeastern United States. *Herpetologica* 43(4):415–423.

Palmer, W.M., and Braswell, A.L. 1995. *Reptiles of North Carolina*. Chapel Hill: University of North Carolina Press.

Palmer, W.M., and Tregembo, G. 1970. Notes on the natural history of the scarlet snake *Cemophora coccinea copei* Jan in North Carolina. *Herpetologica* 26:300–302.

Pearson, P.G. 1955. Population ecology of the spadefoot toad, *Scaphiopus h. holbrooki* (Harlan). *Ecol Monogr* 25:233–267.

Pechmann, J.H.K., Estes, R.A., Scott, D.E., and Gibbons, J.W. 2001. Amphibian colonization and use of ponds created for trial mitigation of wetland loss. *Wetlands* 21(1):93–111.

Peet, R.K., and Allard, D.J. 1993. Longleaf pine vegetation of the southern Atlantic and eastern Gulf Coast regions. *Proc Tall Timbers Fire Ecol Conf* 18:45–81.

Perdue, T., and Green, M.D., eds. 1995. *The Cherokee Removal: A Brief History with Documents*. Boston: Bedford Books of St. Martin's Press.

Petranka, J.W. 1998. *Salamanders of the United States and Canada*. Washington, DC: Smithsonian Institute Press.

Pianka, E.R. 1973. The structure of lizard communities. *Annu Rev Ecol Syst* 4:53–74.

Pianka, E.R. 1975. Niche relations of desert lizards. In *Ecology and Evolution of Communities*, eds. M. Cody and J. Diamond, pp. 292–314. Cambridge: Harvard University Press.

Platt, W.J. 1975. The colonization and formation of equilibrium plant species associations on badger disturbances in a tall-grass prairie. *Ecol Monogr* 45:285–305.

Platt, W.J. 1999. Southeastern pine savannas. In *Savannas, Barrens, and Rock Outcrop Plant Communities of North America*, eds. R.C. Anderson, J.S. Fralish, and J.M. Baskin, pp. 23–51. London: Cambridge University Press.

Platt, W.J., and Schwartz, M.W. 1990. Temperate hardwood forests. In *Ecosystems of Florida*, eds. R. Myers and J. Ewel, pp. 194–229. Orlando: University of Central Florida Press.

Platt, W.J., Evans, G.W., and Davis, M.M. 1988. Effects of fire season on flowering of forbs and shrubs in longleaf pine forests. *Oecologia* 76:353–363.

Platt, W.J., Glitzenstein, J.S., and Streng, D.R. 1991. Evaluating pyrogenicity and its effects on vegetation in longleaf pine savannas. *Proc Tall Timbers Fire Ecol Conf* 17:143–161.

Prouty, W.F. 1952. Carolina bays and their origin. *Bull Geol Soc Am* 63:167–224.

Punzo, F. 1995. An analysis of feeding in the oak toad, *Bufo quercicus* (Holbrook), (Anura, Bufonidae). *Fla Sci* 58(1):16–20.

Pyne, S.J. 1982. *Fire in America, a Cultural History of Wildland and Rural Fire*. Princeton, NJ: Princeton University Press.

Repenning, R.W., and Labisky, R.F. 1985. Effects of even-age timber management on bird communities of the longleaf pine forest in northern Florida. *J Wildl Manage* 49(4):1088–1098.

Reynolds, J.H. 1980. A mark-recapture study of the scarlet snake, *Cemophora coccinea*, in a coastal plain sandhill community. Unpublished master's thesis, North Carolina State University, Raleigh.

Richter, S.C. 1998. The demography and reproductive biology of gopher frogs, *Rana capito*, in Mississippi. Unpublished master's thesis, Southeastern Louisiana University, Hammond.

Richter, S.C., Young, J.E., Seigel, R.A., and Johnson, G.N. 2001. Post-breeding movements of the dark gopher frog, *Rana sevosa* Goin and Netting: Implications for conservation and management. *J Herpetol* 35:316–321.

Robertson, W.B., Jr., and Kushlan, J.A. 1974. The southern Florida avifauna. In *Environments of South Florida: Present and Past*, ed. P.J. Gleason, pp. 414–452. Miami Geological Society Memoir 2, Miami, FL.

Robson, M.S. 1992. Recently extinct Florida red wolf. In *Rare and Endangered Biota of Florida, Volume I, Mammals*, ed. S.R. Humphrey, pp. 29–34. Gainesville: University Press of Florida.

Rogers, L.L. 1999. Black bear. In *The Smithsonian Book of North American Mammals*, eds. D. Wilson and S. Ruff, pp. 157–160. Washington, DC: Smithsonian Institute.

Rostlund, E. 1960. The geographic range of historic bison in the Southeast. *Ann Assoc Am Geogr* 50:395–407.

Rudolph, D.C., Burgdorf, S.J., Conner, R.N., and Schaefer, R.R. 1999. Preliminary evaluation of the impact of roads and associated vehicular traffic on snake populations in east Texas. In *Proceedings of the Third International Conference on Wildlife Ecology and Transportation, September 13–16, 1999*, eds. G. Evink, P. Garrett, and D. Zeigler, pp. 129–136. Missoula, MT.

Savage, H. 1982. *The Mysterious Carolina Bays*. Columbia: University of South Carolina Press.

Scholes, R.J., and Archer, S.R. 1997. Tree–grass interactions in savannas. *Annu Rev Ecol Syst* 28:517–544.

Schurbon, J.M., and Fauth, J.E. 2003. Effects of prescribed burning on amphibian diversity in a southeastern U. S. national forest. *Conserv Biol* 17:1338–1349.

Semlitsch, R.D. 1981. Terrestrial activity and summer home range of the mole salamander (*Ambystoma talpoideum*). *Can J Zool* 59:315–322.

Semlitsch, R.D. 1983a. Structure and dynamics of two breeding populations of the eastern tiger salamander, *Ambystoma tigrinum*. *Copeia* 1983:608–618.

Semlitsch, R.D. 1983b. Burrowing ability and behavior of salamanders of the genus *Ambystoma*. *Can J Zool* 61:616–620.

Semlitsch, R.D. 1998. Biological delineation of terrestrial buffer zones for pond-breeding salamanders. *Conserv Biol* 12(5):1113–1119.

Semlitsch, R.D. 2000. Size does matter: The value of small isolated wetlands. *National Wetlands Newsletter* January–February:5–6, 13.

Semlitsch, R.D., and Bodie, J.R. 1998. Are small, isolated wetlands expendable? *Conserv Biol* 12(5):1129–1133.

Semlitsch, R.D., and Bodie, J.R. 2003. Biological criteria for buffer zones around wetlands and riparian habitats for amphibians and reptiles. *Conserv Biol* 176(5):1219–1228.

Semlitsch, R.D., and Jensen, J.J. 2001. Core habitat, not buffer zone. *National Wetlands Newsletter* 23(4):5–6, 11.

Semlitsch, R.D., and McMillan, M.A. 1980. Breeding migrations, population size structure, and reproduction of the dwarf salamander, *Eurycea quadridigitata*, in South Carolina. *Brimleyana* 3:97–105.

Semlitsch, R.D., and Moran, G.B. 1984. Ecology of the redbelly snake (*Storeria occipitomaculata*) using mesic habitats in South Carolina. *Am Midl Nat* 111:33–40.

Semlitsch, R.D., Brown, K.L., and Caldwell, J.P. 1981. Habitat utilization, seasonal activity, and population size structure of the southeastern crowned snake *Tantilla coronata*. *Herpetologica* 37:40–46.

Semlitsch, R.D., Gibbons, J.W., and Tuberville, T.D. 1995. Timing of reproduction and metamorphosis in the Carolina gopher frog (*Rana capito capito*) in South Carolina. *J Herpetol* 29(4):612–614.

Semlitsch, R.D., Scott, D.E., Pechmann, J.H.K., and Gibbons, J.W. 1996. Structure and dynamics of an amphibian community, evidence from a 16-year study of a natural pond. In *Long-Term Studies of Vertebrate Communities*, eds. M.L. Cody and J.A. Smallwood, pp. 217–248. New York: Academic Press.

Sharitz, R.R., and Gibbons, J.W. 1982. The ecology of southeastern shrub bogs (pocosins) and Carolina Bays: A community profile. U. S. Fish and Wildlife Service, FWS/OBS-82/04. Washington, DC.

Shoop, C.R. 1960. The breeding habits of the mole salamander, *Ambystoma talpoideum* (Holbrook), in southeastern Louisiana. *Tulane Stud Zool* 8:65–82.

Simberloff, D. 1993. Species-area and fragmentation effects on old-growth forests: Prospects for longleaf pine communities. *Proc Tall Timbers Fire Ecol Conf* 18:1–13.

Simpkin, S.M., and Michener, W.K. 2005. Faunal soil disturbance regime of a longleaf pine ecosystem. *Southeast Nat* 4(1):133–152.

Smith, H.M. 1946. *Handbook of Lizards*. Ithaca, NY: Cornell University Press.

Smith, L.L., Smith, K.G., Barichievich, W.J., Dodd, C.K., Jr., and Sorensen, K. 2005. Roads and Florida's herpetofauna: A review and mitigation case study. In *Status and conservation of Florida amphibians and reptiles*, eds. W. Meshaka and K. Babbitt. Malabar, FL: Krieger Press.

Snider, A.T., and Bowler, J.K. 1992. Longevity of reptiles and amphibians in North American collections, 2nd ed. Society for the Study of Amphibians and Reptiles Herpetological Circular No. 21.

Snyder, J.R., Herndon, A., and Robertson, W.B. 1990. South Florida rockland ecosystems: Tropical hammocks and pinelands. In *Ecosystems of Florida*, eds. R. Myers and J. Ewel, pp. 230–274. Orlando: University of Central Florida Press.

Steadman, D.W., and Martin, P.S. 1984. Extinction of birds in the late Pleistocene of North America. In *Quaternary Extinctions, a Prehistoric Revolution*, eds. P.S. Martin and R.G. Klein, pp. 466–477. Tuscon: University of Arizona Press.

Stevenson, D.J., Dyer, K.J., and Willis-Stevenson, B.A. 2003. Survey and monitoring of the eastern indigo snake in Georgia. *Southeast Nat* 2(3):393–408.

Stevenson, H.M., and Anderson, B.H. 1994. *The Birdlife of Florida*. Gainesville: University Press of Florida.

Stickel, W.H., and Cope, J.B. 1947. The home ranges and wanderings of snakes. *Copeia* 1947:127–136.

Stoddard, H.L. 1931. *The Bobwhite Quail: Its Habits, Preservation and Increase*. New York: Charles Scribner's Sons.

Stoddard, H.L. 1962. Use of fire in pine forests and game lands of the deep Southeast. *Proc Tall Timbers Fire Ecol* 1:31–42.

Stoddard, H.L. 1978. Birds of Grady County, Georgia. *Bull Tall Timbers Res Stn* 21:1–175.

Stout, I.J., and Marion, W.R. 1993. Pine flatwoods and xeric pine forests of the southern (lower) Coastal Plain. In *Biodiversity of the Southeastern United States: Lowland Terrestrial Communities*, eds. W.H. Martin, S.G. Boyce, and A.C. Echternacht, pp. 373–446. New York: John Wiley & Sons.

Streng, D.R., Glitzenstein, J.S., and Platt, W.J. 1993. Evaluating the season of burn in longleaf pine forests: A critical literature review and some results from an ongoing long-term study. *Proc Tall Timbers Fire Ecol Conf* 18:2278–263.

Tanner, J.S. 1942. The ivory-billed woodpecker. Research Report No. 1, National Audubon Society.

Tebeau, C.W. 1971. *A History of Florida*. Coral Gables, FL: University of Miami Press.

Tennant, A. 1997. *A Field Guide to the Snakes of Florida*. Houston, TX: Gulf Publishing Co.

Tennant, A., and Bartlett, A.D. 2000. *Snakes of North America Eastern and Central Regions*. Houston, TX: Gulf Publishing Co.

Test, F.C. 1893. The "gopher frog." *Science* 22:75.

Thomas, J.W., Anderson, R.G., Maser, C., and Bull, E.L. 1979. Snags. In *Wildlife Habitats in Managed Forests, the Blue Mountains of Oregon and Washington*, ed. J.W. Thomas, pp. 60–77. USDA Forest Service, Agriculture Handbook 553, Washington, DC.

Timmerman, W.W. 1995. Home range, habitat use, and behavior of the eastern diamondback rattlesnake (*Crotalus adamanteus*) on the Ordway Preserve. *Bull Fla Mus Nat Hist* 328, Pt. I(5):127–158.

Travis, J. 1992. Status undetermined eastern tiger salamander *Ambystoma tigrinum tigrinum* (Green). In *Rare and Endangered Biota of Florida, Volume III, Amphibians and Reptiles*, ed. P.E. Moler, pp. 70–76. Gainesville: University Press of Florida.

Tschinkel, W.R. 1987. Distribution of the fire ants *Solenopsis invicta* and *S. geminata* (Hymenoptera: Formicidae) in northern Florida in relation to habitat and disturbance. *Ann Entomol Soc Am* 81:76–81.

Tuberville, T.D., Bodie, J.R., Jensen, J.B., LaClaire, L.V., and Gibbons, J.W. 2000. Apparent decline of the southern hognose snake (*Heterodon simus*). *J Elisha Mitchell Sci Soc* 116(1):19–40.

Tucker, J.W., Jr., and Robinson, W.D. 2003. Influence of season and frequency of fire on Henslow's sparrows (*Ammodramus henslowii*) wintering on Gulf Coast pitcher plant bogs. *Auk* 120(1):96–106.

Umber, R.W., and Harris, L.D. 1974. Effects of intensive forestry on succession and wildlife in Florida sandhills. *Proc Annu Conf Southeast Assoc Game Fish Comm* 28:686–693.

U.S. Fish and Wildlife Service. 1978. Part 17—Endangered and threatened wildlife and plants; listing of the eastern indigo snake as a threatened species. *Fed Regist* 43(21):4026–4028.

U.S. Fish and Wildlife Service. 2003. Recovery plan for the red-cockaded woodpecker, *Picoides borealis*; 2nd revision. US Fish and Wildlife Service, Atlanta, GA.

Van Balen, J.B., and Doerr, P.D. 1978. The relationship of understory vegetation to red-cockaded woodpecker activity. *Proc Annu Conf Southeast Assoc Fish Wildl Agenc* 32:82–92.

Varner, J.M., and Kush, J.S. 2001. Old-growth longleaf pine forests—filling in the blanks. Proceedings Third Longleaf Alliance Regional Conference, Longleaf Alliance Report 5:204–208.

Vogl, R.J. 1973. Fire in the Southeastern grasslands. *Proc Tall Timbers Fire Ecol Conf* 12:175–198.

Wahlenberg, W.G. 1946. *Longleaf Pine: Its Use, Ecology, Regeneration, Protection, Growth, and Management*. Washington, DC: Charles Lathrop Pack Forestry Foundation.

Walker, J. 1993. Rare vascular plant taxa associated with the longleaf pine ecosystems: Patterns in taxonomy and ecology. *Proc Tall Timbers Fire Ecol Conf* 18:105–125.

Walker, J., and Peet, R.K. 1983. Composition and species diversity of pine–wiregrass savannas of the Green Swamp, North Carolina. *Vegetatio* 55:163–179.

Walsh, P.B, Darrow, D.H., and Dyass, J.G. 1995. Habitat selection by Florida grasshopper sparrows in response to fire. *Proc Annu Conf Southeast Assoc Fish Wildl Agenc* 49:342–349.

Ward, D.B., and Clewell, A.F. 1989. Atlantic white cedar (*Chamaecyparis thyoides*) in the southern states. *Fla Sci* 51(1):8–47.

Ware, S., Frost, C., and Doerr, P.D. 1993. Southern mixed hardwood forest: The former longleaf pine forest. In *Biodiversity of the Southeastern United States*, eds. W.H. Martin, S.G. Boyce, and A.C. Echternacht, pp. 447–493. New York: John Wiley & Sons.

Watts, W.A. 1969. A pollen diagram from Mud Lake, Marion County, north-central Florida. *Geol Soc Am Bull* 80:631–642.

Watts, W.A. 1971. Postglacial and interglacial vegetation history of southern Georgia and central Florida. *Ecology* 52(4):670–690.

Watts, W.A. 1975. A late Quaternary record of vegetation from Lake Annie, south-central Florida.

Geology 3:344–346.
Watts, W.A. 1980. The late Quaternary vegetation history of the southeastern United States. *Annu Rev Ecol Syst* 11:387–409.
Watts, W.A., Hansen, B.C.S., and Grimm, E.C. 1992. Camel Lake: A 40,000-yr record of vegetational and forest history from northwest Florida. *Ecology* 73(3):1056–1066.
Webb, S.D. 1974. Chronology of Florida Pleistocene mammals. In *Pleistocene Mammals of Florida*, ed. S.D. Webb, pp. 5–31. Gainesville: University Press of Florida.
Webb, S.D. 1977. A history of savanna vertebrates in the New World. Part I: North America. *Annu Rev Ecol Syst* 8:355–380.
Webb, S.D. 1990. Historical biogeography. In *Ecosystems of Florida*, eds. R. Myers and J. Ewel, pp. 70–100. Orlando: University of Central Florida Press.
Webb, S.D., and Wilkins, K.T. 1984. Historical biogeography of Florida Pleistocene mammals. In *Contributions in Quaternary Vertebrate Paleontology*, Special Publication No. 8, eds. H.H. Genoways and M.R. Dawson. Pittsburgh, PA.
Webb, S.D., Dunbar, J., and Newsom, L. 1992. Mastodon digesta from north Florida. *Curr Res Pleistocene* 1992:114–116.
Weigl, P.D., Steele, M.A., Sherman, L.J., Ha, J.C., and Sharpe, T.L. 1989. The ecology of the fox squirrel (*Sciurus niger*) in North Carolina: Implications for survival in the Southeast. *Bull Tall Timbers Res Stn* 24:1–93.
Werner, P.A., ed. 1991. *Savanna Ecology and Management: Australian Perspectives and Intercontinental Comparisons*. Oxford, UK: Blackwell Scientific Publications.
White, L.D., Harris, L.D., Johnston, J.E., and Milchunas, D.G. 1975. Impact of site preparation on flatwoods wildlife habitat. *Proc Annu Conf Southeast Assoc Game Fish Comm* 29:347–353.
Wilson, E.O., and Brown, W.L. 1958. Recent changes in the introduced population of the fire ant, *Solenopsis saevissima* (Fr. Smith). *Evolution* 12:211–218.
Wilson, V. 1951. Some notes on a captive scarlet snake. *Herpetologica* 7:172.
Wolfe, S.H., Reidenauer, J.A., and Means, D.B. 1988. An ecological characterization of the Florida panhandle. *US Fish Wildl Serv Biol Rep* 88(12):1–277.
Wolfenden, G.E. 1959. A Pleistocene avifauna from Rock Springs, Florida. *Wilson Bull* 71(2):183–187.
Wood, D.A. 2001. *Florida's Fragile Wildlife: Conservation and Management*. Gainesville: University Press of Florida.
Wright, A.H. 1932. *Life-histories of the Frogs of Okefinokee Swamp, Georgia*. New York: Macmillan Publishing Co.
Wright, A.H., and Wright, A.A. 1949. *Handbook of Frogs and Toads of the United States and Canada*. Ithaca, NY: Comstock Publishing Associates.
Wright, A.H., and Wright, A.A. 1957. *Handbook of Snakes of the United States and Canada*, 2 volumes. Ithaca, NY: Comstock Publishing Associates.
Wright, H.A., and Bailey, W. 1982. *Fire Ecology: United States and Southern Canada*. New York: John Wiley & Sons.
Young, T.S., and Laerm, J. 1993. A late Pleistocene vertebrate assemblage from the St. Marks River, Wakulla County, Florida. *Brimleyana* 18:15–57.

Section III
Silviculture

Chapter 7

Uneven-Aged Silviculture of Longleaf Pine

James M. Guldin

Introduction

The use of uneven-aged silviculture has increased markedly in the past 20 years. This is especially true in the southern United States, where the use of clearcutting and planting is often viewed as a practice whose emphasis on fiber production results in unacceptable consequences for other values, such as those that benefit from maintenance of continuous forest cover over time. Public lands in general, and national forest lands in particular, have become the focal point for the replacement of clearcutting and planting with even-aged and uneven-aged reproduction cutting methods that rely on natural regeneration, and that can better achieve management goals that are defined by residual stand structure and condition rather than by harvested volume.

Land managers in the southern United States are keenly interested in a renaissance for longleaf pine (*Pinus palustris* Mill.) (Landers et al. 1995; Barnett 1999). Of the four southern pines, longleaf pine has experienced the greatest percentage loss in forest area, from 37 million hectares prior to European colonization to approximately 2.2% of that currently (Frost this volume). That scarcity has increased the ecological value of the stands that remain. The scattered tracts of remnant unmanaged longleaf pine stands have high emotional and physical appeal. Managed stands of longleaf pine, especially those in which prescribed fire has been regularly applied, provide exceptional values for endangered species such as the red-cockaded woodpecker (*Picoides borealis* Vieillot), game species such as bobwhite quail (*Colinus virginianus* L.), and fire-dependent species such as wiregrass (primarily *Aristida beyrichiana* Trin. & Rupr.). Then too, many foresters fondly recall the memory of the exceptional quality of lumber that mature stands of longleaf pine were, and are, capable of producing.

The perception exists that many of these values can be provided by management of longleaf pine especially through the use of uneven-aged silviculture. In public debates, this may be supported by little other than the layperson's view that uneven-aged silviculture is the opposite of clearcutting, and thus innately has something to recommend it. Foresters have been a bit more reluctant to wholly embrace the application of uneven-aged silviculture in longleaf pine, citing among other reasons the intolerance to shade of the

---

James M. Guldin • Arkansas Forestry Sciences Laboratory, Southern Research Station, USDA Forest Service, Monticello, Arkansas 71656.

species, the difficulty in obtaining natural regeneration, and the cost.

These factors may in part explain why the focus of discussion regarding uneven-aged silviculture in southern pines is especially prominent on public lands such as national and state forests, and private lands managed as game plantations. These ownership entities share a number of attributes, including a diversity of ownership objectives, and the capability, directly or indirectly, to subsidize timber production with other resource values. For example, the National Forests of Florida have made a commitment to manage longleaf pine using both even-aged and uneven-aged systems. While this is admirable from the perspective of using a diversity of reproduction cutting methods to meet a diversity of forest management objectives, the proposed scale of the practice may outstrip the research that supports widespread application.

There is no reason to suspect that the principles of uneven-aged silviculture cannot be successfully adapted to longleaf pine stands in the lower Atlantic and Gulf Coastal Plains. The method has been successfully applied over time in other southern pines—most notably, mixed loblolly (*P. taeda* L.)–shortleaf (*P. echinata* Mill.) pine stands in the upper West Gulf Coastal Plain (Baker et al. 1996). A review of that history of success and failure will be of value in providing perspective regarding the application of the method in longleaf pine.

Uneven-aged silviculture has been successful in different forest types (Guldin 1996). Success with the method depends on the ability to obtain regeneration of the desired species, and to have that regeneration develop into merchantable size classes. Conversely, failures with uneven-aged silviculture are typically associated with an inability to obtain desired regeneration (Guldin 1996; Guldin and Baker 1998). There is good reason to expect that the details of regeneration establishment and development under an uneven-aged system in longleaf pine stands will be difficult, if the experience associated with the development of the shelterwood method in longleaf pine (Croker and Boyer 1975) is any indication. There is anecdotal evidence to suggest that uneven-aged silvicultural prescriptions can be successful in longleaf pine stands (Farrar and Boyer 1991; Farrar 1996; Moser et al. 2002). There is also considerable debate about the implications of habitat quality for red-cockaded woodpeckers in uneven-aged longleaf pine stands (e.g., Engstrom et al. 1996; Rudolph and Conner 1996). In view of the current situation, the opportunity to develop and refine the application of uneven-aged silviculture in longleaf pine is timely.

This review of the selection method and of the principles that underlie its application for longleaf pine is based on another southern pine species in which experience has been successful over a long period of time—specifically, naturally regenerated stands of loblolly pine with a minor and varying proportion of shortleaf pine in the upper West Gulf Coastal Plain in southern Arkansas. Following that overview, thoughts about the application and modification of the method to longleaf pine will be discussed in detail.

Definitions and Concepts

The goal of any silvicultural system is to advantageously utilize the resources in a given stand for social benefit through the emulation of natural processes of succession and disturbance. Helms (1998) defines a silvicultural system as a planned series of treatments for tending, harvesting, and reestablishing a stand, and a regeneration method as a cutting procedure by which a new age class is created within the stand. Smith (1986) also makes this distinction, using the term "reproduction cutting method" instead of "regeneration method." These terms, "regeneration method" and "silvicultural system," are commonly misapplied in two ways. The first is that they are often mistakenly used interchangeably. The former refers to the short period of time during which a new age cohort of regeneration of the desired species is obtained, and the latter refers to the entire program of treatments for the life of the stand. The second is that they are often mistakenly applied to a forest rather than to the individual stand, which is their intended scope

(Smith 1986; Helms 1998). The confusion is in part because the treatment prescriptions in the silvicultural system are closely related to stand structure at any given point in time, and stand structure is established primarily using the reproduction cutting method at the point of stand or cohort establishment.

The choice of regeneration method determines the scale of disturbance that foresters can imitate (Brockway et al. this volume). Even-aged regeneration methods such as clearcutting, the seed tree method, or the shelterwood method are designed to emulate varying intensities of stand-replacing disturbance events, resulting in a new cohort of regeneration across the entire stand. Unevenaged regeneration methods such as the selection method are designed to emulate a small-scale within-stand disturbance event, resulting in a new age class only in that subset of the stand where the practice was imposed. In either even-aged or uneven-aged methods, the main indication of success in the execution of a regeneration method is whether a new stand of trees is successfully obtained to replace the trees that were removed during the harvest. Thus, the reproduction cutting method is the primary element of the silvicultural system in that the actions that comprise the silvicultural system depend upon the origin of the regeneration, its age distribution, and its spatial distribution over the stand.

The choice of regeneration method also has an inordinate influence on the overall course of silvicultural treatments and the way those treatments are imposed in the stand (Fig. 1). In an even-aged stand, the sequence of silvicultural practices depends upon stand age. The new stand is obtained using a regeneration method that results in a new cohort of regeneration across the entire stand. Subsequent treatments such as site preparation in advance of the new cohort, release of that cohort, and intermediate treatments such as thinning also occur across the entire stand. Each treatment is imposed in a manner that is correlated with the age of the new cohort of regeneration. Eventually, when the stand reaches maturity, a new reproduction cutting method is implemented at the rotation age r, which gives rise to a subsequent stand managed with a subsequent silvicultural system.

Conversely, in an uneven-aged stand, there is no rotation age r. Instead, treatments are based on a cutting cycle c, the basic interval of stand entry, which varies from 5 to 20 years. Cutting cycle harvests are imposed in each stand every c years. However, the trees

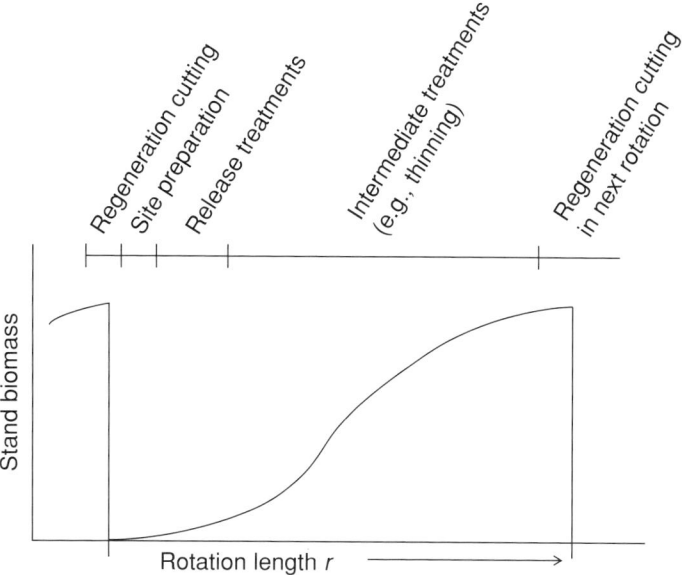

FIGURE 1. Chronosequential application of individual practices of a silvicultural system in evenaged stand. During the rotation age r, treatments are applied across the entire stand to meet silvicultural objectives that are related to tree age.

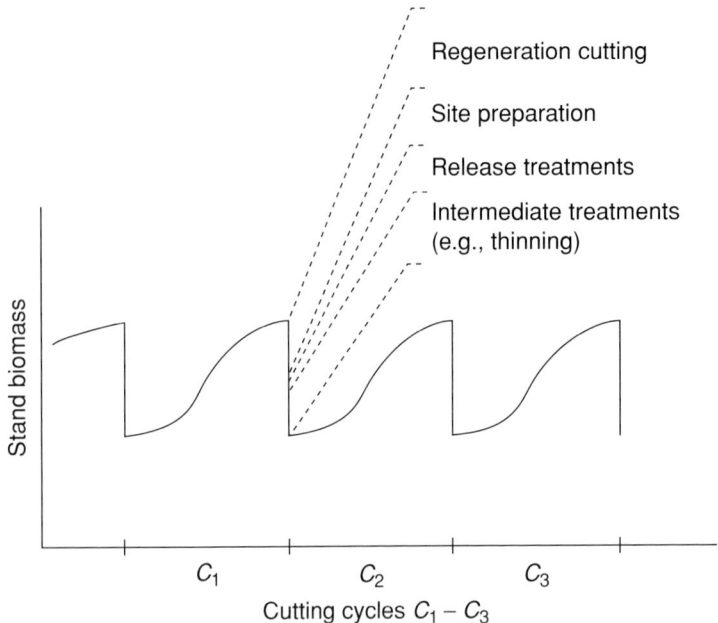

FIGURE 2. Concurrent application of individual practices of an uneven-aged silvicultural system during a cutting cycle harvest in a balanced uneven-aged stand. Treatments are applied to subunits of the stand depending on conditions within each subunit. Each cutting cycle harvest will support similar treatments.

removed during a cutting cycle harvest are taken for different reasons—regeneration cutting, release, or intermediate treatment such as thinning (Fig. 2). As a result, the different silvicultural practices that occur chronosequentially in an even-aged stand and that cover its entire area are conducted at the same time in an uneven-aged stand, but only over a portion of the stand area. This complicates the practical implementation of the method. Uneven-aged systems are often thought to be less intensive than even-aged systems, and that may be true in relation to, say, the degree of site exposure during regeneration cutting or the capital outlay required to establish a new stand. But uneven-aged systems require more attention on the part of the forester, and can be more inefficient to conduct in some ways because the scale of operation is at the scale of subunits within a stand rather than across the entire stand.

The Selection Method—An Overview

Uneven-aged silviculture is implemented using a reproduction cutting method called the selection method, used to regenerate and maintain a multiaged structure by removing trees either singly, or in small groups or strips (Helms 1998). By definition, an unevenaged stand has at least three age classes (Smith 1986; Helms 1998). In practice, age is not measured; stands are managed by controlling either the volume that is harvested, the diameter distribution in reference to a target distribution, or the area within the stand that is cut.

In an uneven-aged stand, growing space is subdivided among trees of all size classes, from regeneration through the largest overstory trees. A starting point to determine the appropriate basal area to maintain in an uneven-aged stand is to apply the basal area found in a mature even-aged stand of the same species that is marginally fully stocked or slightly understocked. A stand in that condition will have a slight amount of growing space available in the understory for the regeneration establishment but will be marginal for regeneration development. If that basal area is translated to an uneven-aged stand, the heterogeneous conditions that typify an uneven-aged stand will result in fully stocked clusters of overstory trees in some areas, and other areas that are sufficiently understocked such that regeneration development can occur. This concept gives rise to two generally accepted methods by which the selection method is implemented—one using very small openings and the other using larger openings.

Single-Tree Selection Method

Singletree selection imitates the smallest scale of disturbance, such as when a single tree falls or dies while standing in the woods. Possible causes of such a disturbance are insects, lightning, disease, windthrow, or some other agent. In unmanaged stands, this process results in gap-phase dynamics indicative of late-successional conditions (White 1979; Runkle 1982; Pickett and White 1985). When a large tree in an uneven-aged stand is removed, the growing space used by that tree is made available to adjacent trees, to smaller trees in the midstory, and to regeneration in the understory. In the smallest gaps, the gap may close before the regeneration can accede into the main canopy, and the regeneration may then persist without further growth or may even succumb to suppression. The occurrence of multiple gaps (where a nearby tree succumbs and creates a second nearby opening, either concurrently with or soon after the first), or expansion of existing gaps (where gap-bordering trees fall) can tip this ecological balance in favor of regeneration survival and development.

In the single-tree selection method, individual trees of all size classes are removed more or less uniformly across the stand (Smith 1986; Helms 1998). This is typically conducted by first identifying the trees that are to remain in the stand. The trees to remove then become obvious because their removal promotes the continued growth and development of the trees that are to remain. The diameter of the tree being removed is directly related both to the silvicultural objective for its removal, and to the retention of stocking levels by size class deemed desirable across the residual diameter distribution. Small-diameter trees, such as pulpwood or small sawlogs, are removed according to classical thinning rationale—to free an immature neighboring tree, presumably one with better form, condition, or other attribute, from the competition provided by the tree being removed. Removal of a large-diameter tree, such as one at or larger than the maximum diameter desired for retention, is intended to create a canopy opening within which regeneration is to become established and to develop.

Group Selection Method

In the group selection method, trees are removed and new age classes are established in small groups (Smith 1986; Helms 1998). Group selection imitates a small-scale natural disturbance that kills a small group or cluster of trees within a stand; examples include mortality of trees from a blowup of a surface fire or an infestation of pine beetles. In theory, the nature of the disturbance should promote a suitable and receptive seedbed for seedling establishment, and the size of the gap that is created by the disturbance is large enough to promote regeneration development. The seed source for that new cohort can be advance growth of regeneration in place prior to creation of the opening, seed from trees that border the gap or stored in the soil beneath the gap, or as seedfall disseminated from the trees being removed within the gap at the time of their removal. The size of the gap affects the species composition of that new cohort. Larger gaps favor species of greater intolerance to shade, while smaller gaps favor shade-tolerant species.

The group selection method is applied or suggested when there is a desire to use an unevenaged silvicultural system in a stand, yet still regenerate shade-intolerant species; hence the interest in the method relative to the southern pines. The maximum size for group openings depends on one's interpretation of ecological literature as modified by prevailing forest management guidelines. It is generally agreed that the upper ecological size limit for tolerant species within a circular group selection opening is one whose radius equals the height of the surrounding trees in the stand (Helms 1998). In trees with a height of 25 m, the ecological upper limit would be on the order of 0.2 ha. Larger openings would then be suggested for intolerant species such as longleaf pine. The suggested maximum group opening size on national forest lands within the Southern Region of the USDA Forest Service is 0.8 ha, which should allow for acceptable regeneration establishment and development of any of the southern pines. In practice, some trees are typically removed from the stand matrix between the groups as well, so as to promote stand health and acceptable basal area levels between group openings, and to prepare trees as seed producers in subsequent group openings.

The decision to locate a group opening within the stand is usually done to alleviate understocking, rectify excessive stem density, or take advantage of existing regeneration. If part of the stand is understocked, creating an opening gives the forester an opportunity to regenerate that group, and thus to restore that part of the stand back to full stocking. At the other extreme, if there is a place within the stand where the trees are all in a surplus size class on the marking tally, creating an opening harvests those trees and helps the forester more quickly achieve the desired residual stand density. Finally, if a species is difficult to regenerate naturally, one should not fail to create openings within the stand where advance regeneration is found.

Group selection has a number of administrative advantages over single-tree selection that contribute to its popularity. Most group openings serve as points of concentration for logging operations in the immediate area; logs are frequently decked in the openings, and haul roads typically run from one group to another. As a result, group openings are often heavily scarified. This is an advantage in promoting pine regeneration, which requires exposed mineral soil for optimal seed germination and establishment. Moreover, the group opening is the only part of the stand in which regeneration is expected. Site preparation or release treatments can thus be restricted to the groups, which is an advantage in that less area must be treated and the specifications for treatment can be made clearer than when a comparable treatment is prescribed in a single-tree selection stand. The advent of geographic positioning system technology adds to the ease of contracting treatments, since the precise location of each group can be specified.

Selective Cutting

The term "selective cutting" is often used to describe harvesting activity that resembles uneven-aged regeneration cutting in that trees remain on the site. But a strict definition of the term is "select some trees to cut and cut them"; it has no commonly accepted professional meaning except a derogatory one. The term does not refer to, and is not synonymous with, the practice of uneven-aged harvests using the selection method. All too often, stands harvested using selective cutting are high-graded—harvest is uncontrolled, the best trees are cut, and the poorest remain. Perhaps the most important silvicultural distinction between the selection method and selective cutting is that under the latter, no provision is made for establishment or development of desired species of regeneration. This is a trap into which improperly applied harvests under the selection method can fall.

Regulation of Uneven-Aged Stands

Regardless of whether single-tree selection or group selection is implemented, the methods for ensuring the regulation of growth and harvest are the same. Two methods of regulation have historically been associated with the selection method—regulation of the sawtimber volume in the stand through harvest of growth, and regulation of the structure of the stand through conformance with a target diameter distribution. With group selection, a third method enters the picture, in which regulation is based on proportion of harvested area during each cutting cycle.

These methods have varying degrees to which they conform to the origins of uneven-aged silviculture (Guldin 1996). The selection method, the most recently developed of the regeneration methods historically, traces its origins to the Dauerwald in Germany, which evolved as a counterpoint to area-based regulation methods. Under the Dauerwald, allowable cut was based on stand volume growth; a volume equal to roughly 15 times the annual growth was retained on the site (Troup 1928, 1952). Other European work at the turn of the century used a negative exponential algebraic relationship among the number of trees by diameter class to quantify the reverse J-shaped curve that approximates an uneven-aged stand (deLiocourt 1898), and that approach was further developed in North America in modern approaches to forest management (Meyer et al. 1961). Little historic evidence exists to support an area-based regulation method in the evolution of uneven-aged silvicultural systems. On the other hand, Smith (1986) points out that regulation method should be independent of silvicultural system; under such logic, any of the following systems might legitimately be applied to regulation of uneven-aged stands.

Regulation of Volume

The most successful application of the selection method in southern pines has been at the Crossett Experimental Forest (EF) in Ashley County, AR, in the mixed loblolly–shortleaf pine forest type of the upper West Gulf Coastal Plain. The Crossett EF, managed by the Southern Research Station of the USDA Forest Service, was established in 1934 and is still active. It supports several long-term research studies and demonstrations, notably the Good and Poor Farm Forestry Forty Demonstration Stands and the Methods of Cutting Study.

The uneven-aged stands of loblolly–shortleaf pine at the Crossett EF were regulated using the volume control-guiding diameter limit (VCGDL) method (Reynolds 1959, 1969; Baker et al. 1996; Guldin 1996, 2002). In this method, sawlog volume and volume growth are used to calculate harvests, and trees are marked based on whether their individual growth rates are sufficient to allow them to maintain acceptable sawlog volume growth.

Implementation of the VCGDL method has four broad steps. First, a current inventory of the stand is taken. That inventory is used to prepare a stand and stock table that quantifies the stem density and sawlog volume before harvest by diameter class per unit area. This requires application of an appropriate local sawlog volume table, so that the sawlog volume by diameter class can be included in the table.

Second, the compound growth rate of the stand must be determined. This is usually done by averaging the growth rate for a number of trees of varying size in the stand, using 10-year radial increment measured from increment cores, and calculated per tree using the formula

$$G = \exp\left[\frac{(V_0/V_p)}{n}\right] - 1$$

where

G = growth rate (%)
V_p = previous tree volume
V_0 = present tree volume
n = growth interval (years)

Alternatively, experience shows that in managed stands, one can use an appropriate compound growth percentage for sawlog volume increment. Values between 6% and 8% are typical in unmanaged and managed loblolly–shortleaf pine stands, respectively, in south Arkansas. An appropriate range for longleaf pine would be based on local experience.

Either approach then requires the forester to determine an after-cut volume for the planned harvest. The after-cut volume is the cumulative volume to which the current stand must be reduced. That level is set by predicting the future cutting cycle length, by selecting the future stand volume sought at that time, and then by calculating the volume to which the current stand must be reduced so that it will grow to the intended future volume at the appropriate rate of growth.

Third, the allowable cut is calculated as the difference between the before-cut volume and the planned after-cut volume. This leads directly to the calculation of the guiding diameter limit (GDL), which is that diameter class that meets the allowable cut if all trees in larger diameter classes are cut. Usually, part of the

GDL class must also be cut to exactly match the allowable cut.

Finally, the field crews are given the GDL class and the percentage of the GDL class to cut. Markers are instructed to retain trees above the GDL if they are growing acceptably, and then to mark an equivalent volume to that retained in diameter classes smaller than the GDL. Marking crews can only do this efficiently by memorizing the appropriate local volume table. The crews then must keep a running tally, either mentally or on a notepad, of cumulative volume retained above the GDL and that removed below the GDL. At the end of the marking, the volume marked below the GDL should balance that retained above the GDL (Fig. 3).

The VCGDL method has a number of advantages. It requires the field crew to examine sawlog component of the stand from the perspective of trees that should be retained. As a result, crews can balance whether to cut and leave trees across a range of diameter classes. It also requires that large high-value trees above the GDL have a compelling record of growth to be retained.

However, there are several limitations of the VCGDL method. The approach does not provide any evidence to the forester about the growth of trees below the sawtimber size class. Foresters must judge in some other way whether regeneration is being established, and whether sub-sawtimber size classes are developing at an acceptable rate. That judgment is typically based on experience rather than objective standards, and that can be a limitation. Second, the method requires a high degree of experience on the part of the field crews who are marking the stand, especially in regard to estimation of the volume of trees above and below the GDL that are being retained and marked, respectively, such that the cumulative volume tally balances when the marking is completed. Finally, because decisions are made in the field about retaining trees above the GDL

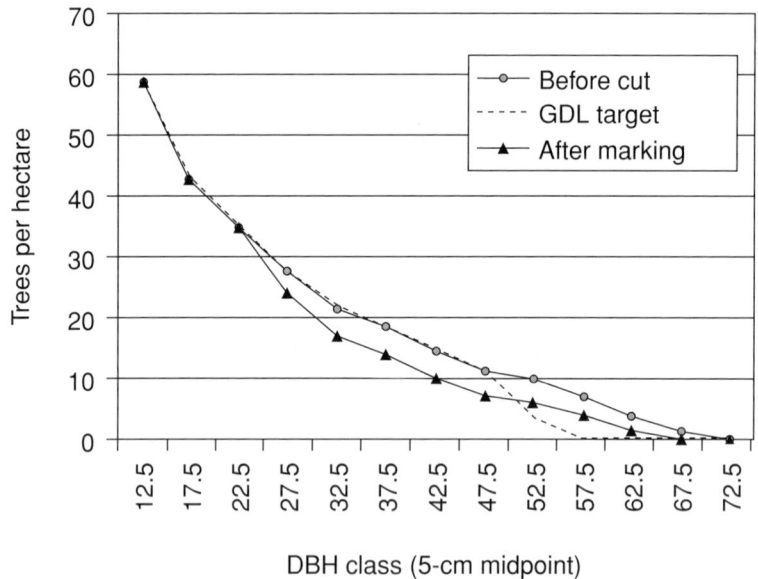

FIGURE 3. The volume control-guiding diameter limit (VCGDL) regulation approach conceptually applied in the 1978 before-cut inventory from the Good Farm Forestry Forty demonstration stand, an uneven-aged loblolly–shortleaf pine stand on the Crossett Experimental Forest in southern Arkansas. Before-cut, GDL target, and after-cut diameter distributions are drawn as curves rather than histograms. The GDL target reflects the allowable cut in sawtimber cubic volume based on 6% growth rate. The after-cut diameter distribution illustrates one possible outcome resulting from retaining trees above the guiding diameter limit, and removing trees below the limit such that the volume of the stand is retained at the guiding level.

and removing trees below the GDL, the use of computer models to predict stand development in advance of harvest is difficult.

The VCGDL regulation approach evolved as a means to regulate single-tree selection stands. However, modifying the approach to regulate a stand being managed using group selection is relatively straightforward. All trees within the group would be marked, and added to the marking tally. Trees smaller than the GDL within a group would have their volume added to the below-GDL cumulative volume tally. That would lead to retaining an equivalent volume of trees at or above the GDL in the matrix between groups.

Regulation of Stand Structure

The regulation of stand structure is based on the notion that the diameter distribution of a balanced uneven-aged stand has an ideal theoretical relationship, which can be compared with the actual stand structure for generating an after-cut residual stand (and indirectly, a marking tally) that carries the existing stand closer to the theoretical ideal. Several approaches can be developed to quantify this ideal theoretical stand, such as use of stand density index (Long 1998) or leaf area index (O'Hara 1996). But the most common in southern pines is based on the assumption of a constant ratio q in the number of trees in adjacent size classes, according to the simple formula

$$q = \frac{t_n}{t_{(n+i)}}$$

where

$q = q$ ratio
$t_n =$ number of trees per unit area in the nth diameter class
$t_{(n+i)} =$ number of trees per unit area in the next larger class of class width i

Thus, q is dependent on diameter class width, and the use of a given q ratio must include reference to the class width i. If the maximum diameter class D of trees to retain in the stand is known, one can use q to construct a negative exponential relationship that can be fit to any desired residual basal area B per unit area in the stand. Specification of B, D, and q thus constitutes a unique solution of diameter distribution. This approach, called the BDq approach, is used to generate the target balanced diameter distribution against which the existing stand structure is compared. The method was developed by Leak (1964), and its practical implementation was described in detail by Marquis (1978). Modifications for uneven-aged stands of loblolly–shortleaf pines in the West Gulf region were described in Baker et al. (1996), and for southern pines generally by Farrar (1996).

Simply stated, selection of the target BDq parameters allows the forester to calculate a unique hypothetical target diameter distribution. This is typically prepared on a unit area (per-hectare) basis. The diameter distribution of the before-cut stand, prepared from a preharvest inventory, is then compared to that target. In an ideal case, the before-cut stand will contain a surplus of trees in every diameter class compared with the target; that surplus then becomes the marking tally. However, far more common is the situation in which some diameter classes in the preharvest stand will contain a surplus of trees compared to the target, and others will contain a deficit of trees. Here, the basal area of those deficits must be calculated, and that basal area deficit must be accounted for by retaining more trees than called for in those diameter classes that have a surplus relative to the target stand (Fig. 4). Ultimately, deficits in a given diameter class will be corrected through ingrowth from smaller diameter classes over time.

When the final tally of trees to cut is determined for each diameter class, the proportion of trees to cut by diameter class is calculated. That information—number of trees to cut, and percentage, by diameter class—is given to the field crews, who use that information as they mark the stand. Field crews will find it easier to base their marking on the proportion rather than the absolute number, since it is easier to think about removing a set percentage of a given diameter class rather than an absolute number of trees in a given diameter class per unit area. That also allows the reinforcement

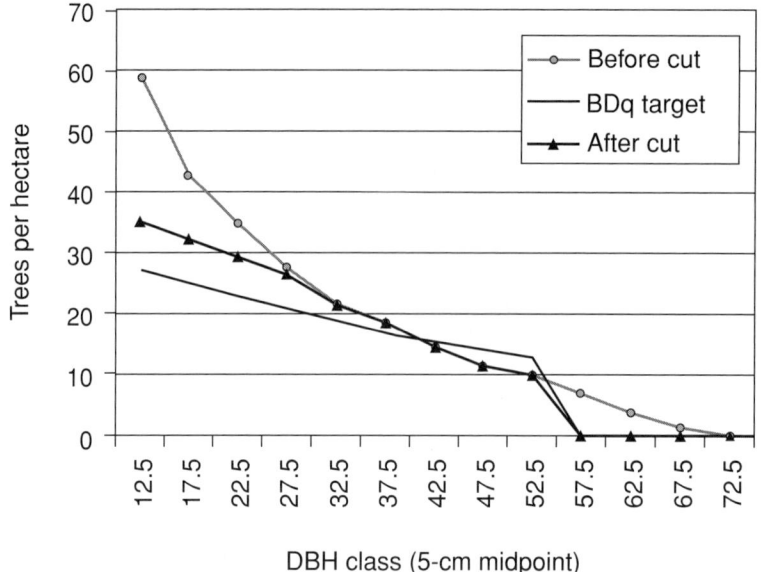

FIGURE 4. Regulation of stand structure using the basal area–diameter–q ratio (BDq) method conceptually applied in the 1978 before-cut inventory from the Good Farm Forestry Forty demonstration stand, an uneven-aged loblolly–shortleaf pine stand on the Crossett Experimental Forest in southern Arkansas. Before-cut, BDq target, and after-cut diameter distributions are drawn as curves rather than histograms. The BDq target reflects a $B = 14\,\mathrm{m}^2\,\mathrm{ha}^{-1}$, $D = 57.5$ cm, and q (5-cm classes) of 1.44. The after-cut diameter distribution illustrates the compensation by basal area according to the q ratio; the basal area in deficit diameter classes is retained in surplus diameter classes such that the target basal area is retained.

to be given that the poorest percentage of trees in the diameter class should be marked for harvest, and the best trees retained. The number of size classes the field crews must work with can be reduced if broader product classes are used. For instance, a fivefold product classification that includes small pulpwood, large pulpwood, small sawtimber, medium sawtimber, and large sawtimber would be very convenient for field crews to apply.

If some method other than the BDq approach is used to generate the target structure, the process for implementation still is most efficient if conducted as described above. Suppose a target diameter distribution is generated using a power function, for example, rather than the negative-exponential BDq approach. Once the target diameter distribution is obtained, there is still need to compare the existing stand to that target, generate a marking guide, and to compensate for diameter class deficits between the preharvest stand and the target, so as not to overcut the preharvest stand, and finally to determine if the projected harvest is operable.

Structural regulation has the advantage of objectivity. When generating a target diameter distribution, target diameter class data can be calculated for submerchantable diameter classes as well, which can provide guidance about whether cutting-cycle harvests are providing acceptable regeneration establishment and development through the submerchantable component and acceptable recruitment into the merchantable component of the stand. This depends on the assumption that the mathematical relationship used to characterize the stand structure is biologically meaningful at the smallest size classes, and this may not be the case for the negative exponential relationship upon which q is based (Baker et al. 1996).

The main disadvantage is that the process for calculating the marking tally is cumbersome, especially in cases in which deficit diameter classes are adjusted according to the mathematical relationship used in the initial target calculation. Spreadsheet programs are available to assist this calculation (Baker et al. 1996). A second disadvantage is that the appropriate B, D, and q parameters for application to longleaf pine have yet to be identified through research or practice.

Area-Based Regulation

Regulation of stands managed by group selection has been advocated using the area-based regulation concept borrowed from even-aged forest management (e.g., McConnell 2002). Under this simple device, the initial decision is to establish a rotation age, r, for the trees in the uneven-aged stand, essentially the age at which trees for the species under management are typically harvested in a comparable even-aged context. The area a of the stand is then divided by the rotation age r, and the quotient represents the proportion of the stand area a to be cut annually. That is converted to a, an area to be cut in a given cutting cycle harvest by multiplying the annual percentage of area to cut times the length of the cutting cycle, according to the simple formula

$$A_c = (a/r)c$$

where

A_c = area to be cut in a cutting-cycle harvest
a = stand area
r = rotation age and
c = planned length of the cutting cycle (years)

The problem in using area-based regulation with group selection is more theoretical than practical. It is difficult to distinguish between group selection and patch clearcutting, the small-opening even-aged variant of the clearcutting method that is also regulated using this approach. The best way to draw a distinction between the area-based regulation of the group selection method versus patch clearcutting is through applications that increase the within-stand heterogeneity of structure. Examples include varying the area cut in any one cutting cycle, varying group size and shape, or placing openings in a pattern that is not geometric or predictable. Use of group opening sizes less than two tree heights in radius, such that the entire group opening is under the ecological influence of the gap-bordering trees, would also provide ecological distinctions with patch clearcutting.

Adaptive Experience in Southern Pines

The long-term studies and demonstrations at the Crossett EF in south Arkansas provide keys to the successful implementation of the method in mixed loblolly–shortleaf pine stands. That background is the best source of experience with the method in the South and serves as a point of departure for considering the application to other forest types.

Regeneration

At the stand level, the first indicator of long-term forest sustainability is whether adequate regeneration is obtained when a regeneration cutting is made in a stand. This is especially true with the selection method, which requires a delicate balance between the stocking and development of the merchantable component of the residual stand versus the stocking and development of seedlings and saplings. It applies also in situations where conversion from even-aged condition to uneven-aged condition is imposed. It is critical to obtain regeneration after a cutting cycle harvest in any conversion, transition, or initial steps in implementation of either the single-tree or group selection method.

Abundant seed crops are an excellent attribute on which to rely when prescribing a reproduction cutting method that depends on natural regeneration. Long-term data on seedfall in loblolly–shortleaf pine stands on the Crossett EF show that, on average, natural regeneration is adequate or better four years in five, and rarely do seed failures occur in two consecutive years (Cain and Shelton 2001). This prolific seedfall is one of the reasons underlying the successful application of either even-aged or uneven-aged reproduction cutting methods that rely on natural regeneration in this forest type and region (Baker and Murphy 1982; Zeide and Sharer 2000; Cain and Shelton 2001).

Conversely, irregular seed crops can reduce the probability of success in obtaining natural regeneration under the selection method. This is important both for initial period of conversion to uneven-aged structure, and in maintenance of that structure. Experience at the Crossett EF suggests that failure to secure a new age class following a given cutting cycle harvest was not in itself an impediment to maintaining desirable uneven-aged structure, but missing two age classes in consecutive cutting cycle harvests is to be avoided (Reynolds 1969). If a given cutting cycle harvest fails to secure a new age cohort of regeneration, supplemental site preparation efforts should be conducted at the next cutting cycle harvest to ensure that regeneration is obtained (Guldin and Baker 1998).

Rehabilitation of Understocked Conditions

The stands on the Crossett EF originated as cutover understocked stands, and were managed in a manner by which stocking was built over time. The stands had been harvested by the Crossett Lumber Company in 1915 to a 38-cm stump limit, roughly equivalent to a 30-cm diameter limit. No management occurred on the area until it was leased to the Forest Service in 1934 (Guldin 2002). These stands were not fully stocked, homogeneous, and even-aged; rather, the stands showed considerable within-stand heterogeneity at the start. The research and demonstration work that began at that point successfully restored understocked and marginally stocked stands back to full stocking through harvest of a portion of growth. Two elements of this work were especially important.

The first was the reaction of these pines to removal of competition. In the upper West Gulf Coastal Plain, both loblolly and shortleaf pine respond to release at advanced age. Data from studies at the Crossett EF (Baker et al. 1996) suggest that pine stems in the 10- to 15-cm diameter class will respond to release if their

FIGURE 5. A loblolly pine sapling on the Crossett EF that meets the minimal size criteria for response to release—a 20% live crown ratio, and diameter outside bark at the base of the live crown of 5 cm. A similar rule of thumb for response to release would be helpful to have in applying the selection method to longleaf pine. (James M. Guldin)

live crown is greater than 20% and if the stem diameter at the base of the live crown exceeds 5 cm (Fig. 5).

The second was the use of herbicides to control hardwoods that were competing with the pines. Effective hardwood control was critical both as site preparation for the establishment and development of new seedlings and also as a release treatment and liberation cutting (Smith 1986) to free established pine saplings and small merchantable stems. Thus, the ability to use herbicides effectively to control hardwoods competing with pines, and to then have the pines respond quickly to the growing space made available, lies at the root of success in using the selection method to manage loblolly–shortleaf pine stands in the West Gulf region.

Developmental Dynamics

Uneven-aged stands exist in a delicate balance between understocking and growth. Baker et al. (1996) describe this balance in the context of shade management. That balance is controlled by three factors: the distribution of basal area retained in the residual stand immediately after a cutting cycle harvest, the length of the cutting cycle, and the operability of the future cutting cycle harvest.

After a cutting cycle harvest, the residual overstory trees grow and some degree of ingrowth into the overstory also occurs. As a result, stocking levels increase in the overstory over time. But this increased stocking serves to increasingly inhibit the development of regeneration. Subsequent cutting cycle harvests will be needed to reduce overstory stocking sufficiently to allow the continued development of the initial regeneration cohort, and to obtain a new cohort. On the other hand, if a given cutting-cycle harvest removes too much basal area, the stand will not grow rapidly enough to allow an operable cutting cycle harvest in the subsequent cutting cycle.

One simple metric for the upper limit of acceptable basal area to carry in an uneven-aged stand is to quantify the residual basal area in a classic low thinning at which accidental regeneration just begins to be suppressed. This point approximates the highest acceptable before-cut basal area in uneven-aged stands. For example, pine regeneration can become established in even-aged Coastal Plain loblolly stands thinned to 16 m^2 ha^{-1}, but will cease to make acceptable height growth if basal area exceeds 17–18 m^2 ha^{-1}. Because overstory tree distribution in uneven-aged stands is more heterogeneous, these basal area levels represent an upper limit to the acceptable basal area range for successful uneven-aged prescriptions.

The lower limit is defined by maintaining acceptable overstory growth over the expected duration of the cutting cycle. In uneven-aged loblolly–shortleaf pine stands at the Crossett EF, the target residual basal area after a cutting cycle harvest is roughly 14 m^2 ha^{-1}, and the stands grow approximately 0.55–0.7 m^2 ha^{-1} in basal area annually. After 5 years, the stands will have grown in basal area to about 17 m^2 ha^{-1}, and the subsequent cutting cycle harvest can then be imposed.

Because basal area can also be related to stand volume, operational feasibility of harvests can be tested. In south Arkansas, loblolly–shortleaf pine stands support after-cut volumes of roughly 26–27 m^3 ha^{-1} of sawlog volume, and the stands grow on the order of 2.2–2.4 m^3 ha^{-1} of sawlog volume annually. After 5 years, the stands will support roughly 37–40 m^3 ha^{-1} of sawlog volume. Operable harvests in the vicinity are about 9 m^3 ha^{-1} of sawlog volume. Thus, sawlog volume growth of the stand can be cut on roughly a 5-year interval with operational harvests, which also maintains regeneration development. Metrics such as these are needed for longleaf pine stands.

Another clue about the appropriate upper limit of basal area is whether acceptable rates of height growth can be maintained in the regeneration component. In loblolly and shortleaf pine stands, height growth of regeneration is a useful indication of maintaining the ability to recover full growth potential. Minimum acceptable annual height growth in these species is 0.15 m. If seedlings or saplings less than 1.3 m in height are not growing at this rate, they will probably not survive.

Finally, the Crossett EF experience suggests a final visual clue for determination of acceptable balance between overstory and understory—the presence of foliage of the desired species at all levels of the canopy profile in the stand. If regeneration is successfully established and making acceptable height growth, and if repeated cutting-cycle harvests are successful in obtaining regeneration, seedlings and saplings will be visible in the stand. The longer the period of successful silviculture under the selection method, the more prominently will foliage of the desired species be found at all levels of the canopy profile (Fig. 6).

FIGURE 6. A view of conditions in an uneven-aged stand of loblolly–shortleaf pine within the Poor Farm Forestry Forty Demonstration Area on the Crossett EF immediately following the cutting cycle harvest in the spring of 2003. Note the presence of foliage of the desired pines at all heights in the canopy profile, a simple visual clue that denotes sustainable regeneration cutting over time in uneven-aged stands. (James M. Guldin)

Marking Rules

During his tenure at Crossett, CEF founding scientist Russ Reynolds explicitly refused to identify his volume-control method as "single-tree" or "group" selection; he called it "selection." Occasionally large openings would occur, occasionally small ones would suffice. Reynolds's key decision was whether the tree being examined while marking was of acceptable form, size, and quality to retain. Reynolds captured this concept in the simple phrase, "cut the worst and leave the best" (Reynolds 1959, 1969). Attention to this simple marking rule ensured that stem quality was gradually improved over time. As practiced on the Crossett EF, the selection method has a reputation as one that produces sawtimber of high quality (Guldin and Fitzpatrick 1991); the long-term application of a marking rule such as this contributes to that reputation.

This rule raises distinctions between regulation by volume under the VCGDL and regulation by structure under the BDq method. It is easier to leave the best trees under the volume control regulation method versus the BDq, because the marking tally in the BDq is specific to a given diameter class whereas the marking tally of the volume control method cuts across diameter classes. Under VCGDL, a residual tree is judged to be part of the population of "best" trees regardless of diameter class. Conversely, in the BDq method, a tree that is retained is judged relative to other trees in that diameter class only, and the proportion to cut changes from one diameter class to another.

For example, suppose that a BDq marking tally requires removal of 1 in 10 trees in the 45-cm class, but half of the trees in the 30-cm class. Field crews will invariably come across a tree in the 20th percentile of quality in the 45-cm class immediately adjacent to a smaller tree of better absolute form and with better developmental potential in the 40th-50th percentile of quality in the 30-cm class, and will complain about marking the better tree and leaving the poorer one. The answer for the field crews in that event is to use common sense, and to mark the poorer tree. Carrying that logic to its conclusion leads to a critical point relative to the BDq method. Of the B, D, and q variables, residual basal area is most important to retain, followed by maximum diameter; q is least important. Some thought has been given to modifications of the BDq method as a BD method (Baker et al. 1996); this would result in essentially a basal area control method implemented in a manner similar to regulation by volume, but in which the basis for compensation among trees being retained is by equivalence of basal area rather than volume.

Reynolds's marking rule also raises distinctions between group selection and single-tree selection. A rule that guides the forester to "cut the worst and leave the best" can be more strictly followed in single-tree selection than in group selection. In most instances, the localized area of the stand within which a group opening is planned will contain trees that under an individualistic evaluation would qualify for retention. That might allow one to further refine the logic for placement of group openings—locate the opening in those parts of the stand where a disproportionate number of the trees within the planned group are of poorer condition than those in other parts of the stand. Such manipulation in the location of group openings is possible if the group selection method is being implemented using regulation by volume or structure. However, improvement in residual stand condition as a result of this marking rule is by definition unlikely to occur under area regulation of group selection, especially if imposed using strict geometric patterns.

The Selection Method in Longleaf Pine

Interest in implementing the selection method in the longleaf pine forest type is driven by a number of considerations. Foremost among them is to develop habitat conditions in longleaf stands that favor the species that inhabit these stands, such as bobwhite quail (Moser et al. 2002) and the red-cockaded woodpecker (McConnell 2002). To a certain extent, arguments about habitat condition that can be developed in uneven-aged stands of longleaf pine are premature without a careful examination of what a sustainable application of the selection method would look like in longleaf pine, using the subjective metrics developed from our understanding of the method in the loblolly–shortleaf pine forest type.

The state-of-the-art treatise on the selection method in longleaf pine (Farrar 1996) is a primary source for managers to consider as the selection method is operationally applied in longleaf pine. Equally important in application to the selection system in longleaf pine is research on longleaf pine autecology that culminated three decades ago on the Escambia EF near Brewton, AL (Croker and Boyer 1975), where

FIGURE 7. A view of the shelterwood method in application to longleaf pine in 1982 on the Escambia Experimental Forest, Brewton, AL. The residual basal area in the overstory was 7 m^2 ha^{-1}, and seedlings have emerged from the grass stage several years following the seed cut. (James M. Guldin)

detailed studies of the reproductive biology and silvics of longleaf pine were fundamental to the development of the even-aged shelterwood method (Fig. 7). A subjective interpretation of these sources suggests that a successful prescription for the selection method in longleaf pine will require attention to regeneration establishment, the pattern of implementation, the approach to regulation, and developmental dynamics. Among the largest challenges will be the integration of prescribed fire as a standard element of the method.

Regeneration

The application of natural regeneration in a selection method for longleaf pine will be difficult. Seed production is much less reliable in longleaf pine, where adequate seed crops only occur between 10 and 20% of the time (Wahlenberg 1946), than in loblolly pine. The degree of silvicultural attention required to make a successful prescription involving natural regeneration will be greater for longleaf pine than for loblolly pine, if for no other reason than the greater infrequency of adequate seed crops. This can be especially problematic in mixed-species southern pine stands that include longleaf pine as part of the mix, because the other pines will be more prolific seed producers.

Careful attention to silvicultural detail is needed to ensure practical success with natural regeneration in longleaf pine. For example, the key to the development of the shelterwood method in longleaf pine was the detailed work by Croker (1973). He reported that over a 7-year period, cone production in a longleaf pine stand reached an optimum when the stand basal area of longleaf pine was 6.88 m² ha⁻¹ and declined as basal area decreased or increased from that level. Greater overstory basal area resulted in less seedfall and reduced numbers of seedlings. A uniform residual overstory of 10.33 m² ha⁻¹ resulted in virtually no surviving saplings over time (Boyer 1993). Fieldcraft such as that described in the development of the shelterwood method (Croker and Boyer 1975) would improve natural seedfall in any selection method applied in longleaf pine stands. Because cone production is a highly inherited trait genetically (Croker 1964), marking crews should include an evaluation of past cone production as a decision element in whether to retain a tree during cutting cycle harvests (Fig. 8).

On national forest lands in the South, another practical approach for management of longleaf pine using the selection method is to plan for natural regeneration, but to use planting as a fallback position to prevent excessive delays in reforestation. There will be two opportunities for successfully obtaining natural regeneration prior to planting. The first chance is that associated with the initial harvest. Foresters with the USDA Forest Service generally allow a logger a multiyear

FIGURE 8. Longleaf pine seed after seedfall in fall 1982, a good seed year in a managed even-aged longleaf pine stand in central Louisiana. A prescribed fire had been used that year to prepare the seedbed for the anticipated seed crop. (James M. Guldin)

window within which to complete a cutting cycle harvest. Hopefully, the period when the stand is actually cut would occur in conjunction with an adequate seed year. However, the administration of sales on National Forest lands precludes the ability to guarantee harvest in conjunction with seed crops and receptive seedbeds. Logging contractors are typically given several years to harvest a timber sale, and cultural work to improve seedbed condition must be programmed a year in advance.

The second chance is to catch the first good seed crop after the sale closes. Site preparation should be conducted to prepare a receptive seedbed when that seed crop occurs. This, too, is constrained by administrative procedures. Site preparation on National Forest lands is funded using proceeds from timber sales under the Knutsen–Vandenberg Act of 1933, which limits expenditures to a 5-year period following sale closure. If longleaf pine produces one good seed crop in 5 years, the chances are that site preparation can be timed to that expected seedfall. However, site preparation dollars must be requested in the fiscal year in advance of that in which they would be spent—and the ability to predict a good seed year is limited to 6 months in advance of seedfall (Croker and Boyer 1975).

Practically, then, a forester with the USDA Forest Service has 4 years after the cutting cycle harvest is concluded to obtain natural regeneration; the fifth year must be devoted to spending the available funds to prepare the site and plant seedlings. Standards should be developed that provide guidance to silviculturists about when natural regeneration difficulties are likely to be profound (such as poor seed producers left on the site, an absence of advance growth, or the lack of adjacent stands that could contribute seed to a cutover area from adjacent mature trees). That information could help foresters identify stands that might be good candidates to plant initially under residual overstories, rather than to tackle the chain of efforts to synchronize site preparation to seedfall.

In the past decade, a focus of regeneration research in longleaf pine has been to examine longleaf seedling establishment, survival, and growth in openings of various size and condition. Results are generally of the opinion that a clumped residual overstory condition, suggestive of the group selection approach, promotes early seedling development when compared with homogeneous overstory conditions. Seedlings initiate height growth primarily in the center of gaps, but height growth is reduced along the borders of gaps or adjacent to residual trees. This border effect is on the order of 12–20 m. The major reasons for this seedling growth pattern are related to increased light intensity in gaps (Grace and Platt 1995; Palik et al. 1997; McGuire et al. 2001; Battaglia et al. 2002; Gagnon et al. 2003) and decreased levels of intraspecific competition for soil resources in gaps (Brockway and Outcalt 1998).

More than any of the other southern pines, longleaf pine will benefit from some applications that involve planting as the source of regeneration. However, research experience with that is limited. Probably a minimum number of seedlings is rationally feasible to plant. If natural regeneration is inadequate, and planting is needed, one should plant enough seedlings to ensure a fully stocked stand of planted seedlings. The number to plant will vary depending on seedling stock and the level of understory competition. For example, 1000–1200 seedlings per hectare may be sufficient using containerized planting stock and a preplant herbicide treatment in a group selection opening. Without these refinements, more seedlings will be needed.

Planting should be done in association with an effective site preparation prescription that promotes survival and height growth of the planted seedlings. Genetic improvement of planting stock has produced seedlings that make rapid early height growth under open conditions with intensive site preparation. Although families selected for rapid growth in the open would probably be successful if planted beneath a residual overstory or within a group opening that is under the ecological influence of the overstory trees that surround the group, there are opportunities to explore the best families to plant in conditions that are subject to partial shade or neighbor influence. However, there is little basis for this at present.

Pattern of Implementation

Because the presence of overstory trees affects longleaf pine seedling establishment and development, most recent research has concentrated on the possibilities associated with use of the group selection method in longleaf pine. Farrar (1996, 1998) suggests that a "modified group selection" approach be used in which groups are not removed until regeneration is established beneath the group. He further suggests that this be integrated with cyclic prescribed burning (Fig. 9). Either the VCGDL or the BDq regulation approaches would be feasible (Farrar and Boyer 1991). Under the BDq, Farrar's suggested target structure for longleaf

FIGURE 9. Longleaf pine seedlings and saplings of natural origin in a group selection opening on the Escambia Experimental Forest in Brewton, AL, during the 1982 growing season. (James M. Guldin)

pine includes a residual basal area target B of 14 m^2 ha^{-1}, maximum retained diameter D of 50 cm, and a q ratio for 5-cm diameter classes of 1.44, and a suggested cutting cycle length of 10 years (Farrar 1996). His uneven-aged marking guidelines contain a description of how to implement either method for longleaf pine. The burning program is required to keep competing hardwoods in check and to keep seedbeds prepared for any seedfall that might occur. When seedlings become established at acceptable densities within an area (local distributions equivalent to 1800–2000 trees ha^{-1}), cutting cycle harvest to remove the overstory trees will allow seedlings to initiate height growth. Subsequent cutting cycle harvests can be used to expand existing groups or to establish new groups, and as a free thinning in the matrix of the stand between the group openings.

Given the success in regenerating longleaf pine naturally using the shelterwood method, another possibility, or perhaps a variant of Farrar's modified group approach, is to adapt the shelterwood prescription for longleaf pine (Croker and Boyer 1975) in the context of a group selection regeneration method, where the groups are treated as small shelterwood openings. Smith (1986) noted that although group selection is usually imposed by removing all trees within the group, groups could certainly be created that retained overstory trees within them as silviculturally appropriate. This method would resemble a group selection with reserves (cf. Helms 1998) except that the reserves are explicitly retained for the silvicultural purpose of obtaining natural regeneration.

A shelterwood-based group selection approach in longleaf pine would use groups within which longleaf pine seed trees are retained at shelterwood (6–7 m^2 ha^{-1}) residual basal area levels. During one cutting cycle harvest, groups would be marked to resemble the seed cut of a shelterwood residual basal area (Smith 1986), using the same decision variables for seed tree retention as described in Croker and Boyer (1975). During the intervening cutting cycle, prescribed fires would be imposed as an element of the prescription, so as to prepare the site for seedfall, and Farrar (1998) offers suggestions to accomplish that. The seed trees in the group openings would eventually maximize their ability to produce cones, seedfall would be optimized 3 to 5 years after the cutting cycle harvest, and at the next adequate seed year seedlings would become established in the shelterwood group opening. The residual seed trees within the group would then be removed in the subsequent cutting cycle harvest to release established seedlings. If the removal cut of the shelterwood residuals within the group opening is not timely, seedling survival would be compromised.

Alternatively, if groups are initially created without residual trees, an area adjacent to the group opening could be retained at shelterwood basal areas such that group expansion would occur in the subsequent cutting cycle harvest, resulting in an expanded group coalescence. The same suggestions regarding marking, use of prescribed fire, seedling development, overstory removal, and so on would apply in this modification of the practice.

It might be that stands larger than 15 ha could be managed this way, and also that the method would be amenable to group openings larger than the 0.81-ha maximum for group selection suggested by the USDA Forest Service. Research would be needed to determine an effective range of group opening size such that longleaf seedlings can initiate height growth in a short period of time. It is likely that the size would be larger rather than smaller. This approach could be regulated using any of the usual regulation methods that apply to uneven-aged stands.

Approach to Stand Regulation

More than in other southern pines, managers in longleaf pine are looking for ways to retain larger trees, in some cases much larger, above the diameter limits that have been established in other uneven-aged experience. This would be done to meet resource attributes, values, or needs for associated species within the longleaf forest ecosystem, such as legacy trees or for nest construction by species such as the red-cockaded woodpecker. There is little theory available in the uneven-aged literature to account for the influence of large trees retained above the maximum diameter or in addition to the desired target residual stand. But even if only a few large trees are retained, their presence can adversely affect the development of the stand, because retaining them prevents growing space from being used by trees of other sizes. The regulation method and marking guides must account for any trees that are retained for special purposes, simply because large trees usurp considerable growing space that if unaccounted for could disrupt stand development.

For example, the BDq method calls for harvest of the worst trees and retention of the best in diameter classes at or below the maximum retained diameter D, but calls for all trees above the D to be cut. There are no active research studies on the Crossett EF that test whether trees larger than D can be retained. As a starting point, the basal area of retained trees should be included in the calculations of stand structure, simply because they create a very large influence in basal area calculations. An 80-cm tree has a basal area of 0.5 m^2, or roughly 4% of the residual basal area target of 13 m^2 ha^{-1} after a typical cutting-cycle harvest. If three trees above D per hectare were retained for special purposes and the stand below D was managed for the after-cut residual basal area target of 13 m^2 ha^{-1}, the actual basal area in the stand would be 14.5 m^2 ha^{-1}, more than 10% higher than the target. Over time that would adversely affect stand development. Similarly, under volume control, the volume and the volume growth of those big trees must be averaged into the calculation used to determine the allowable cut and the guiding diameter limit. Trees above the limit can be retained at the discretion of the marker, provided that an equivalent volume is then marked below the guiding diameter limit.

Growth and Yield

Empirical data on growth and yield of uneven-aged stands of longleaf pine are difficult to find because uneven-aged stands managed for a sufficient length of time are relatively rare (Kush et al. this volume). One of the few papers to cite data on growth and yield directly under the selection system is that of Farrar and Boyer (1991), who describe the growth of two uneven-aged stands on the Escambia EF over a 10-year period in comparison to that of a demonstration stand being managed using the shelterwood method. In all three stands, sawtimber volume growth was comparable at roughly 2 m^3 ha^{-1}, or roughly two-thirds the rate expected in the study. At these rates of volume growth, cutting cycles of 10 years or longer will be required to generate operable harvests.

Farrar and Boyer (1991) also speculated that over time, volume yield from selection stands will likely be less than that produced from a forest of large even-aged stands, because the zone of competition between large and small trees is minimized with large blocks found in even-aged stands (Farrar and Boyer 1991). The use of group selection, which would also have less area within the stand in this zone of competition relative to single-tree selection, might partly compensate for that hypothesized volume shortfall.

In a study of longleaf pine regeneration development under varying residual overstory basal area levels, Boyer (1993) reported that even a few residual longleaf pine parent trees resulted in substantial growth reductions of the new age cohort. The growth of two-aged stands in this study was less than half the growth reported in the naturally regenerated even-aged stands released from overstory competition when young. This provides another estimate of the total merchantable volume growth that might be produced in uneven-aged stands—less than half that found in comparable released even-aged stands.

Given the shortage of long-term data on growth and yield from uneven-aged stands in the literature, among the first priorities for research is to better quantify the growth and yield that one can expect from application of the selection system in the longleaf pine forest type. At a minimum, one should expect reduced rates of volume growth, especially in total merchantable cubic volume, in the uneven-aged stands. This point has been observed elsewhere in application of the method in southern pines (Guldin and Baker 1998).

Developmental Dynamics

If uneven-aged silviculture is applied in longleaf pine stands similar to that in other forest types where the method has been successful, a number of attributes will be apparent. Longleaf pine trees in the sawtimber size class will constitute roughly two-thirds of the residual basal area, but only 25% of the number of stems 10 cm and larger in the stand (Farrar et al. 1984; Baker et al. 1996). Cutting cycle harvests will require two visits by field crews to the stand—one to obtain the before-cut stand inventory upon which regulation calculations are based and to determine operability of the proposed cutting cycle harvest, and a second to actually mark the stand using the marking guidelines that the regulation calculations produce. Marking will follow a pattern of cutting the worst longleaf pines, and leaving the best, during every cutting cycle. Some form of competition control that meets the multiple silvicultural objectives of site preparation, release, and liberation cutting will be required on the order of every other cutting cycle. In longleaf pine stands, that will probably take the form of prescribed fire, with perhaps occasional herbicide use or mechanical felling of competing hardwood species. Regeneration of longleaf pine must be monitored after each cutting cycle harvest. Some regeneration will be expected to become established and to initiate height growth after every cutting cycle harvest, and especially after the first cutting cycle harvests in stands recovering from understocked conditions or being converted from even-aged fully stocked conditions. After several decades of implementation using a 10-year cutting cycle, visual examination of stands will reveal foliage of longleaf pine present at all levels of the canopy profile from the ground up through the main canopy. It will require three or more 10-year cutting cycle harvests to approach an uneven-aged structure, and longer to develop a well-balanced stand structure. Fewer cutting cycles will be needed if the initial stand condition is understocked and if multiple age cohorts are already present; more will be needed if the stand under management

7. Uneven-Aged Silviculture

FIGURE 10. A within-stand view of a new regeneration cohort in a longleaf pine stand managed as a quail plantation in southern Georgia. (James M. Guldin)

was initially a well-stocked even-aged stand at midrotation or later.

This description is at odds with several existing and notable approaches to management of longleaf pine—that proposed for management of the red-cockaded woodpecker, and that reflected in the Stoddard–Neel approach for management of uneven-aged longleaf pine stands for quail habitat. Open understory conditions are sought for both red-cockaded woodpeckers and for bobwhite quail. An understory full of trees in the submerchantable and smallest merchantable size classes are sought in an uneven-aged stand. The question is whether uneven-aged practices can be, or have been, modified so as to simultaneously maintain an open understory condition, while concurrently maintaining the development of regeneration cohorts in sufficient number and distribution within the uneven-aged stand (Fig. 10).

Consider the Stoddard–Neel approach of uneven-aged silviculture, an excellent description of which was recently published by Moser et al. (2002) (see also boxes A and B). In this approach, single-tree selection is used to maintain low residual basal area at 15 m^2 ha^{-1} or less; open midstory conditions are maintained, removals are made from below, and regeneration occurs in patches. The purpose of this variant of the single-tree selection method is to provide habitat for northern bobwhite quail. Graphs of the diameter distributions for several quail plantations (Moser at al. 2002) differ from the reverse J-shaped curve typically associated with uneven-aged silviculture at the stand level, specifically in a lack of trees in the smallest diameter classes. For example, no 1-cm diameter class less than 10 cm in any of the plantations managed using the Stoddard–Neel approach has more than 5% of the total number of pines per hectare. Conversely, the 1995 preharvest inventory on the Good Farm Forestry Forty demonstration stand on the Crossett EF showed 4450 trees ha^{-1} in the submerchantable diameter classes (1.5–8.9 centimeters inclusive), corresponding to 92% of the total trees per hectare in the demonstration stand (Guldin unpublished data). This raises the question of whether regeneration is being recruited at a sufficient rate under the Stoddard–Neel approach to ensure the long-term sustainability of not only uneven-aged stand structure, but also the habitat values for which those stands are justifiably prized.

Similarly, debate is currently active in the literature about the habitat values provided by uneven-aged stands for the red-cockaded woodpecker (Engstrom et al. 1996; Rudolph and Conner 1996; Hedrick et al. 1998). This debate has its roots within the application of the Stoddard–Neel selection system as well, since it is largely upon that experience that habitat descriptions of uneven-aged longleaf pine stands are based. But the debate embraces other situations in which desired habitat for the red-cockaded woodpecker is inconsistent with information available from the literature about uneven-aged stand dynamics. It is not really a question of the conditions that are appropriate for the red-cockaded woodpecker, which are fairly well defined (Conner and Locke 1982; Hooper 1988; Hooper et al. 1991; Conner et al. 1994; Ross et al. 1997). Rather, the question is one in which a regeneration method that provides desired residual stand conditions is sustainable, according to the first rule of sustainability at the stand level—that an imposed regeneration method must result in the successful establishment and development of regeneration.

Summary

The successful practice of uneven–aged silviculture in mixed loblolly–shortleaf pine stands of the upper West Gulf Coastal Plain has been characterized by a number of attributes (Guldin and Baker 1998). Key factors are attention to stand-level regulation, use of appropriate residual basal area levels that approximate those found in slightly understocked mature even-aged stands, establishment and development of regeneration, and attention to a marking rule that cuts the poorest trees and leaves the best across a range of diameter classes.

Longleaf pine shares some silvical attributes with loblolly and shortleaf pines, but not all. The favorable elements will be useful in development or refinement of the selection method in longleaf pine. First, dominant or codominant longleaf pines respond to release, though suppressed trees do not (Boyer 1990). Thus, cutting cycle harvests in which codominant or better longleaf pines are released from competition of others will stimulate a growth response in the residual stand, which promotes continued stand development. This attribute has been a feature of the Stoddard–Neel variant of the selection method in longleaf pine as well (Moser et al. 2002). Second, it can be successfully managed using the shelterwood method (Croker and Boyer 1975), which has been observed in other forest types where the selection method has been applied (Guldin 1996).

Where longleaf differs most prominently from other southern pines is in the periodicity of seed crops and the difficulty in securing natural regeneration. This will require new interpretations of existing knowledge, and the development of new knowledge, to ensure that longleaf pine seedlings can become established and can develop properly following regeneration cutting under the selection method. With respect to natural regeneration, refinements of existing knowledge from the application of the shelterwood method might be promising as a variation under modifications of the group selection method (Farrar 1996). Conversely, should natural regeneration techniques fail or result in unacceptable delays in regeneration establishment or development, technology should be developed for application of planting as an alternative or a preferred method of obtaining establishment of regeneration at acceptable levels. More than any other southern pine, or for that matter any other species in which the selection method has been used, planting seedlings for reforestation of uneven-aged stands will have a prominent place in the successful application of the selection method for longleaf pine.

The biggest question that remains unresolved is the level at which regeneration development can be considered acceptable in the selection method. The Stoddard–Neel selection method points in one direction about this, and experience from the selection method as practiced in mixed loblolly-shortleaf pines, and in other forest types, points in another. The Stoddard–Neel approach differs from most other instances of successful application of uneven-aged silviculture due to the smaller number of stems in the submerchantable class. That difference leads to allied relationships in habitat condition that promotes open midstory conditions in one case, and midstory conditions occluded by development of seedlings and sapling in the submerchantable diameter classes in other conditions. Ultimately, the question relates to the degree to which deviations from the reverse J-shaped structure can be considered to be sustainable at the stand level. This is the most prominent research gap in our understanding of the regeneration dynamics of uneven-aged longleaf pine stands, and one that is critical in order to ultimately evaluate what constitutes sustainability of uneven-aged structure in longleaf pine stands in the long term.

References

Baker, J.B., and Murphy, P.A. 1982. Growth and yield following four reproduction cutting methods in loblolly–shortleaf pine stands—A case study. *South J Appl For* 6(2):66–74.

Baker, J.B., Cain, M.D., Guldin, J.M., Murphy, P.A., and Shelton, M.G. 1996. Uneven-aged silviculture for the loblolly and shortleaf pine forest cover types. USDA Forest Service, General Technical Report SO–118. New Orleans, LA: Southern Research Station.

Barnett, J.P. 1999. Longleaf pine ecosystem restoration: The role of fire. *J Sustain For* 9(1/2):89–96.

Battaglia, M.A., Mou, P., Palik, B., and Mitchell, R.J. 2002. The effect of spatially variable overstory on the understory light environment of an open-canopied longleaf pine forest. *Can J For Res* 32(11):1984–1991.

Boyer, W.B. 1990. Longleaf pine. In *Silvics of North America: 1. Conifers; 2. Hardwoods*. Agriculture Handbook 654, tech. cords. R.M. Burns and B.H. Honkala, vol. 1, pp. 405–412. Washington, DC: USDA Forest Service.

Boyer, W.B. 1993. Long-term development of regeneration under longleaf pine seedtree and shelterwood stands. *South J Appl For* 17(1):10–15.

Brockway, D.G., and Outcalt, K.W. 1998. Gap-phase regeneration in longleaf pine wiregrass ecosystems. *For. Ecol Manage* 106(2,3):125–139.

Cain, M.D., and Shelton, M.G. 2001. Natural loblolly and shortleaf pine productivity through 53 years of management under four reproduction cutting methods. *South J Appl For* 25(1):7–16.

Conner, R.N., and Locke, B.A. 1982. Fungi and red-cockaded woodpecker cavity trees. *Wilson Bull* 94:64–70.

Conner, R.N., Rudolph, D.C., Saenz, D., and Schaefer, R.R. 1994. Heartwood, sapwood, and fungal decay associated with red-cockaded woodpecker cavity trees. *J Wildl Manage* 58:728–734.

Croker, T.C. 1964. Fruitfulness of longleaf trees more important than culture in cone yield. *J For* 65: 488.

Croker, T.C. 1973. Longleaf pine cone production in relation to site index, stand age, and stand density. Research Note SO-156. New Orleans, LA: USDA Forest Service, Southern Forest Experiment Station.

Croker, T.C., and Boyer, W.D. 1975. Regenerating longleaf pine naturally. Research Paper SO–105. New Orleans, LA: USDA Forest Service, Southern Forest Experiment Station.

deLiocourt, F. 1898. De l'amenagement des Sapinières. Bulletin of the Society of Foresters. Franche-Comté Belfort, Besançon, pp. 396–409.

Engstrom, R.T., Brennan, L.A., Neel, W.L., Farrar, R.M., Lindeman, S.T., Moser, W.K., and Hermann, S.M. 1996. Silvicultural practices and red-cockaded woodpecker management: A reply to Rudolph and Conner. *Wildl Soc B* 24(3):334–338.

Farrar, R.M., Jr. 1996. Fundamentals of uneven-aged management in southern pine. Miscellaneous Publication 9. Tallahassee, FL: Tall Timbers Research Station.

Farrar, R.M. 1998. Prescribed burning in selection stands of southern pine: Current practice and future promise. In *Fire in Ecosystem Management: Shifting the Paradigm from Suppression to Prescription*, eds. T.L. Pruden and L. A. Brennan, pp. 151–160. Tall Timbers Fire Ecology Conference Proceedings No. 20. Tallahassee, FL: Tall Timbers Research Station.

Farrar, R.M., Jr. and Boyer, W.D. 1991. Managing longleaf pine under the selection system—Promises and problems. In *Proceedings of the 6th Biennial Southern Silvicultural Research Conference; 1990 October 30–November 1, Memphis, TN*, pp. 357–368. General Technical Report SE–70. Asheville, NC: USDA Forest Service, Southeastern Forest Experiment Station.

Farrar, R.M., Murphy, P.A., and Willett, R. L. 1984. Tables for estimating growth and yield of uneven-aged stands of loblolly–shortleaf pine on average sites in the West Gulf area. Bulletin 874. Fayetteville: University of Arkansas, Arkansas Agricultural Experiment Station.

Gagnon, J.L., Jokela, E.J., Moser, W.K., and Huber, D.A. 2003. Dynamics of artificial regeneration in gaps within a longleaf pine flatwoods ecosystem. *For Ecol Manage* 172:133–144.

Grace, S.L., and Platt, W.J. 1995. Neighborhood effects on juveniles in an old-growth stand of longleaf pine, Pinus palustris. *Oikos* 72(1):99–105.

Guldin, J.M. 1996. The role of uneven-aged silviculture in the context of ecosystem management. *West J Appl For* 11(1):4–12.

Guldin, J.M. 2002. Continuous cover forestry in the United States: Experience with southern pines. In *Continuous Cover Forestry: Assessment, Analysis, Scenarios*, eds. K. von Gadow, J. Nagel, and J. Saborowski, pp. 295–307. Boston: Kluwer Academic Publishers.

Guldin, J.M., and Baker, J.B. 1988. Yield comparisons from even-aged and uneven-aged loblolly–shortleaf pine stands. *South J Appl For* 12:107–114.

Guldin, J.M., and Baker, J.B. 1998. Uneven-aged silviculture, southern style. *J For* 96(7):22–26.

Guldin, J.M., and Fitzpatrick, M.W. 1991. Comparison of log quality from even-aged and uneven-aged loblolly pine stands in south Arkansas. *South J Appl For* 15(1):10–17.

Hedrick, L.D., Hooper, R. C., Krusac, D. L., and Dabney, J. M. 1998. Silvicultural systems and red-cockaded woodpecker management: Another perspective. *Wildl Soc B* 26(1):138–147.

Helms, J.A., ed. 1998. *The Dictionary of Forestry*. Bethesda, MD: The Society of American Foresters.

Hooper, R.G. 1988. Longleaf pines used for cavities by red-cockaded woodpeckers. *J Wildl Manage* 52:392–398.

Hooper, R.G., Krusac, D.L., and Carlson, D.L. 1991. An increase in a population of red-cockaded woodpeckers. *Wildl Soc B* 19:277–286.

Landers, J.L., Van Lear, D.H., and Boyer, W.D. 1995. The longleaf pine forests of the Southeast: Requiem or renaissance? *J For* 93(11):39–44.

Leak, W.B. 1964. An expression of diameter distribution for unbalanced uneven-aged forests. *For Sci* 10:39–50.

Long, J.N. 1998. Multiaged systems in the central and southern Rockies. *J For* 96(7):34–36.

McConnell, W.N. 2002. Initiating uneven-aged management in longleaf pine stands: Impacts on red-cockaded woodpecker habitat. *Wildl Soc B* 30(4):1276–1280.

McGuire, J.P., Mitchell, R.J., Moser, E.B., Pecot, S.D., Gjerstad, D.H., and Hedman, C.W. 2001. Gaps in a gappy forest: Plant resources, longleaf pine regeneration, and understory response to tree removal in longleaf pine savannas. *Can J For Res* 31(5):765–778.

Marquis, D.A. 1978. Application of uneven-aged silviculture on public and private lands. In *Uneven-Aged Silviculture and Management in the United States*, pp. 25–61. USDA Forest Service. Timber Management Research, General Technical Report WO-24.

Meyer, H.A., Recknagel, A.B., Stevenson, D.D., and Bartoo, R.A. 1961. *Forest Management*, 2nd edition. New York: Ronald Press.

Moser, W.K., Jackson, S.M., Podrazsky, V., and Larsen, D.R . 2002. Examination of stand structure on quail plantations in the Red Hills region of Georgia and Florida managed by the Stoddard–Neel system: An example for forest managers. *Forestry* 75(4):443–449.

O'Hara, K.L. 1996. Dynamics and stocking-level relationships of multi-aged ponderosa pine stands. *For Sci* 42(4), Monograph 33.

Palik, B.J., Mitchell, R.J., Houseal, G., and Pederson, N. 1997. Effect of canopy structure on resource availability and seedling response in a longleaf pine ecosystem. *Can J For Res* 27(9):1458–1464.

Pickett, S.T.A., and White, P.S., eds. 1985. *The Ecology of Natural Disturbance and Patch Dynamics*. New York: Academic Press.

Reynolds, R.R. 1959. Eighteen years of selection timber management on the Crossett Experimental Forest. USDA Technical Bulletin 1206.

Reynolds, R.R. 1969. Twenty-nine years of selection timber management on the Crossett Experimental Forest. USDA Forest Service Research Paper SO-40.

Reynolds, R.R., Baker, J.B., and Ku, T.T. 1984. Four decades of selection management on the Crossett Farm Forestry Forties. Arkansas Agricultural Experiment Station Bulletin 872.

Ross, W.G., Kuhlavy, D.L., and Conner, R.N. 1997. Stand conditions and tree characteristics affect quality of longleaf pine for red-cockaded woodpecker cavity trees. *For Ecol Manage* 91:145–154.

Rudolph, D.C., and Conner, R.N. 1996. Red-cockaded woodpeckers and silvicultural practice:

Is uneven-aged silviculture preferable to even-aged? *Wildl Soc B* 24(2):330–333.

Runkle, J.R. 1982. Patterns of disturbance in some old-growth mesic forests of eastern North America. *Ecology* 63:1533–1546.

Smith, D.M. 1986. *The Practice of Silviculture*, 8th edition. New York: John Wiley & Sons.

Troup, R.S. 1928. *Silvicultural Systems*. London: Oxford at the Clarendon Press.

Troup, R.S. 1952. *Silvicultural Systems*, 2nd edition. Edited by E. W. Jones. London: Oxford University Press.

Wahlenberg, W.G. 1946. *Longleaf Pine: Its Use, Ecology, Regeneration, Protection, Growth, and Management*. Washington, DC: Charles Lathrop Pack Forestry Foundation and USDA Forest Service.

White, P.S. 1979. Pattern, process, and natural disturbance in vegetation. *Bot Rev* 45:229–299.

Zeide, B., and Sharer, D. 2000. Good forestry at a glance: A guide for managing even-aged loblolly pine stands. Arkansas Forest Resources Center Series 003. University of Arkansas Division of Agriculture, Arkansas Agricultural Experiment Station.

BOX 7.1
The Stoddard–Neel Approach

A Conservation-Oriented Approach

Steven B. Jack, W. Leon Neel, and Robert J. Mitchell

Joseph W. Jones Ecological Research Center, Newton, Georgia 39870

Uneven-aged systems of forest management have been recommended as good conservation-oriented resource management approaches. A multi-aged approach is thought to be especially appropriate for longleaf pine ecosystems managed for diversity and conservation purposes because it is believed that historical longleaf pine forests contained multiple age classes with many cohorts older than is typically found in today's landscape.

Herbert Stoddard and Leon Neel developed, over the course of several decades, a unique approach (hereafter referred to as the Stoddard–Neel or S-N approach) to uneven-aged management of open pine–grassland ecosystems on the Coastal Plain region of north Florida and south Georgia. Their approach was developed on private hunting preserves in the region, where aesthetics and wildlife considerations were as important, or even more important, than the production of timber volume. The overall objective of the S-N approach was and is to restore and perpetuate the forest and all its components, while at the same time meeting individual land owner objectives, with the production of timber and subsequent economic return as by-products of this focus on long-term stewardship. Properties managed under their guidance have produced substantial timber volumes over many years while perpetuating the longleaf pine ecosystem and its associated diversity.

The focus of the S-N approach is on longleaf pine, but it is not exclusive to this species and is applicable to more than just mature forests.

The S-N approach is unique not because of the uneven-aged focus or in how trees are selected for removal, but rather in its overall guiding philosophy of resource management and what is selected to remain on the site after each harvest entry. The S-N approach is grounded in a strong land ethic, with the maintenance of a healthy and multifaceted forest as the major goal. Rather than a primary focus on timber management with other resources and amenities as secondary or ancillary, the S-N approach seeks to preserve all characteristics of the ecosystem and then determine how much timber is available for harvest. The result is that no one value of the ecosystem is maximized at the expense of other amenities, but rather a balance of ecological, economic, and aesthetic values is maintained.

Under the S-N approach the forest is never terminated, and continual forest cover is maintained with multiple regeneration events that lead to a multi-aged structure. The S-N approach also incorporates a long-term view to forest management rather than a short-term, economically driven model for forestry, with most results realized over many years from cumulative effects rather than from one or two discrete management actions. The S-N approach is also inherently conservative, where the capital in the form of standing volume is protected and only a portion of the growth is harvested in any given stand entry. Natural disturbance events in longleaf pine forests, which tend to be small in extent and of low intensity, are used as a guide to harvest operations. Finally, the S-N approach embraces and creates complexity and diversity in the forest rather than trying to simplify forest structure or homogenizing structure over the landscape.

These guiding principles of the S-N approach lead to several attributes in implementation. First, prescribed fire is crucial and integral to the S-N approach within the pine–grassland ecosystem. Fire addresses several objectives: (1) it controls hardwood encroachment and maintains the open midstory characteristic of the forest type; (2) it helps propagate fire-adapted and fire-dependent species of plants as well as appropriate habitat for wildlife native to the ecosystem; and (3) over time it helps to define the upland areas for pine management. In the S-N approach prescribed fire is objective driven and frequent such that all seasons for burning are potentially used to meet objectives, and variability in fire effects are achieved as objectives change over time. Because prescribed fire is the primary management tool utilized in the approach, silvicultural operations are carried out to ensure the presence of adequate fuels (e.g., pine needles) to conduct frequent burns.

Selection of trees for harvest is similar to that found in other uneven-aged approaches to forest management (i.e., "cut the worst and leave the best" with trees marked for removal based upon their vigor and growth potential) but tends to favor older and larger trees as residuals much more than other approaches. With the S-N approach a mix of age classes is maintained in the forest, including a substantial complement of older trees beyond their "financial maturity." Also, all trees with visible defect are not marked for removal because having some trees with economic defect provides unique wildlife habitat and adds diversity to the forest.

The S-N approach is not quantitative nor easily quantified. No uniform target stand structure is specified to guide the timber marking process as is characteristic of most other uneven-aged approaches, with the result that the stand density and basal area are highly variable in a forest managed under the S-N approach. Natural gaps are utilized and expanded to provide "space" for regeneration rather than targeting some percentage of the forest to be in evenly distributed gaps. While no elaborate calculations are carried out to determine annual yield, periodic inventories are conducted to guide the determination of stand growth and how much volume is available to be harvested at any point in time. Maximizing the growth of individual trees is not a priority of the S-N approach, but individual tree quality is an important consideration and is one of the factors that help drive the economics of the approach.

Regeneration, especially of longleaf pine, is of high concern in the S-N approach, but in comparison to other uneven-aged approaches much less emphasis is placed on obtaining regeneration at each stand entry. This is due to two primary factors: (1) the long time scale of management planning under the S-N approach and the longevity of longleaf pine result in less need to obtain regeneration establishment as frequently as is necessary when managing for fixed maximum age class, tree size, or a specified diameter distribution; and (2) the conservative timber marking and less emphasis on producing timber means that there is a reduced requirement for continual recruitment to replace large numbers of harvested trees. The result is that regeneration tends to occur in small, discrete patches rather than being fairly uniformly distributed throughout the forest, and the understory and midstory appear much more open than is true for many uneven-aged forests.

Because trees are maintained in the forest for a long time and residual trees are carefully selected, harvesting operations are carried out with great care and close monitoring (as is the case with all approaches to uneven-aged management) in order to minimize damage to regeneration, residual adult trees, groundcover, and soil. The S-N approach differs from many other approaches, however, in the high level of utilization of cut stems. Intensive merchandizing of cut trees is

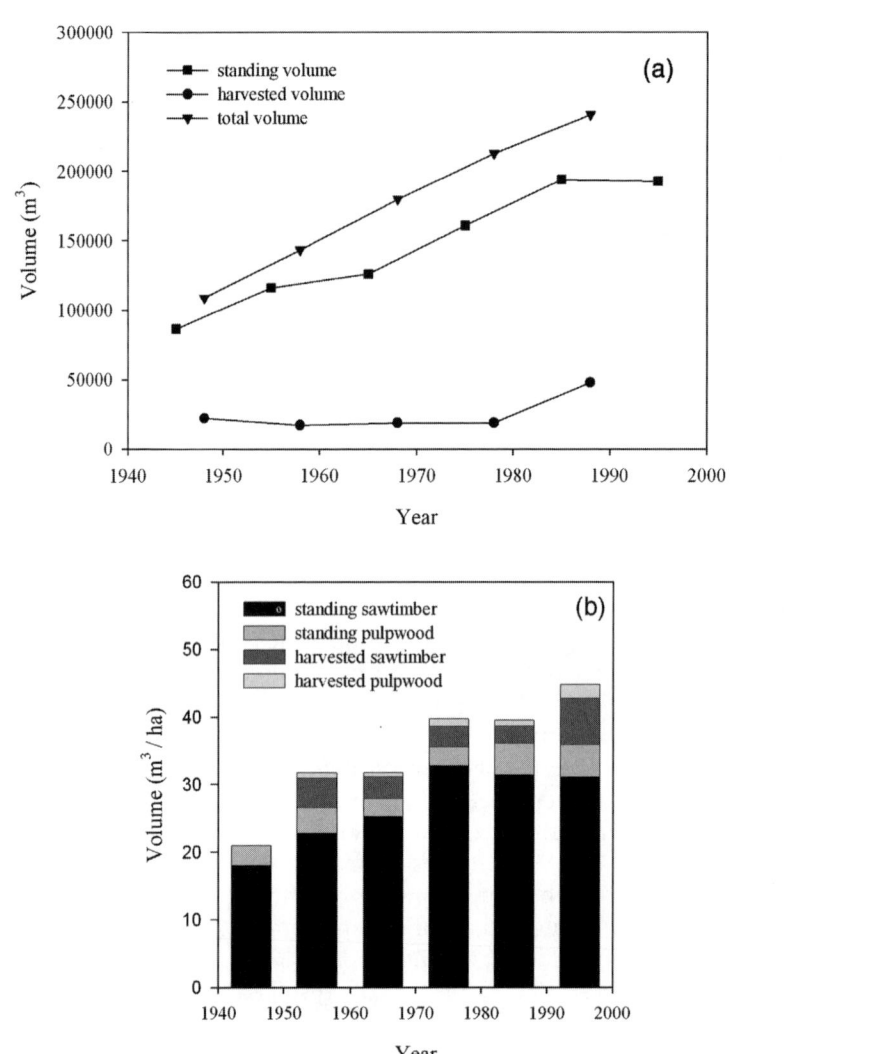

FIGURE 1. Data from Greenwood Plantation located near Thomasville, GA, showing results of management using Stoddard–Neel approach. (a) Total, standing, and harvested volume (for entire property) over 5 decades; (b) standing and harvested per hectare volumes by product class. Data are from periodic (every 10 years) timber inventories from 1945 to 1995 and from timber harvest records for same time period. Original volume data were in board feet (Scribner) per acre and were converted using the factors 1 m^3 = 450 bf (green) and 1 bf/A = 0.00549 m^3/ha.

important to implementing S-N approach because it helps to support the conservative marking of trees. That is, maximizing the value of trees at the landing by careful allocation to different product classes can greatly increase the return to the landowner and offset the lower harvest volumes. The careful merchandizing and the varied structure of the forest also allow for greater flexibility in responding to changing timber markets.

The S-N approach has been questioned and criticized in regard to its long-term sustainability, and has been characterized by some as only tending large trees with no consideration for regeneration so that the system will eventually collapse as all the old trees either die or are harvested (i.e., mining the limited resource). However, an example from Greenwood Plantation near Thomasville, GA, managed using the S-N approach for more than 60 years shows that the forest can be sustained and improved utilizing this form of management (see box by Moser for other examples). Box A Figure 1 shows that over 50 years the total standing volume on the plantation increased (in total and per acre) while still supporting significant timber harvests. In addition, other data (not shown) indicate that there is continual ingrowth to larger tree sizes from regeneration and recruitment. Even though high timber volumes were harvested from the property it did not come at the expense of other resources: excellent quail hunting was maintained, as were the diversity of the understory flora and a large red-cockaded woodpecker population. These results provide strong evidence that the S-N approach is able to achieve significant conservation objectives and maintain and perpetuate the longleaf pine forest with a range of amenities while still producing substantial income from timber harvest.

The S-N approach cannot be fully explained in this overview because it relies heavily on the "art" aspects of silviculture (as opposed to "science" and quantification) and there is no numerical equation or "recipe" that can be distilled for direct application in the field. The S-N approach is a product of the time and place where it was developed, and is not appropriate for all forests or landowner objectives. However, when conservation goals are paramount but some utilization and income are required, the S-N approach to forest management is an excellent alternative to more traditional or prevalent uneven-aged silvicultural approaches, particularly in pine-grasslands of the Southeast.

BOX 7.2
The Stoddard–Neel System
Case Studies

W. Keith Moser

USDA, Forest Service, North Central Research Station, Forest Inventory and Analysis, St. Paul, Minnesota 55108.

In the eighteenth century, the Coastal Plain uplands of what is now the southeastern United States were covered with large, variably stocked longleaf pine (*Pinus palustris* Mill.) stands. Hurricanes and tornadoes caused many areas to develop as even-aged stands. Smaller scale disturbances, such as lightning strikes and fires of varying intensity, caused stands to develop diverse age- and size-class structures (Frost 1993). With fires occurring every 1–10 years, longleaf pine was well-suited to multi-aged forest conditions because it can respond to release at almost any age and because it tolerated fire at all age classes better than its pine and hardwood competitors can (Boyer 1990), although fire will always impact growth (Boyer 2000). Longleaf pine also has some reputation for insect and disease resistance (Moser et al., 2003), further limiting the opportunity for widespread even-aged forests caused by insect and disease attacks.

As described in the accompanying chapter by Guldin, quantifying multi-aged forests has assumed that a balanced forest has a reverse J-shaped diameter distribution that can be described by a constant ratio of succeeding diameter classes. While some may argue that the shape of the curve refers to the diameter distribution of forests, not stands (de Liocourt 1898), many foresters have sought to implement this structure at the smallest scale possible, even in shade-intolerant species.

Attempts to maximize fiber production with such a structure can devalue the very ecological attributes that make longleaf pine ecosystems important. Overall diversity is reduced by competition from a dense overstory and midstory, a necessary result of a constant q-factor (Farrar 1996) in an uneven-aged structure.

The Stoddard–Neel system of uneven-aged management, practiced in the Red Hills region of Florida and Georgia, uses single-tree selection to create a multi-aged forest that provides an open midstory and a great deal of growing space to the understory, thereby maintaining suitable habitat for species of interest, such as northern bobwhite quail, gopher tortoise, and red-cockaded woodpecker (Engstrom et al., 1995) (see the box by Jack et al. for details of this approach). In this system, large old pines are considered valuable both for their influence on the rest of the ecosystem and as habitat for species of special concern. Trees are moved through the midstory in patches as opposed to trying to maintain a widespread midstory. Additionally, in the context of maintaining healthy quail habitat, basal areas were largely kept below 15 m^2 ha^{-1} to ensure enough growing space for the understory plants that provide food for the bobwhite (Moser and Palmer 1997).

Starting in the 1920s, Herbert Stoddard determined the connection between quail population dynamics, prescribed fire, and overstory structure (Stoddard 1931). During World War II, Stoddard devised the "Stoddard system" of selection system management to supply wood for the war effort while still maintaining a forest overstory. Beginning in 1950, Leon Neel started working with Stoddard and continued his management methods after Stoddard retired.

I present below the data from three studies. A 4-ha block of the Wade Tract, referred to as "Area WT" below, an 80-ha ecological preserve where no harvesting has occurred, provides a template of an unmanaged diameter distribution. (Originally collected by W. J. Platt, reported in several articles including

FIGURE 1. Diameter distribution of plantations.

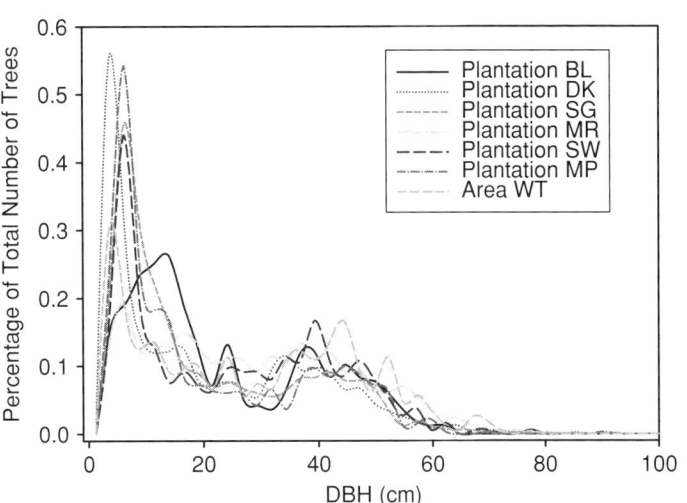

Platt et al., 1988; the data were excerpted from Cressie 1991.) In a second (pilot) study, one section of a quail shooting plantation was sampled for species, heights, and diameters using 0.04-ha plots at an intensity of 10% ("Plantation MP," Moser unpublished data). In the third study, the structure of several Stoddard–Neel forests was examined over 3 years ("Plantations MR, SW, SG, BL, and DK"). This last investigation (Moser et al., 2002) looked at diameter distribution and spatial arrangement at four scales: ownerships, 40-ha blocks, transects (7–8 per block) and 0.04- or 0.08-ha plots (about 7 to 8 of these per transect). The researchers took the forestwide diameter distributions and converted them into proportions for each diameter class. They compared the diameter distributions at the ownership level to the distributions of the constituent blocks, the diameter distributions of the blocks to the constituent transects, and the distributions at the transect level to those of the constituent plots.

Box B Figure 1 shows a summary of the plantation-level diameter distributions from all three studies. Under the Stoddard–Neel system, the number of trees decreases rapidly as the diameter increases, up to a point. Next there is a "bump" or second mode in the larger diameter classes, after which the number of trees again decreases. The open midstory is the result of fire and mechanical control of sapling ingrowth and a preference of the Stoddard–Neel system for retaining large trees. The shape of the diameter distribution of the plantations differs from the shape of the distribution of the classic reverse-J curve. Given the quail-hunting constraints of a low residual basal area and an open midstory, the Stoddard–Neel system is a more appropriate structure in these longleaf pine stands. Plantations MR and BL are the exceptions that validate the rule: they have a higher proportion of loblolly pine, so maintaining the smaller size classes is more difficult with an annual fire regime.

Moser et al. (2002) then compared the relationship between size classes: forestwide versus plantation, plantation versus block, block versus transect, and transect versus plot. They found little significant difference between the forestwide, plantation, or block distributions. Of the 404 plots compared with their parent transects, 161 were different, approximately 40%. There is greater variation from the forestwide distribution at the transect and plot levels than at the block level, and a more distinctly two-storied (bimodal) structure (Moser et al., 2002).

At the plot level, the higher proportion of plots different from their parent transects reflects the preeminence of site-specific and tree-specific decisions. As one goes down in scale, the percentage of management units with distributions significantly different from the forestwide distribution increases. These data suggest that the Stoddard–Neel system focuses on managing plantations and blocks, not stands or smaller units, as uneven-aged entities. The two-storied nature of the diameter distributions at all scales suggests that a forester could manage the smaller units as two-aged (really two-sized) stands.

While the practitioners of the Stoddard–Neel system have stated that they use single-tree selection to create and maintain such forests, like most foresters, they incorporated the practice of negative (improvement) thinning—focusing on removing poorer quality trees to increase the quality of the residual stand. Neel stated that the forest grows at a rate of 4% per year (Leon Neel personal communication) and that approximately 90% of annual growth can be harvested in periodic removals (Neel 1993). What is unique about the system, however, is the emphasis on retaining large, old trees. As there is only a fixed amount of growing space, other parts of the age (size) class distributions must occupy a proportionately smaller area.

Before Stoddard's investigations in the 1920s and 1930s, landowners were unaware of the relationship between overstory structure, fire, and quail; so plantations accumulated volume and new age classes of trees. One can assume that the diameter (and perhaps age) distribution was normal at the inception of the Stoddard–Neel system in the 1940s. A forest manager could then create and maintain the Stoddard–Neel structure with a "low" thinning (thinning the overstory from below). Longleaf pine crowns were not very dense (Gagnon et al., 2004), so the combination of large and frequent sunflecks through the canopy and the openings created by timber harvest or quail management would release growing space to the understory. The value of the Stoddard–Neel system is that it maintains a continuous forest canopy while maintaining an open midstory. A forest maintained under the BDq system with a constant ratio between successive diameter classes would require a very flat distribution (a very low "q") to satisfy these requirements, and would run the risk of insufficient ingrowth at the upper diameter levels due to mortality losses by the smaller trees, particularly under an intensive prescribed burning regime. Longleaf pine is a long-lived species, allowing the forest manager to use the Stoddard–Neel system to maintain a vigorous overstory while "filtering" trees through the midstory at a density and rate compatible with maintaining the hunting and ecological attributes of the Red Hills plantations; often cover blocks or "ring-arounds," areas left unburned to provide nesting and cover habitat for quail (Moser and Palmer 1997), serve the additional function of providing regeneration opportunities for longleaf pine. Like any partial harvesting practice, the Stoddard–Neel system requires management discipline, in that one cannot remove overstory volume at a rate greater than ingrowth. A suitable quantity of overstory is necessary to provide habitat and fuel for prescribed burning.

The key points of the system include maintaining densities below 15 m^2 ha^{-1}, managing the overstory with removals from below, maintaining a reproduction component in the stand, and allowing transition from reproduction to overstory on some small proportion of the area.

References

Boyer, W.D. 1990. *Pinus palustris* Mill. Longleaf pine. In *Silvics of North America. Vol. 1: Conifers*, comps. R.M. Burns and B.H. Honkala, pp. 405–412. USDA Agriculture Handbook 654.

Boyer, W.D. 2000. Long-term effects of biennial prescribed fires on the growth of longleaf pine. In *Proceedings of the Tall Timbers Fire Ecology Conference, No. 21, Fire and forest ecology: Innovative silviculture and vegetation management*, eds. W.K. Moser and C.F. Moser, pp. 18–21. Tallahassee, FL: Tall Timbers Research Station.

Christensen, N.L. 2000. Vegetation of the southeastern Coastal Plain. In *North American Terrestrial Vegetation*, 2nd edition, eds. M.G. Barbour and W.D. Billings, pp. 397–448. London: Cambridge University Press.

Cressie, N.A. 1991. *Statistics for Spatial Data*. New York: Wiley.

de Liocourt, F. 1898. *De l'amenagement des Sapinieres*. Bulletin of the Society of Foresters. Franche-Comte Belfort, Besançon, pp. 396–409.

Engstrom, R.T., Brennan, L.A., Neel, W.L., Farrar, R.M., Lindeman, S.T., Moser, W.K., and Hermann, S.M. 1995. Silvicultural practices and red-cockaded woodpecker management: A reply to Rudolph and Conner. *Wildl Soc Bull* 24(2): 334–338.

Farrar, R.M. 1996. *Fundamentals of Uneven-aged Management in Southern Pine*. Tall Timbers Research Station, Tallahassee, FL. Miscellaneous Publication No. 9.

Frost, C.C. 1993. Four centuries of changing landscape patterns in the longleaf pine ecosystem. In *Proceedings of the Tall Timbers Fire Ecology Conference, No. 18, the Longleaf pine ecosystem: Ecology, restoration and management*, ed. S.M. Hermann, pp. 17–43. Tallahassee, FL: Tall Timbers Research Station.

Gagnon, J.L., Jokela, E.J., Moser, W.K., and Huber, D.A. 2004. Characteristics of gaps and natural regeneration in mature longleaf pine flatwoods ecosystems. *For Ecol Manage* 187(2–3):373–380.

Moser, W.K., and Palmer, W.E. 1997. Quail habitat and forest management: What are the opportunities? In *Forest Landowner Magazine Annual Landowners Manual*, March/April, 56–63.

Moser, W.K., Jackson, S.M., Podrázský, V.V., and Larsen, D. 2002. Examination of stand structure on quail plantations in the Red Hills region of Georgia and Florida managed by the Stoddard–Neel system: An example for forest managers. *Forestry* 75(4):443–449.

Moser, W.K., Treiman, T., and Johnson, E.E. 2003. Species choice and risk in forest restoration: A comparison of methodologies for evaluating forest management decisions. *Forestry* 76(2):137–147.

Neel, W.L. 1993. An ecological approach to longleaf forestry. In *Proceedings of the Tall Timbers Fire Ecology Conference, No. 18, the Longleaf pine ecosystem: Ecology, restoration and management*, ed. S.M. Hermann, p. 335. Tallahassee, FL: Tall Timbers Research Station.

Platt, W.J., Evans, G.W., and Rathbun, S.L. 1988. The population dynamics of a long-lived conifer (*Pinus palustris*). *Am Nat* 131:491–525.

Stoddard, H.L. 1931. *The Bobwhite Quail: Its Habits, Preservation and Increase*. New York: Scribner.

Chapter 8

Longleaf Pine Growth and Yield

John S. Kush, J. C. G. Goelz, Richard A. Williams, Douglas R. Carter, and Peter E. Linehan

Introduction

Across the historical range of longleaf pine (*Pinus palustris* Mill.), less than 10% of lands previously occupied by longleaf ecosystems are currently in public ownership (Johnson and Gjerstad 1999; Alavalapati et al., this volume). The remainder is owned by private entities ranging from the forest industry, to timberland investment organizations, to highly varied nonindustrial private landowners. Any significant recovery of longleaf is therefore dependent on the participation of the private sector. Certainly, for the forest industry, and many other investor-type groups, the need for competitive returns from forest management is extremely important. And although experience has indicated that economic return is often not the primary motivator for nonindustrial landowners, it usually plays some role in management decision-making.

One major area requiring more knowledge is the need for models to reliably project growth and, ultimately, economic value of longleaf pine. Some limited data are available for projecting natural stands of longleaf that may be extrapolated to yield estimates of potential growth in planted stands, but there is a great deal of uncertainty when gains in seedling quality, competition control, fertilization, and other silvicultural techniques are factored in. Much of the reestablishment of longleaf pine taking place today is occurring on old fields and pastures. At least half of that planted is done so using containerized seedlings, usually employing both intensive site preparation and follow-up herbaceous competition control to improve survival and accelerate growth.

Longleaf pine can grow competitively with, or even exceed, the growth of other southern pine species on many sites. If markets continue to award quality wood products, particularly utility poles, with premium prices, longleaf is highly competitive. Private industrial and nonindustrial landowners should therefore respond positively to that possibility and make longleaf a vital part of their portfolio.

John S. Kush • School of Forestry and Wildlife Sciences, Auburn University, Auburn, Alabama 36849. **J. C. G. Goelz** • USDA Forest Service, Southern Research Station, Pineville, Louisiana 71360. **Richard A. Williams** • School of Forest Resources and Conservation, University of Florida, Milton, Florida 32572. **Douglas R. Carter** • School of Forest Resources and Conservation, University of Florida, Gainesville, Florida 32611. **Peter E. Linehan** • The Pennsylvania State University, Mont Alto Campus, Mont Alto, Pennsylvania 17237.

Historical Perspective

The first European explorers to visit the southeastern U.S. Coastal Plain found a vast parkland of low ground cover growing under a relatively open canopy of pine (Bartram 1791; Schwarz 1907). Depending on the geography and/or soils, the dominant tree was longleaf pine.

Schwarz (1907) noted that within pure stands of longleaf pine, certain minor variations existed. The most important variation was in the density of trees. Ordinarily, the stand of trees did not maintain uniformity over more than a few hundred acres; often changing abruptly even within 50 acres. Schwarz (1907) gave this description:

Thus we may enter a stand of mature timber, with trees from 90 to 120 feet in height and with ample spaces here and there in the crown cover, giving entrance to the light from overhead. After walking perhaps only a few hundred paces, we may find the trees suddenly beginning to close up their crown spaces. They grow smaller and more numerous, until presently they form a tolerably dense grove; and then they open up once more into the original stand of mature, tall trees. Occasionally, too, a tract of old trees of fairly uniform height is replaced by one in which the trees show diversity in size, ranging from mere poles to veterans of the forest.

Although an extremely intolerant tree, which will thrive best in even-aged stands, the natural form of longleaf pine tends toward small, even-aged groups of a few hundred square feet. Being naturally resistant to fire, large clearings are never caused by fire. In regions of severe winds, or tornadoes, larger even-aged patches and strips are found, sometimes one-quarter to one-half mile in width, which have come in after blowdown. These are pretty well interspersed with patches or single survivors of the old forest, which have acted as seed trees. Fire always has and always will be an element in longleaf forests, and the problem is not how fire can be eliminated but how it can be controlled so as to secure reproduction; second, to prevent the accumulation of litter and reduce the danger of a disastrous blaze. The factor that probably determines growth and yield is root competition for soil moisture. Longleaf stands are subject to severe droughts. The slow juvenile growth and long taproot of the young tree indicate its adaptation to this condition. Very young stands of longleaf may be quite crowded and remain so for 50 to 80 years. But it was found that such stands, if closely crowded, fell off in growth so badly that there was a distinct loss of production. Trees less than 100 years of age continued to grow vigorously in diameter even in rather dense groups, provided such groups were isolated and did not form a complete stand. But above this age it was found that groups of several trees standing close together would have differentiated themselves into dominant and suppressed trees, one or two trees with large crowns showing continued growth, while the rest were almost stationary. These occurred in groups surrounded by open space and could not be accounted for by struggles for light. Root competition alone can account for the thinning out of mature longleaf forests and the wide spacing of veteran trees. It also accounts, to a greater extent than intolerance, for the absence of seedlings under the open crowns of veteran trees, and their appearance only in openings at some distance from such trees. The indicated management for longleaf pine is to avoid crowding and not to attempt rotations much longer than 100 years or the production of large sizes. The ideal form for young stands of longleaf would be to have them stocked at most with only about twice as many saplings as should be standing in the form of mature trees at the end of the rotation.

It is evident that under natural conditions, even in the presence of repeated fires, the longleaf pine forest renews itself, young trees coming in on areas left blank by the death of old timber. All trees in a stand do not grow equally fast, nor continue to grow at the same rate. In longleaf pine this is especially noticeable. Only the largest trees, with the biggest crowns, continue to grow at a rapid rate after a stand has reached merchantable size. After a longleaf pine stand reaches the age of about 120 years, the loss from red rot, fire, and suppressed growth increases so fast that the net gain in growth on the stand would not pay the

Old Growth Information from the Literature

Based on timber tallies of 162 ha (400 acres) of pure even-aged, old-growth longleaf pine stands in Tyler County, TX, Chapman (1909) found that trees per hectare dropped from 148 to 27 (60 to 11 trees/acre) going from a stand 100 years old to 320 years old with average diameter at breast height (DBH) increasing from 37.0 to 77.9 cm (14.0 to 29.5 in.). Mean annual growth [in thousand board feet (MBF) per acre per year] reached at a maximum at 110 years. This indicated that longleaf pine does not increase much in yield after 120 years with the increase in mortality due to decay, fire, and other disturbances. The increase in total yield is very slow up to 250 years and then diminishes. Trees less than 100 years old will grow vigorously in diameter even in rather dense stands provided groups are isolated and do not form a complete stand.

Wahlenberg (1946) noted that old-growth forests were aggregations of even-aged stands covering areas from a few hundred square feet to several acres. Well-stocked stands had 74 to 247 merchantable trees/ha (30 to 100 trees/acre), and poorer sites as few as 5 to 7 trees/ha (2 to 3 trees/ha). Using data from Forbes and Bruce (1930), Wahlenberg (1946) observed that over wide regions of the South, the forest contained many age and size classes of trees, with only small areas usually limited to a single age class.

Growth and Yield of Natural Stands

There has been a great deal of research on the growth and yield of longleaf pine. Farrar (1979a) published the first growth and yield equations for thinned stands of even-aged natural longleaf pine. Uneven-aged stands are more complex structurally and thus more difficult to model.

Other attempts at predicting growth have involved the use of empirical yield tables, which are direct estimates of growth based on stand structure and volume tables. Davis (1966) indicated that good estimates of growth should provide growth directly in hectares, include the fewest possible variables, require a minimum of field data, provide estimates in cubic foot volume, do not use age as a primary variable, and treat height growth differently than diameter growth.

Spurr (1952) proposed a method to account for diameter growth and height growth in projecting volume growth. One must first be able to predict basal area and tree height growth. Growth in basal area is largely a function of stand density while height growth is primarily a function of site quality and stand age. Once basal area and height measurements are estimated and volume calculated for the present stand, then estimates of volume for some future stand can be calculated. Assuming the stand form factor remains unchanged, the relationship between the product of basal area and height (for a given volume) will hold for the future period, i.e.,

$$\frac{PBA \times PHt}{P\ Vol} = \frac{FBA \times FHt}{F\ Vol}$$

where

PBA is the present basal area
PHt is the present height
P Vol is the present volume
FBA is the future basal area
FHt is the future height
F Vol is the future volume

The difference between future volume and present volume is the increment of growth for the period.

Another approach is to use a stand, or stock table, projection. The essence of a stand table projection is to estimate the future stand based on the present one. The problem is what diameter increment to use and how to apply the expected growth to the diameter class. Diameter increment data could be added to the midpoint of the diameter class. This method assumes that all trees in a given diameter class are at the midpoint and grow at the same rate. This assumption reveals the limitation of this method of predicting stand growth as all trees in a given diameter class will not be the same size and all trees will not be at the midpoint of the diameter class.

A second method applies the average diameter growth to all trees within a class while recognizing dispersion of individuals within the same class. This method uses a movement ratio technique to predict what percentage of trees within a diameter class move up into larger diameter class, and what percentage will remain in the same diameter class. The movement ratio is defined as

$$M = (g/i) \times 100$$

where

M = movement ratio
g = diameter growth increment
i = diameter class interval

For example, assume the average diameter growth was 5.3 cm (2.1 in.) over a period of time, and the diameter class interval is 5 cm (2 in.), then the movement ratio is calculated to be 105. This means that 5% of the trees move up two diameter classes while 95% of the trees move up only one diameter class.

Growth and yield models have been developed for uneven-aged stands of loblolly and shortleaf pines (Farrar et al. 1984; Murphy and Farrar 1985). Some general inferences from these studies conducted in Arkansas were that loblolly and shortleaf pine average annual growth could be expected to be around 0.3 m² (3 ft²) of basal area and 2.4 to 3.3 m³ (84 to 116 ft³) of merchantable volume growth. Volume growth on the Mississippi study locations showed slightly lower production compared to the Arkansas sites (Baker et al. 1996).

Farrar (1996) provided guidelines for the uneven-aged management of longleaf pine. Nature managed longleaf pine as small patches of even-aged stands across an uneven-aged landscape. The main drawback with uneven-aged management with longleaf pine is that a lot of work is required to keep up with how a stand is growing. The fact that it takes a considerable amount of work to manage longleaf pine under an uneven-aged system has discouraged past management of longleaf pine in this way. However, recently public agencies such as state forest divisions or departments and the USDA Forest Service are attempting to manage longleaf pine using the uneven-aged silvicultural approach.

The best estimate of longleaf pine growth and yield for natural stands can be found in Farrar (1979b). The USDA Forest Service established a regional longleaf pine growth study (RLGS) in the mid-1960s in southwest Georgia, northwest Florida, southern Alabama, and southern Mississippi. The data from this study and the subsequent formulas represent the net volume growth and yield one might expect in the absence of adverse influences such as weather, insects, and disease. The equations for estimating growth and yield (inside bark) of thinned natural longleaf pine are given below (the symbols and letters follow throughout all of the formulas).

Basal area is calculated as

$$BA = 0.2296 B_2$$
$$B_2 = e[(A_1/A_2)\ln(B_1) + 6.0594(1 - A_1/A_2)]$$

where

BA = projected basal area at the end of the period in square meters per hectare
B_2 = projected basal area at the end of the period in square feet per acre
e = exponential function
A_1 = initial stand age in years (beginning of the period)
A_2 = stand age at the end of the period in years
$\ln$ = natural logarithm
B_1 = initial basal area in square feet per acre

Total volume is given by

$$TVIM_2 = 0.06997\ TVI_2$$
$$TVI_2 = e[2.6776 + 0.015287(S)$$
$$- 21.909/A_2 + (A_1/A_2)\ln(B_1)$$
$$+ 6.0594(1 - A_1/A_2)]$$

where

$TVIM_2$ = projected stand total volume, inside bark, in cubic meters per hectare at the end of the period, DBH ≥ 1.5 cm (0.6 in.), 6.1-cm (0.2 ft) stump, S = site index in feet, base age of 50 years

TVI_2 = projected stand total volume, inside bark, in cubic feet per acre at the end of the period

When $A_1 = A_2$ and $B_1 = B_2$ or the growth period is 0, then the above equation can be reduced to

$$TVIM = 0.06997\ TVI$$
$$TVI = e[2.6776 + 0.015287(S)\\ - 21.909/A + \ln(B)]$$

where A, S, and B are current stand age, site index, and basal area, respectively, TVIM is the predicted current stand volume in cubic feet per hectare, and TVI is the predicted current stand volume in cubic feet per acre.

The equation for calculating the predicted stand total volume (outside bark) is

$$TVOM = 0.06997\ TVO$$
$$TVO = TVI\,[1 + e\{-0.1785 + 43.629/S\\ + 1108.6/(SB) - 0.42802(\ln(A))\\ - 360.87(\ln(A))/(SB)\}]$$

where

TVOM = cubic meters per hectare
TVO = volume in cubic feet per acre

The above equation for calculating the total stand volume (outside bark) can be used for either the beginning or initial stand condition or the final condition using the appropriate TVIM, TVI, age, site index, and basal area figures.

Farrar (1979b) also published formulas to determine merchantable volumes for both present and future stands. The merchantable volume formula is

$$V4IM = 0.6997\ V4I$$
$$V4I = TVI/[1 + e\{2.623 + 316.77/S\\ + (SB) - 2.8248(\ln(A))\\ - 3326.7(\ln(A))/(SB)\}]$$

where

V4IM = predicted stand merchantable volume in cubic meters per hectare, inside bark, DBH ≥ 9.1 cm (3.6 in.), top DOB (diameter outside bark) ≥ 7.6 cm (3 in.), 6.1-cm (0.2 ft) stump

V4I = predicted stand merchantable volume in cubic feet per acre

These formulas have their limitations. First, they were derived from the first 5 years of the study. Actual growth beyond 5 years has not been used to adjust formula coefficients. Indeed, the study report indicates that the estimates of total cubic foot volume provided the most reliable results while the formula for estimating merchantable volume was the least reliable. A second limitation is that these equations were derived for trees growing in stands that were thinned. These equations used trees from thinned stands, a 5-year growth period, and low mortality. Future yield predictions were also not reliable, and the estimates became unrealistically large for unthinned stands.

Even though the equations developed by the Forest Service and reported by Farrar (1979b) have their limitations, they do provide a useful estimate of a stand's volume and an estimate of the stand's future volume and basal area. Stand predictions of growth and yield are useful to landowners attempting to manage their forests.

Quicke et al. (1994) used the RLGS database and produced an individual tree basal area increment (BAI) model for longleaf pine. The model is an intrinsically nonlinear equation, which is constrained so that it performs within the bounds of biologically reasonable outputs for any combination of values for the independent variables. All parameters in the equation were estimated simultaneously. This is a departure from the more traditional potential-times-modifier approach in which parameters for a potential growth function are estimated from a sample of trees exhibiting the fastest growth. Independent variables used to describe BAI are stand basal area, the competitive position of an individual tree within the stand calculated as the sum of the basal areas of all trees larger than the subject tree, mean age of dominant and co-dominant trees, and individual tree diameter outside-bark at breast height. Noticeably absent from the model is an independent variable that explicitly characterizes site differences.

Further work by Quicke et al. (1997) created an individual tree annual survival rate model. Variables used in the model were predicted diameter increment and diameter at breast height (DBH). Predicted annual survival rates ranged from 0.92 for a tree with a 2.54 cm (1 in.) DBH and an annual diameter increment of 0.13 cm, to over 0.99 for any tree larger than 15 cm in DBH. Stand level verification was based on 102 comparisons of observed and predicted trees per acre. Mean residuals, expressed as a percentage of observed final trees per acre, were 3% and 6% for projection periods of 5 and 10 years, respectively. The model predicts noncatastrophic mortality. In conjunction with a basal area increment model, it can be used to predict changes in the structure of longleaf pine stands. Meldahl et al. (1997) used the RLGS dataset to calculate needle fall, standing biomass, net primary productivity, and projected leaf area. In addition, climatic variables were included in tree and stand models.

Another study by Saucier et al. (1981) developed weight, volume, board-foot, and cord tables for major southern pine species, including longleaf pine. Data for this study were derived by felling sampled trees and measuring diameter, total height, and height to various merchantability limits.

The equation for predicting the total tree green weight using DBH and total height is

$$YM = 0.4536\,Y$$
$$Y = -44.418879 + 0.20297(D^2 Th)$$

where

YM = total tree weight in kilograms
Y = total tree weight in pounds
D = DBH in inches
Th = total height in feet

The total tree green weight to a 10.2-cm (4 in.) top is given by

$$YM = 0.4536\,Y$$
$$Y = -36.83043 + 0.15608(D^2 Th)$$

The study gives the cubic foot volume of the stem to a 10.2-cm (4 in.) top as

$$VM = 0.2832\,V$$
$$V = -0.84281 + 0.02216(D^2 Th)$$

where

VM = volume in cubic meters
V = volume in cubic feet

For longleaf pine, the paper also gives green weight to 17.8-cm (7 in.) and 22.9-cm (9 in.) tops; green weight of wood, bark, and foliage; wood volume to 17.8- and 22.9-cm tops; board foot (Scribner) volumes; green weight of sawtimber per MBF; and pulpwood weights and volumes.

Growth and Yield of Planted Stands

The most broadly based system of stem profile equations (and hence volume) is provided by Clark et al. (1991). Clark et al. (1991) include equations to predict stem diameter at any height, given diameter at breast height and at Girard's Form Height (5.3 m or 17.3 ft), total height, and the height at which diameter is to be predicted. The equations can also be used to estimate height at a given minimum diameter (merchantable height), and volume in cubic feet to any minimum diameter, or volume between a maximum and minimum diameter (such as pulpwood volume above sawtimber volume). There also are equations to use if height to a top diameter, rather than total height, is known. If diameter at form height is not measured, Clark et al. give an equation to predict it from diameter at breast height and total height. Bark thickness at breast height can also be estimated from an equation that predicts diameter inside bark. Clark et al. furnish parameter estimates for longleaf pine that represent South-wide estimates as well as subregions, namely, Coastal Plain (Atlantic coast and eastern Gulf—Alabama and points east and north), Piedmont (all states), and Deep South (Texas, Louisiana, and Mississippi).

Thomas et al. (1995) provide biomass and taper equations for longleaf pine in thinned and unthinned plantations in Louisiana and Texas. Their taper equation predicts upper stem diameters as a function of relative height, diameter at breast height, and plantation age. Volume is obtained by the integral of the taper equation, between limits of merchantability. Weight of the bole is obtained by doubly integrating (across diameter and height) a function for specific gravity. However, their equation for specific gravity was unique for the ages of the sampled trees (35, 45, or 50 years), and thus has somewhat restricted applicability. Thomas et al. compared their taper equations for four classes of stands: (1) an unthinned Louisiana plantation; (2) a thinned Louisiana plantation; (3) a different thinned Louisiana plantation and two thinned Texas plantations; (4) natural and plantation-grown longleaf from various stands in Alabama. These four classes produced taper equations that differed; the Alabama taper curve was particularly different from the others. This suggests that site and management differences can have large effects on stem taper. If stem taper varies so much from stand to stand, this suggests that regional volume equations will be poor estimators for any given stand, unless the volume/taper equation includes some measurement of form beyond simply measuring diameter at breast height and total height. This suggests that Clark et al. (1991), with diameter at form height determined from local data, would be preferable to regional volume equations. Clark et al. (1991) can be applied to natural and planted longleaf, and they have estimated their equation for all common species or species groups that are associates of longleaf pine.

Baldwin and Polmer (1981) used data from some of the same plantations as Thomas et al. (1995). They fit three different taper equations for different classes of crown ratio (less than 36%, between 36% and 50%, and greater than 50%). Crown ratio potentially can reflect the differences in taper among trees within and among stands. Even when crown ratio is not measured on every tree, it should be possible to assign crown ratio class reasonably accurately to trees as height is measured. These taper and volume equations should be useful for estimating volume in longleaf pine stands in the western Gulf states (Texas and Louisiana). Brooks et al. (2002) provide taper and cubic foot volume equations for young plantations in southwest Georgia; they compared their equations to the equations of Baldwin and Polmer (1981) and Baldwin and Saucier (1983), but they did not measure crown ratio, and so could not fully utilize Baldwin and Polmer's (1981) equations. The taper equation of Brooks et al. (2002) was slightly superior to that of Baldwin and Polmer (1981), and their volume equation was slightly superior to those of Baldwin and Polmer (1981) and Baldwin and Saucier (1983); however, this is expected because they used the same data to fit and test their equation, while the Baldwin equations were fit to other data. It is reasonable to expect bias when an existing equation is applied to a new dataset that is geographically distinct from the data on which the equation was estimated. However, the bias was not large, and when the bias was made equal to the bias of the equation of Brooks et al. (2002), the taper equation of Baldwin and Polmer (1981) would have produced lower absolute error than the equation of Brooks et al. (2002) (calculations by the second author). This suggests that Baldwin and Polmer (1981) might be more broadly applicable than Brooks et al. (2002).

Baldwin and Saucier (1983) provide aboveground weight and volume estimators for unthinned planted longleaf pine, using 111 of the 113 trees sampled in Baldwin and Polmer (1981). Rather than using a taper equation, they used a combined variable equation (Clutter et al. 1983) given by

$$\log(Y) = b_0 + b_1 * \log(D^2 H)$$

where Y is either volume or some biomass component, D represents diameter at breast height, and H represents total height. To estimate volume to some minimum top diameter, they estimate a volume ratio equation that is multiplied by the value for total volume. As they did not use any variable, such as crown ratio, that could explain stand-to-stand and tree-to-tree variability in taper, we suggest the taper equations of Baldwin and Polmer (1981) would be preferable for estimation of volume. However, Baldwin and Saucier's (1983) biomass equations are probably the only regional equations to estimate green and dry weight of wood, bark, branches, and foliage, given the limitations in application of the bole biomass equations of Thomas et al. (1995). There are biomass equations for longleaf in very young natural stands (presumably natural stands, the publication only indicates they are even-aged; Edwards and McNab 1977) and an old natural stand (Taras and Clark 1977). The results of Taras and Clark (1977) are included in USDA (1984) tables.

To estimate board foot volume, there are a few different approaches. The first approach is to use volume tables derived from natural stands, such as those found in USDA (1929). These volume tables use diameter at breast height and total height to predict board foot volume by several different log rules. The second approach employs form class volume tables, such as Mesavage and Girard (1946). Wiant (1986; Wiant and Castaneda 1977) created equations that approximate the Mesavage and Girard (1946) tables for form class 78 (form class is diameter inside bark at 17.3 feet height divided by diameter outside bark at breast height). Use of the equations may be more efficient for some individuals than looking up the values in a table, although the table look-up can be programmed. Wiant (1986) assumed a 3% change in volume for each point of form class change from 78 (higher volumes would be obtained with higher values of form class). A landowner may have a good idea of the form class for his holdings, perhaps as a function of diameter and height of the tree. Alternatively, Clark et al. (1991) have an equation for diameter inside bark at 5.3 m (17.3 ft). Because of the structure of the equation, form class is a constant for a given total height; that is, diameter at breast height does not affect form class. This is counterintuitive, as it would seem likely that form class should depend on diameter and height, but the relationship seems to hold for the data of Clark et al. (1991). The corresponding form class produced by their equation is provided in Table 1. On the other hand, Parker (1998) suggests that taper could be constant within diameter classes rather than height classes. The taper equations of Thomas et al. (1995) suggest that form class varies in response to diameter, height, and age, and does so differently for thinned and unthinned stands. Age has a relatively small difference on taper, but DBH and height have

TABLE 1. Form class (in percent) by height relationships for three subregions and South-wide, as calculated from an equation of Clark et al. (1991)

Region	Total height (ft)								
	40	50	60	70	80	90	100	110	120
South-wide	66	73	77	79	81	82	83	83	84
Coastal Plain (AL to VA)	67	73	77	79	80	81	82	82	83
Piedmont	65	73	77	80	81	82	83	84	84
Deep South (TX, LA, MS)	62	72	77	80	83	84	85	86	86
Region	Total height (m)								
	12.2	15.2	18.3	21.3	24.4	27.4	30.5	33.5	36.6
South-wide	66	73	77	79	81	82	83	83	84
Coastal Plain (AL to VA)	67	73	77	79	80	81	82	82	83
Piedmont	65	73	77	80	81	82	83	84	84
Deep South (TX, LA, MS)	62	72	77	80	83	84	85	86	86

TABLE 2. Form class (in percent) at age 40 related to diameter (DBH) and total height for thinned and unthinned stands in Louisiana and Texas, calculated from taper equation of Thomas et al. (1995) and bark thickness equation of Clark et al. (1991)

DBH (in.)	Height (ft)	Thinning	Form class
9	50	Thinned	74
9	80	Thinned	81
12	50	Thinned	75
12	80	Thinned	82
9	50	Unthinned	73
9	80	Unthinned	79
12	50	Unthinned	74
12	80	Unthinned	80
DBH (cm)	Height (m)	Thinning	Form class
22.9	15.2	Thinned	74
22.9	24.4	Thinned	81
30.5	15.2	Thinned	75
30.5	24.4	Thinned	82
22.9	15.2	Unthinned	73
22.9	24.4	Unthinned	79
30.5	15.2	Unthinned	74
30.5	24.4	Unthinned	80

a larger effect (Table 2). There are relatively small differences in absolute numbers between Tables 1 and 2. However, there are fundamental differences in the choice of the factors that affect form class. This could be very important when contrasting different silvicultural practices. If silvicultural practices affect form class, the chosen equation might not reflect those differences and thus the real difference among treatments might not be apparent.

The values in Table 1 can be used in concert with Mesavage and Girard's (1946) tables or Wiant's (1986) equation. Borders and Shiver (1995) used the taper equation of Clark et al. (1991) to produce board foot tables for loblolly, slash, and shortleaf pines. Their tables suggested greater volume than Mesavage and Girard (1946). This might suggest that Mesavage and Girard (1946) underestimate board foot volume for longleaf pine as well. Borders and Shiver (1995) present the final procedure for calculating board foot volume. They used a taper equation to calculate inside bark diameters at the scaling diameter of each log of fixed length, and directly applied their chosen log rule to calculate volume of each log. Any taper equation could be used in this way, although it is tedious if the calculations are done by hand rather than programmed. A program to calculate board foot volume in this way is available from the second author (jcgoelz@fs.fed.us) or from the programmer, Daniel Leduc (dleduc@fs.fed.us).

Poles are a high-value product, greatly exceeding the value of sawtimber per board foot, and thus it is critical to determine yield of poles. Any of the taper equations can also be used to determine whether a tree of a given diameter at breast height and total height possesses the minimum dimensions for top diameter and length of poles. ANSI (1987) provides specifications for dimensions of poles for 10 quality classes for pole lengths of 6.1 to 38.1 m (20 to 125 ft), in increments of 1.5 m (5 ft). The specifications are in terms of circumference (diameter times π) at 1.8 m (6 ft) from the butt of the pole. The taper equation can be solved for diameter at a height of 2.2 m (7.3 ft; counting stump and sawkerf). Hawes (1947) assumed a constant ratio of diameter inside bark at a height of 1.8 m (6 ft) to be 0.88 times diameter outside bark at breast height. Using taper and bark thickness equations would probably be more accurate for a specific tree. Quicke and Meldahl (1992) used a taper equation for natural longleaf to create such tables. Any taper equation and bark thickness equation for planted longleaf could be used in a similar fashion. The differences between the tables of Quicke and Meldahl (1992) and Hawes (1947) were greatest for long poles. Busby et al. (1993) found that 90% of trees in their plantations of longleaf pine in Louisiana were sufficiently free of defect to produce a pole if they met the diameter and length requirements.

Evaluating Site Quality

Site quality is a critical component of most growth and yield models. Lohrey and Bailey (1977) created a growth and yield model for unthinned plantations of longleaf pine. In it, they used the site index equations produced by Farrar (1973). Farrar (1973) developed his equations to reproduce the graphical site curves provided for natural second growth

TABLE 3. Number of trees (A) per acre and (B) per hectare to be planted to achieve thinning thresholds of GLSDI of 50% of maximum or 4/3 of crown closure for different levels of quadratic mean diameter and survival (to age of first thinning)

DBH (in)	90% survival		70% survival		50% survival	
	1/2 maximum	4/3 crown closure	1/2 maximum	4/3 crown closure	1/2 maximum	4/3 crown closure
4	1355	1014	1742	1303	2439	1824
6	604	463	777	595	1088	834
8	341	264	438	340	613	475
10	218	170	280	219	393	307
12	151	119	195	153	273	214
DBH (cm)						
10.2	3348	2506	4305	3220	6027	4507
15.3	1493	1144	1920	1470	2689	2061
20.4	843	652	1082	840	1515	1174
25.4	539	420	692	541	971	759
30.5	373	294	482	378	675	529

longleaf pine in USDA (1929). It should be noted that while the USDA (1929) curves indicate zero height before age 5, Farrar's (1973) curves are not conditioned in this way, and do not adequately represent the USDA (1929) curves at ages below 15 years. Apparently, Lohrey and Bailey (1977) found the curves for natural longleaf pine stands (USDA 1929; Farrar 1973) adequately represented the plantation grown longleaf data that they had available. Goelz and Leduc (2003) provided preliminary site index curves using the data of Lohrey and Bailey (1977), as well as additional measurements on the same plots, and supplemental plots that were not available to Lohrey and Bailey (1977). The curves of Goelz and Leduc (2003) are very similar to the USDA (1929) curves for site index of 70 at base age of 50 (Fig. 1). However, the curves for site index 50 and 90 are considerably different from the USDA (1929) curves. The anamorphic curves represented by USDA (1929) do not change shape as site index changes. There is a very common phenomenon that arises in polymorphic site index curves where the curves for the lower sites tend to be more linear while the curves for the higher sites tend to be more curved, achieving a higher proportion of the asymptotic height at younger ages (Goelz and Burk 1996). This expected pattern is missed in the anamorphic curves, but is obtained in the polymorphic curves of Goelz and Leduc (2003). Brooks (2004) describes an equation to predict dominant stand height for young longleaf plantations in southwest Georgia.

Boyer (1980, 1983) suggested that planting site (old field, unprepared cutover, mechanically prepared cutover) and stand density (survival at 10 years) affected early growth of longleaf pine. Boyer used a simple Schumacher equation:

$$\log_{10}(H) = b_o + b_1 \frac{1}{A}$$

to fit a common guide curve for all site conditions, where H and A represent height and age, respectively, and the b_i are parameters. Then, he expanded the equation by making b_1 a linear function of surviving trees per unit of land and height at age 15 (height at age 15 was only included for the two cutover sites). This structure produced site curves that are anamorphic for old-field situations and polymorphic for the two cutover situations and having a common asymptote for all combinations of site index and site condition. Although he labeled his curves by height at age 25, by including height at age 15 as a predictor variable, he was essentially creating base age specific site curves with 15 as the base age.

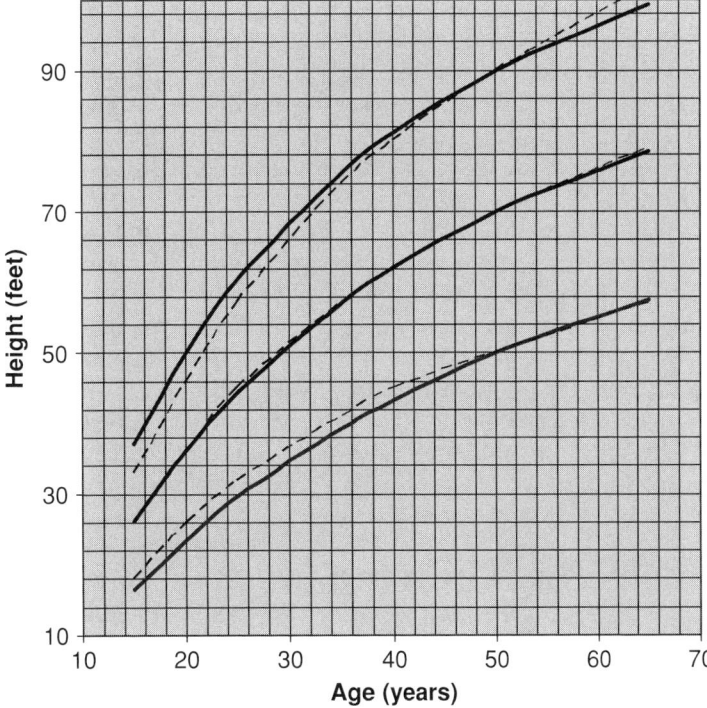

FIGURE 1. Comparison of site index curves from USDA (1929), dashed lines, and Goelz and Leduc (2003), solid lines, for site index (base age 50) of 50, 70, and 90 feet.

Goelz and Leduc (2004) suggested that a base age of 50 years provided more reliable estimates of site index than a base age of 25 years. The trade-off seems to be to either model this early height growth directly, or lose some of the capacity to define site quality, or to ignore the early, largely random variation and concentrate on the intrinsic productivity of the site. We suspect that the different bias of the perspectives of Boyer (1980, 1983) and Goelz and Leduc (2004) arises from the different natures of their datasets, predominantly 15 years or younger versus predominantly 16 years and older.

Shoulders and Tiarks (1980) produced regression models for planted longleaf pine in Louisiana and Mississippi relating height at age 20 to rainfall, slope, and available soil moisture (1/3 atmosphere percent minus 15 atmosphere percent) in the B2 horizon. Longleaf pine height was greatest where rainfall was least 122 cm (48 in.) per year (evenly split between warm season, April through September, and cool season), on very modest slopes (1.6%), and where moisture-holding capacity suggested loams or sandy loam soils (6–9%). Rainfall is clearly related to geography, and thus the effect of rainfall is confounded with soil differences and perhaps difference in constituents of the competing vegetation. Moisture-holding capacity affects competing vegetation, and thus the effect may be due to competition rather than moisture availability per se. An earlier analysis at age 15 (Shoulders and Walker 1979) suggested that the effect of rainfall was different on sites with droughty soils compared to wet or intermediate sites. Rainfall was positively related to height for droughty soils while negatively related to height for wet and intermediate sites. This suggests that aeration may be limiting to growth where soils may become saturated, but not on inherently dry sites. To estimate site index, a user could estimate height at age 20, using the equation of Shoulders and Tiarks (1980), then apply this height to a site index equation or curve.

Harrington (1990) produced an expert system to predict site index of both natural and planted longleaf pine from 25 soil and physiographic variables. Most of the variables need only be given qualitative values, or assigned to classes of continuous variables. A user can make reasonable guesses for the more difficult to measure variables, such as percent phosphorus, from soil survey information or regional data. The system can predict base age 50-site index within 1.7 m (5.5 ft). It is presented as a stand-alone computer application, but the source code could be extracted and incorporated into larger growth and yield systems. As Shoulders and Tiarks (1980) developed their equation from data arising only from Louisiana and Mississippi, Harrington's (1990) system might be more suitable outside those two states.

Evaluating Growing Stock and Stand Density of Longleaf Pine Stands

Stand density affects growth and shape of individual trees as well as understory plant communities and affects habitat quality for wildlife (Grelen and Lohrey 1978; Clutter et al. 1983; Haywood et al. 1998). Stand density can be described as simple measurements of number of trees, basal area, or volume per acre. Or the variables may be combined into a stand density index. Most stand density indices are functions of two or more of (1) basal area per acre, (2) trees per acre, or (3) average tree size (in terms of quadratic mean diameter, volume, or weight). Most stand density indices are independent of stand age and site quality, except as those variables influence tree size and mortality. Reineke's (1933) stand density index (SDI) is $N(10/D_q)^{-1.605}$, where N represents number of trees per acre, D_q represents quadratic mean diameter, and -1.605 is an empirically derived constant for all species. The Reineke relationship arose out of the observation that there seemed to be a limiting straight-line relationship when the logarithm of quadratic mean diameter was plotted versus the logarithm of trees per acre. Reineke (1933) suggested a maximum stand density index of 400 for longleaf pine, which is the same as the maximum SDI of shortleaf and slash pines, but less than the maximum for loblolly pine (450). The southern variant of the Forest Vegetation Simulator (Donnelly et al. 2001) uses a maximum Reineke stand density index of 390 for longleaf pine, based on their data.

The actual maximum density line has limited applicability to management since most forests would be maintained at densities much less than the maximum. However, a line can be defined that is parallel to the limiting density line that represents the threshold of significant density-dependent mortality (Drew and Flewelling 1979; Dean and Baldwin 1993). This line typically represents 50% to 55% of the maximum density. Under typical management, a stand would be maintained at or below this level of density.

In Fig. 2, we plot a limiting relationship for longleaf pine plantations from a large database

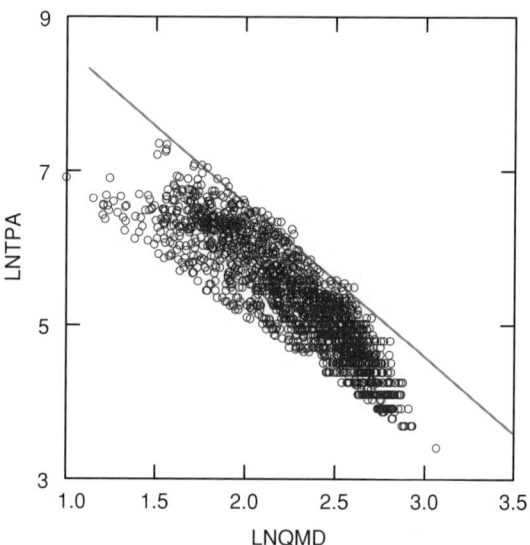

FIGURE 2. A limiting density relationship for longleaf pine plantations. Only plots with a Reineke stand density index of 100 or greater were used. The line was fit by minimizing the function: $loss = (observed - predicted)^2 \left(N\left(\frac{D_q}{10}\right)^{-1.992}\right)^{12}$. This weighting ensures that the line will approach the limit of the data. Note that there are no plots with large quadratic mean diameter near the limiting line; plots with the largest quadratic mean diameter were thinned fairly heavily.

(Goelz and Leduc 2001). However, we found the exponent to be −1.992, rather than Reineke's −1.605. If the exponent was −2, this would indicate that the limiting density relationship would represent a maximum basal area for all levels of quadratic mean diameter. As −1.992 is very close to −2, the maximum basal area is 49.1 m²/ha ± 0.19 (214 ft²/acre±2) across a wide range of quadratic mean diameter from 6.4 to 66 cm (2.5 to 26 in.). The southern variant of the Forest Vegetation Simulator (FVS) employs a maximum basal area of 48.9 m²/ha (213 ft²/acre) (Donnelly et al. 2001). This is very similar to the maximum indicated by the data of Goelz and Leduc (2001). However, this should not be surprising, as Donnelly et al. (2001) incorporated the data of Goelz and Leduc (2001) with data they had available from natural stands.

We used the limiting density relationship to build a density management diagram for longleaf pine plantations. As the exponent is not −1.605, we call the index the Goelz–Leduc stand density index (GLSDI) for longleaf pine, rather than the Reineke stand density index. It may be calculated as $N(D_q/10)^{-1.992}$, with the maximum SDI calculated to be approximately 393. Three lines are present on Fig. 3. There is a maximum density line, a line representing 50% of maximum, or the threshold for significant density-dependent mortality, and a line that represents crown closure, as determined by the equations of Smith et al. (1992) for crown diameter of open-grown longleaf pine. As the GLSDI is based on an exponent very close to two, and as Smith et al. (1992) predict open-grown crown diameter as a linear function of diameter, basal area is nearly constant across a broad range of diameter. Thus, maximum basal area is approximately 44.7 m²/ha (215 ft²/acre), the threshold for significant density-dependent mortality is 22.2 m²/ha (107 ft²/acre), and crown closure occurs at approximately 14.5 m²/ha (63 ft²/acre). Appropriate levels of basal area would vary depending on management objectives. However, a basal area of 14.5 m²/ha (63 ft²/acre) would provide high rates of individual tree growth while not sacrificing much whole stand growth. Higher levels of basal area would produce slower individual tree growth, but somewhat greater whole stand growth, somewhat higher log quality, and losses to mortality would be slight if stands were thinned before they exceeded 24.6 m²/ha (107 ft²/acre). Although the data of Goelz and Leduc (2001) included some plots with greater than 49.4 m²/ha (215 ft²/acre) of basal area, the plots were small (roughly 0.04 ha, 0.1 acre) and were not likely indicative of larger plots (0.4 ha, 1.0 acre) or stands.

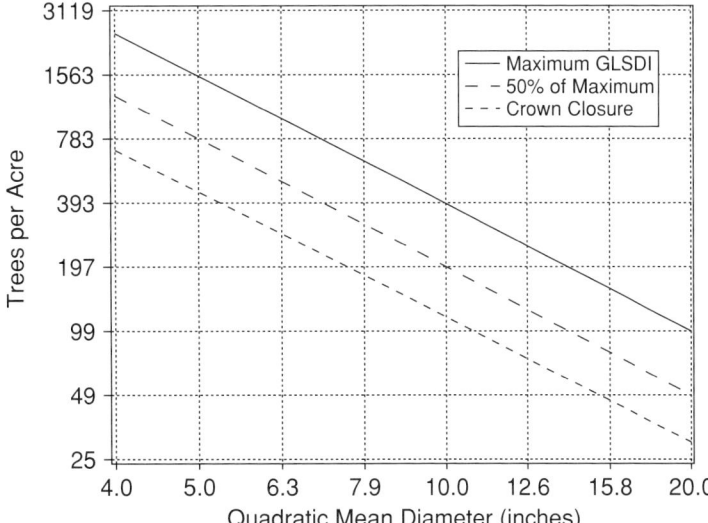

FIGURE 3. Density management diagram for longleaf pine plantations. Lines are drawn for maximum density, 50% of maximum, and for crown closure. Note the axes are scaled logarithmically.

The GLSDI also can be used to help guide initial planting density. We will consider two thresholds for thinning—50% of maximum GLSDI and 4/3 times the density at crown closure. The former threshold implies removal of one-third to one-half of the basal area while the latter implies removal of about one-fourth to one-third of the basal area in the thinning. Merchantability of a thinning will be influenced by size of the trees that are harvested. We consider first thinnings at quadratic mean diameters between 10.2 and 30.5 cm (4 and 12 in.). We explore three levels of survival to first thinning—50%, 70%, and 90%. Note this is survival to first thinning, not initial survival. Finally, our calculations assume there are no desirable volunteer trees in the plantation. If quadratic mean diameter is 10.2 cm (4 in.), and survival to first thinning is 90%, 3033 trees/ha (1355 trees/acre) must be planted to produce 24.6 m²/ha (107 ft²/acre). At a quadratic mean diameter of 15.2 cm (6 in.), 1492 trees/ha (604 trees/acre) are required, and at 20.3 cm (8 in.), 138 surviving trees/ha (341 trees/acre) are needed. So, if thinning a stand with a quadratic mean diameter of 15.2 cm (6 in.) is practicable, and the manager sought to maximize total volume growth, initial planting density should be about 1483/ha (600/acre), assuming survival of 90%. If survival were 70%, more than 1903 trees/ha (770 trees/acre) would be planted. The consequence of lower than expected survival means that thinning could be delayed a few years until quadratic mean diameter is greater. For example, if 90% survival were anticipated, but 50% was achieved, thinning would be delayed until quadratic mean diameter was at least 5.1 cm (2 in.) larger than anticipated, which is predicted to be about 5 years later using the model of Lohrey and Bailey (1977). Besides the delay to first thinning, there would likely be a reduction in log quality. A stand that is not sufficiently dense to thin until it has a quadratic mean diameter from 25.4 to 20.5 cm (10 to 12 in.) will likely have more defects due to knots and persistent dead branches, unless pruning was performed.

Acknowledgments

Drs. Tom Croker and William Boyer are acknowledged for the work they did over the decades at the Escambia Experimental Forest and for their teachings on how to naturally regenerate longleaf pine and manage it. We extend our sincere thanks to Dr. Robert M. Farrar, Jr. It was his tenacity that has provided much of the data available today for modeling longleaf pine growth and yield.

References

ANSI. 1987. American national standard for wood poles—specifications and dimensions. ANSI 05.1–1987.

Baker, J.B., Cain, M.D., Guldin, J.M., Murphy, P.A., and Shelton, M.G. 1996. Uneven-aged silviculture for the loblolly and shortleaf pine forest cover types. General Technical Report SO-118. USDA Forest Service, Southern Research Station.

Baldwin, V.C., Jr., and Saucier, J.R. 1983. Aboveground weight and volume of unthinned, planted longleaf pine on West Gulf forest sites. Research Paper SO-191. USDA Forest Service, Southern Forest Experiment Station, New Orleans, LA.

Baldwin, V.C., Jr., and Polmer, B.H. 1981. Taper functions for unthinned longleaf pine plantations on cutover West Gulf sites. In *Proceedings of the First Biennial Southern Silvicultural Research Conference*, ed. J.P. Barnett, pp. 156–163. November 6–7, 1980, Atlanta, GA. General Technical Report SO-34. USDA Forest Service, Southern Forest Experiment Station.

Bartram, W. 1791. *Travels through North & South Carolina, Georgia, East & West Florida, the Cherokee Country, the Extensive Territories of the Muscogulges, or Creek Confederacy, and the Country of the Chactaws*. Published as the Francis Harper's naturalist edition. Athens: University of Georgia Press. 1998.

Borders, B.E., and Shiver, B.D. 1995. Board foot volume estimation for southern pines. *South J Appl For* 19: 23–28.

Boyer, W.D. 1980. Interim site-index curves for longleaf pine plantations. Research Note SO-261. USDA Forest Service, Southern Forestry Experiment Station, New Orleans, LA.

Boyer, W.D. 1983. Variations in height-over-age curves for young longleaf pine plantations. *For Sci* 29: 15–27.

Brooks, J.R. 2004. Predicting and projecting stand dominant height from inventory data for young longleaf pine plantations in southwest Georgia. In *Proceedings of the Twelfth Biennial Southern Silvicultural Research Conference*, ed. K. Connor, pp. 187–188. Biloxi, MS. February 24–28, 2003. General Technical Report SRS-71. Asheville, NC: USDA, Forest Service, Southern Research Station.

Brooks, J.R., Martin, S., Jordan, J., and Sewell, C. 2002. Interim taper and cubic-foot volume equations for young longleaf pine plantations in southwest Georgia. In General Technical Report SRS-48, pp. 467–470. Asheville, NC: USDA Forest Service, Southern Research Station.

Busby, R.L., Thomas, C.E., and Lohrey, R.E. 1993. Potential product values from thinned longleaf pine plantations in Louisiana. In *Proceedings of the Seventh Biennial Southern Silvicultural Research Conference*, ed. J. C. Brissette, pp. 645–650. Mobile, AL, November 17–19, 1992. General Technical Report SO-93. USDA Forest Service, Southern Forest Experiment Station, New Orleans, LA.

Carmean, W.H. 1975. Forest site quality evaluation in the United States. *Adv Agron* 27:209–267.

Chapman, H.H. 1909. A method of studying growth and yield of longleaf pine applied in Tyler County, Texas. *Proceedings, Society of American Foresters* 4: 207–220.

Clark, A., III, Souter, R.A., and Schlaegel, B.E. 1991. Stem profile equations for southern tree species. Research Paper SE-282. USDA Forest Service, Southeastern Forest Experiment Station, Asheville, NC.

Clutter, J.L., Fortson, J.C., Pienaar, L.V., Brister, G.H., and Bailey, R. L. 1983. *Timber Management: A Quantitative Approach*. New York: John Wiley & Sons.

Davis, K.P. 1966. *Forest Management: Regulation and Valuation*, 2nd edition. New York: McGraw–Hill.

Dean, T.J., and Baldwin, V.C., Jr. 1993. Using a density-management diagram to develop thinning schedules for loblolly pine plantations. Research Paper SO-275. USDA Forest Service, Southern Forest Experiment Station, New Orleans, LA.

Donnelly, D., Lilly, B., and Smith, E. 2001. The southern variant of the forest vegetation simulator. USDA Forest Service, Forest Management Service Center, Ft. Collins, CO.

Drew, J.T., and Flewelling, J.W. 1979. Stand density management: An alternative approach and its application to Douglas-fir plantations. *For Sci* 25: 518–532.

Edwards, M.B., and McNab, W.H. 1977. Biomass prediction for young southern pines. *J For* 77: 291–292.

Farrar, R.M., Jr. 1973. Southern pine site index equations. *J For* 71: 696–697.

Farrar, R.M., Jr. 1979a. Growth and yield predictions for thinned stands of even-aged natural longleaf pine. Research Paper SO-156. USDA Forest Service, Southern Forest Experiment Station.

Farrar, R.M., Jr. 1979b. Volume production of thinned natural longleaf. In *Proceedings Longleaf Pine Workshop*, ed. W.E. Balmer, pp. 30–48. Mobile, AL, October 17–19, 1978. USDA Technical Publication SA-TP3. USDA Forest Service, SE Area, State & Private Forests, Atlanta, GA.

Farrar, R.M., Jr. 1996. Fundamentals of uneven-aged management in southern pine. Tall Timbers Research Station Miscellaneous Publication 9.

Farrar, R.M., Jr., Murphy, P.A., and Willett, R.L. 1984. Tables for estimating growth and yield of uneven-aged stands of loblolly-shortleaf pine on average sites in the West Gulf area. Bulletin 874. Fayetteville: University of Arkansas, Arkansas Agricultural Experiment Station.

Forbes, R.D., and Bruce, D. 1930. Rate of growth of second-growth southern pines in full stands. U.S. Department of Agriculture Circular 124.

Goelz, J.C.G., and Burk, T.E. 1996. Measurement error causes bias in site index equations. *Can J For Res* 26: 1585–1593.

Goelz, J.C.G., and Leduc, D.J. 2001. Long-term studies on development of longleaf pine plantations. In *Proceedings of the Third Longleaf Alliance Regional Conference, Forests for Our Future*, ed. J.S. Kush, pp. 116–118. Alexandria, LA, October 16–18, 2000. Auburn, AL: Longleaf Alliance, Auburn University.

Goelz, J.C.G., and Leduc, D.J. 2003. A model for growth and development of longleaf pine plantations. In *Proceedings of the Fourth Longleaf Alliance Regional Conference*, ed. J.S. Kush. Southern Pines, NC, November 17–20, 2002. Auburn, AL: Longleaf Alliance, Auburn University.

Goelz, J.C.G., and Leduc, D.J. 2004. Reproducibility and reliability: How to define the population of trees that represent site quality for longleaf pine plantations. In *Proceedings of the Twelfth Biennial Southern Silvicultural Research Conference*, ed. K. Connor, pp. 189–195. Biloxi, MS, February 24–28, 2003. General Technical Report SRS-71. USDA Forest Service, Southern Research Station, NC: Asheville.

Grelen, H.E., and Lohrey, R.E. 1978. Herbage yield related to basal area and rainfall in a thinned longleaf plantation. Research Note SO-232. USDA Forest Service, Southern Forest Experiment Station, New Orleans, LA.

Harrington, C.A. 1990. PPSITE—A new method of site evaluation for longleaf pine: Model development and user guide. General Technical Report SO-80. USDA Forest Service, Southern Forest Experiment Station, New Orleans, LA.

Hawes, E.T. 1947. A method of determinating southern pine pole classes from d.b.h. *J For* 45: 204–205.

Haywood, J.D., Boyer, W.D., and Harris, F.L. 1998. Plant communities in selected longleaf pine landscapes on the Catahoula Ranger District, Kisatchie National Forest, Louisiana. In *Proceedings of the Ninth Biennial Southern Silvicultural Research Conference*, ed. T.A. Waltrop, pp. 86–91. Clemson, SC, February 25–27, 1997. General Technical Report SRS-20. USDA Forest Service, Southern Research Station, Asheville, NC.

Johnson, R., and Gjerstad, D. 1999. Restoring the longleaf pine forest ecosystem. *Alabama's Treasured Forests* Fall: 18–19.

Lohrey, R.E., and Bailey, R.L. 1977. Yield tables and stand structure for unthinned longleaf pine plantations in Louisiana and Texas. Research Paper SO-133. USDA Forest Service, Southern Forest Experiment Station, New Orleans, LA.

Mesavage, C., and Girard, J.W. 1946. *Tables for Estimating Board-Foot Volume of Timber*. Washington, DC: USDA Forest Service.

Means, D.B. 1996. Longleaf pine forest, going, going, ... In *Eastern Old-Growth Forests: Prospects for Rediscovery and Recovery*, ed. M. B. Davis, pp. 210–229. Washington, DC: Island Press.

Meldahl, R.S., Kush, J.S., Rayamajhi, J.N., and Farrar, R.M., Jr. 1997. Productivity of natural stands of longleaf pine in relation to competition and climatic factors. In *The Productivity and Sustainability of Southern Forest Ecosystems in a Changing Environment*, ed. R. A. Mickler and S. Fox, pp. 231–254. New York: Springer-Verlag.

Murphy, P.A., and Farrar, R.M., Jr. 1985. Growth and yield of uneven-aged shortleaf pine stands in the interior highlands. Research Paper SO-218. USDA Forest Service, Southern Forest Experiment Station.

Parker, R.C. 1998. Field and computer application of Mesavage and Girard form class volume tables. *South J Appl For* 22: 81–87.

Platt, W.J., Evans, G.W., and Rathbun, S.L. 1988. The population dynamics of a long-lived conifer (*Pinus palustris* Mill.). *Am Midl Nat* 131(4):491–525.

Quicke, H.E., and Meldahl, R.S. 1992. Predicting pole classes for longleaf pine based on diameter breast height. *South J Appl For* 16:79–82.

Quicke, H.E., Meldahl, R.S., and Kush, J.S. 1994. Basal area growth of individual trees: A model derived from a regional longleaf pine growth study. *For Sci* 40(3):528–542.

Quicke, H.E., Meldahl, R.S., and Kush, J.S. 1997. A survival rate model for naturally regenerated longleaf pine. *South J Appl For* 21(2):97–101.

Reineke, L.H. 1933. Perfecting a stand-density index for even-aged forests. *J Agric Res* 46:627–638.

Saucier, J.R., Phillips, D.R., and Williams, J.G., Jr. 1981. Green weight, volume, board-foot, and cord tables for the major southern pine species. Georgia Forest Research Paper No. 19. Georgia Forestry Commission, Research Division.

Schwarz, G.F. 1907. *The Longleaf Pine in Virgin Forest: A Silvical Study*. New York: John Wiley & Sons.

Shoulders, E., and Tiarks, A.E. 1980. Predicting height and relative performance of major southern pines from rainfall, slope, and available moisture. *For Sci* 26:437–447.

Shoulders, E., and Walker, F.V. 1979. Soil, slope, and rainfall affect height yield in 15-year-old southern pine plantations. Research Paper SO-153. USDA Forest Service, Southern Forest Experiment Station, New Orleans, LA.

Smith, W.R., Farrar, R.M., Jr., Murphy, P.A., Yeiser, J.L., Meldahl, R.S., and Kush, J.S. 1992. Crown and basal area relationships of open-grown southern pines for modeling competition and growth. *Can J For Res* 22:341–347.

Spurr, S.H. 1952. *Forest Inventory*. New York: Ronald Press.

Taras, M.A., and Clark, A., III. 1977. Aboveground biomass of longleaf pine in a natural sawtimber stand in southern Alabama. Research Paper SE-162. USDA Forest Service, Southeastern Forest Experiment Station, Asheville, NC.

Thomas, C.E., Parresol, B.R., Le, K.H.N., and Lohrey, R.E. 1995. Biomass and taper for trees in thinned and unthinned longleaf pine plantations. *South J Appl For* 19:29–35.

USDA. 1929 [revised 1976]. Volume, yield, and stand tables for second-growth southern pines. USDA Miscellaneous Publication 50. Washington, DC.

USDA. 1984. Tables of whole-tree weight for selected U.S. tree species. General Technical Report WO-42. USDA Forest Service, Washington, DC.

Wahlenberg, W.G. 1946. *Longleaf Pine: Its Use, Ecology, Regeneration, Protection, Growth, and Management*. Washington, DC: Charles Lathrop Pack Forestry Foundation.

Wiant, H.V., Jr. 1986. Formulas for Mesavage and Girard's volume tables. *North J Appl For* 13:147–148.

Wiant, H.V., Jr., and Castaneda, F. 1977. Mesavage and Girard's volume tables formulated. Resource Inventory Notes BLM 4. USDI, Bureau of Land Management. Denver Service Center.

Section IV

Restoration

Chapter 9

Restoring the Overstory of Longleaf Pine Ecosystems

Rhett Johnson and Dean Gjerstad

Introduction

Restoring longleaf pine trees to the southeastern landscape is a daunting task, because more than 97% of the original area has been lost to other uses (Landers et al. 1995; Frost this volume). However, many of the disincentives and difficulties in managing for longleaf pine have been addressed and solved or exposed as misconceptions, and landowners across the region are expressing renewed interest in returning this once-dominant southern pine to their lands. Several recent publications providing information to landowners and natural resource managers on longleaf pine restoration and management have appeared (Earley 1997, 2002; Franklin 1997; Kush 1997, 1999, 2001, 2003; Johnson 1999; Mitchell et al. 2000).

The economic and silvicultural benefits of longleaf pine are generally well-known: high-quality products; fire tolerance; insect and disease resistance; wind firmness; the ability to grow and thrive on harsh sites; and the ability to respond to thinning at virtually any time in its long life (Johnson 1999). The burgeoning pine straw market promises to yield even more financial benefit to landowners who manage their forests for its production.

Less well quantified but no less appreciated are the ecological benefits of longleaf forests, particularly when they are restored to ecological function (Frost 1990). The grassy and herbaceous understory is rich in species, some present only in fire-maintained longleaf ecosystems (Brockway and Lewis 1997; Glitzenstein et al. 1998; Kush et al. 1999a). Wildflowers abound, particularly in the fall, and are attended by scores of butterflies. Wildlife responds well to these forests. Bobwhite quail (*Colinus virginiana*) do particularly well in longleaf forests, as do fox squirrels (*Sciuris niger*). Wild turkeys (*Meleagris gallapavo*), deer (*Odocoileus virginiana*), and gopher tortoises (*Gopherus polyphemus*) find good habitat there and many songbird species are regular residents. Several threatened or endangered species of birds, reptiles, and plants are dependent on burned longleaf forests for their existence.

Interested landowners face two distinct but related tasks in their efforts to restore longleaf pine to their lands. One, the establishment of the trees themselves, is challenging but there are tested and reliable strategies for most situations (Franklin 1997; Boyer 1999; Hainds 1999, 2001, 2003a,b). The second, restoration

---

Rhett Johnson • Solon Dixon Forestry Education Center, School of Forestry and Wildlife Sciences, Auburn University, Andalusia, Alabama 36420.   **Dean Gjerstad** • School of Forestry and Wildlife Sciences, Auburn University, Auburn, Alabama 36849.

of the entire longleaf pine forest community, or at least enough of it to provide all of the functions necessary to provide the benefits that make it unique, is considerably more difficult and may be very expensive. It will certainly be a long-term effort.

This chapter describes 10 forest conditions common to the longleaf pine region today and suggests ways to begin the overstory restoration process. These situations, in general order of the level of difficulty in restoring ecosystem structure and function from most to least difficult, include: (1) agricultural fields and pastures; (2) "off-site" upland hardwoods; (3) abandoned cutover forest land; (4) mixed hardwood–pine or pine–hardwood forest; pine stands without a significant longleaf component and (5) no fire history or (6) with a recent history of fire; pine stands with greater than 4.5 m^2 basal area (ba) per hectare of longleaf pine in the overstory and (7) no recent fire history or (8) with a recent history of fire; stands that are predominately longleaf pine and (9) with no recent fire history or (10) with a recent history of fire.

Regeneration Options

Artificial Regeneration

Direct Seeding

The success of planting from seed, called "direct seeding" by foresters, depends on creating a well-prepared seedbed, sowing sufficient quantity of seed and having favorable weather conditions for seed germination and seedling early growth (Barnett and Baker 1991). Creating a well-prepared seedbed can range from a low-cost prescribed fire to intensive mechanical and chemical site preparation for sites with thick sod or heavy hardwood competition. The purpose of site preparation is to provide exposed mineral soil for seed germination and to reduce competing vegetation that can impact the survival and early growth of seedlings (Lowery and Gjerstad 1991). A major deterrent to direct seeding is that the cost of seed necessary for successful regeneration is higher than the cost of seedlings required for successful regeneration. Seed loss to predation and seedling mortality following seed germination make it necessary to sow from 1.7 to 3.4 kg of seed per hectare (Mann 1970; Barnett and Baker 1991). In addition, to prevent excessive predation by birds and rodents, the seed must be treated with repellents. Successful regeneration using this technique is dependent on adequate rainfall in the days and weeks following sowing, otherwise germination will be poor and mortality of newly germinated seedlings will be high (Williston and Balmer 1971).

Seedlings

In situations where there are no residual longleaf pines to provide seed, the most viable option is to plant longleaf seedlings. There are two types of seedlings available on the market today, bareroot and containerized (Barnett et al. 1990). Both types of seedlings can be planted successfully; however, a regional survey of practitioners found that average survival was 20% higher for container stock (85%) than for bareroot stock (Boyette 1996).

Bareroot seedlings are grown for approximately 1 year from seed in nursery beds and then lifted with no soil on the roots immediately before planting (Mexal and South 1991). The lifting process causes most fine roots to be lost and these fine roots must be regenerated after outplanting to enable the seedling to survive. Left unchecked in the nursery bed, longleaf lateral and taproots can grow to 1 m or longer in length, making them unwieldy to transplant. Thus, roots are usually pruned several weeks prior to lifting to produce a more compact root system and to stimulate fine root growth (Mexal and South 1991). The pruned roots are easier to plant and the residual fine roots will eventually regenerate new roots when outplanted. Root pruning should only be done by trained personnel in the nursery seedbed and never by tree planters prior to outplanting. Typically, root pruning by tree planters results in a significant increase in seedling mortality (Wakeley 1953; Mexal and South 1991). Characteristics of quality bareroot seedlings include healthy green foliage, root collar diameters of at least 12 mm, and several healthy lateral roots.

Containerized seedlings are grown in a variety of hard-walled vessels or in peat pots from seed. The entire plant, with root mass intact in the potting medium, is removed from the container, placed in shipping boxes, and shipped to the planting site (Brissette et al. 1991). In this case, the fine feeder roots are intact at the time of outplanting and frequently begin root growth immediately. Characteristics of quality containerized seedlings include healthy foliage, root collar diameters of at least 6 mm, firm plugs, and light brown or white root tips (indicating root growth) visible along the sides of the plug (Barnett et al. 2002; Hainds and Barnett 2003).

Site Preparation Alternatives

Preparation of a site for planting can take many forms and vary considerably in intensity of site disturbance (Table 1). Primary reasons to perform site preparation include better planting access and control of unwanted vegetation, thus enhancing the survival and early

TABLE 1. Advantages and disadvantages of mechanical and chemical site preparation techniques.

	Mechanical	
Technique	Advantages	Disadvantages
Bulldozing (windrow, rootrake)	1. Complete vegetation removal 2. Easy to administer 3. Machine plant	1. Short-term vegetation control 2. Negative impact on herbaceous plants 3. High cost 4. Potential loss in soil productivity and erosion 5. Loss of area in windrows 6. Potential water quality and sedimentation problems
Shear and pile	1. Less soil and site disturbance, therefore less potential for erosion and site degradation 2. Easy to administer 3. Machine plant	1. Short-term vegetation control 2. High cost
Disking	1. Easy to administer 2. Effective for natural regeneration 3. Machine plant	1. High cost 2. Potential erosion on steep slopes 3. Short-term vegetation control 4. Negative impact on herbaceous plants
Bedding	1. Improved survival on wet sites 2. Machine plant 3. Easy to administer	1. Expensive 2. Short-term vegetation control 3. Negative impact on herbaceous plants
3-in-1 (combines plowing, subsoiling, and bedding in one pass)	1. Easy to plant 2. Concentration of organic matter in bed tincreases soil fertility near seedling 3. Reduces woody plant competition in bedded area	1. Expensive 2. Competing vegetation between beds not affected 3. Likely severe impact on herbaceous plants in bedded zone 4. Beds need to settle prior to planting
Drum chopping	1. Fewer water quality and erosion problems 2. Minimal impact on herbaceous plants 3. Machine plant (if obtain good burn) 4. Usually provides for a good fire 5. Easy to administer	1. Considerable hardwood resprouting commonly occurs 2. Must be used in combination with fire to clear debris 3. Short-term vegetation control

(Cont.)

TABLE 1. (Continued)

Technique	Mechanical Advantages	Mechanical Disadvantages
Sub soiling (ripping)	1. Breaks up plow pan allowing improved drainage and root growth	1. Do not plant in rip as seedlings tend to sink resulting in mortality; plant to the side of the rip
Scalping	1. Removes sod, pest, and pathogens in planting zone on old agricultural fields 2. Low cost 3. Easy to administer	1. Must follow terrain to avoid erosion
Burning	1. Low cost 2. Positive impact on herbaceous plants 3. Reduction of fuels 4. Controls many small woody plants	1. Temporary air pollution 2. Short-term vegetation control 3. Possibility of escape 4. Normally used in combination with other methods
Total utilization	1. If done correctly, little needed for site prep. 2. Easy to plant	1. Rarely accomplished 2. Short-term vegetation control 3. Often considerable hardwood resprouting

	Chemical Advantages	Chemical Disadvantages
	1. Lower capital expenditures 2. Higher productivity (more acres per day) 3. Less resprouting of hardwoods 4. Can be used on more rugged terrain 5. Less soil disturbance 6. Growth response of crop trees	1. Narrow window of application (day/season) 2. Must train personnel 3. Need follow-up burn to facilitate planting 4. Controversial 5. May reduce herbaceous cover and species richness 6. Difficult to administer

growth of newly established seedlings (Lowery and Gjerstad 1991). In pastures and agricultural fields, scalping and subsoiling provide some measure of both. In improved pastures, a herbicide treatment is recommended prior to scalping to control introduced grasses like fescue (*Festuca arundinacea*), Bermuda grass (*Cynodon dactylon*), and Bahia grass (*Paspalum notatum*). These grasses must be controlled prior to planting because their well-developed perennial root systems will outcompete longleaf seedlings for soil water and result in poor longleaf pine survival. Additionally, there are no effective post plant treatments to control pasture grasses. Subsoiling or ripping is intended to break up the hardpan or plow pan common to many agricultural soils and allow drainage and root growth. Scalping can provide a weed-free zone immediately after planting, lessen disease and insect damage potential, and improve soil moisture conditions.

In situations following a timber harvest, coarse woody debris can be a problem for tree planters. Treatments to reduce or redistribute tree tops and other debris may be used to make the site more accessible for replanting (Lowery and Gjerstad 1991). Controlling unwanted vegetation on the site prior to planting will improve access and reduce competition for the new stand. To that end, techniques such as chopping, subsoiling, bedding plows, and herbicides, all used alone or in combination and in conjunction with fire, can damage rootstocks of residual woody vegetation and allow the new stand greater access to the site's resources (Lowery and Gjerstad 1991). Subsoiling and bedding treatments can prepare a good rooting environment for the new trees as well as control competing vegetation. Herbicides provide cost-effective control of unwanted vegetation prior to planting in practically every situation. Modern herbicides are commonly

TABLE 2. Herbicides commonly used to manage longleaf pine forests.

Common name of active ingredient	Herbicide family	Uses[a]	Species controlled	Resistant species
Glyphosate	Unclassified	SP, CS	Most plants when absorbed through foliage	Plants not receiving foliar spray
Imazapyr	Imidazolinone	SP, CS, release, HWC	Grasses, broadleaf, hardwoods	Legumes, pines, elms, blackberry, wax myrtle
Triclopyr	Pyridine	SP, CS	Hardwoods, broadleaf weeds	Grasses
Sulfometuron methyl	Sulfonyl urea	SP, HWC	Many grasses and broadleaf weeds	*Andropogon* sp., Bermudagrass, trumpet creeper
Hexazinone	Triazine	SP, CS, BS, release, HWC	Hardwoods and herbaceous species	Some grasses, *Vaccinium* sp.
Picloram	Pyridine	SP, CS	Hardwoods, pine, broadleaf weeds	grasses

[a] SP = site preparation; CS = cut surface; BS = basal soil; release = controlling hardwoods in pine stands; HWC = herbaceous weed control.

selective in their effect on various plant species (Table 2). For example, some herbicides selectively kill grasses (sethoxydim) while others (imazapyr) do not kill legumes and still others (triclopyr, hexazinone) kill only broadleaf plants (Minogue et al. 1991). This trait allows knowledgeable managers the opportunity to almost surgically extract undesired species and create room for desirables. The choice of herbicides or *prescription* should be made by a trained professional to best suit the vegetation to be controlled.

By the same token, intensive site preparation has the potential to destroy desirable components of the understory or to hamper their recovery (Bengston et al. 1993). If understory restoration is a primary consideration for a landowner, a balance must be struck between maximizing site preparation intensity to increase tree growth and survival and minimizing the impact on other ecosystem components. Optimally, fire alone does little to retard understory development and, in fact, is likely to enhance it. In most situations where sites are being converted back to longleaf pine, however, fire is not sufficient to allow optimal longleaf pine growth and survival. Selective herbicides, used in conjunction with fire, offer some of the benefits of site control while sparing some of the desirable plant community. The landowner/manager should have a clear idea of what understory species are desirable and a good inventory of what species are present prior to choosing a chemical.

Chopping, a mechanical treatment which employs a heavy ribbed metal drum pulled across the site, causes minimal soil disturbance and seedbed disruption, and has little impact on herbaceous plants, but tends to stimulate sprouting of woody growth if no other treatment is applied (Lowery and Gjerstad 1991). An effective follow-up treatment to control woody sprouts is a second chop in late summer or early fall when root carbohydrate reserves are low.

As in most forest management situations, cost is a factor in any decision and will vary with the intensity of site preparation. Cost-effectiveness can only be determined if desired outcomes and priorities are clearly understood and stated.

Planting

Longleaf pine can be either hand planted or planted with a planting machine. No matter which method is used, planting depth is a critical factor in determining the survival and early growth of longleaf seedlings. Research conducted by The Longleaf Alliance (Hainds 2003a,b) has indicated that planting containerized seedlings with the terminal bud and root collar slightly above the soil surface results in excellent survival. Seedlings planted with the terminal bud covered by soil suffer high mortality and slow growth. Maintaining an optimal planting depth is particularly important with containerized seedlings and may be difficult to obtain with a planting machine. It is easier to hand plant containerized seedlings than bareroot seedlings. Bareroot seedlings should be planted so that the root collar and terminal bud are right at or within 6 mm of the soil surface after the soil settles (Barnett et al. 1990). Machine-planting usually results in a slight berm around the seedlings created by the packing wheels. That berm typically subsides over time, so that seedlings planted using a machine should be planted with the terminal bud slightly below the soil berm surface. Most longleaf planters find it easier and more efficient to plant bareroot seedlings by machine because of the long coarse lateral roots found on many longleaf seedlings (Barnett et al. 1990). Hand-pruning these roots leads to high mortality and stuffing them into the small slit typical of hand planting can cause root binding or "J-rooting." Most tree planters recommend that as much of the root mass that will fit be inserted into the planting hole, and the remainder left out to air prune. Because containerized seedling roots are in a compact plug, that problem is avoided.

Planting with bareroot seedlings should take place any time in the late fall or early winter when the soil is moist. It is generally preferable to plant in November or December, but all planting should be complete before March (Long 1991). Bareroot longleaf pine seedlings do not store as well, even in cold storage, as do other pines and should be planted as quickly as possible after lifting (Barnett et al. 1990). Containerized seedlings can be planted at any time of the year as long as there is moisture in the soil and can begin growing immediately on outplanting (Hainds 2003b). Very good results have been obtained by planting in November and December, giving the seedlings a chance to develop a root system while competitors are dormant. This head start can be very important in the spring, when grasses, forbs, and woody competitors begin early growth. Planting containerized seedlings in the late spring and summer has also proven to be successful as long as there is adequate soil moisture. Some managers use late planting to replace mortality from earlier plantings.

Planting with bareroot seedlings generally calls for more care than is required for containerized seedlings as success is more dependent on uncontrollable factors such as weather and the window for planting is much narrower. Bareroot seedlings must regrow the fine feeder roots lost in the lifting process before seedling growth can take place. Experienced individuals and organizations can be successful planting bareroot seedlings, but the margin of error is smaller than with containerized seedlings and any misstep can lead to low survival and/or poor early growth. Cost can also be a factor in the choice of seedling type. Currently, bareroot seedlings cost about 40–50% less than containerized seedlings. However, the cost of seedlings is a small part of the total cost of reforestation and replanting after a failure can be very expensive. Simply put, containerized seedlings cost more, but offer insurance against failure as well as improved early growth.

Release Treatments

Because longleaf responds poorly to competition, it is often desirable to "release" it by controlling undesirable vegetation. The most important competitors for longleaf seedlings typically are grasses. Aggressive grass competition can cause high seedling mortality and retard growth of surviving seedlings (Hainds 2003b). The longer seedlings stay in the grass stage, the more susceptible they are to brown spot needle blight (*Mycosphaerella dearnessii*), an important disease of longleaf pine seedlings. Woody competition has a more limited effect on early survival of longleaf pine seedlings, but can also retard growth. Woody competition can be controlled after the longleaf is planted with fire, herbicides, or a combination of both. Grasses and forbs, on the other hand, are generally only top-killed by fire and can rebound aggressively. The most important competition takes place below ground for growing space, soil moisture, and nutrients. Burning grasses does little to reduce that competition. Treating herbaceous plants

9. Restoring the Overstory

FIGURE 1. Shelterwood systems employ a series of cuts over several years and can result in even- or two-aged stands. Trees marked with an "X" are to be harvested in the indicated cutting cycle.

a. The first cut removes nonlongleaf pines unless they are needed to provide fuel for prescribed fires and all hardwoods. Approximately 11.5–16 m^2/ha of basal area of well-distributed longleaf pine should be left across the site.

b. The second cut comes after several years and several fires to control underbrush. This cut will leave a residual basal area of 5.5–7 m^2/ha of well-distributed, well-formed longleaf pines, all proven cone and seed producers.

c. A winter or spring burn before seed fall a year when there is a good cone crop prepares a seed bed for the new germinants. If stocking is adequate (greater than 11,000/ha), the seed trees can be removed in the second or third year following seed fall to yield an even-aged stand. Seed trees should be removed while the seedlings are in the grass stage. If a two-aged stand is desired, some or all of the seed trees may be retained. Fire should be excluded for the first year after seed fall.

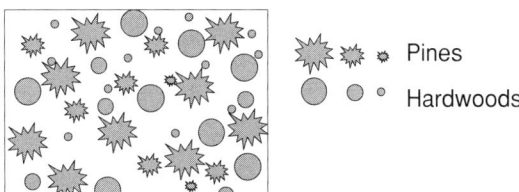
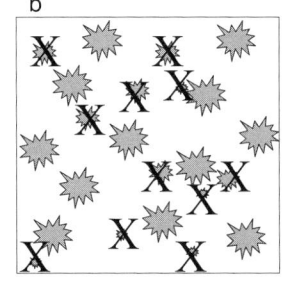
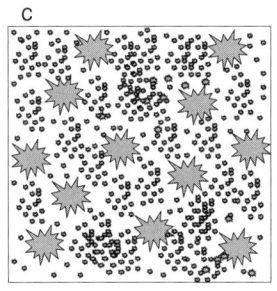

with herbicides provides seedlings with a competition-free rooting zone, allowing them to establish healthy and competitive root systems. Selective herbicides can be used to favor some plants while controlling others. Banded applications only treat the immediate area around the seedling rows, preserving desirable native vegetation between bands. The effective period for most of the chemicals used for release is one growing season or less.

Natural Regeneration

Even-Aged Systems

Where adequate seed sources exist, longleaf pine may be regenerated from natural seed (Dennington and Farrar 1983; Barnett and Baker 1991). Trees should be at least 30 cm in diameter at breast height (DBH) and preferably 40 cm or larger before being considered as reliable seed producers. If enough longleaf trees of sufficient size are present and evenly distributed across the site, natural regeneration is an option. The longleaf shelterwood system, as developed at the Escambia Experimental Forest in Alabama, is a modified seed tree system adapted to allow for the heavier seed of longleaf pine (Croker and Boyer 1975; Dennington 1990). In this shelterwood system, a longleaf stand is first thinned to a basal area (BA) of about 12 m^2/ha or about 125–150 trees at least 30 cm in diameter per hectare (Fig. 1; Boyer and White 1990). The resulting stand is left to grow while the tree canopies expand into the gaps between them. Approximately 5 years later, a second cut is made to reduce the number of trees to 60–75 trees/ha

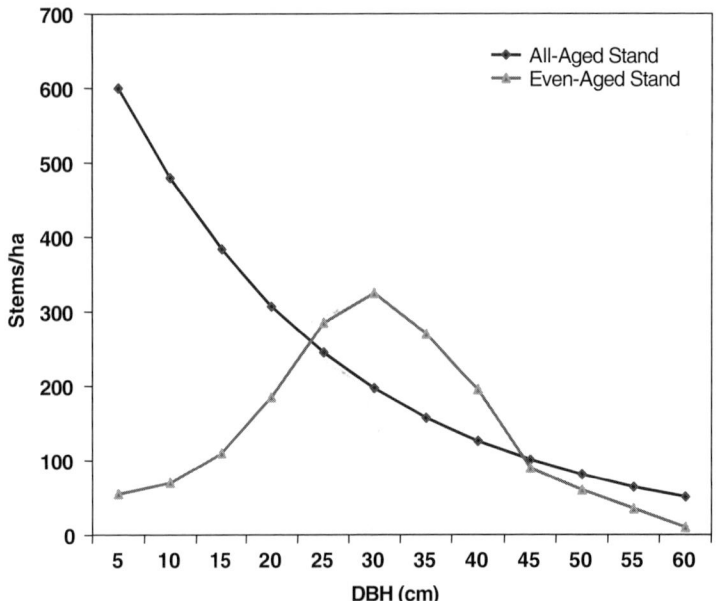

FIGURE 2. Stand structures and appearances of traditional uneven-aged and even-aged mature longleaf stands are shown. Regulated uneven-aged stands have a stem diameter distribution curve that resembles a reverse "J," while even-aged stands typically have a bell-shaped distribution curve.

(BA of 5.5–7 m²/ha). The residual trees should be selected for desirable qualities, because they will provide the seed for the new forest. During this period, prescribed fire should be used frequently to keep the woody brush in the understory under control. Longleaf cones take 2 years to mature, so it is possible to anticipate good seed crops by monitoring flower production more than a year before actual seed fall (Croker 1971). Failing that, green cones are easily spotted in the late spring and early summer of the actual seed year. It is recommended that an average of 2500 cones/ha be present to achieve successful regeneration. With each cone producing approximately 50 seeds, 2500 cones will produce nearly 125,000 seeds/ha. Typically, 90% of the seed and newly germinated seed are lost to predation by wildlife or die due to adverse environmental conditions. The desired goal is that 7500 to 15,000 seedlings/ha be present before the overstory is removed. A prescribed burn in the winter or spring before seed fall will greatly enhance seed "catch" and germination. When the new stand is established and at least 1 year old, the overstory should be removed to achieve maximum growth. Some seedlings will be damaged by the logging activity, but in virtually every incidence more than adequate numbers of seedlings will be present after the overstory removal. Generally, 2500 to 3500 seedlings/ha in height growth and free of overhead competition are considered a successfully established stand. Over time such stands will develop into an even-aged stand with bell-shaped diameter distribution (Fig. 2).

Uneven-Aged Systems

Uneven-aged longleaf pine stands with a reverse J diameter distribution can be created through several approaches (Figs. 2 and 3). If an uneven- or all-aged stand is desired, the quickest and easiest approach is a modified shelterwood system where all or a portion of the overstory trees are retained. This approach will create a two-aged stand with the first seed fall and, over time and successive seed crops, can come to contain several age classes. However, such stands are not commonly found on the landscape. The trade-off is reduced growth of the new seedlings due to competition from the overstory trees (Boyer 1993; Palik et al. 1997, 2003). Another option is to create gaps of 0.1–0.2 ha in the canopy across the stand by removing several trees in clumps (Fig. 4; McGuire et al. 2001). This approach mimics natural gaps resulting from overstory mortality due to lightning strikes, insect infestations, wind throw, and other causes of mortality.

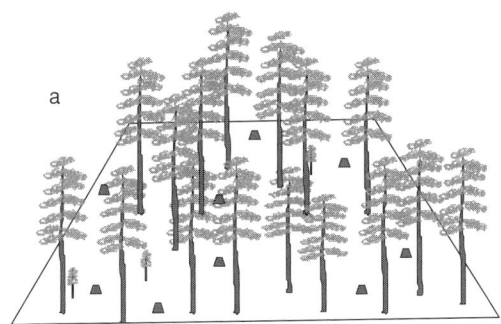
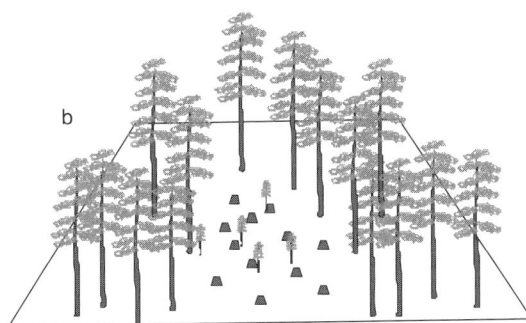
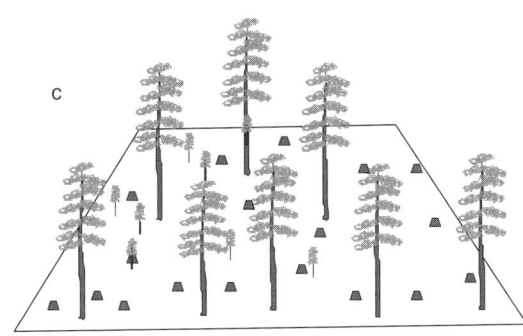

FIGURE 3. There are various approaches to uneven-aged management. (a) Single tree selection, (b) group selection, and (c) modified shelterwood cuts are visualized above.

Again, fire must be used to retard invasion by undesirable hardwoods and woody brush. As the gaps seed in with longleaf, they are gradually enlarged by removing trees around the margins. The gaps soon begin to resemble small "domes," with the oldest, tallest, and fastest growing saplings in the middle and the youngest, slowest growing, and smallest at the edges or margins of the openings. Over time and successive seed crops and cuts, these gaps merge into each other and trees representing many age classes occupy the site. If the gaps are too small, the residual trees compete vigorously with the new seedlings and can prevent successful establishment (McGuire et al. 2001). If the gaps are too large, the heavy longleaf seed might not reach the middle and it can be very difficult to get fire to burn across the gap due to the lack of continuous pine needle litter to carry the fire. This can result in establishment of undesirable woody species in the gaps. In open stands, single-tree removal can create adequate openings for regeneration, but the removal of several trees in clumps at each entry may be more cost-effective and provide better sites for seedling growth (Fig. 5).

An even-aged stand can be converted to an all-aged or all-sized stand over a series of harvests using the BDq method (Fig. 6). The first harvest typically spares the smallest size classes and most of the largest except where stems exceed the predetermined diameter limit. Most of the harvest takes place in the middle diameters, where stem counts exceed those indicated by the desired distribution curve. Typically, to retain the desired basal area, some of the larger trees are retained and removed in later cuts. Over several harvests and seed crops, the stand distribution will begin to resemble the desired reverse J shape and the all-aged condition reached.

A version of this technique is possible when there is no natural longleaf seed source if fire can be used in the existing stand. Cutting gaps into existing pine stands and planting with containerized seedlings can be successful. Experience has suggested that at least minimal site preparation and fairly large (greater than 0.1 ha) openings are necessary for satisfactory results. Again, repeating the process over time with several cuts should result in an all-aged longleaf stand. This approach may result in growth reductions (Farrar and Boyer 1991; Boyer 1993), but does allow the retention of mature trees on the site at all times.

Situations Requiring Artificial Regeneration

Agricultural Fields and Pastures

Although old agricultural fields and pastures can appear to be "blank slates," they typically

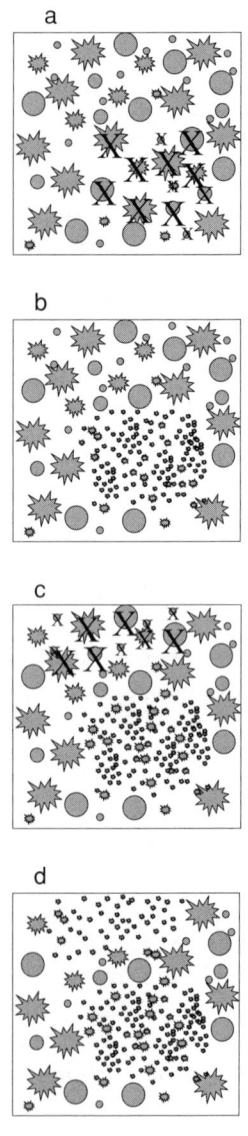

FIGURE 4. Small group selection harvests can yield uneven-aged longleaf pine stands. Trees marked with an "X" are to be removed. Symbol size indicates the relative size of the tree represented.

a. The first cut removes hardwoods and pines in clusters of 0.1 to 0.2 ha in area and at least 30 m wide. Seed-producing longleaf surrounding these clusters are retained. Prescribed fire must be utilized to keep these openings free of hardwoods and pines other than longleaf until regeneration with longleaf is achieved.

b. After skipping a year of fire to protect the new seedlings, burning is resumed. Mature trees adjoining the openings may suppress growth of the new seedlings and may be removed.

c. Another series of clusters can be selected and removed on the desired cutting cycle, following the same regime as above. The seedlings from the first clusters are now small saplings.

d. Continued removal of mature trees in small groups or clusters will eventually release the younger trees, the first of which are now entering larger size classes. An all-aged stand or forest is developing, with trees of several age and size classes.

are the most difficult of all sites to restore to longleaf pine ecosystem function (Hainds 2003b). The pasture grass and agricultural weed complex is aggressive and most of these sites have little or no residual seed bed from the early native forest. Reforestation, of course, must be through planting. Direct seeding techniques are seldom successful. If seed is broadcast, germination is uncertain and predation by birds and small mammals can be severe. Planting individual seeds by hand or drilling the seed into the soil is labor intensive and success is usually spotty.

Preparing the site for planting is critical. If pasture grasses are present, they should be killed before planting is attempted (Hainds 2003b). Killing grasses like Bermudagrass, fescue, and Bahiagrass is difficult and usually requires herbicides. An effective technique includes mowing, allowing the grass to grow back to about 15 cm in height, and then applying glyphosate, imazapyr, or a combination of the two herbicides. Glyphosate can be applied anytime the grass is growing. Imazapyr is most effective in late spring or early summer. Follow-up spot treatments will likely

FIGURE 5. Single tree selection harvests can also yield an all-aged condition, although growth of younger age classes is likely suppressed. Trees marked with an "X" are to be harvested. Symbol size represents tree size.

a. Individual trees are selected for removal throughout the stand, choosing suppressed trees, trees with bad form, and trees of species other than longleaf for harvest. Some hardwoods may be retained for wildlife purposes and some nonlongleaf pines may spared to provide continuous fuel for prescribed fire.

b. Seedlings should accumulate in small patches throughout the stand, particularly in the small gaps created by the earlier harvest. A second round of tree removals based on tree quality, position in the canopy, and stand density may be performed.

c. The thinning process continues on regular intervals, retaining the most desirable individuals and removing some to release nearby seedlings and saplings.

d. Prescribed fire is continuously applied. Thinning continues to be used to improve stand quality and to free established regeneration from competition. An all-aged stand comprised of trees of several age and size classes is developed.

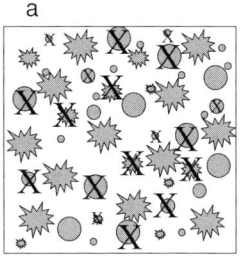

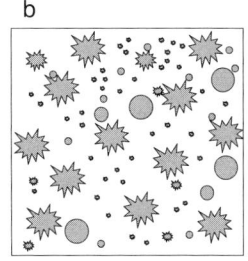
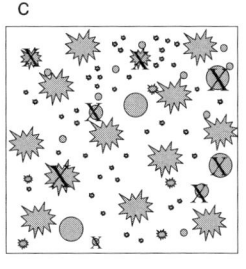
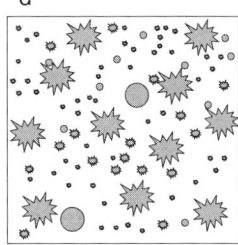

be necessary to control these tough competitors, especially Bermudagrass.

In both pastures and old fields, the next step is to scalp the site (Hainds 2001). Scalping peels back the upper 5–7.5 cm in old fields and 7.5–10 cm in sod pastures in a furrow 60–90 cm wide. Scalping implements are available commercially or may be constructed from modified fire plows. There are commercial vendors in many areas who will provide scalping services. It is extremely important to rigorously follow contours when scalping. Soil movement along the scalped furrow can not only doom the planting effort but damage water quality and ruin the productivity of the field for future generations as well. Unless the site has been subsoiled within the past 2 years, it is usually a good idea to pull a ripper bar along the furrow to break up plow pans caused by previous agricultural tillage. Depth of the rip should be from 45 to 60 cm. If planting is to be done using a mechanical planter, it is recommended that

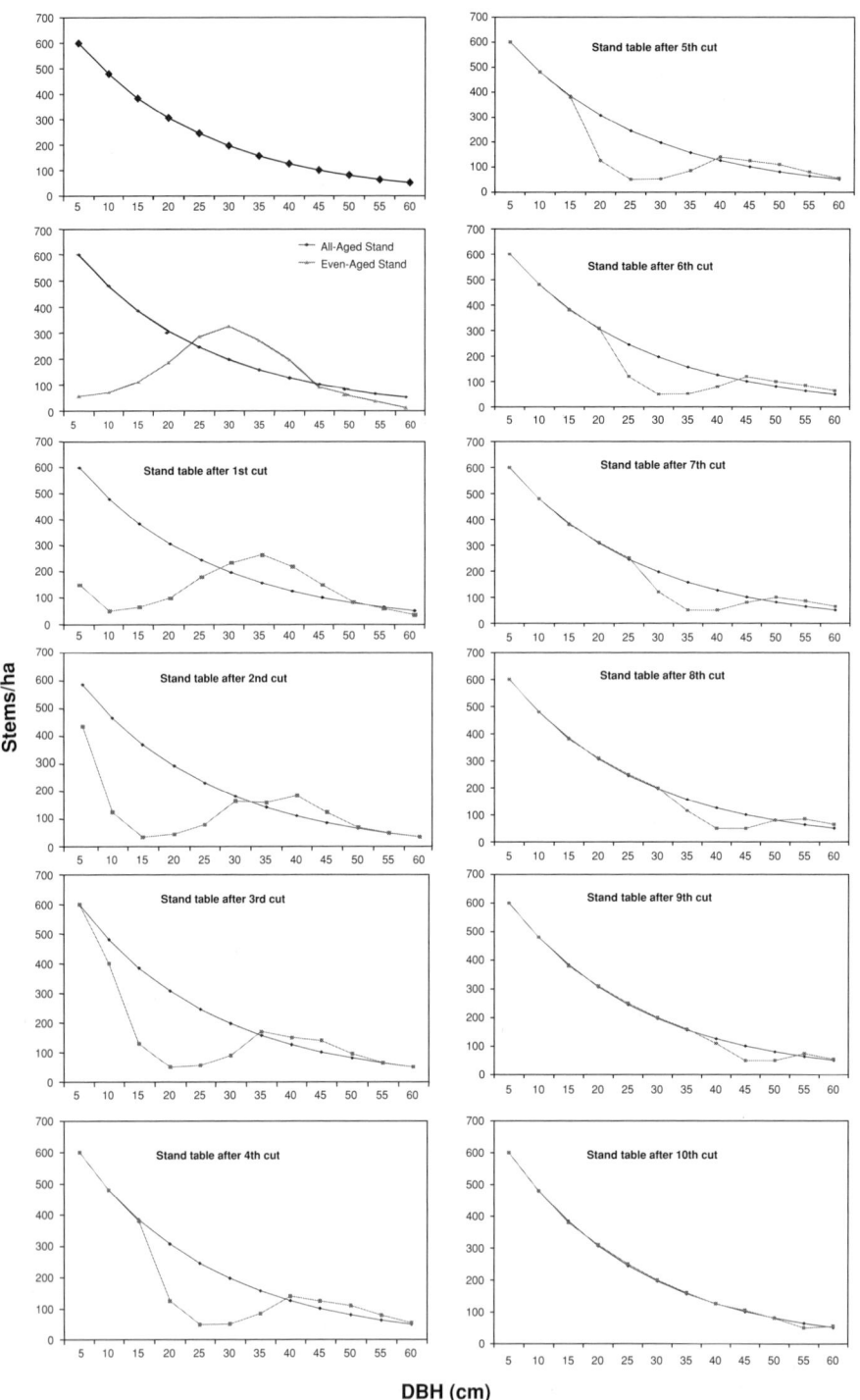

FIGURE 6. An even-aged stand can be converted to an all-aged or all-sized stand over a series of harvests using the BDq management method. Existing and desired stand distribution curves are developed and superimposed. The first harvest typically spares the smallest size classes and most of the largest except where stems exceed the predetermined diameter limit (D). Most of the harvest takes place in the middle diameters, where stem counts exceed those indicated by the desired distribution curve. Typically, to retain the desired basal area (BA), some of the trees above the "line" are retained and removed in later cuts. Over several harvests and seed crops, the stand distribution should begin to resemble the desired reverse "J" shape and the all-aged condition reached. In this example, a cycle of 10 harvests is depicted.

the rip be oriented approximately 30 cm off center in the scalped area. This is done so that the seedlings may be planted in the center of the furrow and not in the ripped area. If trees are to be planted by hand, they should not be planted in the rip, but 15 cm or more to one side. This is because seedlings planted in the ripped zone tend to sink several centimeters into the soil, resulting in mortality. Scalping greatly improves survival in agricultural areas (Barnard et al. 1995; Hainds 2001). The furrow typically remains virtually weed free for much of the first growing season, reducing competition for moisture and nutrients. In addition, there is evidence that scalping reduces loss to a number of pests and diseases, including white fringed beetle (*Graphognathus* spp.) grubs and charcoal root rot (*Rhizoctonia* spp.) (Barnard et al. 1995).

Planting should be accomplished as soon in the fall as there is adequate moisture in the soil (Hainds 2003b). All planting should be completed before Christmas if at all possible. Planting earlier in the fall gives the seedlings a chance to initiate root growth before entering dormancy and gives them a head start the following spring. It is recommended that planting be done with containerized seedlings in these situations. Recent research indicates that seedlings should be planted so that 1.2 to 2.5 cm of the plug is exposed, especially in scalped areas (Hainds 2003b). Soil tends to move into these areas and seedlings with terminal buds covered by soil are not likely to survive. In the spring following planting, herbaceous weed control treatment might be desirable. One frequently recommended treatment is 0.15–0.2 kg of Oust™ per hectare in March or April. The addition of 0.75 kg of Velpar DF™ per hectare gives broader spectrum control and may extend the effectiveness of the treatment (Hainds 2003b). A premixed version of these chemicals, Oustar™, can also be applied at a rate of 0.75 kg per treated hectare. These treatments are applied directly over the top of the seedlings and generally in a band 1–1.5 m wide or in spots of 1.5-m diameter over each seedling. In banded treatments, total volume of spray (chemical plus water) should be at least 95 liters/ha. Before spraying, one or two seedlings should be excavated and checked for new root growth. If the seedlings have not begun to initiate new root growth outside the plug, delay spraying until that occurs. Also, areas where soil pH is high (greater than 6.0) should not be treated with Oust™. Oust™ is a root growth inhibitor, and high-pH soils tend to exacerbate its effectiveness. If root growth on new seedlings is retarded by the chemical, April and May droughts may cause high mortality.

Another herbicide combination that has proven successful in these situations is Arsenal™ and Oust™ (Hainds 2003b). Arsenal™ should not be applied over longleaf seedlings earlier than 4–6 weeks after foliage growth initiation in the spring. One treatment combines an early treatment with 0.15 or 0.2 kg of Oust™ per treated hectare in March or April followed by a subsequent treatment, if needed, with 0.35 or 0.4 kg of Arsenal™ per treated hectare in May or later. The Arsenal™ and Oust™ may be applied together in one treatment 4–6 weeks after initiation of foliage growth in the spring. Late germinants such as crabgrass can prove disastrous to old-field and pasture plantings if only the early treatment is used.

Since these sites generally have no remnants of the native forest understory vegetation and little or no seedbed to draw on, the use of chemicals to control vegetation does little damage to the total restoration effort. Once the trees are established on agricultural sites, they can be burned almost immediately. In fact, it may be difficult to use fire effectively on many of these sites because there are often few fine fuels on the ground to carry the fire. Old-field weeds like dog fennel (*Eupatorium capillofolium*), ragweed (*Ambrosia* spp.), and goldenrod (*Euthamia* spp.) do not burn well and do not carry fire. Broomsedge (*Andropogon* spp.) is a better fuel and will carry a fire if there is a wind of 5 km/h or more. The trees are fire tolerant throughout their life, but the stage when they are 15 cm to 1 m tall is the most vulnerable. Trees in this size class should not be burned during the growing season when they are "candling" (when the terminal bud is elongating and is tender and white). Young longleaf may be burned successfully with strip headfires, although other types of fires may be successful as well. Mowing between rows may be helpful for access and to help carry fire, but care should be taken to avoid damaging seedlings.

Off-site Upland Hardwoods

Many longleaf sites become occupied by hardwood species if clearcut or other heavy harvest is followed by fire suppression and no attempt is made at longleaf pine regeneration. Typically, light seeded species like sweetgum (*Liquidambar styraciflua*) and loblolly pine (*Pinus taeda*) are the first tree species to colonize the site. Without disturbance, particularly fire, more shade-tolerant hardwood species come to dominate the site, in the process providing enough shade to preclude regeneration by the residual pines. These hardwood species in the longleaf pine region typically include oaks (*Quercus* spp.) of various species, hickories (*Carya* spp.), and maple (*Acer* spp.). Without further disturbance, the site will soon be crowded with woody brush and mixed hardwood regeneration, resulting in a shaded forest floor with little or no herbaceous vegetation. On mesic sites, this can be a fairly rich mix of species. On xeric or very dry sites, the dominant species are typically scrub oaks like turkey oak (*Quercus laevis* Walt.), bluejack oak (*Quercus incana* Bartr.), sand post oak (*Quercus margaretta* Ashe), and blackjack oak (*Quercus marilandica* Muench.). The effect, if left undisturbed for long periods, is similar; no pines in the overstory or understory, a mixed woody midstory and understory, and very few grasses or forbs. Restoring longleaf pine and longleaf function to this forest type is slow, but can be accomplished. Longleaf pine restoration can only be accomplished by artificial means, i.e., planting.

If the hardwoods on the site are of commercial size, the first step is to remove them with a timber harvest. Examination of the site after harvest will determine the presence of desirable understory components. If there is no significant desirable native vegetation on the cutover site, site preparation can proceed. Because hardwoods are vigorous resprouters, it is usually desirable to control regrowth with a herbicide. These chemicals can be applied aerially, with ground-mounted sprayers, or by hand with backpack sprayers. The chemical may be broadcast, banded, or directed at sprout clumps. It is usually recommended that the site lay out for one growing season after harvest to allow sufficient resprouting to aid herbicide efficacy.

Choice of chemical should be based on species to be controlled as well as species to favor (Cantrell 1985; Minogue et al. 1991). Herbicides differ in their mode of action and which species they control. For instance, chemicals whose active ingredients include glyphosate (e.g., Roundup™ or Accord™) are only foliar active and affect only the plants that are sprayed. They control grasses and forbs as well as woody plants, but are much less effective on waxy-leafed plants. Products containing hexazinone as an active ingredient, on the other hand, are soil active. That is, they are taken up through the plant's root system and work internally in the plant. Hexazinone products, like Velpar™, Pronone™, and ULW ™, are ineffective on many grasses like the common *Andropogon* species (e.g., broomsedge, bluestems) and wiregrasses (e.g., *Aristida* spp., *Muhlenbergia* spp.) at low rates (Brockway and Outcalt 2000). This may be an important factor where understory restoration is an objective. Likewise, desirable wildlife species such as American beautyberry (*Callicarpa americana*) and many *Vaccinium* species (e.g., blueberry, sparkleberry, huckleberry) are also tolerant of hexazinone. Products like Arsenal™ or Chopper™, which contain imazapyr as an active ingredient, afford broad scale control of many species, but have little effect on nitrogen-fixing legumes which can also provide important food for many wildlife species (Minogue et al. 1991). Tank mixes of herbicides afford broader spectra of control, but care should be taken to ensure that they are compatible. In short, prescriptions for herbicides should be tailored carefully to fit both species that are to be controlled and species that are to be spared.

In the case of mesic off-site hardwood conversion, there is likely to be little desirable vegetation under the stand to protect and the choice of site preparation method is more likely to depend on what species are to be controlled prior to planting (Haywood et al. 2004). Herbicide labels are a good source of information on efficacy of the chemical on target species. It is advisable to obtain professional assistance, however, before choosing and applying an herbicide.

Mechanical site preparation in combination with fire is another choice for effective off-site hardwood conversion (Table 1). Shearing, raking, plowing, and bedding in preparation for planting are commonly applied practices (Lowery and Gjerstad 1991). All of these techniques, however, are likely to destroy existing native ground cover and make its reestablishment more difficult (Bengston et al. 1993). This delay in herbaceous recovery can delay the reintroduction of prescribed fire. Chopping with a drum chopper is effective where debris is light and can be done with minimal impact on native understory. These mechanical treatments typically result in many hardwood sprouts from roots and stumps and may threaten the survival and growth of longleaf pine seedlings unless controlled.

In the case of scrub oaks (*Quercus* spp.) on xeric sites, mechanical treatments or herbicides used in combination with fire or fire alone will often stimulate vegetative growth by desirable grasses and forbs (Brockway and Outcalt 2000). Before longleaf pine can be reestablished, however, the site must be made accessible for the planters and the oaks must be reduced in number to minimize competition with longleaf pine seedlings and allow the development of herbaceous fuels to carry later fires. These oaks are fairly fire-tolerant and are best killed with a herbicide. Choices include a hexazinone product applied as a broadcast or banded treatment, a directed or broadcast treatment with an imazapyr product, or stem injection with a variety of chemicals. Again, choices should be based on existing competition and desirable vegetation and prescriptions should be prepared by knowledgeable professionals. Planting with containerized or bareroot longleaf pine seedlings can then usually proceed. Many native species typically have a diminished presence on these sites and may recover without further assistance beyond prescribed fire, but depleted seed banks may require planting of some desired species.

Mesic sites occupied by vigorous hardwood species can be restored to longleaf pine in a two-step process. After clearing the hardwoods through harvest and controlling the sprouts and other woody competitors with chemicals, a "nurse crop" of loblolly or slash pine might be planted and fire introduced as early as possible in the rotation, usually between ages 8 and 12, and kept in the system until final harvest of the pines, either as pulpwood in 12–15 years or as solid wood products, with thinnings, at 25–30 years. This will help control unwanted woody vegetation through the early shading provided by the faster growing loblolly or slash pine and provide pine fuel for subsequent fires. Keeping fire in the system on a 2- to 4-year rotation, with occasional growing season fires as part of the mix, should prepare the site for planting with longleaf and encourage any native species to assert themselves prior to establishment of the longleaf pines.

Abandoned Cutover Forest Land

Cutover woodland sites typically do not have a longleaf pine seed source present, so they must be regenerated by planting if longleaf pine is to be reestablished. Depending on the length of time since longleaf occupied the site, site preparation for planting can take various forms. If the site has been cut over and abandoned for several years, it is very likely to be occupied by woody brush, hardwood seedlings and saplings, grapevine (*Vitis rotundifolia*) and other vines, and even loblolly pine, an aggressive pioneer species. There is typically little herbaceous understory if the site has been cut over for very long. In this case, site preparation is usually most successful if it is done with herbicides. The chemical(s) chosen would depend on the species mix and the mode of application on the density and height of the vegetation. If herbicides are chosen for site preparation, they should be followed in a timely fashion by fire to augment their effectiveness and to make access for planting easier. Fire alone is seldom effective in these situations due to a lack of fine fuels and the lack of total kill on hardwood rootstocks. If the cutover is recent, then fire might be effective assuming there is sufficient fine fuel, usually grasses and forbs, to carry the fire across the site. It is usually desirable to follow up the fire with directed herbicide treatments either before or after planting.

When the site is prepared for planting, containerized or bareroot longleaf seedlings should be planted as early in the fall as feasible with the limiting factor being adequate soil moisture (Hainds 2003b). A spring herbaceous release treatment with selective herbicides carefully applied will enhance survival and growth of the newly planted seedlings. Fire may be introduced as early as the first year postplanting and a mixture of dormant and growing season fires should encourage existing native vegetation (Walker 1998). Caution is advised when the young seedlings are in the candling stage and before they are 1.5–2 m tall. If the desired native plant community does not appear within the first 2 or 3 years of burning, it may be necessary to reintroduce it or supplement remnant populations with seed or seedlings (Glitzenstein et al. 1998).

Mixed Hardwood–Pine or Pine–Hardwood Forest

Many historical longleaf pine sites are occupied today by forests composed of loblolly and/or shortleaf pine (*Pinus echinata* Mill.) and mixed hardwoods. In much of the natural longleaf range, this is the typical forest condition following a total harvest if no attempt was made to reforest. These forests can produce valuable timber, excellent wildlife habitat, and are often aesthetically pleasing. Still, they occupy sites that once supported longleaf pine ecosystems and are well suited to that forest type. For a variety of reasons, including optimizing ecological and economic value, many managers desire to restore longleaf pine to these sites. Doing so requires the planting of longleaf pine seedlings. One technique that has been applied successfully begins with the commercial harvest of the hardwood component, followed by a growing season fire, preferably in late March through mid-May. Clearing logging slash away from the boles of the residual pines manually or mechanically is a wise precaution. The intense heat these piles or tops can generate can damage or kill the roots and cambium layer and scorch the crowns of nearby trees (Swezy and Agee 1991; Hanula et al. 2002). Continuing to burn on a 2- or 3-year rotation, using growing season and dormant season burns, should control the hardwood sprouting and most of the pine regeneration. When the woody understory is diminished, the next step is to remove the nonlongleaf pine component in a final harvest. Following the harvest and prior to replanting, site preparation is typically needed to remove unmerchantable hardwoods and pines. The method of site preparation should be chosen with the level of residual competition and the presence or absence of desired ground-cover species in mind. If the fire has been effective, the woody competition should be minimal and any residual ground cover should have begun to express itself. Postplanting control of hardwood sprouts might be desirable and can be achieved by directed spray with foliar active herbicides or spot treatments with hexazinone products (Minogue et al. 1991). Restoration of understory vegetation is likely achievable through repeated use of fire, particularly growing season fire, throughout the rotation unless the site was in cultivation for an extended period prior to becoming reforested (King et al. 1997). If desired understory recovery does not materialize within 5 years or less, it may be necessary to artificially supplement it with seed or seedlings.

Another approach to restore mixed pine–hardwood stands is to leave a portion (no more than 2.3 m^2/ha of basal area (BA)) of the loblolly, slash (*Pinus elliottii*) or shortleaf pine stand intact after the pine harvest and underplant with longleaf seedlings. This would create a two-aged stand and would help provide fuel for subsequent fires. There are several cautions, however. There is evidence that even as little as 2.3 m^2/ha BA can significantly retard growth of the new stand (Boyer 1993). More importantly, there may not be adequate pine fuel to carry sufficiently hot fires to control the loblolly, slash, or shortleaf pine seedlings resulting from this technique. These pine seedlings can be important competitors for the longleaf pine and choke out desirable understory vegetation if not controlled. The likely outcome of this approach is the creation of a mixed species pine stand.

Pine Stands without a Significant Longleaf Component

There are many essentially pure stands of pines of species other than longleaf pine on former longleaf pine sites. Some of these are the result of earlier removal of longleaf pine and recolonization by loblolly, shortleaf, or slash pine. Many more are the result of deliberate replacement of longleaf pine by planting either loblolly or slash pine. These sites must be reforested to longleaf by planting as there is no available seed source. The task of restoring the ground cover is directly linked to the previous history of fire in the stand and prior land use (e.g., agricultural uses) (Frost 1993).

Pine Stands without Longleaf in the Overstory and No Recent History of Fire

To restore the form and function of a longleaf pine ecosystem to this forest type requires the reintroduction of fire into the system first. If the period without fire has been long (greater than 10 years) and fuel accumulations are high (greater than 8000 kg/ha), some type of fuel treatment may be necessary to avoid damage to the existing stand (Haywood et al. 2004). These treatments might include mowing and/or herbicides prior to the burn to reduce "laddering" of fuels and get fuels down onto the ground where they can deteriorate prior to the fire. In any event, if fuel buildup is a concern, the first burn should be a cool one. For example, the burn should be in the dormant season on a cool day with moderate humidity (30–50%), fuel moisture, and wind (e.g., 10–20 km/h) to carry the heat out of the canopy quickly. The type of fire can vary, but strip headfires with 20 m or less between strips is a good compromise between a slow backfire and hotter fires (Wade and Lunsford 1988). On-site observations are necessary to prescribe fire correctly. It may be necessary to follow this initial fire with a second dormant season fire within 2 years under similar conditions. The first fire will reduce fine fuels on the forest floor, but only top kill woody vegetation, creating a potential problem for the next fire. A second dormant season fire will "knock down" and begin to consume some of that dead fuel. An assessment of fuel conditions should be made after the second fire to determine if fuel conditions will allow a growing season fire. A thinning might be performed at this point to encourage ground cover plants. If a thinning is performed, planning for the next fire should take into consideration the resulting slash and protect the residual trees (Haywood et al. 2004). After the thinning and the first growing season burn, fire should be continued on at least 3-year and preferably 2-year intervals for at least two cycles. Ideally, at least two successive growing season burns should be performed prior to harvest of the overstory in preparation for reforestation because hardwood brush rootstocks are persistent and will occupy growing space with overstory removal unless otherwise killed.

The regeneration harvest can take two forms. The first approach is to clearcut the existing stand and, after site preparation, replant with longleaf seedlings. If burning has been effective, site preparation can be minimal, consisting of reducing or removing the woody debris to allow the planters access to the site. Spot herbicide treatments might be desirable to control persistent woody clumps, but broadcast chemicals should not be necessary. Optimal growth of the seedlings will require control of the herbaceous vegetation with chemicals, but care should be taken to protect desirable native understory species.

A second technique involves retaining 5–8 m^2/ha BA of the existing pine stand intact as a fuel source for future fires and to maintain the appearance of a forest while the new stand is developing. Harvesting should be done so as to create scattered gaps throughout the stand. Underplanting with containerized seedlings can be accomplished in the newly created gaps. Growth and survival of the new stand will likely be affected negatively (McGuire et al. 2001; Palik et al. 2003). Competition by regeneration from seed from the residual trees poses a greater problem for the longleaf pine seedlings. Fire will be necessary to control that regeneration while the longleaf pine becomes established. Eventually, the remnant pine overstory should be removed in a series of steps, creating or enlarging existing gaps for underplanting. Gap size is important. If gaps sizes are greater than 0.1 ha, it is difficult to get fire to carry into the centers to control invading pines or hardwoods. Smaller gaps allow few resources for the new seedlings and make survival and growth problematic (Palik et al. 2003).

In all cases, fire must be used regularly to maintain the stand and to encourage desirable native understory plants. The new seedlings can be safely burned within a year of planting (Wahlenberg 1946). When the seedlings are just emerging from the grass stage (0.2–1.5 m tall), they are vulnerable to fire and should be burned very carefully or not at all to avoid damage. Once the trees are 1.5 m tall or taller, they can be burned fairly safely unless they are candling. Once they are 2 m tall, they can be burned safely in any season. One recommended technique employs narrow strip headfires for the early burns (Mobley et al. 1978). Later fires can take many forms, but growing season fires should be part of the mix. If the understory does not recover, it may be because the site was farmed prior to the previous stand, not uncommon in the Southeast.

Pine Stands without Longleaf in the Overstory and with a Recent Fire History

The task of restoration is made much easier in this situation. At least two growing season fires with a 2-year interval should be performed prior to harvesting the overstory pines. The next steps can be similar or identical to the scenarios described above, beginning with the harvest of the overstory. As above, the harvest can take either the form of a clearcut or a thinning to create gaps in which to plant longleaf seedlings.

Situations in Which Natural Regeneration Is an Option

Pine Stands with BA > 4.5 m^2/ha Longleaf Pine in the Overstory

Some natural pine stands in longleaf pine's range contain a significant longleaf pine component in the overstory. Some of these stands offer the option of natural regeneration if there is an adequate longleaf pine seed source and that seed source is well distributed. Longleaf pine seeds are the heaviest of the southern pine seeds and do not usually disperse farther from the parent tree than the tree's height, typically 30 m or less (Boyer and Peterson 1983). Consequently, it is recommended that no fewer than 50 seed trees/ha greater than 40 cm in diameter at breast height (DBH) be present for adequate coverage of the area with seed (Barnett and Baker 1991). The only way to determine if this is true is a good ground inventory or "cruise." Plot tallies should be kept separately because an average "trees per acre" figure is of limited use if the trees are not well distributed.

If there is an evenly distributed and adequate longleaf pine seed source available, then it might be possible to regenerate the stand with longleaf pine naturally via some form of the shelterwood method. In the classic shelterwood method used in longleaf pine systems, a preparatory cut is made to obtain a BA of approximately 14 m^2/ha to allow the crowns of potential seed trees to expand, therefore producing more cones (Boyer and White 1990). When the crowns have expanded into the canopy gaps, a second cut is made to reduce the BA to approximately 7 m^2/ha, favoring good cone producers well distributed across the site. Fire is used throughout this process to control competing vegetation, particularly woody shrubs, and to encourage a desirable herbaceous understory.

Once the desired overstory stocking level is obtained, the cone crop must be monitored in anticipation of a good seed year. Longleaf pine produces good seed years sporadically, on the average of every 6 years across most of its range, although it may vary from one every 3 years to one every 30 years (Croker and Boyer 1975; Boyer and White 1990; Boyer 1999). Prescribed fire is necessary while waiting on a good seed crop to maintain control of the understory. Because longleaf has a "2-year" cone, inventory of the potential seed crop can begin in the year previous to seed fall. Binocular counts of female flowers the year prior are an index of the coming seed crop. In the spring and summer of the next year, binocular surveys of tree crowns can be conducted to estimate the number of maturing cones (Croker 1971). When the seed trees have an average of 30 cones/tree or 2500 cones/ha, a burn should be performed to prepare a seedbed. This burn should take place in late winter or early spring (March–April) to control vegetation and reduce the litter layer, allowing the seed to fall on mineral soil. Sufficient time should be allowed for litter and vegetation regrowth to cover the soil and conceal the seed from seed-eating birds and small mammals until it can germinate. Longleaf pine seed typically falls in October and November and germinates as soon as it encounters adequate moisture (Boyer and White 1990). If seeds fall on heavy litter, they may germinate but suffer from May droughts or be killed in the first fire because their roots are not in mineral soil (Boyer and White 1990).

Classic shelterwoods are even-aged regeneration systems, and the seed trees are removed when an adequately stocked new stand is achieved. The seed trees may be retained, at the owner's discretion, to create a two-aged stand. There is evidence that the retention of as little as 2.3 m^2 BA/ha can significantly retard growth of seedlings, however (Boyer 1993). If the stand is a mixed species pine stand with longleaf as a minor component, it may be necessary to use fire or other means to control seedlings of the other species while the shelterwood process is underway. It will probably be necessary to leave some trees of other pine species in the first shelterwood cut to provide fuel and shade to control invading woody brush. In the second cut, however, it is unwise to leave any loblolly or other pines to provide seed to compete with the longleaf pine seedlings. Because these species are annual and prolific producers of widely dispersed seed, they are excellent colonizers of unoccupied sites. It is preferable to leave gaps in the stand than to retain loblolly seed trees. Once loblolly seedlings attain 2 m or more in height, typically in 3 or 4 years, they are tolerant of fire and are difficult to control that way.

There is always the option to clearcut and plant with longleaf pine seedlings after establishing a burning regime and controlling the woody understory. This is a more reliable and quicker way to reestablish the longleaf pine component in these stands, but sacrifices the appearance of the forest for the short term and creates an even-aged stand. Mature forest obligate species such as red-cockaded woodpeckers will be eliminated by this practice. In some cases the management goal is to maintain the native genetic population. In such situations, choices include natural regeneration or collecting seed from the site for use in artificial regeneration. The new stand should still be managed with fire as described below, with caution advised when the seedlings are most vulnerable.

Stands with No Recent History of Fire

Fire must be a part of the restoration process, but must be reintroduced very carefully if it has been long excluded. Excessive fuel buildups can result in damaging or lethal fires, even in mature stands (Haywood et al. 2004; Outcalt and Wade 2004). In stands with a pine overstory other than longleaf, forest floor fuel depths may not be as great, because loblolly and shortleaf pine needles are shorter and decay more rapidly than do longleaf pine needles. Still, if fuel accumulations are very deep, feeder roots may have grown up into the litter and can be damaged or killed by a hot fire. In addition, litter buildups composed of bark, cones, and needles surrounding the base of the boles of standing timber can smolder for hours and even days, damaging the cambium where the bark is thin. In addition, long-term fire exclusion usually results in a woody mid- and understory. These fuels can feed hot fires, especially if needles drape on vines and shrubs and "ladder" flames up into the canopy.

The first fire in a long unburned pine stand should be as cool as possible, burning on a cold, breezy day with moderate humidity (30–60%) and relatively high fuel moisture. In extreme cases, fuel treatments such as mowing or raking around the boles of overstory trees might be employed. If longleaf pine is to be favored in these stands, special attention should be paid to the bases of existing longleaf pine. Raking or wetting the fuel around the trees might lessen the danger of damage from fire if scale of operation allows. Strip headfires move the heat through the stand relatively quickly without allowing intensity to build to dangerous levels (Wade and Lunsford 1988). The second fire might also be a dormant season strip headfire to further reduce fuel loads, encourage root growth into deeper soils, and consume or topple woody shrubs top-killed by the first fire. The interval between the first and second fire should be no more than 3 years. Subsequent fires should take place at 2- or 3-year intervals, depending on fuel conditions. The third or fourth fire should be a growing season fire, ideally in late March through early May. From this point on, at least every third burn should be a growing season burn, extending the window into early summer. As fuel loads are reduced, hotter fires can be safely used.

The understory should begin to respond following the first growing season fire and continue to develop with successive burns. If this is not the case, the process might be "jump-started" by planting or seeding desired species.

Stands with a Recent History of Fire

Pine stands with a recent fire history may be restored in the same manner, but it is much more likely that the woody understory and accumulated forest floor fuel will not require special treatment. Burning should begin with a growing season fire in late March or April and follow with both growing season and dormant season burns until the stand is regenerated. At that point, fire should be continued as often as needed to control fuels and to encourage restoration of the desired understory.

Stands that Are Predominately Longleaf Pine

There are several options for regeneration and maintenance of stands dominated by longleaf pine. Regeneration of these stands can be by artificial or natural means and can take place all at once, in several stages, or continuously. Even-aged stands may be maintained in an even-aged condition by regenerating by clearcutting and replanting or by use of the shelterwood method. The existing stand can usually be retained for a relatively long time (200 years or more) given the long life span of longleaf pine (Matoon 1922; Wahlenberg 1946).

If the objective is an uneven- or all-aged stand or forest, there are several techniques available. Creating gaps in the canopy of an even-aged mature longleaf pine forest allows longleaf pine regeneration to occur when seed crops are available. If the gaps are created prior to a seed crop, fire must necessarily be used to keep those gaps from becoming occupied with woody brush or hardwoods like turkey oak (*Quercus laevis*) or water oak (*Quercus nigra*). Another approach includes creating gaps after seed fall opening holes in the canopy after new seedlings have established themselves and giving them space to grow into. Optimum gap size is a subject of some discussion among managers and researchers, but there are some generally accepted trade-offs. Smaller gaps decrease seedling survival and growth rate (McGuire et al. 2001). The removal of single trees throughout a longleaf pine stand can work with longleaf pine management systems if the stand is fairly open in nature (less than 7 m^2/ha) and growth rates are not particularly important. Small group selection harvests of 0.1 to 0.2 ha are adequate for longleaf pine regeneration (McGuire et al. 2001; Palik et al. 2003). Typically, subsequent cuts are used to expand these openings in anticipation of new seedlings and perpetuating the range of age classes of an all-aged forest. "Domes" of young longleaf pine saplings and seedlings are created, with the oldest and tallest saplings in the center of the gaps and the height and age decreasing with distance from the center. Eventually, as the openings are enlarged, they merge into each other and a forest with groups of trees representing many seed years results. This stand structure is considered typical of much of the presettlement longleaf pine forest (Matoon 1922; Wahlenberg 1946). Patch size undoubtedly varied in presettlement forests, because natural forces of varying intensity created them. Lightning strikes, fires, tornados, and insects likely caused small gaps in the forest, while hurricanes and catastrophic fires had the potential to flatten large areas. It is postulated that constant frequent fire relegated slash and loblolly pines and hardwoods to wetter, less fire-prone areas, leaving the field, so to speak, to longleaf pine (Schwarz 1907; Chapman 1932). Recurring fires probably kept invading species at bay, while the longleaf pine gradually reoccupied large areas from seed and seedlings in place and residual seed trees that survived the catastrophic event (Heyward 1939).

Longleaf Stands with No Recent Fire History

Longleaf pine forests are more likely than other pine forests to accumulate high levels of forest floor litter. Longleaf needles are larger and more decay resistant than those of other pines and can build up litter depths of 25 cm or more (Hermann 2001). Fire must be reintroduced into these situations very cautiously. Fuel treatments such as raking around existing trees, wetting areas immediately around existing trees, and mowing or otherwise removing standing fuels are suggested techniques to avoid or reduce mortality of the overstory. Longleaf pine is fire-tolerant, not fire-proof. Feeder roots frequently grow into the litter layer in long unburned stands and can be severely damaged by even moderately hot surface fires. Because it is common for fuel to accumulate around the base of large longleaf pine in fire-excluded situations, fires can smolder in these piles of bark, foliage, and dead wood for long periods, damaging the cambium layer of the bole at a location where the insulating bark is often thin to begin with and killing or damaging feeder roots, making them more vulnerable to invading diseases (Kush et al. 1999b; Hanula et al. 2002). High mortality levels in mature longleaf pine stands are not unusual when fire is reintroduced in long unburned stands. If scale permits, extraordinary efforts are warranted to prevent this mortality. If it does not, the first fires should be conducted on cool or cold days, with moderate humidity (40% or greater) and moderate to high fuel moisture levels. Wind speeds should be moderate (e.g., 8–12 km/h) and steady. Strip headfires are recommended, with strips 30 m or less apart (Wade and Lunsford 1988). This allows the fires to move fairly rapidly through the fuel without building excessive intensity and without igniting the smoldering fires so dangerous to longleaf pine survival. As the fuel is gradually reduced with successive fires, more latitude with fire intensity and season is allowed. Eventually, when most feeder roots are in mineral soil and the fuel load is reduced, growing season fires may be introduced into the fire regime to encourage the herbaceous understory typical of longleaf pine forests. When regenerating longleaf pine stands naturally, it is important to remember that longleaf pine seed should fall on mineral soil or very light litter to achieve best germination and survival and fires should be timed accordingly. It is equally important to remember that longleaf pine seedlings are extremely vulnerable to fire in the first year after germination and mortality is high on seedlings with root collar diameters less than 0.6 cm (Boyer 1979). Once seedlings start height growth, fires should be applied cautiously or withheld when a majority of the seedlings are between 15 cm and 1.5 m in height, as this size class is vulnerable to fire, particularly when the terminal bud is in the "candle" stage (Grelen 1982).

Longleaf Stands with a Recent History of Fire

The key to managing longleaf pine stands with a recent fire history is to maintain the fire regime and decide on the method of regeneration. If the understory is still occupied by woody shrubs, a switch to growing season burns will encourage the proliferation of grasses and forbs typically indicative of restored longleaf pine forests (Lewis and Harshbarger 1976; Brockway and Lewis 1997; Glitzenstein et al. 1998; Haywood et al. 2001; Kush 2001). The choice of regeneration method will shape the stand's structure for the future. Both even- and uneven-aged conditions are natural in longleaf forests (Croker and Boyer 1975; Boyer 1979, 1999; Dennington and Farrar 1983; Boyer and White 1990; Barnett and Baker 1991; Farrar 1996). By maintaining fire in this system, the common characteristic is a fire-maintained plant and animal community that can for the foreseeable future be maintained.

References

Barnard, E.L., Dixon, M.N., and Ash, E.C. 1995. Benomyl root dip and scalping improves performance of longleaf pine on pest-infested agricultural croplands. *Tree Planters Notes* 46:93–96.

Barnett, J.P., and Baker, J.B. 1991. Regeneration methods. In *Forest Regeneration Manual*, eds. M.L. Duryea and P. M. Dougherty, pp. 35–50. Norwell, MA: Kluwer Academic Publishers.

Barnett, J.P., Lauer, D.K., and Brissette, J.C. 1990. Regenerating longleaf pine with artificial methods. In *Proceedings of the Symposium on the Management of Longleaf Pine, April 4–6, 1989, Long Beach, MS*, pp. 72–93. General Technical Report SO-75. USDA Forest Service, Southern Forest Experiment Station.

Barnett, J.P., Hainds, M.J., and Hernandez, G.A. 2002. Interim guidelines for growing longleaf seedlings in containers. General Technical Report SRS-60. Southern Research Station, USDA Forest Service.

Bengston, G., DuPre, J., Twomey, W., and Hooper, R. 1993. Longleaf ecosystem restoration in the wake of Hurricane Hugo. In *Proceedings of the Tall Timbers Fire Ecology Conference, No. 18*, ed. S. M. Hermann, pp. 339–347. Tall Timbers Research Station, Tallahassee, FL.

Boyer, W.D. 1979. Regenerating the natural longleaf pine forest. *J For* 77:572–575.

Boyer, W.D. 1987. Volume growth loss: A hidden cost of periodic prescribed burning in longleaf pine? *South J Appl For* 11(3):154–157.

Boyer, W.D. 1993. Long-term development of regeneration under longleaf pine seedtree and shelterwood stands. *J For* 17:10-15.

Boyer, W.D. 1999. Longleaf pine: Natural regeneration and management. *Alabama's Treasured Forests* 18:7–9.

Boyer, W.D., and Peterson, D.W. 1983. Longleaf pine. In *Silvicultural Systems for the Major Forest Types of the United States*. Agricultural Handbook 445. pp. 153–156. Washington, DC: US Department of Agriculture.

Boyer, W.D., and White, J.B. 1990. Natural regeneration of longleaf pine. In *Proceedings of the Symposium on the Management of Longleaf Pine, April 4–6, 1989, Long Beach, MS*, pp. 94–113. General Technical Report SO-75. USDA Forest Service, Southern Forest Experiment Station.

Boyette, W.G. 1996. 1995 survey of longleaf pine restoration efforts in the South. North Carolina Division of Forest Resources, Raleigh.

Brissette, J.C., Barnett, J.P., and Landis, T.D. 1991. Container seedlings. In *Forest Regeneration Manual*, eds. M.L. Duryea and P.M. Dougherty, pp. 117–141. Norwell, MA: Kluwer Academic Publishers.

Brockway, D.G., and Lewis, C.E. 1997. Long-term effects of dormant-season prescribed fire on plant community diversity, structure and productivity in a longleaf pine wiregrass ecosystem. *For Ecol Manage* 96:167–183.

Brockway, D.G., and Outcalt, K. W. 2000. Restoring longleaf pine wiregrass ecosystems: Hexazinone application enhances effects of prescribed fire. *For Ecol Manage* 137:121–138.

Cantrell, R.L. 1985. A guide to silvicultural herbicide use in the southern United States. School of Forestry and Wildlife Sciences, Auburn University, AL.

Chapman, H.H. 1932. Is the longleaf type climax? *Ecology* 8(4):328–334.

Croker, T.C., Jr., 1971. Binocular counts of longleaf pine strobili. SO-127. New Orleans, LA. USDA Forest Service, Southern Forest Experiment Station.

Croker, T.C., Jr., and Boyer, W.D. 1975. Regenerating longleaf pine naturally. SO-105. New Orleans, LA. USDA Forest Service, Southern Forest Experiment Station.

Dennington, R. W. 1990. Regenerating longleaf pine with the shelterwood method. R8-MB47. Atlanta, GA. USDA Forest Service, Southern Region.

Dennington, R. W., and Farrar, R. M., Jr. 1983. Longleaf pine management. R8-FR3. Atlanta, GA. USDA Forest Service, Southern Region.

Earley, L.S., ed. 1997. *A Working Forest: A Landowner's Guide for Growing Longleaf Pine in the Carolina Sandhills*. Sandhills Area Land Trust, Southern Pines, NC.

Earley, L.S., ed. 2002. *Managing the Forest and the Trees: A Private Landowner's Guide to Conservation Management of Longleaf Pine*. Zebulon, NC: Theo Davis Sons Inc.

Farrar, R.M., Jr. 1996. Fundamentals of uneven-management in southern pines. Miscellaneous Publication No. 9. Tall Timbers Research Station, Tallahassee, FL.

Farrar, R.M., Jr., and Boyer, W.D. 1991. Managing longleaf pine under the selection system—Promises and problems. In *Proceedings of the Sixth Biennial Southern Silvicultural Research Conference, October 30–November 1, 1990*, Memphis, TN, pp. 357–368. General Technical Report 70, Asheville, NC: USDA Forest Service Southeastern Forest Experiment Station.

Franklin, R.M. 1997. Stewardship of longleaf pine forests: A guide for landowners. Longleaf Alliance Report No. 2. The Longleaf Alliance, Solon Dixon Forestry Education Center, Andalusia, AL.

Frost, C.C. 1990. Natural diversity and status of longleaf pine communities. In *Forestry in the 1990s—A Changing Environment*, eds. G. Youngblood and D.L. Frederick, pp. 26–35. Regional Technical Conference. Pinehurst, NC: Society of American Foresters.

Frost, C.C. 1993. Four centuries of changing landscape patterns in the longleaf pine ecosystem. In *Proceedings of the Tall Timbers Fire Ecology Conference, No. 18*, ed. S. M. Hermann, pp. 17–43. Tall Timbers Research Station, Tallahassee, FL.

Glitzenstein, J.S., Streng, D.R., Wade, D.D., and Platt, W.J. 1998. Maintaining and restoring species diversity in longleaf pine groundcover: Effects of fire regimes and seed/seedling introductions. In *Proceedings of the Longleaf Pine Ecosystem Restoration Symposium*, Society for Ecological Restoration 9th Annual International Conference, ed. J.S. Kush, pp. 72–75. Longleaf Alliance Report No. 3. The Longleaf Alliance, Andalusia, AL.

Grelen, H.E. 1982. May burns benefit survival and growth of longleaf pine seedlings. In *Proceedings of the Second Biennial Southern Silvicultural Research Conference*, ed. E. P. Jones, pp. 70–73. General Technical Report SE-24. Asheville, NC: USDA, Forest Service, Southeastern Forest Experiment Station.

Hainds, M.J. 1999. Successfully planting longleaf pine. *Alabama's Treasured Forests* 18:10–11.

Hainds, M.J. 2001. Scalping aids survival of longleaf. *Alabama's Treasured Forests* 20(3):24–27.

Hainds, M.J. 2003a. Determining the correct planting depth for container-grown longleaf pine seedlings. In *Proceedings 4th Biennial Regional Longleaf Alliance Conference*, pp. 66–68. November 17–20, 2002, Southern Pines, NC.

Hainds, M.J. 2003b. Establishing longleaf seedlings on agricultural fields and pastures. In *Proceedings 4h Biennial Regional Longleaf Alliance Conference*, pp. 69–73. November 17–20, 2002, Southern Pines, NC.

Hainds, M.J., and Barnett, J.P. 2003. Container-grown longleaf pine quality. In *Proceedings 4th Biennial Regional Longleaf Alliance Conference*, pp. 63–65. November 17–20, 2002, Southern Pines, NC.

Hanula, J.L., Meeker, J.R., Miller, D.R., and Barnard, E.L. 2002. Association of wildfire with tree health and numbers of pine bark beetles, reproduction weevils and the associates in Florida. *For Ecol Manage* 170:233–247.

Haywood, J.D., Harris, F. L., and Grelen, H. E. 2001. Vegetation response to 37 years of seasonal burning on a Louisiana longleaf pine site. *South J Appl For* 25(3):122–130.

Haywood, J.D., Bauman, T.A., Goyer, R.A., and Harris, F.L. 2004. Restoring upland forests to longleaf pine: Initial effects on fuel load, fire danger, forest vegetation, and beetle populations. In *Proceedings of the 12th Biennial Southern Silvicultral Research Conference*, ed. K. F. Connor, pp. 299–303. General Technical Report SRS-71. Asheville, NC: USDA Forest Service, Southern Research Station.

Hermann, S.M. 2001. A brief overview of fire and season of burn in native longleaf pine ecosystems. In *Forest for Our Future. Restoration and Management of Longleaf Pine Ecosystems: Silvicultural, Ecological,*

Social, Political, and Economic Challenges. Proceedings of the 3rd Longleaf Alliance Regional Conference. October 16–18, 2000, comp. J.S. Kush, pp. 53–57. Alexandria, LA. Longleaf Alliance Report No. 5.

Heyward, F. 1939. The relation of fire to stand composition of longleaf pine forests. *Ecology* 20(2):287–304.

Johnson, R. 1999. Restoring the longleaf pine forest ecosystem. *Alabama's Treasured Forests* 18(4):18–19.

King, S. E., Grace, S. L., and Hedman, C.W. 1997. The importance of site history on ground cover species of longleaf pine systems: Implications for restoration. In *Longleaf Pine: A Regional Perspective of Challenges and Opportunities*, Proceedings of the First Longleaf Alliance Conference, September 17–20, 1996, comp. J.S. Kush, pp. 102–103. Mobile, AL. Longleaf Alliance Report No. 1.

Kush, J.S., comp. 1997. *Longleaf Pine: A Regional Perspective of Challenges and Opportunities*, Proceedings of the First Longleaf Alliance Conference, September 17–20, 1996, Mobile, AL. Longleaf Alliance Report No. 1.

Kush, J.S., comp. 1999. *Longleaf Pine: A Forward Look*, Proceedings of the 2nd Longleaf Alliance Conference, November 17–19, 1998, Charleston, SC. Longleaf Alliance Report No. 4.

Kush, J.S., comp. 2001. *Forest for Our Future. Restoration and Management of Longleaf Pine Ecosystems: Silvicultural, Ecological, Social, Political, and Economic Challenges*, Proceedings of the 3rd Longleaf Alliance Regional Conference, October 16–18, 2000, Alexandria, LA. Longleaf Alliance Report No. 5.

Kush, J.S., comp. 2003. *Longleaf Pine: A Southern Legacy Rising from the Ashes*, Proceedings of the 4th Longleaf Alliance Regional Conference, November 17–20, 2002, Southern Pines, NC. Longleaf Alliance Report No. 6.

Kush, J.S., Meldahl, R. S., and Boyer, W.D. 1999a. Understory plant community response after 23 years of hardwood control treatments in natural longleaf pine (*Pinus palustris*) forests. *Can J For Res* 29:1047–1054.

Kush, J.S., Varner, J.M., and Meldahl, R. S. 1999b. Slow down, don't burn too fast... Got to make that old-growth last. In *Longleaf Pine: A Forward Look*, Proceedings of the 2nd Longleaf Alliance Conference, comp. J. S. Kush, pp. 109–111. Longleaf Alliance Report No. 4.

Landers, J.L., Van Lear, D.H., and Boyer, W.D. 1995. The longleaf forests of the Southeast: Requiem or renaissance? *J For* 93(11):39–44.

Lewis, C.E., and Harshbarger, T.J. 1976. Shrubs and herbaceous vegetation after 20 years of prescribed burning in the South Carolina coastal plain. *J Range* Manage 29(1):13.

Long, A.J. 1991. Proper planting improves performance. In *Forest Regeneration Manual*, eds. M. L. Duryea and P. M. Dougherty, pp. 303–321. Norwell, MA: Kluwer Academic Publishers.

Lowery, R. F., and Gjerstad, D.H. 1991. Chemical and mechanical site preparation. In *Forest Regeneration Manual*, eds. M.L. Duryea and P. M. Dougherty Norwell, MA: Kluwer Academic Publishers.

Mann, W.F., Jr. 1970. Direct-seeding longleaf pine. USDA Forest Service Research Paper SO-57. Southern Experiment Station.

Matoon, W.R. 1922. Longleaf pine. US Department of Agriculture Bulletin No. 1061. Washington, DC.

McGuire, J.P., Mitchell, R. J., Moser, E.B., Pecot, S. D., Gjerstad, D. H., and Hedman, C.W. 2001. Gaps in a gappy forest: Plant resources, longleaf pine regeneration, and understory response to tree removal in longleaf pine savannas. *Can J For Res* 31:765–778.

Mexal, J.G., and South, D.B. 1991. Bareroot seedling culture. In *Forest Regeneration Manual*, eds. M.L. Duryea and P. M. Dougherty, pp. 89–116. Norwell, MA: Kluwer Academic Publishers.

Minogue, P. J., Cantrell, R. L., and Griswold, H. C. 1991. Vegetation management after plantation establishment. In *Forest Regeneration Manual*, eds. M. L. Duryea and P. M. Dougherty. Norwell, MA: Kluwer Academic Publishers.

Mitchell, R.J., Neel, W.L., Hiers, J.K., Cole, F. T., and Atkinson, J.B., Jr. 2000. A model management plan for conservation easements in longleaf pine-dominated landscapes. Joseph W. Jones Ecological Research Center, Newton, GA.

Mobley, H.E., Jackson, R.S., Balmer, W.E., Ruziska, W.E., and Hough, W.A. 1978. A guide for prescribed fire in southern forests. USDA Forest Service, State and Private Forestry, Atlanta.

Outcalt, K.W., and Wade, D.D. 2004. Fuels management reduces tree mortality from wildfires in southeastern United States. *South J Appl For* 28(1):28–34.

Palik, B., Mitchell, R.J., Houseal, G., and Pederson, N. 1997. Effects of canopy structure on resource availability and seedling responses in a longleaf pine ecosystem. *Can J For Res* 27:1458–1464.

Palik, B., Mitchell, R.J., Pecot, S., Battaglia, M., and Pu, M. 2003. Spatial distribution of overstory retention influences resources and growth of longleaf pine seedlings. *Ecol Appl* 13:674–686.

Robbins, L.E., and Myers, R.M. 1992. Seasonal effects of prescribed burning in Florida: A review. Tall Timbers Research, Inc. Miscellaneous Publication No. 8. Tallahassee, FL.

Schwarz, G.F. 1907. *The Longleaf Pine in Virgin Forest—A Silvical Study*. New York: John Wiley & Sons.

Swezy, D.M., and Agee, J.K. 1991. Prescribed-fire effects on fine-root and tree mortality in old-growth ponderosa pine. *Can J For Res* 21(5):626–634.

Wade, D.D., and Lunsford, J.D. 1988. A guide for prescribed fire in southern forests. Technical Publication TP11. Atlanta, GA. USDA Forest Service, Southern Region.

Wahlenberg, W.G. 1946. *Longleaf Pine: Its Use, Ecology, Regeneration, Protection, Growth, and Management*. Washington, DC: Charles Lathrop Pack Foundation.

Wakeley, P.C. 1953. Planting the southern pine. Government Printing Office, Washington, DC. USDA Agricultural Monograph No. 18.

Walker, J. 1998. Ground layer vegetation in longleaf pine landscapes: An overview for restoration and management. In *Proceedings of the Longleaf Pine Ecosystem Restoration Symposium*, Society for Ecological Restoration 9th Annual International Conference, Longleaf Alliance Report No. 3, comp. J.S. Kush, pp. 2–13. The Longleaf Alliance, Andalusia, AL.

Williston, H.L., and Balmer, W.E. 1971. Direct seeding of southern pines—A regeneration alternative. Forest Management Bulletin. US Forest Service, State and Private Forestry, Atlanta, GA.

Chapter 10

Restoring the Ground Layer of Longleaf Pine Ecosystems

Joan L. Walker and Andrea M. Silletti

The longleaf pine ecosystem includes some of the most species-rich plant communities outside of the tropics, and most of that diversity resides in the ground layer vegetation. In addition to harboring many locally endemic and otherwise rare plant species (Peet this volume) and enhancing habitat for the resident fauna (Costa and DeLotelle this volume), the ground layer vegetation produces fine fuel needed to carry low-intensity surface fires that perpetuate the ecosystem. Ecosystem restoration requires the restoration of both the ground layer plant community and the pine canopy.

Ground layer restoration in longleaf pine communities is an area of active investigation, through both adaptive management projects and formal research. However, there is no restoration manual for the longleaf pine community. Instead, restoration practitioners develop their action plans based on an ecological reference model and project goals, and achieve their objectives using conventional natural resources management and horticultural methods. Given the natural heterogeneity of the longleaf pine ecosystem at multiple scales and the differences imposed by a varied land use history, one could argue that there never will be a manual to adequately describe or prescribe restoration protocols for all situations; however, we believe there are general patterns within this ecological system that can guide restoration protocol development. In addition, restorationists have practical experience that is not documented in the peer-reviewed literature at this time but nevertheless converges on some necessary steps for successful restoration. In this chapter we summarize the general lessons learned from ongoing restoration efforts, and discuss ecological aspects of the ground layer vegetation that guide us in extrapolating this information to other sites. Our purpose is to share information so that we might advance the restoration of the ground cover in longleaf pine communities by minimizing avoidable mistakes and by identifying critical information needs.

In the first section we provide an overview of the ground cover vegetation in the longleaf pine ecosystem. We then describe extant conditions in longleaf pine sites often targeted for restoration, including the ways they differ from reference conditions and their derivation from widespread historical land uses. The next two sections summarize lessons learned from research and restoration projects that emphasize (1) altering canopy structure to favor ground-layer restoration or (2) starting new populations of ground cover species. We close

---

Joan L. Walker and Andrea M. Silletti • USDA Forest Service, Southern Research Station, Clemson, South Carolina 29634.

with an assessment of information needed to advance ground cover restoration in the longleaf pine ecosystem.

Our review of existing restoration projects shows that they are being conducted on a relatively narrow subset of possible longleaf pine habitats. Significant projects we know of are concentrated in the Atlantic and Gulf Coastal Plains; mesic savannas and flatwoods, loamy upland sites, and xeric to subxeric sites in the Fall-line Sandhills are represented. We note the absence of projects in the middle Atlantic Coastal Plain, where few examples of remnant vegetation remain, in the mountain longleaf pine communities of the Blue Ridge and Cumberland Plateau, and in the longleaf pine–bluestem communities. Information presented in this chapter draws on projects conducted by researchers and restoration practitioners in Florida, South Carolina, and Georgia. Although many of the same issues exist in the underrepresented areas, unique species, habitats, spatial contexts, and historical land uses are bound to generate some restoration challenges that we have not addressed.

Ground Layer Vegetation in Longleaf Pine Landscapes: An Overview

Throughout the range of longleaf pine the general picture of a frequently burned high-quality natural area shows a predominantly herbaceous ground layer dominated by grasses with a diverse mixture of forbs. Woody species, if present, are short and inconspicuous. Most of the common species are sun-loving perennials with an ability to resprout after fire. Fire typically stimulates the flowering and seed production by many characteristic species, and there are apt to be species flowering at most any time during the growing season.

Grasses, legumes, and composites are the most common plant families in these burned habitats (Harcombe et al. 1993; Peet and Allard 1993; Drew et al. 1998). Other common families include the sedges, especially the beak rushes (*Rhynchospora* spp.), and lilies. More unusual plants include orchids and carnivorous species, often associated with wet, nutrient-poor sites.

Locally, ground layer composition and structure vary with fire frequency and soil conditions, typically characterized by soil texture and interpreted as variation in soil moisture status (Peet and Allard 1993). Overall, frequently burned sites have more species at small spatial scales than sites where fire has been eliminated; and intermediate-to-wet sites support more species than very dry sites (Peet this volume).

In spite of these general patterns, there is considerable compositional variation from one part of the region to another. Most herbaceous species have geographic ranges that are much smaller than that of longleaf pine. Species that have more or less restricted geographic ranges are known as endemic species, and the longleaf pine ecosystem has many subregional and local endemic species (Estill and Cruzan 2001; LeBlond 2001; Sorrie and Weakley 2001). As the geographic limit of a species' range is reached, it drops out of the local flora but may be replaced by an ecologically similar species. This results in changing species composition in the ground layer. Species with very small geographic distributions (narrow endemics) are prone to extinction and include some of the ground layer species that are federally listed as Endangered or Threatened (Walker 1998). Because there are important differences among sites, describing the ecologically appropriate composition for restoration must be done carefully.

Reference Models and Goals for Ground Layer

Restoration practitioners use "reference models" to describe the ecological potential for a project site. A reference model is a description of the restoration site as it may have looked and functioned in the past, before negative changes had occurred. Ideally the description answers questions about composition, structure, and function. Intact remnant patches of the target ecosystem, such as nearby "natural areas," are sometimes identified as "reference sites." They are selected to match the restoration project site with respect to geography and physical environment and are believed to represent the historic or contemporary potential conditions. Nearby environmentally similar sites are the best reference sites, but the condition of any proposed reference site may have been influenced by unknown stochastic events in the past (such as extreme weather or disturbance events) so that its condition may not represent the project site potential.

Besides using reference sites, restoration ecologists use other kinds of "reference information" to develop reference models (Table 1; White and Walker 1997). Ideally the project planner would conduct a site assessment to gather current information about the site to be restored, including a description of the underlying environmental conditions, and search for accurate historical information about the same site. Desirable historical information includes historical photographs, written descriptions, plant and animal species lists, frequency of burn in the area and under what conditions, and/or reports of significant disturbances or past land uses.

Practitioners sometimes use historical or contemporary information from other sites, or from less specific geographic areas. Though such information may be useful, it is important to remember that information about places is generally place- and time-specific. The more distant or more general the information source, the less likely it will accurately represent a specific project site and the less useful it will be for setting feasible objectives (Table 1). Egan and Howell (2001) recommend that restorationists use a combination of site analysis (same time, same place), historic information from the project site (different time, same place), and information from contemporary reference sites (same time, different place).

TABLE 1. Examples of reference information that can be used to develop a reference model.[a]

Time/space	Restoration project site	Different site or general location
Contemporary (Observed directly; change can be monitored)	(Site analysis) • Physical environment. Examples: soil type, fertility, hydrology, topographic position, etc. • Biotic environment. Examples: (1) canopy—composition, age, size class distribution, origin; (2) other vegetation—composition, species abundance, presence of exotic species • Disturbance evidence. Examples: fire scars, plow lines	(*Reference site* if it matches the conditions and geography of site to be restored) • Same as for site analysis data The nearer the site to the project site, the more likely that information can be used directly in reference model • General location information Examples: county species lists, herbarium records
Historical (Snapshot of past; cannot observe change or know effect of stochastic events in the past)	*HIGH VALUE for reference model* (Site history) • Photographs, with dates • Written descriptions of physical and biotic conditions, past land uses or disturbances (sources: deeds, explorers' accounts, diaries and letters of previous owners) *HIGH VALUE*	*HIGH-MODERATE VALUE* (Historical information from different sites) • Similar to site history data • Fire scar data in general landscape or region • Regional land use history • Pollen data (prehistorical) *LOW VALUE*

[a] Reference information can be classified into four categories based on geographic source of data (the site to be restored versus a different or general location) and whether the data represent current (contemporary) or historical conditions. Modified from White and Walker 1997 with permission from Blackwell.

A variety of restoration goals may be compatible with the site's ecological potential, for example, establishing fine fuels to facilitate fire use in timber management, improving habitat for a bobwhite quail, or creating an aesthetically desirable setting. Goals like these examples may be viewed as restoring a subset of the composition, structure, and function that historically characterized a site, in contrast to the ambitious goal of restoring the entire complement of species and their interactions. In practice, such "partial restoration" goals are far more likely to be achieved than "complete restoration of biodiversity" (Lockwood and Pimm 1999). The feasibility of project goals must be examined in light of the ecological capability of the restoration site including limitations imposed by spatial scale and context (White 1996), as well as the resources that are available to do the work and to maintain it (White and Walker 1997; Ehrenfeld 2000).

Recent Land Uses and Legacies: Starting Points for Restoration

Altered fire regimes, plantation establishment, and conversion of forest lands to agriculture have resulted in loss of the ground cover diversity throughout the longleaf pine range (Wear and Greis 2002). Ongoing and completed restoration projects that we reviewed all fall into one of these recent land use history classes. Of these classes pine plantations are the most heterogeneous. They differ in canopy species (primarily loblolly or slash pine), in age, and in methods of establishment. Also, pre-planting site histories vary, most significantly in whether they have a history of modern cultivation in contrast to continuously forested or lightly cultivated. Finally, they may have experienced a period of fire suppression. As a result of diverse management histories, plantations may resemble both agricultural sites and sites with altered fire regimes.

Although these land use groups do not represent mutually exclusive conditions, we think it useful to consider them because they differ from reference conditions in different ways and thus represent somewhat distinct challenges to restoration (Fig. 1). These largely anthropogenic disturbances have generated very different starting points in terms of physical conditions and especially of biotic legacies, which are the remnant components of the undisturbed longleaf ecosystem. Effort needed to restore a site will vary inversely with the amount of biotic legacy remaining (Fig. 2).

Altered Fire Regimes

The historical fire regime has been described as one of frequent, low-intensity surface fires. The extent of individual fire events and return interval are likely to have varied with topography, thus among different parts of the longleaf range (Frost 1998). It is assumed that most acres of longleaf pine habitat ignited by lightning burned during the early to mid-growing season, but Native American ignitions spanned the seasons (Robbins and Myers 1992). Over the last 60 years land managers have reduced the spatial extent of fires, shifted the predominant season to winter burning (which may be associated with lower intensity fires owing to high fuel moisture and low air temperatures), and reduced fire frequency or eliminated fires altogether. These practices are associated with increased densities and expanded distributions of woody species (Platt et al. 1991; Robbins and Myers 1992; Waldrop et al. 1992; Streng et al. 1993; Glitzenstein et al. 1995; Gilliam and Platt 1999; Drewa et al. 2002) and decreased abundance of herbs (Walker and Peet 1983; Peet and Allard 1993).

The hardwood component increases with fire exclusion or reduced fire frequency, but the specific composition varies both geographically and with site conditions within a landscape (Gilliam et al. 1993; Harcombe et al. 1993; Liu et al. 1997; Gilliam and Platt 1999; Varner et al. 2003). The losses in the ground cover after long periods without fire (more than two decades) are so profound that studies of extant old growth with significant fire exclusion focus almost exclusively on the woody species component (e.g., Gilliam and Christensen 1986; Gilliam et al. 1993; Gilliam

10. Restoring the Ground Layer

FIGURE 1. Different starting conditions for longleaf pine community restoration: (a) Xeric site where fire was excluded for about 30 years; (b) slash pine plantation on a mesic site once occupied by longleaf pine. Note increased turkey oak with fire exclusion, in contrast to absence of hardwoods in the plantation where hardwoods were controlled. Both have abundant pine straw or leaf litter, but lack a diverse herb layer.

and Platt 1999; Varner et al. 2000). Descriptions of these sites note low richness and sparse cover of herbaceous species, as in the Gilliam et al. (1993) description of the Boyd Tract, an old-growth remnant in the North Carolina sandhills: "sparse and relatively species-poor, typical of pine forest herb layers under chronic no-fire conditions." The most abundant species

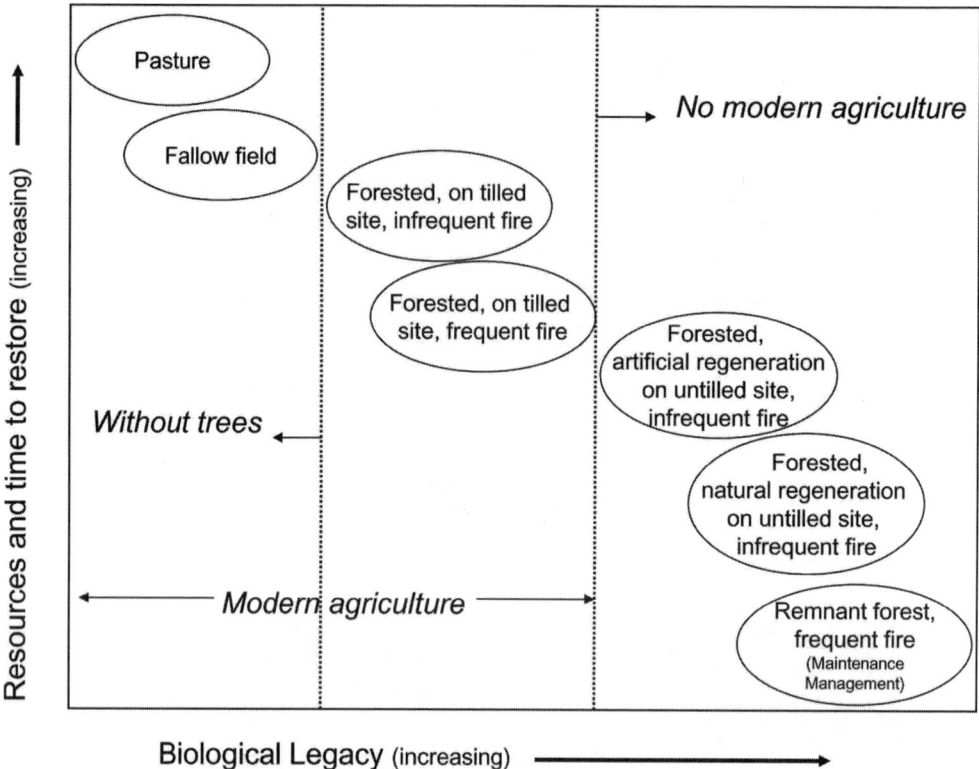

FIGURE 2. Relationship between time and resources needed for restoration and the abundance of remnant biota (biological legacy) on the site to be restored. Ovals indicate the relative position of common starting conditions as defined by land use history. Sites subjected to modern agricultural methods (repeated machine tilling) are distinct from sites that have remained in forest or escaped intense agriculture. Nonforest sites require planting trees, and all previously tilled sites are likely to require species additions. Infrequently burned forests may need hardwood removal or other canopy manipulations. The abundance of remnant biota on forested sites varies inversely with the intensity of any site preparation used for pine regeneration.

in the ground layer were seedlings of pine and oaks and other hardwoods (such as flowering dogwood, mockernut hickory, and black gum). After 45 years of fire exclusion in an old-growth site in Escambia County, Alabama, a single herb, *Acalypha virginica*, was present prior to restoration fire treatments (Varner et al. 2000). Results of fire frequency experiments, generally sampled through several fires over a decade or less, indicate that increased woody species dominance can occur in relatively short periods of time (Mehlman 1992; Beckage and Stout 2000; Glitzenstein et al. 2003).

Reduced fire frequency leads to scale-dependent decreases in herbaceous species richness: decreases in richness are most evident at small spatial scales (less than or equal to 1.0 m² plots) and less evident at larger scales (greater than 600 m²). This pattern has been shown both in mesic productive savannas (Walker and Peet 1983; Glitzenstein et al. 2003) and in xeric sites (Walker 1998). Species retained at larger scales may provide on-site seed sources for restoring the ground layer via natural dispersal and establishment. How long species will persist is not known; if retention is short-lived, opportunities for restoring residual populations with fire alone will diminish with time.

Rates of species loss associated with reduced fire frequency vary with plant groups. Among the species most likely to be lost or significantly reduced in mesic to wet longleaf pine savannas are the dominant rhizomatous and bunch grasses (Walker and Peet 1983; Glitzenstein et al. 2003), species present as basal rosettes (many composites), many sedges and other small monocots, and insectivorous species. Mehlman (1992) found predictable patterns of species loss, or conversely of persistence, with fire exclusion in drier upland longleaf pine forests. He identified more species persisting in high-frequency than in low-frequency, burned or unburned sites. While all three groups of stands included ruderal and "climax" longleaf pine associates, legumes were significantly associated only with burned groups, and only woody species were significantly associated with fire exclusion. Based on work in prairies, which resemble longleaf pine ground cover vegetation in their dominance by bunch grasses, and abundance of composites and legumes, Leach and Givnish (1996) confirmed higher than expected losses of nitrogen-fixing legumes and small-seeded species, and that losses were more pronounced on more productive sites. Regionally rare species were lost at a rate more than twice the average for all species. We expect similar patterns for longleaf pine savannas. We do not know whether species associated with high fire frequencies persist as a result of fire-associated vigor, or if persistence requires continued seedling establishment in fire-created "safe sites" (*sensu* Harper 1977).

The most obvious and consistent effects of season of burning are effects on woody stems, with growing season fires being more effective at reducing both size and density compared to dormant season burning (Robbins and Myers 1992 and references therein). Drewa et al. (2002) reported shrubs sprouted more vigorously following dormant season fires than growing season fires, and further, that repeated growing season fires reduced the size but not the number of stems of established shrubs. Others have suggested that trees once established are not easily removed by fire (Rebertus et al. 1989; Platt et al. 1991; Glitzenstein et al. 1995), even after 30 years of annual or biennial summer burning (Waldrop et al. 1992).

Despite reports that growing season burning stimulates flowering and increases synchrony of flowering (Platt et al. 1988; Streng et al. 1993; Brewer and Platt 1994) and that dominant grasses flower only infrequently without growing season burning (Robbins and Myers 1992), there have been no convincing changes in abundance and composition directly related to season of burning (Streng et al. 1993; Brockway and Lewis 1997). But, fire season may affect the herbaceous community indirectly through changes in canopy structure and consequent changes in the environment (reduced resource availability, particularly light and water) for herbs (Harrington and Edwards 1999).

In summary, fire frequency is likely to be more important than season of burning for maintaining longleaf pine communities; sites with a history of frequent dormant season fire are likely to have retained most of the species found in a nearby reference site. If a substantial period of fire exclusion has occurred, however, it is not likely that simply restoring the fire regime will restore the herb layer.

Plantation Establishment and Management

The condition of the ground layer in plantations varies with land use history, site preparation methods, stand age, treatments applied during stand development, and site type. A history of machine tilling is likely to have the greatest adverse impacts on the ground layer. For example, Hedman et al. (2000) showed that as little as 2 years of cultivation prior to site preparation and planting resulted in sites with reduced species richness and cover compared to reference longleaf pine stands in southern Georgia. Effects of other factors (stand age, canopy composition, site preparation, recent fire history) were small by comparison.

Site preparation effects vary with the type of method and with intensity. Mechanical methods include treatments such as drum chopping (crushing with a roller) and leaving the vegetation, or shearing (cutting at the ground level) and piling the organic materials. Intensity may be increased, for example by weighting the chopper and rolling the site more than once, for more complete competition control. Mechanical methods generally reduce the cover of both woody and herbaceous species directly, but many individuals survive to resprout. All mechanical methods expose some mineral soils, and may inadvertently redistribute topsoil with nutrients and organic matter.

Herbicides (chemical methods) can be used to target specific plant groups, and are effectively used to reduce woody species with presumably low impacts on herbaceous species. Treatments can be broadcast or applied to individual stems, further increasing the specificity of applications. Chemical treatments do not disturb soil, reducing opportunities for weeds to become established. Both mechanical and chemical methods are often coupled with fire, which maintains pine dominance and benefits herbaceous species, especially grasses.

Regardless of the method, the general objective of site preparation is to favor the establishment and early growth of planted pines, and often results in increased herbaceous cover in the first few years. Residual herbaceous species may increase, and exposed mineral soil provides a seed bed for both weedy and desirable climax species (species of undisturbed longleaf communities). Because many climax species do not have adaptations for rapid dispersal and establishment, the short-lived flush of herbaceous growth following site preparation is relatively enriched with ruderal species and depleted of climax herbs (Swindel et al. 1986; Glitzenstein 1993).

Herbaceous cover and richness tend to decline with plantation age, and without intervening fire treatments herb cover can decline significantly by age six (Zutter and Miller 1998) while woody species increase. Additional silvicultural treatments may reduce herbaceous species, for example the reduction of pineland threeawn with fertilization (White 1977); or invigorate the herb layer, as by thinning (Grelen and Enghardt 1973; Means 1997; Harrington and Edwards 1999). Over a range of site conditions, high herbaceous cover has been associated with frequent burning and inversely related to basal area of the canopy trees, suggesting the potential benefits of burning and thinning in plantations to restore the ground layer (Hedman et al. 2000). The ground layer in plantations (on untilled sites) often includes a surprisingly large number of the species found in remnant forests on similar site types (Hedman et al. 2000; Smith et al. 2001). A study of xeric communities in the Carolina Sandhills National Wildlife Refuge showed that 40-year-old plantations had nearly identical species-presence lists as remnant longleaf pine stands (Walker and van Eerden unpublished data). Importantly, however, *Aristida stricta*, the dominant bunch grass, and *Gaylussacia dumosa*, the second most abundant ground cover species in remnant sites, were essentially eliminated from plantations. Compared to the xeric sandhills sites, more productive sites are likely to lose herb species and cover relatively quickly, and to provide fertile ground for weedy species (Smith et al. 2001).

In summary, except on old agricultural sites, plantations are likely to support many characteristic native species, and thus might be restored without species additions. However, the loss of grasses and dominant ground cover species may limit the effectiveness of fire as a restoration tool. We do not know how long populations of nonweedy species can persist in longleaf pine plantations; thinning may increase their longevity, but early postestablishment stands are most likely to have residual populations to "rescue." In this way they resemble sites where reduced fire frequency was the primary disturbance. In other plantations desirable trees may be present, but characteristic herbs missing, indicating the need for modifying canopy structure and adding characteristic herbaceous species. The need to establish or augment herbaceous species populations makes them similar to conditions in agricultural sites.

Agricultural Sites

Established pastures and recently cultivated fields present a predictable condition nearly devoid of any vestiges of the former ground cover. An agricultural history can have long-lasting impacts on the vegetation, soils, and microorganisms of other forest types (Foster et al. 2003), but there is little information available about the long-term effects of agricultural use on longleaf pine systems and how those effects impact restoration efforts. Much of what is known comes from project reports and proceedings of regional restoration conferences, mainly from sites where the primary agricultural use was management of improved pasture. These sites are characterized by the complete absence of longleaf pine, a depauperate to nonexistent native species pool, and domination by cultivated grasses such as bahia (*Paspalum notatum*) or early successional old-field weeds. Consequently, most of the time, energy, and money invested in restoring such sites go to eliminating the nonnative and undesirable vegetation and establishing new populations of longleaf pine ground cover species.

Restoration Tasks

Based on conditions described in the previous sections, we recognize some general conditions requiring management action (Table 2). The canopy may be dominated by species other than longleaf pine, and at altered densities (often greater than the reference model) and distributions (more regular in pine plantations than reference conditions). Similarly, the composition and structure of the ground layer vegetation may be changed, including species richness and relative abundance of species. Common or rare species may be absent; native ruderal species and exotics species may

TABLE 2. Summary of ecosystem changes that may have to be treated to achieve restoration goals. The necessity to treat any of these depends on specified project goals.

Condition	Alternatives: general treatments	Can fire alone fix it?	Does it matter for biodiversity conservation and sustainability?
Woody species			
Canopy/subcanopy density higher than reference	Prescribed fire Mechanical treatments Chemical treatments	Depends: yes, if fuels adequate and long time; no, if no fuels and short time constraint	Yes
Canopy composition altered	Regenerate to longleaf pine; approaches may vary from clearcutting through progressive thinning and patch regeneration	No	Yes, for long-term success
Herbs			
Absence or scarcity of dominant or common species	Prescribed fire Direct seeding Plant plugs	No; possibly increase sparse population, but probably take a long time	Yes
Absence or scarcity of rare species	Prescribed fire Direct seeding Plant plugs	Probably not	Sometimes; depends on objectives for site
Presence of persistent weeds (natives)	Prescribed fire Hand/mechanical "weeding" Chemical treatments	Depends on identity of weeds and available time; some are really difficult	Sometimes; depends on nature of "weed," but probably not
Presence of exotic species	Prescribed fire Hand/mechanical "weeding" Chemical treatments	Depends, but for noted species in longleaf pine systems, they seem to tolerate fire	Sometimes; depends on nature of "weed," but probably should be remedied
Site conditions			
Hydrology altered	Restore drainage	No	Yes
Soil structure/fertility altered	Burn off excess organic capital	Maybe	Yes

be present. Finally, the physical condition of the site itself may have been altered, for example by attempts to drain wet sites or to alter microsites for pine seedling establishment. All changes in structure and composition may be combined with the elimination of fire.

It is not necessary to remedy all altered conditions in order to achieve some restoration goals. The addition of rare species may be optional, and should be pursued if restoring the biodiversity in a nature preserve is the goal; or if their establishment supports a rare species conservation goal, and then only if postrestoration management can maintain high-quality habitat conditions (Gordon 1994). The presence of native weeds may go untreated if they are not aggressively displacing desired climax species (D'Antonio and Meyerson 2002), but we favor eliminating exotic species where possible as ecological ramifications may not be known at this time.

Except for restoring altered physical conditions, such as site hydrology changed by drainage ditches, prescribed burning is essential for rectifying and maintaining the restored condition of nearly all aspects of community change (Table 2). Additional possible treatments can be grouped based on two overall goals: (1) restore canopy structure to a condition that promotes ground layer establishment and vigor and (2) start new populations or augment existing populations of native ground layer species. This chapter focuses on actions required to restore the ground layer vegetation, and does not address establishing the longleaf pine. However, approaches to restoring the longleaf component can affect ground layer development, and successful restoration will require coordinating longleaf and ground layer restoration.

Changing Canopy Structure to Enhance Ground Layer Vegetation

Aside from establishing longleaf pine in the canopy, the most common objective for canopy management is to reduce a hardwood and shrub component. Fire, mechanical methods (e.g., felling, girdling, drum chopping, shearing), and chemical (herbicide) methods are used. All of these treatments, alone or in combinations, both reduce and control trees and shrubs, and affect existing ground layer vegetation to varying degrees.

Fire is promoted as a "natural" method with positive benefits and most restoration practitioners and researchers concur that in some cases fire alone may restore canopy structure and favorable conditions for ground cover recovery, but that restoration will require multiple fires over relatively long times (Robbins and Myers 1992; Waldrop et al. 1992; Glitzenstein 1993; Streng et al. 1993). Factors that limit the capacity for fire to restore structure include a lack of fine fuels, presence of ladder fuels that may promote crown damage, and thick duff that resists burning when moist and kills trees when it does burn. The problem is particularly vexing when the site contains desirable old trees with heavy duff accumulations at their bases (Varner et al. 2000; Kush et al. 2004). Compared to the presumed historical fire regime, a prescribed fire regime for restoration may differ in seasonality, frequency, and intensity, or be combined with pretreatments to protect desirable biological legacies like remnant old trees or trees with red-cockaded woodpecker cavities (see Box 10.1). An initial series of cool, winter burns may effectively reduce duff accumulations and protect old trees in fire-suppressed stands (Kush et al. 2004).

The effectiveness of fire for changing canopy structure can be enhanced by combining burning with mechanical and/or chemical treatments (Tanner et al. 1988; Outcalt 1994; Walker and van Eerden 1998; Provencher et al. 2001; Kush et al. 2004; Walker et al. 2004). In general, mechanical or chemical treatments reduce hardwoods and subsequent fires consume fuels and maintain hardwoods as basal sprouts.

The best-documented study of the effects of treatments to restore canopy structure in a longleaf pine ecosystem was conducted in a large-scale experiment in the sandhill communities at Eglin Air Force Base in the

Florida Panhandle (Provencher et al. 2001). The study plots were second-growth longleaf pine stands that had a long history of fire suppression. Consequently, the midstory had become dominated by a variety of oak species and there was a very sparse understory, with mats of hardwood leaf litter interspersed with bare ground. The goal of the study was to use management techniques commonly employed to reduce hardwood midstory in longleaf pine systems, and to document their effects on both target (oak) and nontarget (herbaceous) species, thereby testing the hypothesis that restoration of the habitat structure would be sufficient to return the understory vegetation to reference conditions. Three hardwood reduction treatments were used: spring burning, application of a hexazinone herbicide, and mechanical felling/girdling of hardwoods. The herbicide and felling/girdling treatments were followed by fuel reduction burns in the year after treatment. These plots were compared to both untreated controls and reference plots to determine the effect of oak reduction treatments on herbaceous species richness and densities. They predicted an increase in plant species richness and in densities of herbaceous plants that qualitatively tracked increasing levels of hardwood reduction.

All treatments were effective for reducing oaks; however, 4 years after treatment, results suggested that fire alone was the least effective hardwood reduction method, but yielded the greatest increases in ground cover species richness and densities. Brockway and Outcalt (2000) similarly reported that hexazinone followed by burning more effectively reduced turkey oak and shrub density and enhanced ground layer recovery in an oak-dominated site than did burning alone. Provencher et al. (2001) concluded that if gradual reduction of hardwood densities were acceptable, fire was an effective and cost-efficient means of hardwood control and would benefit the ground layer vegetation. Chemical and mechanical control were recognized to be viable options for situations where hardwood reduction is needed immediately, but they cost up to eight times more than burning, showed less understory improvement, and were judged to be effective for restoring community structure in the long term only if followed by prescribed burning.

Among mechanical options, drum chopping has been widely applied to reduce hardwoods, especially small oaks (*Quercus laevis, Q. incana, Q. margaretta*), and other woody species such as saw palmetto (*Serenoa repens*) (Tanner et al. 1988). Light chopping treatments (single passes with an empty drum chopper) have short-lived impacts on dominant bunch grasses in dry sites (*Aristida beyrichiana* in Florida flatwoods [Grelen 1959] and *A. stricta* in South Carolina sandhills [Walker and van Eerden 1998; Walker et al. 2004]), but more intensive treatments are likely to substantially reduce or eliminate the dominant wiregrass (Grelen 1962; Moore 1974).

Mechanical treatment effects vary with season of treatment, and when applied in the same year as prescribed fire. In a field experiment at the Carolina Sandhill National Wildlife Refuge prescribed fire (growing season, dormant season, and no-burn treatments) was combined with light drum chopping (growing season, dormant season, and no-chop treatments), and treatment effects on wiregrass and turkey oak recovery were evaluated. Wiregrass recovered to pretreatment levels within two growing seasons in all growing season burn treatments, regardless of chopping treatment; dormant season burn plots recovered more slowly, the impact exacerbated by extreme drought (Walker et al. 2004; Fig. 3). The plots showing the slowest recovery were chopped after dormant season burning, perhaps as a result of chopping without a layer of pine straw to protect wiregrass roots and crowns. Other tools that cut or crush understory trees without intense ground disturbance are available and we would expect similar effects. Chemical applications can effectively reduce hardwoods within pine stands on a range of site conditions. Hexazinone formulations are often used on oaks and shrubs typically found on mesic to dry sites; treatments for mesic sites are more variable including glyphosate, triclopyr, and 2,4-D formulations (Litt et al. 2001).

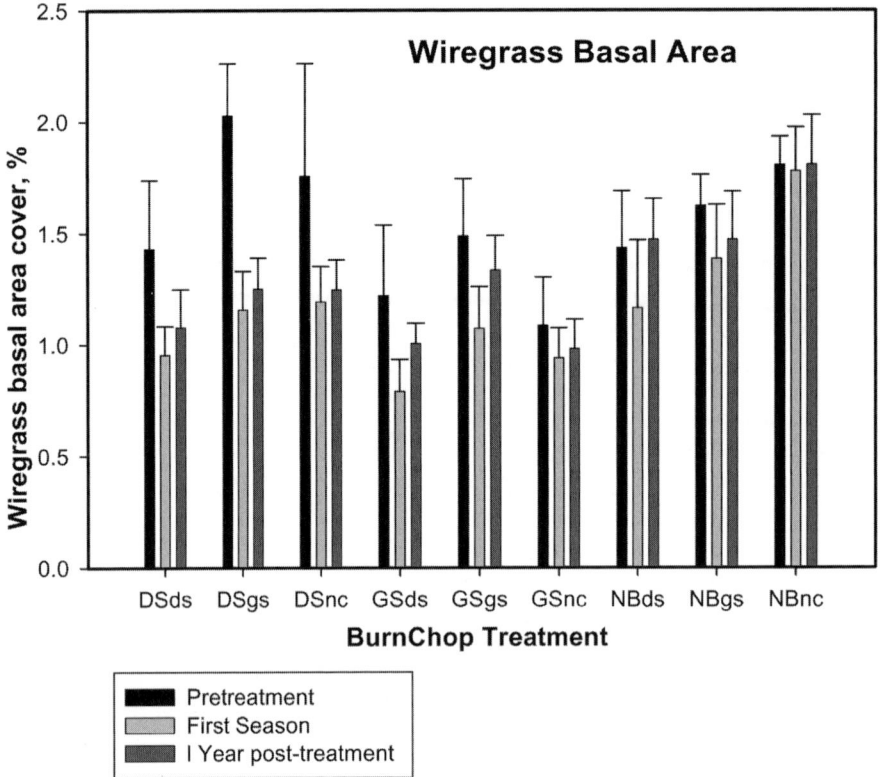

FIGURE 3. Basal area of wiregrass (*Aristida stricta*) before and following experimental burning and drum chopping at the Carolina Sandhills National Wildlife Reserve, South Carolina. Study sites were longleaf pine saw timber or pine-scrub oak stands on Alpin soils. Three burn and three chopping (single pass, unweighted drum) treatments were defined by season: dormant season (DS), growing season (GS), and not treated (that is, not burned [NB] or not chopped [NC]). Treatments: Uppercase letters indicate burn season; lowercase indicate chopping season. Measurements shown are pretreatment, after one growing season, and after two growing seasons. NB treatments did not change through time; GS burns recovered to pretreatment levels; DS burns had not recovered to pretreatment levels in two seasons (see Walker et al. 2004).

Although target woody species are successfully controlled with chemical applications, the effects on nontarget species of the ground layer are not fully understood. A recent review of herbicide application studies to determine effects on native, nontarget species (Litt et al. 2001 and references therein) found that extremely variable treatments and their application mostly in plantations rather than natural stands make it difficult to evaluate herbicide use for restoration of ground layer vegetation. The effects of herbicide use on the ground layer varied with habitat and with the specific herbicide, or combination of herbicides, used. Hexazinone herbicides were most widely used as they are especially effective against common midstory hardwood species such as oaks, sweetgum, and sumacs.

In flatwoods habitats, all herbicides used reduced species richness and cover of herbaceous and woody ground layer plants. Decreases ranged from 5.1% in herbaceous species richness compared to control using a form of hexazinone, to 71.8% in total species richness using a mixture of three herbicides. The only study to document vegetative cover reported declines in the cover of both herbaceous (27.2%) and woody (58.6%) vegetation after hexazinone application.

In sandhills habitats, the effects were more varied. Woody cover and density were reduced 10.3% to 55.9% by hexazinone application, but woody biomass increased 105.3% with use of 2,4-D. Graminoid density and cover generally increased with hexazinone application, but the response of nongraminoid herbaceous plant cover ranged from a 49.8% increase to a 33% decrease with hexazinone use. The response of ground layer species richness to herbicide use in sandhills was dependent on the type of herbicide used and the application rate. Total species richness increased anywhere from 6.4% to 81%, while herbaceous species richness was shown to increase 55.2% in the one study for which it was reported.

Among pine plantation studies, treatments were especially variable with respect to both herbicide or herbicide combination used and application rate, making it difficult to describe any general response patterns. Herbicide treatments generally increased herbaceous species richness (10.5% to 84.7%), and reduced woody species richness to varying degrees, never exceeding a 17.2% decline. Graminoid species richness increased by 30.8% in one study, decreased by 16.7% in another, and showed intermediate responses in the rest. Herbicides tended to decrease total species richness in plantations, with declines as much as 11.2% reported. However, triclopyr and glyphosate herbicide application increased total species richness by 10.9 and 8.7%, respectively. In plantation studies competition control was the motivation for herbicide use; however, differences between control of desirable ground layer plants and weeds are not reported. Thus, it is impossible to determine the contribution of each group to the reported changes in species richness.

Very few studies have reported herbicide effects on individual species of concern, such as wiregrass or other herbaceous species. With respect to wiregrass, study results range from increases of up to 7480% to decreases of 142%, depending on the specific chemical and application rate used. Even studies using the same herbicide have shown a wide range of responses, and responses within the same study can vary widely from year to year. At this point in time, it is difficult, if not impossible, to draw any general conclusion about the effects of herbicide use on wiregrass, simply because the little information that we do have shows no consistent pattern. The same is true for any individual species of interest, although it is important to note that responses of three threatened species in Mississippi to herbicide use were all negative.

In summary, herbicide use is generally successful in reducing mid- and understory hardwoods in all systems; however, there remain significant unknowns about impacts on native species, especially those in the herbaceous ground layer. Additional well-designed studies in natural systems would provide much needed information and would be advised before large-scale application of herbicides is used as a restoration method, especially in those areas with remnant native plant populations.

Most studies of restructuring the canopy to restore diverse ground layer vegetation have been conducted in comparatively dry sites. It seems likely that higher productivity sites might differ in the following ways: need for more frequent retreatment; need for more intensive initial treatments relative to the period of fire exclusion; greater challenge from exotic species; more profound species losses because mesic sites will develop more intense competitive species interactions, mesic sites have more species to lose, and mesic sites have more relatively rare species which are prone to elimination (Leach and Givnish 1996). As a result of more species losses, we suggest that mesic sites are more likely to require species reintroductions.

Plantation Restoration Strategies

Several research groups have proposed strategies for restoring plantations. Based on the results from an experiment to study the relative importance of light and soil water availability and litterfall in limiting herbaceous density and cover in longleaf pine plantations, Harrington and Edwards (1999) concluded that conventional silvicultural treatments, including thinning, herbicides, and prescribed fire, can be used to create a stand structure that favors herb layer diversity and production. Although the study was conducted in sites where the herbaceous composition differed substantially from an undisturbed ground layer, they suggest similar conditions would favor climax species. That assertion may be generally true, but we suspect that once established, climax species would respond more slowly than old-field species to changing conditions. Harrington and Edwards (1999) caution that prescribed fire (every 2 to 3 years) and thinning must be applied periodically to maintain an open structure that favors herbs. Alternatively, managing the stand for herbaceous layer diversity and productivity could begin with site preparation, using herbicides (for control of woody species) and prescribed fire to benefit pine seedlings and existing herbs. Missing herbaceous species may be added at this time to restore the composition of the herbaceous community. (See related information in the section on Direct Seeding.)

Kirkman and Mitchell (2002) describe a progressive thinning strategy to restore even-aged slash pine plantations to multi-aged longleaf pine communities with diverse ground cover. The work is being conducted in an upland site in southwest Georgia and in a flatwoods site in the Florida Panhandle. Gradual thinning leaves pines producing litter to support surface fires and providing for future timber harvest, while creating conditions that favor herbaceous species. This research group is investigating the effects of gap size, and of different methods (including herbicide and mowing treatments) to control woody species growth and promote a grassy ground layer. Treatments also include seeding *Aristida beyrichiana* in experimental gaps. Researchers will monitor herb layer development with burning to determine the need for additional species introductions. No results are published yet, but this approach has promise for restoring longleaf plantations as well as for restoring longleaf to sites currently planted in other pines. Methods such as these could be especially helpful to land managers with responsibilities to recover red-cockaded woodpecker populations challenged to restore both the canopy and ground layer of existing plantations (U.S. Fish and Wildlife Service 2001). A gradual conversion and restoration would retain the value of the plantation as woodpecker foraging habitat, while developing future habitat.

Altering Species Composition

Species composition may differ from reference conditions by the absence of common or rare species, or the presence of weedy natives or exotic species. In this section we focus on starting new populations of native species, although exotic species effects can pose significant problems for ecological restoration in the longleaf pine system as elsewhere (Hobbs and Humphries 1995; D'Antonio and Meyerson 2002). For example, cogon grass (*Imperata cylindrica*), a well-studied exotic rhizomatous grass invasive in the southern part of the longleaf pine range, can displace native grasses and alter the fire regime because it burns more intensively than the native bunch grasses (Lippincott 2000; Jose et al. 2002). By changing the fire regime, cogon grass has the potential to alter patterns of species recruitment and persistence through time.

Exotic species clearly challenge restoration efforts, but an exhaustive treatment of the topic is beyond the scope of this chapter. For more details we refer the reader to the rapidly expanding literature on exotic species, including excellent sources of information for identification and control of exotic plant species (e.g., Miller 2003). Especially helpful are websites devoted to management of non-native plants, including a site with information from U.S. federal and state governments (http://www.invasivespecies.gov), and from the Nature Conservancy's Invasive Species Initiative (http://tncweeds.ucdavis.edu).

TABLE 3. Comparison of direct seeding and outplanting options.

Direct seeding	Outplanting
Advantages	*Advantages*
Economical ($3K/acre)	Can choose individual target species
Simultaneously introduce multiple species known to co-occur	No need to disrupt existing conditions
	No special planting tools
Can create custom seed mixes by varying timing and methods of collection	Can be done on slopes where seeding equipment cannot be used safely
Can be done concurrent with site preparation	Conducive to volunteer assistants
	Good success for many species
Can be done in winter, before frost, when competition for labor is lower	Reduced susceptibility to drought at early stages
	Appropriate for rare species
Mechanized approaches can treat large areas	Few seeds are needed to ensure establishment objectives
Genetically diverse seeds can be used so that site conditions "select" most suitable individuals	Stock can be propagated any time when seed is available
	Shorter period of competition control needed in many cases
Disadvantages	*Disadvantages*
Unreliable establishment requires large seed supplies	Expensive (up to $10K/acre)
	Introduce only one species at a time
Not as useful for rare species	Available stock may be limited by the need for hand-collecting seed and size of nursery
Special care needed to create seed mixes	
Seeding rates difficult to determine to ensure outcome	Germination and initial establishment in greenhouse conditions; may favor genotypes less suitable for future establishment in field conditions
Competition control essential	

Options for Starting New Populations

Options for starting new populations include direct seeding, out-planting nursery stock, or transplanting wild stock (Guerrant 1996). In the context of biodiversity conservation the latter approach is generally regarded as a last resort, reserved for rescuing native plants from sites destined for destruction, and will not be addressed further. Both direct seeding and outplanting nursery stock have been used successfully in longleaf pine restoration projects and have advantages and disadvantages (Table 3). Economic considerations give direct seeding a clear advantage over planting plugs. Costs for using plugs, which include seed collection, nursery personnel, site preparation, and planting, can run from $3000 (van Eerden, unpublished report to North Carolina Department of Agriculture) to as high as $12,000/acre (Seamon, in Disney Wilderness Preserve 2000); cost estimates for direct seeding were estimated at $155–650 and $300–400/acre at the same sites, respectively, and include maintenance of the seed collection site, seed collection, site preparation, and seeding. Machine planting options for direct seeding make it possible to treat large areas, and when seed mixes (mixed species, or mixed collections from more than one site) are used, established individuals will represent genotypes that are successful as seedlings in field conditions rather than greenhouse conditions. Outplanting approaches allow for selecting individual species (e.g., rare species), controlling the genetic composition of the new population, and for establishing plant cover quickly, but may be best suited for small areas. Native seed is not available commercially, so seed may have to be supplied to a grower for seedling production by special order.

Key issues associated with seeds include what species to plant; where, how, and when to collect seed; how to clean and store native seed; seed viability, germination requirements, and factors that affect seedling establishment.

Species Selection

Criteria for species selection include: (1) the species' habitat is similar to the restoration site conditions; (2) the restoration site is within the natural distribution range of the target species; (3) the species is needed to meet the project goal. For examples, restoring fine fuel production, restoring the diversity of vascular plants to a site, and restoring habitat for a rare butterfly require different suites of species. In order to maintain a restoration project, burning must be possible, and we recommend fuel production, through the establishment of dominant perennial grasses and retention of onsite fuel sources (e.g., pine straw), as an objective for all restoration projects. If restoration goals do include the introduction of uncommon species, there is as yet no general consensus about whether to add these species at the beginning of the restoration process (Weber 1999), or to wait to add them until after a matrix of dominant native species is established (Packard and Mutel 1997). Gordon (1994) developed a dichotomous key to support or guide management decisions to introduce (or not) a native species. Although this tool highlights issues associated with individual species, especially "at-risk" species, some of them are directly applicable to restoration, such as considering genetic and environmental suitability of the donor site, considering impacts on any remnant populations on the recipient site, and the potential for managing the site after the species introductions.

Seed Sources

Abundant seed production is expected in sites with abundant flowering, often resulting from burning in the current or previous growing season (Platt et al. 1988; Robbins and Myers 1992; Streng et al. 1993). However, viable seed production in native plant populations is a complex process dependent on successful pollination, fertilization, seed development, and seeds escaping predation or destruction by pathogens. In theory, all of these processes could be affected through various mechanisms by season of burning. Independent of recent fire history, year-to-year and site-to-site variation in seed production is typical of natural plant populations (Fenner 1985). Thus, abundant flowering does not necessarily predict abundant seed production.

There are very few direct measurements of the magnitude of seed production in natural populations of longleaf pine associates. In *Pityopsis graminifolia* (Brewer and Platt 1994) and *Aristida stricta* (van Eerden 1997) viable seed production (seeds/plant) following growing season burns was significantly greater than after dormant season fires, while Hiers et al. (2000) reported that effects of burn season on seed production in legumes varied with species. Greater losses of some legume species' seed to predators occurred after winter compared to growing season fires, but that also varied among species (Hiers et al. 2000).

Genetic Considerations

The genetic composition of populations at the seed donor site can affect the success of the new population by providing genotypes suitable for the environmental conditions at the restoration sites. Donor composition can also affect the genetic structure of residual populations in or near the restoration site by introducing new genes and creating novel genotypes via genetic recombination. To minimize potentially adverse consequences, collection sites should be as near as possible and as similar as possible with respect to physical environment to the planting site.

Widespread species, such as some of the dominant grasses, may harbor considerable genetic diversity across their ranges (Hamrick et al. 1991; Millar and Libby 1991). Based on morphological, geographic, and ecological factors, Peet (1993) divided *A. stricta* (*sensu* Radford et al. 1968) into a more northerly species, *A. stricta*, and more southerly taxon, *A. beyrichiana*. *Aristida stricta* and *A. beyrichiana* dominate the ground cover in many longleaf pine communities, and despite disagreement as to the taxonomic status (Walters et al. 1994; Kesler et al. 2003), the bunchgrass is undoubtedly variable within its geographic range. Further, species with wide habitat tolerances within the same landscape may exhibit ecotypic differentiation, as Kindell et al. (1996) demonstrated for *A. beyrichiana*. They reported a differential performance of seedlings from different habitats (xeric sandhills versus mesic flatwoods in north Florida) in common gardens and reciprocal plantings, such that individuals grew better in sites similar to their habitat of origin. Brewer (1995) showed that for the widespread *Pityopsis graminifolia*, individuals from different locations, with potentially different ecologically limiting conditions, responded differently to fire.

Matching environmental conditions of donor and recipient sites may be more important for species that have limited potential for gene flow among populations, including shorter-lived rather than long-lived perennials, animal rather than wind pollinated, and species with no adaptations for widespread dispersal (Hamrick et al. 1991). In such species, populations tend to be genetically distinct (compared to species with ample gene flow among populations), and consequently more finely adapted to local environmental conditions. If no good collection site match is available, Huenneke (1991) suggests that collecting from multiple suitable sites may be advantageous in producing a genetically diverse propagule mix, and thereby increasing the likelihood of including a suitable environmental match to the restoration site.

Removing seed from a donor site may have adverse effects at the donor site, particularly if the persistence and structure of the community rely on frequent establishment from sexual reproduction, or if collected species provide critical food resources for native fauna. Because most species are perennial in this system, and there have been few observations of seedling establishment, effects of periodic seed removal on donor site composition are not expected to be significant. Out of concern for potential adverse effects of collection, collectors often follow informal "rules" such as: take less than 50% of a strong perennial or less than 10% of an annual; take only what you are prepared to handle responsibly; avoid trampling; collect as close to the restoration sites as is practical (Apfelbaum et al. 1997). The Center for Plant Conservation developed collection guidelines for preserving the diversity of rare plant species (Center for Plant Conservation 1991).

Seed Collection and Handling

Timing

Because each species has a specific phenology, there is no one best time to collect seeds. Plants that are ready for harvest have full-sized seeds with seed coats changing color, usually from green to a darker hue, and dry stems (Apfelbaum et al. 1997). Baskin and Baskin (1998) recommend harvesting when the seeds would naturally disperse. Not only are there differences among species, but seeds of the same species can mature both at different times across the range of longleaf pine systems due to differences in climate and topography, and at different times from year to year due to variations in weather. Seed maturation may also be affected by season of burning, but effects likely vary with species. Some guidelines for seed collection have been published, such as those by Pfaff et al. (2002), which list collection dates for several species of grasses and forbs. But, because of the variations listed above, these types of recommendations should be used with caution and paired with direct observation of the maturity of the plants. More is known about the timing of seed collection for wiregrass than for most other species. Although wiregrass seeds seem to ripen in midfall (October) there is an "after-ripening" effect, such that seeds collected later in the fall and into winter have higher germination rates. van Eerden (1997) found that *Aristida stricta* seeds collected in December had higher germination rates than seeds collected from the same North and South Carolina sandhill sites in November. Similar results were found for *A. beyrichiana* collected in Georgia sandhills (Walker and Silletti unpublished data).

Seed Harvest, Cleaning, Storage

Methods for collecting, cleaning, and storing seed for prairie restorations (Apfelbaum et al. 1997; Clinebell 1997) are mostly applicable for seed handling for longleaf pine restoration projects (Glitzenstein et al. 2001). Generally, seed can be collected by hand, which is especially useful for rare or infrequent species or when individual species are needed to enrich an existing site, or mechanically, which is very effective for collecting seed mixtures (Fig. 4). Hand collection methods vary from simply stripping individual seed heads by hand, to collecting entire infructescences with clippers, to collecting small seeds with a hand-held vacuum.

Several types of seed harvesting machines are available, but often prohibitively

FIGURE 4. Bulk seed collection with an ATV mounted seed stripper in a remnant upland site (a) and emptying the collection hopper into a storage bin (b). Seed collections include seed from all species in fruit at the collection time, e.g., common large grasses and composites. (Photo courtesy of Lin Roth.)

expensive. Sharing the cost of equipment may provide a feasible option, but requires cooperation in collection efforts. Green silage cutters harvest and collect all aboveground plant material. The resulting "green" harvest contains seeds as well as other vegetation and must be distributed quickly so as to avoid seed-destroying mildew as the plants decompose. Pull-type and front-end mounted seed strippers harvest seeds plus accessory plant parts (mostly dry) with rotating brushes. Seed stripper types, available in sizes suitable for mounting on four-wheelers and larger, have been used in longleaf pine systems, but smaller models are most convenient for harvesting seed from sites with trees. Pfaff et al. (2002) provide more details on seed harvester equipment and sources.

Seed harvests are processed to varying degrees depending on how long they are to be stored, how pure the seed must be, and such practical considerations as how much space is available for storing (see Apfelbaum et al. 1997 for details; Baskin and Baskin 1998). Except for fleshy fruits, such as blueberries and huckleberries common to the longleaf system, harvests are usually dried before processing further. [Fleshy fruits require special handling to separate seed and pulp. See Phillips (1985) for suggestions.] Experience supports that collecting seeds on low-humidity days and spreading them out of the weather in a warm place provides adequate drying. Collections generally include a variety of other plant parts and are cleaned in several stages (threshing, scalping, final cleaning), depending on the desired final condition. Methods can be simple such as hand sorting, to more complex screening, and milling of various forms. (See Packard and Mutel 1997 for details and references.)

Seeds of many longleaf-associated native species stored indoors in paper or grass seed bags retain viability for at least a year. Glitzenstein et al. (2001) report acceptable germinability for 2 years, but much reduced viability and deformed seedlings after 2 years of storage at room temperatures. Storing seeds in dry unheated areas (e.g., unheated storage shed) will expose seed to temperature variations similar to field conditions, and may be useful for seed collected in the fall and intended for planting the following season (Glitzenstein et al. 2001; Pfaff et al. 2002). However, Pittman and Karrfalt (2000) report that viability of wiregrass seed drops rapidly after 8 months of storage at ambient temperatures, and they used annually collected seed for seedling production.

Factors Affecting Germination and Establishment

Properly collected seeds of many longleaf pine associates readily germinate without elaborate pretreatments. Germination rates across common plant families are similar and highly variable, ranging from zero to greater than 80% in laboratory, greenhouse, or outdoor trays exposed to ambient environmental variations (Pfaff and Gonter 1996; van Eerden 1997; Glitzenstein et al. 2001; Pfaff et al. 2002). Results of a study of 42 species characteristic of Atlantic coastal plain savannas indicate that germination rates within a species vary from site to site and year to year, but are not related to time of burning or time since burning (Glitzenstein et al. 2001). In that study, most trials exceeded 30% germination. Glitzenstein and colleagues compared germination rates in laboratory trials with germination in flats exposed to outdoor conditions and found that for some species, especially fall-seeding composites, field germination exceeded lab trials.

Several treatments have been reported to increase germination rates in some common longleaf pine associates. Cold stratification increases germination in fall-fruiting composites such as *Liatris* spp. (Pfaff and Gonter 1996), and perennial grasses including *Andropogon gerardii, Schizachyrium scoparium, Ctenium aromaticum, Erianthus giganteus, Aristida beyrichiana* (Glitzenstein et al. 2001), and *A. stricta* (van Eerden 1997). A period of after-ripening reportedly benefits germination rates in *Aristida beyrichiana*, with germination increasing for up to 5 months in dry storage (Pittman and Karrfalt 2000). Finally, heat treatments, which can be as simple as pouring boiling water over seeds and allowing them to cool slowly, increase germination in various legume species (Cushwa et al. 1968; Pfaff and Gonter 1996; Baskin and Baskin 1998). Testing germinability is the most reliable basis for calculating seeding rates and for determining timing of harvest; but as a general rule, mature seeds should be planted at the same time they are naturally dispersed (Baskin and Baskin 1998). Delays in planting some species result naturally in induced dormancy (Baskin and Baskin 1998). Consult references in this section for more information about native seed germination and growing native species.

Although seedlings of many native species have been established and grown under greenhouse and nursery conditions (Pfaff and Gonter 1996; Glitzenstein et al. 2001; Dagley et al. 2002; Pfaff et al. 2002) and even commercially produced (Pittman and Karrfalt 2000), there is little information about factors that affect seedling establishment either from naturally dispersed seed in intact longleaf communities or from seed introduced into field conditions. Experimental results suggest that seedling establishment in intact longleaf pine communities is rare, and that general failure of seedling establishment can be attributed to competition from established dominant species (Brewer et al. 1996; van Eerden 1997; Glitzenstein et al. 2001); higher establishment in mesic compared to xeric sites suggests that competition for water may be the specific cause of mortality (van Eerden 1997; Glitzenstein et al. 2001).

In a garden experiment using a variety of species, Glitzenstein et al. (2001) found that each species was most successful in soil and drainage conditions that most closely matched the environments where it grows naturally. Thus, matching species and probably matching seed source habitats for species found on a broad environmental gradient (especially soil moisture in the longleaf pine system) will most surely enhance establishment success.

The presence of litter likely affects germination and establishment (Fowler 1986; Facelli and Pickett 1991) and species-specific responses to experimental litter treatments were observed at the Carolina Sandhills National Wildlife Refuge (Walker unpublished data; Fig. 5). The effects of pine straw litter on the germination and establishment of ten common sandhills species were monitored for one growing season. Germination of common grasses and composites benefited from light litter. Heavy litter seemed to benefit germination among legumes, but overall germination rates were low and at the end of the first season only one seedling of *Baptisia cinerea* remained in a no-litter plot. In summary, available information suggests that planting in an appropriate site with respect to a moisture gradient, with low competition, and low litter loads favors seedling establishment under field conditions. Similar conditions may be achieved via site preparation for direct seeding projects.

Site Preparation and Sowing Treatments

The challenges for controlling competition vary markedly with site history. As a general rule, in previously forested sites where exotic herbaceous species are not dominant in the ground layer, site preparation suitable for planting longleaf pine seedlings will also favor ground cover establishment, as long as a sufficient amount of bare soil is available for plant establishment. Experimental evidence suggests that more complete competition control is likely to benefit seedling establishment. However, we observe acceptable wiregrass establishment (about two clumps per square meter; from about 15.4 kg cleaned seed/ha; Walker and Silletti unpublished data) from broadcast seeding soon after planting trees in sandy sites where piling harvest slash left a mosaic of bare soil, litter, and residual plant cover. Using a cultipacker to press seed into the soil did not increase establishment success on these uneven forest site surfaces. Acceptable establishment was similarly achieved on mesic savanna sites at Fort Stewart, Georgia (Dena Thomson personal communication).

Pasture lands and abandoned agricultural fields present the dual challenges of removing existing vegetation, including nonnative perennial pasture grasses, and reducing the numbers of weed seeds present in the soil

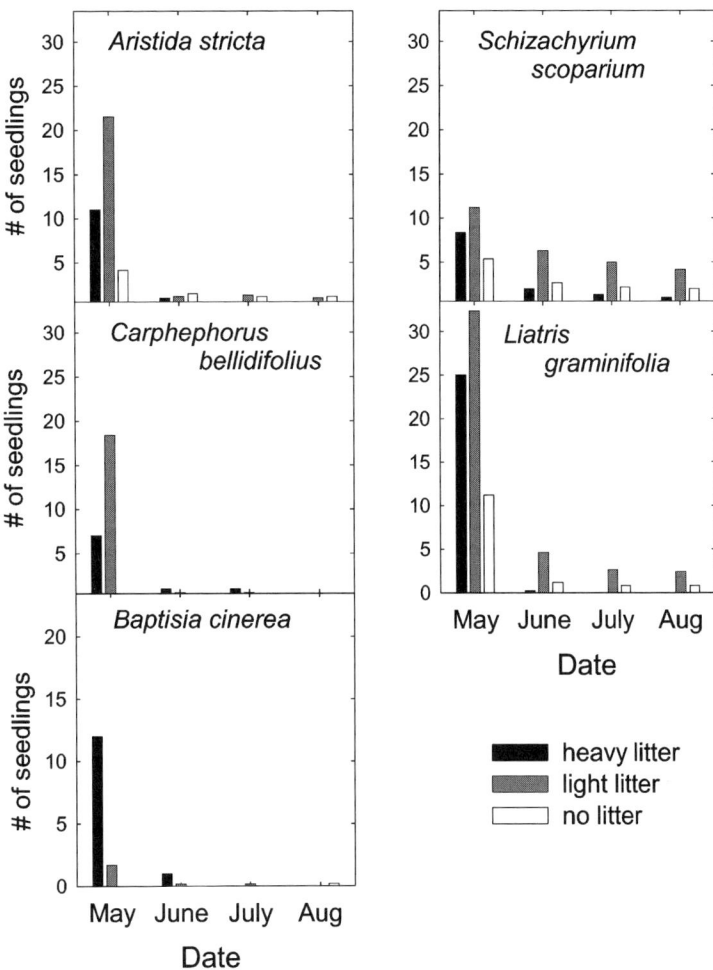

FIGURE 5. Effect of pine straw litter on seedling germination and establishment in selected common sandhills species. Small garden plots at the Carolina Sandhills National Wildlife Refuge were treated with a low litter (comparable to litter deposited in the first year after burning) and heavy litter (twice the low level) application. Bars represent the mean of five replicates per species per treatment. Seedlings were counted in the same plots in four successive months during a single growing season. Number of seeds varied among species, but was constant within a species; thus internal comparisons can be made, but conclusions about species differences cannot be drawn. (Unpublished data).

(the seed bank). Recent projects have demonstrated that removal of unwanted species takes at least 1 year of treatment before planting native species. The generally recommended protocol requires herbicide to remove all vegetation from the site (Disney Wilderness Preserve 2000) followed 3 to 4 weeks later with disking to expose weed seeds allowing them to germinate. Disking is repeated every 4 to 6 weeks for about 6 months prior to planting. Immediately before sowing desired species, the soil is compacted by rolling, and a final herbicide treatment is applied 2 to 3 weeks before planting. Variations on this protocol were shown to effectively prepare sites once dominated by pasture grasses or with cogon grass (*Imperata cylindrica*). If populations of native plants are quickly established after this treatment protocol, additional weeds can be controlled through periodic spot application of herbicide and eventually reduced as they are outcompeted by natives.

Comprehensive studies of sowing treatment effects on establishment of wiregrass and other species were conducted at Apalachicola Bluffs and Ravines Preserve in north Florida (Hattenbach et al. 1998; Seamon 1998; Cox et al. 2004). They examined the effects of eight treatments in a three-factor experiment: sowing native seed alone or with winter rye as a cover crop, rolling the seed in after sowing, or not and adding supplemental water for the first 4 months after sowing or not. They found that neither supplemental water nor sowing an annual cover crop increased ground layer species richness or density. Rolling seeds

FIGURE 6. Standard hay blower distributing wiregrass seed at Fort Gordon, GA. This equipment is effective and widely available.

in immediately after sowing, however, significantly increased wiregrass establishment and survival, as others have reported (Pfaff and Gonter 1996; Bissett 1998).

Additions of fertilizer and mulch to sites after sowing are not recommended. Neither treatment significantly increases establishment of native species under most conditions, but both favor the growth of native and exotic weeds that tend to outcompete natives (Bissett, in Florida Institute of Phosphate Research 1996; Clewell, in Florida Institute of Phosphate Research 1996; Pfaff and Gonter 1996; Jones and Gordon unpublished report to Florida Department of Transportation; Jenkins et al. 2004).

Direct Seeding: How Much, How, When?

Several options for sowing seed including hand broadcasting, hydroseeders, cultipackers, fluffy-seed drills, and fertilizer spreaders have been tested with varying results. The device most often recommended for quick, efficient distribution of native seed, especially wiregrass, is a standard hayblower (Fig. 6), which allows for relatively even distribution of seed over a large area with some control over seed placement (Disney Wilderness Preserve 2000; Jones and Gordon unpublished report to Florida Department of Transportation). See Pfaff et al. (2002) for a more detailed discussion of seeding methods.

The question of at what rate to spread seed (or seed bearing material) highlights the lack of consensus in longleaf pine restoration projects. At a discussion of restoration methods at the Disney Wilderness Preserve Conference on Uplands Restoration, the range of seeding rates for seed stripper collected seed was 25 kg material/ha. The participants agreed on a recommendation of at least 56 kg/ha of material that is 10–11% wiregrass seed by weight, and 8–10% other seed by weight. Their goal was at least three established wiregrass clumps per square meter. For material collected with a green silage cutter, they estimated that approximately 1500 kg/ha would yield the desired establishment rates. One estimate for distributing clean seed was 2.2–2.8 kg/ha (Disney Wilderness Preserve 2000). Other rates found in the literature include 133 kg/ha (Seamon 1998) for stripped material, 4.4–8.8 kg/ha of hand-collected and cleaned seed (van Eerden unpublished report to North Carolina Department of Agriculture), 3.3–4.4 kg seed/ha (Pfaff et al. 2002), and 91 kg/ha (Hattenbach et al. 1998) which yielded 5 to 7 plants/0.5 m^2. Clearly, further experimental trials of seeding rates and resulting yields for different materials are needed.

Adequate soil moisture during early establishment is essential for native plant species. Planting should occur just prior to the season of most reliable moisture (Pfaff and Gonter 1996); in most cases this is during the winter rainy season, from November to February (Pfaff and Gonter 1996; van Eerden 1997). In projects where soil moisture stress is severe, and the site is small enough to make it manageable, irrigation can be applied for the first 3 to 4 months after planting to increase seedling establishment (Jones and Gordon unpublished report to Florida Department of Transportation; Jenkins et al. 2004).

Seedling Plugs: How Many, When, Where?

Seedlings for outplanting are best grown under conditions that ensure adequate growth and survival, while maintaining an environment that is stressful enough to select for stress-tolerant plants and natural root-to-shoot ratios. Glitzenstein et al. (2001) discuss considerations for cultivation including germination and growth media (this can include horticultural media or soil taken from the restoration site, which has the added advantage of providing mycorhizal innoculum), watering regime, and overwintering of seedlings, all of which prepare seedlings for successful outplanting. The results of several studies suggest that, at least for wiregrass, plugs should be at least 6 months old before they are outplanted because younger, smaller seedlings are more susceptible to the effects of drought (van Eerden 1997; Outcalt et al. 1999) and competition (Mulligan and Kirkman 2002a). There is no consensus on the best time of year to plant seedlings; reported planting times ranged from April (Outcalt et al. 1999) to November (van Eerden 1997, unpublished report to the North Carolina Department of Agriculture) for wiregrass and plantings of other species occurred throughout the year (Glitzenstein et al. 2001). Seedling density is also species dependent, but for wiregrass three plugs per square meter is commonly used for experimental purposes (Mulligan and Kirkman 2002a).

In general, survivorship and growth of outplanted plugs in field situations are high, with survivorship rates of 90% (Glitzenstein and Streng, in Florida Institute of Phosphate Research 1996) and 60% (Outcalt et al. 1999) after one growing season, and 80% after two (Glitzenstein and Streng, in Florida Institute of Phosphate Research 1996). Both survivorship and growth are reduced by competition from neighboring plants (Outcalt et al. 1999; Mulligan and Kirkman 2002a). Regarding underplanting seedlings to restore plantations, seedling performance is likely to be maximized when planted in large canopy openings with minimal root competition from both woody and other herbaceous species (Dagley et al. 2002) and low inputs of pine straw litter (van Eerden 1997; Dagley et al. 2002).

Post-planting Management

Prescribed fire is essential to encourage flowering in many species and to control the growth of woody and exotic species. Clewell (in Florida Institute of Phosphate Research 1996) advocates burning as soon as the site is able to carry a fire; however, burning too early can kill young wiregrass plants and slow the growth of those that survive (Outcalt et al. 1999; Mulligan and Kirkman 2002b). Additionally, reports indicate that wiregrass seeds may remain dormant for a year after sowing and germinate in the second season (Seamon 1998; Mulligan and Kirkman 2002b; Cox et al. 2004). It has therefore been recommended that new plants be given at least one to two complete growing seasons and as long as four to five seasons (Outcalt et al. 1999) before prescribed fire is introduced. After that, a 2- to 3-year burn cycle has been suggested, as competition begins to negatively impact wiregrass plants after 3 years (Glitzenstein and Streng, in Florida Institute of Phosphate Research 1996).

Population Establishment: Does It Work?

Population establishment can be considered a success when it results in a self-maintaining population with sufficient genetic diversity for long-term persistence (Pavlik 1996). While few studies have been in place long enough to reach this ultimate goal, there are many encouraging results thus far. Seedling recruitment has been observed in populations of both outplanted wiregrass plugs (Mulligan et al. 2002) and in plots that were direct-seeded (Bissett, in Florida Institute of Phosphate Research 1996). Glitzenstein et al. (2001) monitored six species of outplanted grasses and forbs, including both wiregrass and the rare forb *Parnassia caroliniana*, for 5 years and are optimistic about their chances of long-term success. At the Nature Conservancy Apalachicola Bluffs and Ravines Preserve seeds were collected from 4- and 5-year-old direct-seeded populations, and a direct seeding study for the Florida DOT (Jones and Gordon unpublished report to the Florida Department of Transportation) resulted in a stand that was within the natural range of species cover and able to carry fire in 3 years.

Filling Information Gaps: Adaptive Management and Research

Knowledge about restoring the ground layer in longleaf pine communities has increased substantially in recent years, and results from established projects promise a bright future for restoration. Restoration projects and research efforts underway in various places through the region will yield still more information in the near future (for example, see Box 10.2). We expect that continued knowledge development would benefit from increased collaboration and coordination among research and restoration trials, and further that a widely accessible outlet for developing information will generate even more landowner interest in ground cover restoration.

In addition to the need for increased communication, we have identified some specific information gaps. Restoration research or adaptive restoration projects conducted in other locations within the range of longleaf pine would advance the restoration cause, as well as contribute to understanding the natural variability of the ecosystem. In the absence of more specific information, research projects designed to understand the variation in ecosystem functions across gradients, especially a productivity gradient, may be useful for targeting the most difficult and most pressing restoration needs. There is always a need for more species-specific information about the biology and habitat requirements of both common and rare species, especially regarding reproductive biology. Because species reintroductions are needed for many sites, a more comprehensive understanding of population processes in experimental as well as natural species matrices is essential. Information about persistent native seed banks is scarce (Cohen 1998; Jenkins 2003), but would be especially useful in developing restoration protocols. Finally, while there are suggestions that small fragments of this diverse herbaceous community can persist (Heuberger and Putz 2003), the effects of fragmentation and isolation on the persistence of the ground cover of longleaf are not well known; such knowledge could ensure that feasible restoration goals are established and that restoration resources are targeted where they can be successful.

References

Apfelbaum, S.I., Bader, B.J., Faessler, F., and Mahler, D. 1997. Obtaining and processing seeds. In *The Tallgrass Restoration Handbook*, eds. S. Packard and C.F. Mutel, pp. 99–126. Washington, DC: Island Press.

Baskin, C.C., and Baskin, J.M. 1998. *Seeds: Ecology, Biogeography, and Evolution of Dormancy and Germination.* San Diego: Academic Press.

Beckage, B., and Stout, J.I. 2000. Effects of repeated burning on species richness in a Florida pine savanna: A test of the intermediate disturbance hypothesis. *J Veg Sci* 11:113–122.

Bissett, N.J. 1998. Direct seeding wiregrass, *Aristida beyrichiana*, and associated species. In *Proceedings of the Longleaf Pine Ecosystem Restoration Symposium*, The Longleaf Alliance Report No. 3, pp. 59–60. Fort Lauderdale: The Longleaf Alliance.

Brewer, J.S. 1995. The relationship between soil fertility and fire-stimulated floral induction in two populations of grass-leaved golden aster, *Pityopsis graminifolia*. *Oikos* 74:45–54.

Brewer, J.S., and Platt, W.J. 1994. Effects of fire season and herbivory on reproductive success in a clonal forb, *Pityopsis graminifolia*. *J Ecol* 82:665–675.

Brewer, J.S., Platt, W.J., Glitzenstein, J.S., and Streng, D.R. 1996. Effects of fire-generated gaps on growth and reproduction of golden aster (*Pityopsis graminifolia*). *Bull Torrey Bot Club* 123:295–303.

Brockway, D.G., and Lewis, C.E. 1997. Long-term effects of dormant-season prescribed fire on plant community diversity, structure and productivity in a longleaf pine wiregrass ecosystem. *For Ecol Manage* 96:167–183.

Brockway, D.G., and Outcalt, K. W. 2000. Restoring longleaf pine wiregrass ecosystems: Hexazinone application enhances effects of prescribed fire. *For Ecol Manage* 137:121–138.

Center for Plant Conservation. 1991. Genetic sampling guidelines for conservation collections of endangered plants. In *Restoring Diversity*, eds. D.A. Falk, C.I. Millar, and M. Olwell, pp. 225–238. Washington, DC: Island Press.

Clinebell, R.R., II. 1997. Tips for gathering individual species. In *The Tallgrass Restoration Handbook*, eds. S. Packard and C.F. Mutel, pp. 127–134. Washington, DC: Island Press.

Cohen, S. 1998. Assessment of the restoration potential of severely disturbed longleaf pine sites. Master's thesis. North Carolina State University, Raleigh.

Cox, A.C., Gordon, D.R., Slapcinsky, J.L., and Seamon, G.S. 2004. Understory restoration in longleaf pine sandhills. *Nat Areas J* 24:4–14.

Cushwa, C.T., Martin, R.E., and Miller, R.L. 1968. The effects of fire on seed germination. *J Range Manage* 21:250–254.

Dagley, C.M., Harrington, T.B., and Edwards, M.B. 2002. Understory restoration in longleaf pine plantations: Overstory effects of competition and needlefall. In *Eleventh Biennial Southern Silvicultural Research Conference*, ed. K.W. Outcalt, pp. 487–489. Asheville, NC: USDA Forest Service, Southern Research Station.

D'Antonio, C., and Meyerson, L.A. 2002. Exotic plant species as problems and solutions in ecological restoration: A synthesis. *Restor Ecol* 10:703–713.

Disney Wilderness Preserve. 2000. *Proceedings of the Upland Restoration Workshop*. Kissimmee, FL.

Drew, M.B., Kirkman, L.K., and Gholson, A.K. 1998. The vascular flora of Ichauway, Baker County, Georgia: A remnant longleaf pine/wiregrass ecosystem. *Castanea* 63:1–24.

Drewa, P.B., Platt, W.J., and Moser, E.B. 2002. Fire effects on resprouting of shrubs in headwaters of southeastern longleaf pine savannas. *Ecology* 83:755–767.

Egan, D., and Howell, E.A. 2001. *The Historical Ecology Handbook*. Washington, DC: Island Press.

Ehrenfeld, J.G. 2000. Defining the limits of restoration: The need for realistic goals. *Restor Ecol* 8:2–9.

Estill, J.C., and Cruzan, M.B. 2001. Phytogeography of rare plant species endemic to the southeastern United States. *Castanea* 66(1–2):3–23.

Facelli, J.M., and Pickett, S.T. A. 1991. Plant litter: Its dynamics and effects on plant community structure. *Bot Rev* 57:1–32.

Fenner, M. 1985. *Seed Ecology*. London: Chapman & Hall.

Florida Institute of Phosphate Research. 1996. Proceedings of the Ecosystem Restoration Workshop, Sponsored by the Florida Institute of Phosphate Research and the Society for Ecological Restoration. Lakeland, FL.

Foster, D., Swanson, F., Aber, J., Burke, I., Brokaw, N., Tilman, D., and Knapp, A. 2003. The importance of land-use legacies to ecology and conservation. *BioScience* 53:77–88.

Fowler, N.L. 1986. Microsite requirements for germination and establishment of three grass species. *Am Midl Nat* 115:131–145.

Frost, C. 1998. Presettlement fire frequency regimes of the United States: A first approximation. *Tall Timbers Fire Ecol Conf Proc* 20:70–81.

Gilliam, F.S., and Christensen, N.L. 1986. Herb-layer response to burning in pine flatwoods of the lower coastal plain of South Carolina. *Bull Torrey Bot Club* 113:42–45.

Gilliam, F.S., and Platt, W.J. 1999. Effects of long-term fire exclusion on tree species composition and stand structure in an old-growth *Pinus palustris* (longleaf pine) forest. *Plant Ecol* 140:15–26.

Gilliam, F.S., Yurish, B.M., and Goodwin, L.M. 1993. Community composition of an old growth longleaf pine forest: Relationship to soil texture. *Bull Torrey Bot Club* 120:287–294.

Glitzenstein, J.S. 1993. Panel discussion: Silvicultural effects on groundcover plant communities in longleaf pine forests. In *Tall Timbers Fire Ecology Conference No. 18*, ed. S.M. Hermann, pp. 357–370. Tallahassee: Tall Timbers Research Station.

Glitzenstein, J.S., Platt, W.J., and Streng, D.R. 1995. Effects of fire regime and habitat in three dynamics in north Florida longleaf pine savannas. *Ecol Monogr* 65:441–476.

Glitzenstein, J.S., Streng, D.R., Wade, D.D., and Brubaker, J. 2001. Starting new plants of longleaf pine ground-layer plants in the outer coastal plain of South Carolina, USA. *Nat Areas J* 21:89–110.

Glitzenstein, J.S., Streng, D.R., and Wade, D.D. 2003. Fire frequency effects on longleaf pine (*Pinus palustris* Miller) vegetation in South Carolina and northeast Florida, USA. *Nat Areas J* 23:22–37.

Gordon, D.R. 1994. Translocation of species into conservation areas: A key for natural resource managers. *Nat Areas J* 14:31–37.

Grelen, H.E. 1959. Mechanical preparation of pine planting sites in Florida sandhills. *Weeds* 7:184–188.

Grelen, H.E. 1962. Plant succession on cleared sandhills in northern Florida. *Am Midl Nat* 67:36–44.

Grelen, H.E., and Enghardt, H.G. 1973. Burning and thinning maintain forage in a longleaf pine plantation. *J For* 71:419–425.

Guerrant, E.O., Jr. 1996. Designing populations: Demographic, genetic, and horticultural dimensions. In *Restoring Diversity*, eds. D.A. Falk, C.I. Millar, and M. Olwell, pp. 171–208. Washington, DC: Island Press.

Hamrick, J.L., Godt, M.J.W., Murawski, D.A., and Loveless, M.D. 1991. Correlations between species traits and allozyme diversity: Implications for conservation biology. In *Genetics and Conservation of Rare Plants*, eds. D. A. Falk and K.E. Holsinger, pp. 75–86. Oxford: Oxford University Press.

Harcombe, P.A., Glitzenstein, J.S., Knox, R.G., Orzell, S.L., and Bridges, E.L. 1993. Vegetation of the longleaf pine region of the west gulf coastal plain. In *Tall Timbers Fire Ecology Conference No. 18*, ed. S.M. Hermann, pp. 83–104. Tallahassee: Tall Timbers Research Station.

Harper, J. 1977. *Population Biology of Plants*. New York: Academic Press.

Harrington, T.B., and Edwards, M.B. 1999. Understory vegetation, resource availability, and litterfall responses to pine thinning and woody vegetation control in longleaf pine plantations. *Can J For Res* 29:1055–1064.

Hattenbach, M.J., Gordon, D.R., Seamon, G.S., and Studenmund, R.G. 1998. Development of direct-seeding techniques to restore native groundcover in a sandhill ecosystem. In *Proceedings of the Longleaf Pine Ecosystem Restoration Symposium*, The Longleaf Alliance Report No. 3, pp. 64–70. Fort Lauderdale: The Longleaf Alliance.

Hedman, C.W., Grace, S.L., and King, S.E. 2000. Vegetation composition and structure of southern coastal plain pine forests: An ecological comparison. *For Ecol Manage* 134:233–247.

Heuberger, K.A., and Putz, F.E. 2003. Fire in the suburbs: Ecological impacts of prescribed fire in small remnants of longleaf pine (*Pinus palustris*) sandhill. *Restor Ecol* 11:72–81.

Hiers, J.K., Wyatt, R., and Mitchell, R.J. 2000. The effects of fire regime on legume production in longleaf pine savannas: Is a season selective? *Oecologia* 125:521–530.

Hobbs, R., and Humphries, S.E. 1995. An integrated approach to the ecology and management of plant invasions. *Conserv Biol* 9:761–770.

Huenneke, L.F. 1991. Ecological implications of genetic variation in plant populations. In *Genetics and Conservation of Rare Plants*, eds. D.A. Falk and K.E. Holsinger, pp. 31–44. Oxford: Oxford University Press.

Jenkins, A.M. 2003. Seed banking and vesicular-arbuscular mycorrhizae in pasture restoration in central Florida. Master's thesis. University of Florida, Gainesville.

Jenkins, A.M., Gordon, D.R., and Renda, M.T. 2004. Native alternatives for nonnative turfgrasses in central Florida: Germination and responses to cultural treatments. *Restor Ecol* 12:190–199.

Jose, S., Cox, J., Miller, D.L., Shilling, D.G., and Merritt, S. 2002. Alien plant invasions: The story of cogongrass in southeastern forests. *J For* 100:41–44.

Kesler, T.R., Anderson, L.C., and Hermann, S.M. 2003. A taxonomic reevaluation of *Aristida stricta* (Poaceae) using anatomy and morphology. *Southeast Nat* 2:1–10.

Kindell, C.E., Winn, A.A., and Miller, T.E. 1996. The effects of surrounding vegetation and transplant age on the detection of local adaptation in the perennial grass *Aristida stricta*. *J Ecol* 84:745–754.

Kirkman, L.K., and Mitchell, R.J. 2002. A forest gap approach to restoring longleaf pine-wiregrass ecosystems (Georgia and Florida). *Ecol Restor* 20:50–51.

Kush, J.S., Meldahl, R.S., and Avery, C. 2004. A restoration success: Longleaf pine seedlings established in a fire-suppressed, old-growth stand. *Ecol Restor* 22: 6–10.

Leach, M.K., and Givnish, T.J. 1996. Ecological determinants of species loss in remnant prairies. *Science* 273:1555–1558.

LeBlond, R.J. 2001. Endemic plants of the Cape Fear Arch region. *Castanea* 66: 83–97.

Lippincott, C.L. 2000. Effects of *Imperata cylindrica* (cogongrass) invasion on fire regime in Florida sandhill (USA). *Nat Areas J* 20:140–149.

Litt, A.R., Herring, B.J., and Provencher, L. 2001. Herbicide effects on ground-layer vegetation in southern pinelands, USA: A review. *Nat Areas J* 21:177–188.

Liu, C., Harcombe, P.A., and Knox, R.G. 1997. Effects of prescribed fire on the composition of woody plant communities in southeastern Texas. *J Veg Sci* 8:495–504.

Lockwood, J.L., and Pimm, S.L. 1999. When does restoration succeed? In *Ecological Assembly Rules*, eds. E. Weiher and P. Keddy, pp. 363–392. Cambridge: Cambridge University Press.

Means, D.B. 1997. Wiregrass restoration: Probable shading effects in a slash pine plantation. *Restor Manage Notes* 15:52–55.

Mehlman, D.W. 1992. Effects of fire on plant community composition of North Florida second growth pineland. *Bull Torrey Bot Club* 119:376–383.

Millar, C.I., and Libby, W.J. 1991. Strategies for conserving clinal, ecotypic, and disjunct population diversity in widespread species. In *Genetics and Conservation of Rare Plants*, eds. D.A. Falk and K.E. Holsinger, pp. 149–170. Oxford: Oxford University Press.

Miller, J.H. 2003. *Nonnative Invasive Plants of Southern Forests: A Field Guide for Identification and Control*. GTR SRS–62. Asheville: USDA Forest Service, Southern Research Station.

Moore, W.H. 1974. Some effects of chopping saw-palmetto–pineland threeawn range in south Florida. *J Range Manage* 27:101–104.

Mulligan, M.K., and Kirkman, L.K. 2002a. Competition effects on wiregrass (*Aristida beyrichina*) growth and survival. *Plant Ecol* 163:39–50.

Mulligan, M.K., and Kirkman, L.K. 2002b. Burning influences on wiregrass (*Aristida beyrichiana*) restoration plantings: Natural seedling recruitment and survival. *Restor Ecol* 10:334–339.

Mulligan, M.K., Kirkman, L.K., and Mitchell, R.J. 2002. *Aristida beyrichiana* (wiregrass) establishment and recruitment: Implications for restoration. *Restor Ecol* 10:68–76.

Outcalt, K.W. 1994. Seed production of wiregrass in central Florida following growing season prescribed burns. *Int J Wildland Fire* 4:123–125.

Outcalt, K.W., Williams, M.E., and Onokpise, O. 1999. Restoring *Aristida stricta* to *Pinus palustris* ecosystems on the Atlantic coastal plain, USA. *Restor Ecol* 7:262–270.

Packard, S., and Mutel, C.F. 1997. *The Tallgrass Restoration Handbook*. Washington, DC: Island Press.

Pavlik, B.M. 1996. Defining and measuring success. In *Restoring Diversity*, eds. D.A. Falk, C.I. Millar, and M. Olwell, pp. 127–156. Washington, DC: Island Press.

Peet, R.K. 1993. A taxonomic study of *Aristida stricta* and *A. beyrichiana*. *Rhodora* 95:25–37.

Peet, R.K., and Allard, D.J. 1993. Longleaf pine vegetation of the southern Atlantic and eastern Gulf Coast regions: A primary classification. In *Tall Timbers Fire Ecology Conference*, no. 18 ed. S.M. Hermann, pp. 45–81. Tallahassee: Tall Timbers Research Station.

Pfaff, S., and Gonter, M.A. 1996. Florida native plant collection, production, and direct seeding techniques: Interim report. Brooksville, FL: USDA, Natural Resources Conservation Service, Plant Materials Center.

Pfaff, S., Gonter, M.A., and Maura, C. 2002. Florida Native Seed Production Manual. Brooksville, FL: USDA, Natural Resources Conservation Service, Plant Materials Center.

Phillips, H.R. 1985. *Growing and Propagating Wild Flowers*. Chapel Hill: University of North Carolina Press.

Pittman, T., and Karrfalt, R.P. 2000. Wiregrass propagation. *Native Plants J* 1:45–47.

Platt, W.J., Evans, G.W., and Davis, M.M. 1988. Effects of fire season on flowering of forbs and shrubs in longleaf pine forests. *Oecologia* 76:353–363.

Platt, W.J., Glitzenstein, J.S., and Streng, D.R. 1991. Evaluating pyrogenicity and its effects on vegetation in longleaf pine savannas. *Proc Tall Timbers Fire Ecol Conf* 17:143–161.

Provencher, L., Herring, B.J., Gordon, D.R., Rogers, H.L., Galley, K.E. M., Tanner, G.W., Hardesty, J.L., and Brennan, L.A. 2001. Effects of hardwood reduction techniques on longleaf pine sandhill vegetation in northwest Florida. *Restor Ecol* 9:13–27.

Radford, A.E., Ahles, H.E., and Bell, R.C. 1968. *Manual of the Vascular Flora of the Carolinas*. Chapel Hill: University of North Carolina Press.

Rebertus, A.J., Williamson, G.B., and Moser, E.B. 1989. Longleaf pine pyrogenicity and turkey oak mortality in Florida xeric sandhills. *Ecology* 70:60–70.

Robbins, L.E., and Myers, R.L. 1992. Seasonal effects of prescribed burning in Florida: A review. Miscellaneous Publication No. 8, Tall Timbers Research, Inc., Tallahassee, FL.

Seamon, G. 1998. A longleaf pine sandhill restoration in northwest Florida. *Restor Manage Notes* 16:46–50.

Smith, G.P., Shelburne, V.B., and Walker, J.L. 2001. Structure and composition of vegetation of longleaf pine plantations compared to natural stands occurring along an environmental gradient at the Savannah River Site. In *Eleventh Biennial Southern Silvicultural Research Conference*, ed. K.W. Outcalt, pp. 481–486. Ashville, NC: USDA Forest Service, Southern Research Station.

Sorrie, B.A., and Weakley, A.S. 2001. Coastal plain vascular plant endemics: Phytogeographic patterns. *Castanea* 66:50–82.

Streng, D.R., Glitzenstein, J.S., and Platt, W.J. 1993. Evaluating effects of season of burn in longleaf pine forests: A critical literature review and some results from an ongoing long-term study. In *Tall Timbers Fire Ecology Conference*, ed. S.M. Hermann, pp. 227–263. Tallahassee: Tall Timbers Research Station.

Swindel, B.F., Conde, L.F., and Smith, J.E. 1986. Successional changes in *Pinus elliottii* plantations following two regeneration treatments. *Can J For Res* 16:630–636.

Tanner, G.W., Terry, W.S., and Kalmbacher, R.B. 1988. Mechanical shrub control on flatwoods range in south Florida. *J Range Manage* 41:245–248.

U.S. Fish and Wildlife Service. 2001. Red-cockaded woodpecker recovery plan (*Picoides borealis*). Atlanta: U. S. Fish and Wildlife Service.

van Eerden, B.P. 1997. Studies on the reproductive biology of wiregrass (*Aristida stricta* Michaux) in the Carollina sandhills. Master's thesis, University of Georgia, Athens.

Varner, J.M., III, Kush, J.S., and Meldahl, R.S. 2000. Ecological restoration of an old-growth longleaf pine stand utilizing prescribed fire. In Fire and Forest Ecology: Innovative *Silvicultural and Vegetation Management*, Tall Timbers Fire Ecology Conference Proceedings No. 21, eds. W. K. Moser and C.E. Moser, pp. 216–219. Tallahassee: Tall Timbers Research Station.

Varner, J.M., III, Kush, J.S., and Meldahl, R.S. 2003. Vegetation of frequently burned old-growth longleaf pine (*Pinus palustris* Mill.) savannas on Choccolocco Mountain, Alabama, USA. *Nat Areas J* 23:43–52.

Waldrop, T.A., White, D.L., and Jones, S.M. 1992. Fire regimes for pine-grassland communities in the southeastern United States. *For Ecol Manage* 47:195–210.

Walker, J.L. 1998. Ground layer vegetation in longleaf pine landscapes: An overview for restoration and management. In *Proceedings of the Longleaf Pine Ecosystem Restoration Symposium*, The Longleaf Alliance Report No. 3, pp. 2–13. Fort Lauderdale: The Longleaf Alliance.

Walker, J., and Peet, R.K. 1983. Composition and species diversity of pine–wiregrass savannas of the Green Swamp, North Carolina. *Vegetatio* 55:163–179.

Walker, J.L., and van Eerden, B.P. 1998. Effects of drum-chopping on wiregrass and other herbaceous species in xeric sandhill sites. In *Fire in Ecosystem Management: Shifting the Paradigm from Suppression to Prescription*, Tall Timbers Fire Ecology Conference Proceedings No. 20, eds. T.L. Pruden and L.A. Brennan, pp. 118. Tallahassee: Tall Timbers Research Station.

Walker, J.L., van Eerden, B.P., Robinson, D., and Hausch, M. 2004. Burning and chopping for woodpeckers and wiregrass? In *Red-cockaded Woodpecker: Road to Recovery*, Proceedings of the Red-cockaded Woodpecker Recovery Symposium IV, eds.R. Costa and S.J. Daniels, pp. 683–685. Savannah, GA, January 27–31, 2003. Hancock House Publishers, Blaine, WA.

Walters, T.W., Decker-Walters, D.S., and Gordon, D.R. 1994. Restoration considerations for wiregrass (*Aristida stricta*): Allozymic diversity of populations. *Conserv Biol* 8:581–585.

Wear, D.N., and Greis, J.G. 2002. Southern Forest Resource Assessment. General Technical Report SRS-53. Asheville, NC: USDA Forest Service.

Weber, S. 1999. Designing seed mixes for prairie restorations: Revisiting the formula. *Ecol Restor* 17:196–201.

White, L.D. 1977. Forage production in a five-year-old fertilized slash pine plantation. *J Range Manage* 30:131–134.

White, P.S. 1996. Spatial and biological scales in reintroduction. In *Restoring Diversity*, eds. D.A. Falk, C.I. Millar, and M. Olwell, pp. 49–86. Washington, DC: Island Press.

White, P.S., and Walker, J.L. 1997. Approximating nature's variation: Selecting and using reference information in restoration ecology. *Restor Ecol* 5:338–349.

Zutter, B.R., and Miller, J.H. 1998. Eleventh-year response of loblolly pine and competing vegetation to woody and herbaceous plant control on a Georgia flatwoods site. *South J Appl For* 22:88–95.

BOX 10.1
Prescribed Burning for Understory Restoration

Kenneth W. Outcalt
*Southern Research Station,
USDA Forest Service, Athens,
Georgia 30602*

Role of Prescribed Burning. Because the longleaf ecosystem evolved with and is adapted to frequent fire, every 2 to 8 years, prescribed burning is often useful for restoring understory communities to a diverse ground layer of grasses, herbs, and small shrubs. This restoration provides habitat for a number of plant and animal species that are restricted to or found mostly in longleaf pine communities. Burning can also be used to reduce the midstory layer, which catches shed needles and serves as a ladder to carry understory fires into the crowns of the trees resulting in catastrophic wildfires that can kill vast areas of pines. Prescribed burning also recycles nutrients by releasing those tied up in litter and duff and significantly reduces brown spot needle blight, which attacks longleaf seedlings.

Terms and Techniques. Prescribed burning is the application of fire by trained professionals following a well-developed plan to obtain desired management objectives. Restoration is often done with understory burning or underburning, which is prescribed burning under a forest canopy (McPherson et al. 1990). The fuel for these fires is the understory rough that consists of the accumulated living and dead grasses, forbs and shrubs plus draped needles and the litter layer. The litter layer, the top layer of the forest floor, is composed of recently fallen and largely intact dead needles, leaves, twigs, and branches. A duff layer, composed of partially decomposed litter or fermentation layer and decomposed humus, lies between the litter and mineral soil.

Underburning can be done using heading, backing, flanking, or spot fires, or a combination of these techniques. Heading fires are fire fronts ignited to spread with the wind while backing fires are ignited so the fire front spreads against the wind. Flanking fires are ignited in a line into the wind and thus spread at approximately right angles to wind direction. Spot fires are a series of separate ignition points that are allowed to spread in all directions and thus contain heading, backing, and flanking fires at each spot. Both heading and backing fires can be set as a series of strip fires. Strip heading fires are used to control how fast the fire spreads and thereby the fireline intensity, i.e., the rate of heat energy release (Box A Fig. 1). Placing strips closer together reduces the rate of spread and intensity. Backing fires have low intensities but move slowly and therefore require a lot of time to burn each unit. In addition, backing fires under certain conditions may be quite severe, i.e., cause much of damage to the site, because of excess duff consumption. Internal firebreaks can be constructed for strip backing fires to significantly reduce time to complete the burn. An alternative is to use flanking or spot fires to reduce intensity but speed up the burn without internal fire breaks. These techniques require considerable experience, especially spot firing as you must continually adjust both the spacing and the timing between spots to obtain the desired intensity with changing fuel and weather conditions (Wade and Lunsford 1989).

Burning Prescriptions

Sandhills. Prescribed burning can be used for restoration across the range of sites that longleaf can occupy from dry sandhills to wet savannas. Burning prescriptions depend on the ecosystem type and its current condition. Reduced fire frequency in many xeric and subxeric sandhills longleaf areas has resulted in the development of a midstory layer of native scrub oaks: turkey (*Quercus laevis*), bluejack (*Q. incana*), sand

10. Restoring the Ground Layer

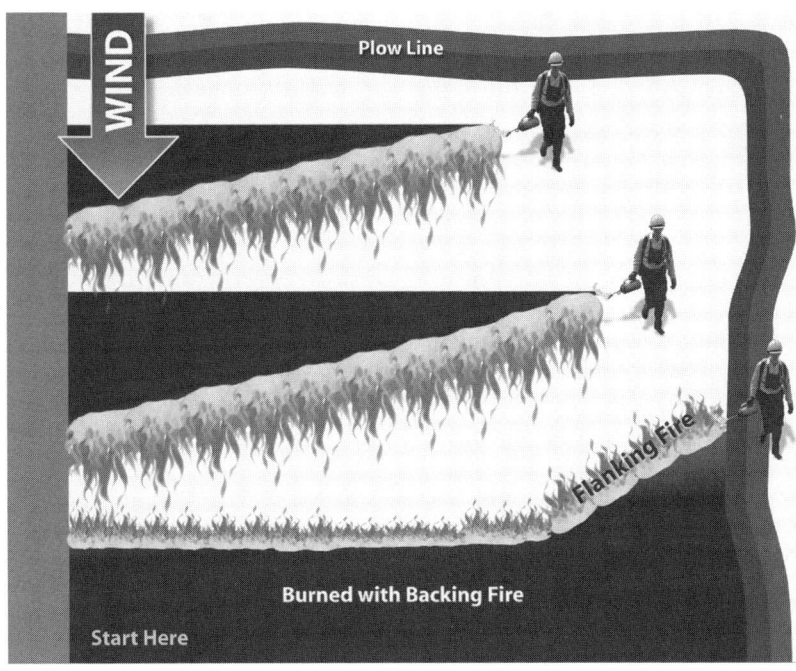

FIGURE 1. Schematic of typical strip heading fire showing securing downwind side with backfire followed by sequential ignition of strip heading fires with a flanking fire to widen fuel free zone along plow line.

live oak (*Q. virginiana* var. *geminata*), and sand post oak (*Q. stellata* var. *margaretta*). Because these sites are very droughty and nutrient limited, even in the absence of frequent burning, they do not develop a continuous closed canopy of midstory hardwoods. Therefore, although greatly reduced, some of the understory grasses do survive. These grasses along with needle litter from longleaf pines furnish sufficient fuel to carry at least a patchy prescribed burn. Repeated applications of prescribed fires during the growing season, i.e., beginning in March, in southern latitudes, and ending in July, can be used to restore these sites by gradually reducing the density of the midstory scrub oaks (Glitzenstein et al. 1995) and promoting the growth of understory grasses and herbs. Managers have found that fire causes wounds on the stems of hardwoods, which are enlarged by subsequent fires, and eventually the top breaks or the stem is girdled and the top dies. Sprouts emerge from the roots of many top-killed stems, but these can be kept in check by subsequent periodic burns.

Flatwoods. On flatwoods and wet lowland pine types restoration means increasing understory diversity in longleaf communities that have been captured by woody species and in many cases developed a substantial midstory layer. The goal is to reduce woody understory and midstory species and allow the grasses and forbs to increase and eventually become at least co-dominant. Prescribed fire can be used to accomplish this transition. Research shows that although growing season burns are sometimes more effective, dormant season burns can also be used to readjust understory composition (Waldrop et al. 1987). For areas not burned for 10 years or more, a couple of dormant season burns should be used to reduce fuel loads before switching to growing season burns. In addition, it is usually best to have these burns close together, i.e.,

2 years or less, to minimize fuel accumulations between burns. Miller and Bossuot (2000) recommend these initial burns be conducted when the drought index is below 250. On sites dominated by saw palmetto (*Serenoa repens*), if burning alone is going to be used for understory restoration, then a series of closely spaced prescribed burns is required. Frequency of burns is more important than season with annual burns the most effective but biennial burns will reduce palmetto-dominance and increase the herbaceous component. It is important not to miss a burn, as this can result in a significant regrowth of palmetto.

Uplands. There also exist upland longleaf types, mostly in Alabama, Mississippi, Louisiana, and Texas, and montane sites in Alabama and Georgia, which have developed unnaturally dense hardwood midstories, and suppressed and impoverished understories. Because these are the most productive longleaf sites, they change the most rapidly, quickly developing midstory layers in the absence of frequent fire. In addition to a very dense midstory and a shrub-dominated understory, these sites also accumulate significant quantities of fuel. As noted for other longleaf ecosystem types, a series of dormant season burns is often necessary to gradually reduce fuel levels before switching to a growing season regiment. However, frequent and multiple growing season burns will be required to reduce the hardwood rootstocks (Boyer 1990), thereby providing conditions favorable to understory grasses and forbs.

Potential Negative Impacts. As with all burning, there is the potential for negative impacts. The most obvious damage is direct tree mortality that can result from excessively hot burns that are too intense and kill tree crowns, including the buds. Tree mortality can also occur with low-intensity but high-severity fires that slowly consume accumulated forest floor duff, and because of their long residence time heat root and stem cambial cells beyond the lethal temperature. Trees suffering from such injury often retain a healthy-looking green crown for some time following the burn, but will eventually die. Longleaf communities needing restoration burning rarely have many seedlings. If significant numbers of seedlings are present, however, excessive seedling mortality can result from burning during the bolting stage. If burning must be done with seedlings at this stage, then burning should be done in the dormant season or early spring prior to the candle stage when seedlings would be most susceptible to fire-caused damage and mortality.

Precautions. In all longleaf types that have not been burned for 10 years or more, there is an excessive buildup of litter and duff around the base of trees. Reintroduction of burning in these stands without excess mortality is best accomplished by a series of dormant season burns. Apply burns when only the litter is dry enough to burn and the duff is too wet to ignite. On upland or sandhills sites, fast-moving heading or flanking fires pushed by a good wind are better than slow backing fires that may dry the duff layer and promote smoldering combustion. Flatwoods sites with palmetto-dominated understories should be burned with heading fires, but will also require a light wind, cool temperature, and higher humidity for the first burn. The objective on all sites is to consume the dry top litter layer while the wet lower duff layer will protect the roots and root collar. Space burns as closely together as fuel to carry the fire will allow. Be cautious in your prescriptions because a patchy burn is preferable to a more complete hot burn that could result in excessive tree mortality. The objective is to gradually reduce the duff layer at the base of trees over a cycle of four or five fires and keep tree mortality at an acceptable level. Once the excess duff layer is removed, apply a growing season burn, again as soon as there is sufficient fuel to carry a good fire.

The most important factor in accomplishing the goal of successful restoration with fire while minimizing the negative consequences is experience. Only through training and practice can you become proficient at selecting proper conditions of temperature, humidity, wind, fuel moisture, and firing techniques keyed to existing fuel types and loads that are likely to produce the desired outcomes. This means obtaining a contract burner from a consulting forestry business or an experienced crew from a nonprofit or government agency until you gain knowledge and experience needed to be a certified burner. A source of information is the U. S. Department of Agriculture, Forest Service, Southern Research Station (http://www.srs.fs.usda.gov), which has a number of relevant publications. Considerable advice and guidance is also available from state forestry agencies, forestry units of southern universities, and local extension agents.

References

Boyer, W.D. 1990. Growing-season burns for control of hardwoods in longleaf pine stands. USDA Forest Service, Southern Forest Experiment Station, Research Paper SO-256, New Orleans, LA.

Glitzenstein, J.S., Platt, W.J., and Streng, D.R. 1995. Effects of fire regime and habitat on tree dynamics in north Florida longleaf pine savannas. *Ecol Monogr* 65(4):441–476.

McPherson, G.R., Wade, D.D., and Phillips, C.B. 1990. *Glossary of Wildland Fire Management Terms Used in the United States.* Washington, DC: Society of American Foresters.

Miller, S.R., and Bossuot, W.R. 2000. Flatwoods restoration on the St. Johns River Water Management District, Florida: A prescription to cut and burn. In *Proceedings of the 21st Tall Timbers Fire Ecology Conference*, pp. 212–215. Tall Timbers Research Station, Tallahassee, FL.

Wade, D.D., and Lunsford, J.D. 1989. A guide for prescribed fire in southern forests. USDA Forest Service, Southern Region, Technical Publication R8-TP 11, Atlanta, GA.

Waldrop, T.A., Van Lear, D.H., Lloyd, F.T., and Harms, W.R. 1987. Long-term studies of prescribed burning in loblolly pine forests of the Southeastern Coastal Plain. USDA Forest Service, Southeastern Forest Experiment Station, General Technical Report SE-45, Asheville, NC.

BOX 10.2
Restoring the Savanna to the Savannah River Site

Don Imm and John Blake
Savannah River Site, USDA Forest Service, New Ellenton, South Carolina 29809

Background

The Savannah River Site (SRS) is a 80,128-ha Department of Energy facility that lies within the upper Coastal Plain of South Carolina and is adjacent to the Savannah River. The area is characterized as a warm temperate climate with moderately well to excessively drained sandy to sandy loam surface soils underlain by sandy loam to clay loam subsoils at varying depths. When the SRS was established in 1951, the land was classified as 40% agriculture and 60% forestland. However, a large portion of the forestland was cut over (~20,000 ha) and had an average stocking of only one-fifth of similar lands in the vicinity. The oldest pine stands dated from about 1890 following Civil War land abandonment.

In 1951 the USDA Forest Service was engaged to reforest the agriculture lands and cut over forest areas. This project was largely complete by the 1970s. During this same period an ecological baseline was established through various research institutions (University of Georgia, University of South Carolina, and the Philadelphia Academy of Sciences) to document the status of the aquatic and terrestrial flora and fauna. In 1974, the Savannah River Site was established as the first National Environmental Research Park. In the late 1970s, coincident with implementation of the first recovery plan for the endangered red-cockaded woodpecker, the prescribed burning was increased to improve habitat conditions for that species, and reduce fuel loading. In the early 1990s, a formal decision was made to link the red-cockaded woodpecker recovery plan with restoration of the native savanna communities (Gaines et al. 1995).

Savanna Restoration Research

Savanna restoration research was targeted at addressing important questions tied to natural resource management decisions. Does historical evidence and current vegetation indicate the presettlement fire savanna composition and distribution? How has land use since European settlement impacted those communities? Can we identify the native savanna species, communities, and habitat relationships? What management practices are effective for restoring and sustaining these communities? Parallel efforts were made to expand field surveys, monitor sensitive species, map remnant communities, and organize existing ecological databases.

Research studies using a combination of historical plats, diaries, and field surveys established that fire savanna communities dominated about 78% of the upland landscape (Frost 1997). These fire-influenced communities ranged from canebrake wetlands and open woodland, to scrub-oak stands, and they contained an array of grass and forb species characteristic of the upper Coastal Plain (Peet and Allard 1995; Duncan and Peet 1996). Cultivation, subsistence hunting, dam construction, and selective tree cutting coupled with fragmentation of the original landscape and fire protection had major ecological impacts on native flora and fauna (White and Gaines 2000). The absence of periodic burning and earlier crop agriculture reduced the abundance of native grasses and forbs, as well as diagnostic species such as wiregrass (*Aristida beyrichiana*) to a few locations (W. T. Batson personal communication), and in doing so reduced the amount of available seed for future establishment beneath newly planted forests. Today, a large percentage of the upland forest is dominated by old field grasses and forbs (Smith 2000; Imm and McLeod 2005). Due to the exclusion of periodic burning through most of the twentieth century, early and midsuccession hardwood species co-dominate with southern pines.

Ecological studies of intact communities (Duncan and Peet 1996) and old fields (Smith 2000) demonstrated that composition and diversity of the grass and herbaceous community follow gradients that are strongly influenced by soils, topography, drainage, and geographical location. While providing clues to ecological controls, they were not precise enough to predict "presettlement" composition for specific parcels. Routine monitoring and surveys for sensitive species, and studies of remnant savanna fragments (Frost unpublished) are further enhancing our knowledge of the original landscape distribution. Numerous studies of habitat relationships for vertebrate species have been conducted. For vertebrate communities (birds, herpetofauna, and small mammals), early succession stands were found to support a wide array of species characteristic of savannas (Grant et al. 1994; Dunning et al. 1995; Yates et al. 1997; Kilgo et al. 2000).

Latent savanna communities at SRS can be grouped into three general categories: (1) old field pine with limited savanna species, (2) wet pine–hardwood forests along wetland margins and swamps, and (3) longleaf pine or pine–hardwood fragments (<1 acre to several acres) with significant remnant populations of savanna species in the understory. For category "1" areas, vegetation research was conducted on ordination factors of old-field communities relative to remnant stands (Smith 2000); overstory conditions and soil tillage effects on seed and seedling establishment for wiregrass (Outcalt et al. 1999); and survival, development, dispersal, and establishment of founder populations along an environmental gradient (Foster and Imm unpublished). For category "2" areas, the SRS established a large experiment on restoration and habitat relationships in the wetlands and upland margins of periodically burned and nonburned Carolina bays. For category "3" areas, studies include an evaluation of thinning and herbicide use on existing understory communities (Harrington and Edwards 1999); an assessment of midstory removal (mechanical, chemical) on existing populations of savanna species (Glitzenstein, unpublished); and a test of the influence of establishment conditions on seed and seedling survival for different species (Primack and Walker 2003).

Landscape Restoration Strategies

Given the land use impacts and uncertainty about original species composition, as well as SRS national defense missions, the stewardship goal is to restore and sustain the native savanna species characteristic of the region. Operational experience at SRS has demonstrated that prescribed burning alone is rarely sufficient to redevelop the native species composition or structure. Periodic burning of large areas in South Carolina is constrained by smoke management regulations. When combined with fuel moisture and weather restrictions, it is extremely difficult to effectively apply periodic burning to large acreage. In addition, the total absence or low numbers of native flora and fauna (e.g., gopher tortoise) severely limits natural recolonization, and woody vegetation competition further suppresses plant development. The operational strategies taking shape at the landscape scale combine prescribed burning, mapping, monitoring, reintroductions of native species, and various silvicultural technologies to achieve the stewardship goals.

The current recovery plan for the red-cockaded woodpecker was realigned to facilitate prescribed burning to reduce competition and litter and duff thickness (Edwards et al. 2003). The primary recovery area is easier to burn, and the soils, topography, and vegetation relationships are consistent with areas historically dominated by savanna communities (Duncan and Peet 1996; Frost 1997; Imm and McLeod 2004). Early succession or regeneration areas are being maintained by reducing planting densities and by initiating burning within a year or two of planting. These areas act as surrogates for savanna habitat because they provide analogous structure required (Johannsen 1998; Krementz and Christie 1999). The current reintroduction of the gopher tortoise at SRS is focused on these areas.

Within the landscape, detailed mapping of sensitive species and remnant fragments is being used to define appropriate silvicultural and burning techniques. For remnant forest fragments with residual savanna species, silvicultural efforts include (1) increased burning frequencies, (2) reductions in tree densities through thinning, (3) mechanical and chemical removal of midstory shrubs, saplings, and overstory hardwoods, and (4) the conversion of off-site species to longleaf pine. For old-field sites, dominated by weedy species a cost-effective strategy for plants may be establishment of founder populations of selected species in heavily thinned, or clear-cut gaps that are periodically burned. With the exception of species such as wiregrass, seed introductions have not been very successful (Primack and Walker 2003). Establishment of native plants by planting small container stock has been promising.

References

Duncan, R.P., and Peet, R.K. 1996. A template for the reconstruction of the natural fire-dependent vegetation of the Fall-line Sandhills, southeastern United States. SRI 96-23R, US Forest Service-Savannah River, New Ellenton, SC.

Dunning, J.B., Jr., Borgella, R., Jr., Clements, K., and Meffe, G.K. 1995. Patch isolation, corridor effects, and colonization by a resident sparrow in a managed pine woodland. *Conserv Biol* 9:542–550.

Edwards, J.W., Smathers, W.M., Jr., LeMaster, E.T., and Jarvis, W.L. 2003. Savannah River Site red-cockaded woodpecker management plan. US Forest Service-Savannah River, New Ellenton, SC.

Frost, C.C. 1997. Presettlement vegetation and natural fire regimes of the Savannah River Site. SRI 97-10-R, US Forest Service-Savannah River, New Ellenton, SC.

Gaines, G.D., Franzreb, K.E., Allen, D.H., Laves, K.S., and Jarvis, W.L. 1995. Red-cockaded woodpecker management on the Savannah River Site: A management/research success story. In *Red-Cockaded Woodpecker: Recovery, Ecology, and Management*, eds. D.L. Kulhavy, R.G. Hooper, and R. Costa, pp. 81–88. Center for Applied Studies, College of Forestry, Stephen F. Austin State University, Nacogdoches, TX.

Grant, B.W., Brown, K.L., Ferguson, G.W., and Gibbons, J.W. 1994. Changes in amphibian biodiversity associated with 25 years of pine forest regeneration: Implications for biodiversity management. In *Biological Diversity: Problems and Challenges*, eds. S.K. Majumbar, F.J. Brenner, J.E. Lovich, J.F. Schalles, and E.W. Miller, pp. 354–367. Easton: The Pennsylvania Academy of Science.

Harrington, T.B., and Edwards, W.B. 1999. Understory vegetation resource availability, and litterfall responses to pine thinning and woody vegetation control in longleaf plantations. *Can J For Res* 29:1055–1064.

Imm, D.W., and McLeod, K.W. 2005. Vegetation types. In *Ecology and Management of a Forested Landscape: Fifty Years of Natural Resource Stewardship on the Savannah River Site*, eds. J.C. Kilgo and J.I. Blake, pp. 85–131. (In Review).

Johannsen, K.L. 1998. Effects of thinning and herbicide application on vertebrate communities in young longleaf pine plantations. MS thesis, University of Georgia, Athens.

Kilgo, J.C., Miller, K.V., and Moore W.F. 2000. Coordinating short-term projects into an effective research program: Effects of site preparation methods on bird communities in pine plantations. *Stud Avian Biol* 21:144–147.

Krementz, D.G., and Christie, J.S. 1999. Scrub-successional bird community dynamics in young and mature longleaf pine–wiregrass savannas. *J Wildl Manage* 63:803–814.

Outcalt, K.W., Williams, M.E., and Onokpise, O. 1999. Restoring *Aristida stricta* to *Pinus palustris* ecosystems on the Atlantic Coastal Plain, USA. *Restor Ecol* 7(3):262–270.

Peet, R.K., and Allard, D.A. 1995. Longleaf pine vegetation of the South Atlantic and Eastern Gulf Coast regions: A preliminary classification. *Proc Tall Timbers Fire Ecol Conf* 18:45–81.

Primack, R.J.L., and Walker, J. 2003. Dispersal and disturbance as factors limiting the distribution of rare plant species at the Savannah River Site and the Carolina Sandhills National Wildlife Refuge. SRI 03-29-R, US Forest Service-Savannah River, New Ellenton, SC.

Smith, G.P. 2000. Structure and composition of vegetation on longleaf pine plantation sites compared to natural stands occurring along an environmental gradient at the Savannah River Site. MS thesis, Clemson University, Clemson, SC.

White, D.L., and Gaines, K.F. 2000. The Savannah River Site: Site description, land-use and management history. *Stud Avian Biol* 21: 8–17.

Yates, M.D., Loeb, S.C., and Guynn, D.C., Jr. 1997. The effect of habitat patch size on small mammal populations. *Proc Annu Conf Southeast Assoc Fish Wildl Agenc* 51:501–510.

Chapter 11

Reintroduction of Fauna to Longleaf Pine Ecosystems

Opportunities and Challenges

Ralph Costa and Roy S. DeLotelle

Introduction

In this chapter, we discuss reintroduction, via translocation, of native fauna into longleaf pine forests. We focus on rare species, including those considered "sensitive," "of special concern," or "candidates" for listing by conservation groups, or state or federal agencies, and on species federally listed as either "threatened" or "endangered" under the Endangered Species Act of 1973, as amended. We cover our topic in four main sections. First, we provide a brief background on the broad concept of reintroductions and translocations as a species-conservation tool. Second, we use the red-cockaded woodpecker (*Picoides borealis*) as a case study to outline how this regional program has succeeded. This section provides a literature review of all relevant issues regarding translocation of red-cockaded woodpeckers. Our goal in using the red-cockaded woodpecker as a case study is to provide an empirical framework so the reader can apply the basics of what we have learned about this species' reintroduction and translocation potential to other species. In our third section we provide brief summaries about other species that may benefit from a reintroduction program. Finally, in our fourth section, we provide a model that may be useful for planning future reintroductions of fauna in longleaf pine forests.

Reintroduction and Translocation: A Conservation Strategy Review

The International Union for Conservation of Nature and Natural Resources/Species Survival Commission, Re-introduction Specialist Group (IUCN/SSC RSG) provides general guidelines for reintroductions considered appropriate internationally (see IUCN/SSC RSG 1995). Their goal is to "...help ensure that the re-introductions [hereafter referred to as reintroduction(s)] achieve their intended conservation benefit, and do not cause adverse side-effects of greater impact." The guidelines' target audiences are managers and scientists who are directly responsible for planning, approving, and carrying out reintroductions. The guidelines contain six important sections: (1) definition of terms, (2) aims and objectives

---

Ralph Costa • U.S. Fish and Wildlife Service, Clemson Field Office, Department of Forestry and Natural Resources, Clemson University, Clemson, South Carolina 29634   **Roy S. DeLotelle** • DeLotelle and Guthrie, Inc., 1220 SW 96th Street, Gainesville, Florida 32607.

of reintroduction, (3) multidisciplinary approach, (4) preproject activities (biological, and socioeconomic and legal requirements), (5) planning, preparation, and release stages, and (6) postrelease activities.

The IUCN/SSC RSG defines "reintroduction" as "an attempt to establish a species in an area which was once a part of its historical range, but from which it has been extirpated or become extinct." They define "translocation" as a "deliberate and mediated movement of wild individuals or populations from one part of their range to another." In this chapter, "augmentation" refers to the translocation of red-cockaded woodpeckers from a donor population to another population to increase its size; "reintroduction" refers to the translocation of individuals from a donor population to unoccupied habitat that was probably occupied historically. We will use "translocation" as a generic term for most discussions related to augmentations and/or reintroductions. In this chapter, we will primarily discuss augmentations, because only one reintroduction of red-cockaded woodpeckers has been accomplished as of 2003 (Hagan and Costa 2001). In the case of red-cockaded woodpeckers, for all practical purposes there is little difference from either a planning or implementation perspective between augmentations and reintroductions.

Griffith et al. (1989) provided the first comprehensive analysis of the status of translocation as a conservation tool. They summarized information on 93 species of birds and mammals (90% game species, 86% success; 7% threatened, endangered, or sensitive, 46% success) translocated from 1973 to 1986 in Australia, Canada, Hawaii, New Zealand, and the United States. Almost 700 translocations were conducted annually, using both captive and wild stock. They found success improved with increased habitat quality, larger numbers of released animals, translocations into the core of a species' historical distribution versus on the periphery or outside the range, and translocations into areas without competitors of similar life form or a congeneric potential competitor. Due to limited survey question responses they were unable to evaluate effects of age, sex, and genetics on success. Griffith et al. (1989) concluded that "...without high quality habitat..." translocations would have poor success regardless of how many animals are released.

Wolf et al. (1996) conducted a follow-up survey to Griffith et al. (1989) and examined 421 bird and mammal translocation programs in Australia, New Zealand, and North America. They documented increases from the results of Griffith et al. (1989) in three potentially important parameters related to successful programs. The median number of animals translocated per program (31.5 to 50.5), the median duration of releases (2 to 3 years), and the proportion of projects releasing greater than 30 individuals (46% to 68%), all increased. Like Griffith et al. (1989), Wolf et al. (1996) also found that high-quality, i.e., "good to excellent," habitat, releases into the core of the species' historical range, and large numbers of translocated individuals positively affected translocation success. In contrast to Griffith et al. (1989), Wolf et al. (1996) found no significant relationship between translocation success and either first age of reproduction ("early or late breeder" in Griffith et al. 1989) or number of offspring ("size of clutches" in Griffith et al. 1989). Both studies found translocation programs for mammals were more successful than those for birds.

Gordon (1994) developed a dichotomous key to provide natural resource managers a tool to help assess the biological and genetic needs and impacts of translocating species. The key is useful for both plants and animals and considers the following factors: (1) degree of threat to species being considered, (2) dispersal from release site, (3) genetic risks, (4) cause of threat, (5) donor source, (6) competitive interactions, (7) consumptive interactions, (8) contamination risks, and (9) release site management. Gordon (1994) suggests that prior to any translocations, decisions should be well documented (the key provides the means to do so) and sufficient postrelease monitoring planned and implemented.

Scott and Carpenter (1987) argued that a lack of reliable published information existed to adequately evaluate success of endangered bird translocation programs. Without such accounts, they concluded that it was difficult to accurately assess the species' status or evaluate success of specific procedures being employed. Furthermore, they concluded that because of the high cost of endangered bird translocation programs, we must ensure high survival probabilities of individual birds and meaningful contributions to the gene pool. To address these issues and concerns they proposed the following guidelines (this partial list only includes those factors related to wild stock): (1) document bird preparation and release methodologies, e.g., capture techniques, holding cages, and transportation methods, (2) record release conditions (see Ellis et al. 1978), and (3) frequently monitor movements and activities of released birds through the first breeding season. Scott and Carpenter (1987) summarized by suggesting that improved documentation of translocation programs applies not only to endangered birds but also to nonendangered birds and numerous other taxonomic groups.

Lessons learned from this review of the translocation literature include: (1) preplanning with documentation is critical, (2) high-quality habitat at the recipient site is mandatory, (3) programs should be restricted to historic range, (4) large numbers of individuals should be moved, (5) a review of literature relevant to one's species or taxa must be conducted, and (6) monitoring postrelease success so that programs can be improved is crucial. Readers are encouraged to review the summary literature discussed above and other current, relevant work related to particular species or taxa before engaging in translocation programs (e.g., see Jones, S. R., ed. 1990. *Endangered Species Update*, Volume 8, No. 1).

Reintroduction of Fauna in Longleaf Pine Ecosystems

Longleaf pine (*Pinus palustris*) occurred in monospecific stands over a larger area than any other tree species in the presettlement landscape of the United States (Platt et al. 1988a; Abrahamson and Hartnett 1990; Frost this volume). Longleaf pine is key to maintaining forest structure in fire-climax flatwoods and pine savannas important to many protected wildlife species (Stout and Marion 1993). These pine forests have been extensively altered by cattle grazing, ditching, fire suppression, conversion to agricultural lands, logging, and naval stores extraction. Today, few pine forests exist that are representative of presettlement conditions. Remaining pine forests differ from those of the presettlement era in having longer fire return intervals, a more even-aged structure, and a denser understory with greater shrub cover and less herb cover. These alterations have led to habitat fragmentation, reduced faunal diversity, and extirpation of many vertebrate species in most longleaf pine landscapes.

The community's dominance in the landscape was facilitated by a suite of pyrogenic traits that favored longleaf pine's survival over hardwoods or other species of pine (Stout and Marion 1993). The physiognomy of classic pine forest is characterized by an emergent tree layer of pines and a ground cover that, although appearing to be species-poor, is one of the most species-rich on the planet (Walker and Peet 1983; Platt et al. 1988b; Peet this volume). Fire is important in maintaining diversity of ground cover and the structural aspects necessary for many wildlife species (Engstrom et al. 1984; Robbins and Myers 1992). In addition to fire, seasonal variation in water availability, low and flat topography, and depression areas also influence structure of pine forests and community types.

Several types of depressional wetlands (e.g., cypress domes, woodland ponds) are found within the matrix of the heterogeneous landscape, as are isolated fragments of scrub, and larger areas of scrubby flatwoods or oak ridges. Size of these embedded communities varies greatly; however, collectively they can provide considerable coverage and diversity. Vegetation of these embedded habitats within the pine forest matrix varies with edaphic conditions, hydroperiod, and basin topography (Abrahamson and Hartnett 1990). A relatively rich fauna is associated with these forests (Harris and Vickers 1984; Means this volume). All amphibians and many reptiles

TABLE 1. List of potential fauna for reintroduction into longleaf pine ecosystems.

Common name	Scientific name	Status[a]	Reintroduction potential[b]
Amphibians			
Mississippi gopher frog	Rana capito sevosa	E	No Recovery plan (more info required)
Flatwoods salamander	Ambystoma cingulatum	T	No Recovery plan (more info required)
Reptiles			
Eastern indigo snake	Drymarchon corais couperi	T	Recovery plan (more info required)
Gopher tortoise	Gopherus polyphemus	T	Recovery plan (yes; disease problems)
Blue-tailed mole skink	Eumeces egregius lividus	T	Recovery plan (no; more info required)
Sand skink	Neoseps reynoldsi	T	Recovery plan (no; more info required)
Louisiana pine snake	Pituophis ruthveni	C	More info required
Birds			
Mississippi sandhill crane	Grus canadensis pulla	E	Recovery plan (yes; captive bred)
Red-cockaded woodpecker	Picoides borealis	E	Recovery plan (yes; wild stock, artificial cavity installation)
Bald eagle	Haliaeetus leucocephalus	T	Recovery plan (yes; wild stock, nest site construction)
Southeastern american kestrel	Falco sparverius paulus	N3	Yes; nest box installation
Bachman's sparrow	Aimophila aestivalis	N3	More info required
Mammals			
Red wolf	Canus rufus	E	Recovery plan (yes; captive bred)
Florida panther	Felis concolor coryi	E	Recovery plan (yes; captive bred)
Florida black bear	Ursus americanus floridanus	N2	More info required
Lousiana black bear	Ursus americanus luteolus	T	Recovery plan (no; more info required)
Sherman's fox squirrel	Sciurus niger shermani	N2	More info required

[a] Status is based on Endangered Species Act listing (E = endangered, T = threatened, C = candidate) or if not federally listed, The Nature Conservancy and the Natural Heritage Network rankings (N = national status; 2 = imperiled, 3 = vulnerable to extirpation or extinction).

[b] Potential is based on information in an approved recovery plan ("yes" or "no" indicates whether reintroduction/augmentation is discussed as a recovery activity) or the authors' evaluation that more information (info) is required for unlisted species and for listed species with and without recovery plans, even if species with plans do not specifically list reintroductions/augmentations as a recovery activity.

of the longleaf pine ecosystem are obligatorily linked with water (e.g., breeding and feeding sites) or upland ridges and, therefore, maintaining a healthy link between uplands and wetlands with varying hydroperiods will result in higher abundance and diversity of species within these taxonomic groups (Vickers et al. 1985).

As a conservation strategy, reintroducing extirpated and augmenting existing small populations of fauna in longleaf pine forests is in its infancy. Red-cockaded woodpeckers are the only resident species with a significant history of translocations, while several species have been supported with limited reintroductions. These include red wolf (Canus rufus), Mississippi sandhill crane (Grus canadensis pulla), and southeastern American kestrel (Falco sparverius paulus) programs. Numerous additional listed, rare, or sensitive species may be potential candidates for similar initiatives (Table 1). Significantly, these potential candidates cover all terrestrial vertebrate taxa, including reptiles, amphibians, birds, and mammals. The list of potential candidates is substantial, covers many life forms over a broad geographic area, includes multiple physiographic provinces, ecoregions, and longleaf pine habitats, and represents the analyses and opinions of numerous federal and state agencies and nongovernmental organizations.

Reintroducing or augmenting native fauna in longleaf pine ecosystems should be considered for the following reasons: (1) recovery of federally or state listed threatened or endangered species, (2) conservation of federal and state candidate and sensitive species, and (3) conservation of rare or keystone species.

In a broad sense, these specific reasons address larger conservation goals, including enhancing long-term survival of a species, maintaining or restoring biodiversity, and increasing conservation awareness (IUCN/SSC RSG 1995). Additionally, specific reasons address legal mandates, e.g., achieving recovery, political, and administrative issues, e.g., precluding the need to list a species, cultural and social issues, e.g., informing and educating the public about the value of saving rare species, and finally, ecological issues, e.g., reintroducing keystone species, which in turn benefits biodiversity by positively affecting a host of other species.

Translocation Case Study: Red-cockaded Woodpecker

This section covers seven major topics: (1) need for translocations, (2) threats to small populations, (3) why translocations work, (4) translocation success, (5) translocation concerns, (6) translocation strategies, and (7) the future. The red-cockaded woodpecker is found in several primary pine forest types within its range, including longleaf pine, shortleaf pine (*Pinus echinata*), slash pine (*Pinus elliottii*), and loblolly pine (*Pinus taeda*). Although forest types within a particular property can be essentially "pure" on a stand-by-stand basis, mixed stands, with one pine species comprising the majority of the overstory, are more typical throughout the species range. Translocations of red-cockaded woodpeckers have taken place in all forest types. To date, no studies have indicated that the forest type affects the outcome (see Edwards and Costa 2004). Therefore, we include literature and data from studies in all forest types and assume that our discussion of red-cockaded woodpecker translocations is relevant to all appropriate southern pine forest types.

Need for Translocations
Prevent Extirpations

Historically, as many as 920,000 groups of red-cockaded woodpeckers may have inhabited the longleaf pine forests from Virginia south to Florida, and west to Texas (Costa 2001). In 2003, 5800 occupied clusters of cavity trees were known throughout the species range, an estimated 99% decline (Costa and Jordan 2003). Initial dramatic population declines were a result of large-scale, extensive habitat loss, primarily from logging, fire suppression, and the naval stores industry (see U.S. Fish and Wildlife Service 2003 for details and additional references).

Once historic populations were fragmented into smaller, disjunct populations, other local factors adversely impacted persistence of many local populations. Factors including population isolation, fragmentation, and size, hardwood midstory encroachment, lack of suitable cavity trees, and demographic and environmental stochasticity caused further declines (Costa and Escano 1989; Conner and Rudolph 1991; Crowder et al. 1998; Letcher et al. 1998; Walters et al. 2002). In combination, these local factors contributed to the sometimes rapid and steady declines, and in numerous cases extirpations, of remnant populations.

Walters (1991) provided a further interpretation of red-cockaded woodpecker population dynamics and its implications for dealing with population decline, i.e., preventing extirpations. His findings support the principle that combating limiting habitat factors, particularly cavity availability, is key to saving and restoring small populations of red-cockaded woodpeckers. In addition, well-planned and -executed translocations, adequate quantities of good-quality nesting and foraging habitat, and an aggressive prescribed burning program will all be necessary to save very small populations from extirpation (e.g., see Brown and Simpkins 2004; Drumm et al. 2004; Hedman et al. 2004; Stober and Jack 2004).

Achieve Recovery

Recovery of the federally listed red-cockaded woodpecker (listed as "endangered" in 1970, 35 Federal Register 16047) depends on establishing sufficient numbers of varying size populations, well-distributed throughout their historic landscape, to enable the species to counteract threats inherent to survival of small populations. Specifically, recovery, or "delisting," of the red-cockaded woodpecker involves establishing at least 29 populations in 11 recovery units (i.e., ecoregions) with the following sizes (size = potential breeding groups): 1 with 1000, 10 with 350, 9 with 250, 3 with 100, and 6 with 40 (see U.S. Fish and Wildlife Service 2003 for details). The 29 "populations" are comprised of 65 individual properties on private, state, and federal lands. Individual properties may or may not be contiguous to other properties comprising a population.

Any population or noncontiguous property harboring fewer than 30 potential breeding groups is eligible to receive translocated birds. Of the 29 recovery populations, 15 (52%) contain fewer than 30 potential breeding groups. Of the 65 properties, 38 (58%) harbor fewer than 30 potential breeding groups; 32 (49%) are noncontiguous. In addition to these populations directly involved in delisting, management and conservation of numerous other populations, i.e., "significant" support populations (see U.S. Fish and Wildlife Service 2003), is critical to help achieve recovery. Many of these support populations also qualify for, and are participating in, translocation programs.

Factors necessitating a translocation program include the fragmented nature of the Southeast's longleaf pine forests, wide and isolated distribution of populations, and inadequate dispersal to increase and sustain demographic health of many small populations. Small, isolated populations are subject to multiple threats and the only way to combat these threats is by increasing population size. Translocation is a powerful management tool capable of relatively rapidly increasing population size to a more self-sustaining level. Once populations harbor 30 or more potential breeding groups, continued population growth can be achieved without reliance on interpopulation translocations (see U.S. Fish and Wildlife Service 2003). However, occasional interpopulation translocations may be necessary to maintain or improve genetic viability (Haig et al. 1993).

Besides preventing extirpation of local populations and achieving recovery for listed species, translocation of other fauna into longleaf pine ecosystems could have various objectives. These include, to: enhance the long-term survival of a species, reestablish a keystone species, maintain and restore natural biodiversity, promote conservation awareness, or a combination of these (IUCN/SSC RSG 1995). Prior to any fauna translocation program in longleaf pine forests, a feasibility study, including a species status review and thorough review of available literature on the species, must be completed. In addition to identifying the need for the program, this background research will help identify and, thereby, focus the stated objectives of the translocation. Once identified, objectives must be clearly stated, understood by all involved parties, and capable of being evaluated on a regular basis to ensure they are being accomplished.

Threats to Small Populations

Threats to small populations include demographic and environmental stochasticity, genetic uncertainty (both inbreeding and genetic drift), and natural catastrophic events (Shaffer 1981, 1987). Based on recent research, population sizes and spatial configurations of red-cockaded woodpecker groups within populations necessary to withstand potential extirpation risks associated with demographic and environmental stochasticity are better understood (Crowder et al. 1998; Letcher et al. 1998; Walters et al. 2002).

Demographic Constraints

Research has demonstrated that the probability of maintaining red-cockaded woodpecker populations is dramatically improved as population size increases and territories are maximally aggregated (e.g., 1 group per 200 acres) (Crowder et al. 1998; Letcher et al. 1998). Since most territories are large, higher population densities can be difficult to attain (Hooper et al. 1982; DeLotelle et al. 1987). Additionally, Crowder et al. (1998) suggested that proximity of adjacent populations might increase the persistence of small populations (also see DeLotelle et al. 2004). Importantly, results of both Crowder et al. (1998) and Letcher et al.

(1998) support the concept of maintaining existing and establishing new populations, even if available habitat limits the ultimate population size to 10–25 groups. It is vital that these small populations be dense and maximally aggregated within suitable habitat (Walters et al. 2002).

Neither the Crowder et al. (1998) nor the Letcher et al. (1998) models allowed for immigration. Letcher et al. (1998) acknowledged that immigration might help support population growth and stability in small populations. However, they noted that the effect of the population's spatial distribution on immigration remains untested. In the absence of immigration, via dispersal, translocation provides a management tool to help maintain small population viability.

Environmental Stochasticity

Environmental stochasticity can affect red-cockaded woodpecker populations in several ways. For example, an exceptionally severe winter could result in high adult mortality, or a prolonged drought during the nesting season could result in poor nestling survival and, therefore, lower recruitment. These types of annual population-level effects have the potential, at least over the short term, to affect population parameters such as group size and recruitment of helpers. In small populations these effects can be even more pronounced leading to population declines (DeLotelle et al. 1995).

Walters et al. (2002) conducted a population viability analysis for red-cockaded woodpeckers by incorporating environmental stochasticity parameters, e.g., probabilities of producing different numbers of fledglings each year, into the demographic model developed by Letcher et al. (1998). Their study results produced two important conclusions. First, density and distribution of groups have as significant an effect on maintaining viable populations through time (that is population fitness) as population size itself. Second, small, highly aggregated populations were relatively stable. Walters et al. (2002) attribute their findings to the "...buffering effect of helpers on population dynamics." That is, when environmental stochasticity results in annual variation in mortality or reproduction, the effect is essentially absorbed by the helper class and, therefore, the overall number of occupied territories remains unaffected. However, ability of the helper class to function as an effective buffer against threats of environmental stochasticity is a function of not only the population's size, but also critically, its density and aggregation (Walters et al. 2002). Significantly, our ability to strategically, via translocations, locate new groups where needed to maximize density and connectivity within populations is and will remain a key management objective as we build small populations to and beyond the threshold sizes necessary to withstand natural demographic (100 potential breeding groups) and environmental (250 potential breeding groups) threats.

Genetic Considerations

Both the short- and long-term survival of small- to moderate-size red-cockaded woodpecker populations will require genetic management. Haig and Nordstrom (1991) provided an excellent primer and guide on genetic management of small populations. Ultimately, in small populations, loss of genetic diversity results in a species' inability to adapt to changing conditions (Selender 1983). Haig and Nordstrom (1991) pointed out that "...loss of genetic diversity is manifested by more severe effects resulting from..." random genetic drift and an increase in levels of inbreeding. While genetic drift can threaten even large populations, inbreeding depression threatens only small populations. The effect of genetic drift, i.e., rate of loss of genetic variation, increases with shrinking population size and mutation rate. Inbreeding depression is one effect of genetic drift (Lacy 1987).

Inbreeding affects fecundity, fertility, development rate, age of sexual maturity, and number of offspring, i.e., factors that control survival and productivity of individuals (Ralls et al. 1988). Inbreeding depression has been demonstrated in red-cockaded woodpecker populations (Daniels and Walters

2000a; Daniels et al. 2000). Daniels et al. (2000) found that populations of at least 40, but perhaps as many as 100, potential breeding groups are necessary to avoid inbreeding depression and concluded that viability of small isolated and declining populations is seriously threatened by inbreeding depression. However, they also calculated that an immigration rate of two or more migrants per year would likely protect against inbreeding depression. Where immigration rates are insufficient (less than two per year), Daniels et al. (2000) recommended translocation as a possible tool to reduce inbreeding.

Stangel et al. (1992), examining genetic variation and population structure, found that most small populations exhibited normal levels of heterozygosity. Based on their findings, they suggested that small populations: (1) should not be considered "lost causes," (2) are valuable reservoirs of unique genetic combinations, and (3) serve as "steppingstones" for gene flow among larger populations. They recommended management and conservation of both small and large populations to maintain the species' historic genetic population structure. Because they found evidence that "... genetic distance increases with geographic distance..." they recommended that donor populations be as near as possible to recipient populations to minimize "disrupting locally adapted populations."

Haig et al. (1994) found no population-specific, diagnostic genetic markers among the 101 red-cockaded woodpeckers sampled from 14 populations rangewide. Significantly, both Stangel et al. (1992) and Haig et al. (1994) concluded that population-specific alleles were not useful for differentiating populations. This suggests that although the current distribution of red-cockaded woodpeckers consists of many, relatively small, fragmented populations, these populations have not been isolated long enough to result in the loss or gain of specific genes within a population (Haig et al. 1994). Based on their overall findings, and in agreement with Stangel et al. (1992), Haig et al. (1994) concluded that nearby populations, as opposed to distant ones, should serve as donors.

Haig et al. (1996), sampling small populations in south Florida, found that birds were no more differentiated or isolated than other populations throughout their range and that there were no genetic differences among birds in longleaf and slash pine habitats. They concluded, "... habitat type does not constrain dispersal ... among geographically proximate populations in south Florida," but recommended translocating red-cockaded woodpeckers from nearby populations and using donors with similar habitat types. However, Edwards and Costa (2004) suggested that, when nearby donor populations of the same habitat type are limited to other small populations, such as in south Florida, translocating birds from larger, nearby populations of different habitat types is acceptable and not likely to be deleterious.

Generally, genetic diversity increases while threats associated with genetic drift decrease with population size. Therefore, population growth strategies must be integral components of small population conservation and management programs. Translocations provide one important tool to grow larger populations and, thereby, increase genetic diversity. Similarly, deleterious effects of genetic drift, including inbreeding depression, can be minimized by natural immigration into small populations and "managed" by translocation of individuals into populations. All genetic research and analyses to date indicate that small populations of red-cockaded woodpeckers are worth saving and likely can be genetically managed via translocations. Similarly, we believe the "small populations are valuable" paradigm may hold true for many longleaf pine ecosystem fauna, although this theory remains untested for most species.

Catastrophic Events

Two types of natural catastrophes potentially threaten red-cockaded woodpecker populations: (1) severe winds (hurricanes, downbursts, and tornados) and (2) southern pine beetle epidemics. While impacts from tornados and downbursts will typically be localized, hurricanes have the potential to affect entire populations. Effects of Hurricane Hugo on the red-cockaded woodpecker population and its habitat on the Francis Marion National Forest, South Carolina, were devastating (Hooper et al. 1990). Because most populations are located in the Atlantic and Gulf Coastal Plains they face a substantial risk from hurricanes (Hooper and McAdie 1995). Reducing the threat of hurricanes at the species level can be accomplished by distributing numerous populations throughout the species range, including insular populations.

Southern pine beetles are not a population-level threat to red-cockaded woodpeckers in longleaf pine forests. Although longleaf pine has high natural resistance to southern pine beetles, loblolly and shortleaf pine are not nearly as resistant, especially under stress (for example, growing on poor soils, at high stocking, severe drought conditions). The rapid and dramatic loss of over 40,000 acres of shortleaf, pitch (*Pinus rigida*), and Virginia pine (*Pinus virginiana*) in the Daniel Boone National Forest, Kentucky, and the subsequent human-induced "extirpation" of the red-cockaded woodpecker population illustrates the catastrophic potential of southern pine beetles (Mills et al. 2004). Proper forest management, including maintaining properly stocked stands and restoring longleaf pine to appropriate sites, can improve a forest's resistance to catastrophic beetle epidemics. Fortunately, at a smaller scale, the loss of individual cavity trees can be mitigated with artificial cavity replacement (Copeyon 1990; Allen 1991).

Summary of Threats to Small Populations of Longleaf Pine Ecosystem Fauna

Managers and biologists considering reintroductions or augmentations of fauna in longleaf pine ecosystems must thoroughly understand the threats to small populations of the target species. This is critical because small populations will remain at risk for a relatively long period of time. Therefore, it is vital to know at what population sizes the risks of demographic, environmental, and genetic stochasticity are overcome. If species-specific data on population sizes and associated resilience to threats are not available, information for similar taxa should be researched. It is imperative that research identifies and understands the potential for population loss and reduction from natural catastrophes. Additionally, it is critical to design translocation programs to withstand the possible effects of catastrophic events to the maximum extent practicable. Proceeding with translocation programs with population goals (sizes, locations, and configurations) that are set without knowledge and consideration of threats to small populations is inefficient and will jeopardize the program's success.

Why Translocations Work

Red-cockaded Woodpecker Demographics and Sociobiology

The red-cockaded woodpecker is a nonmigratory, territorial, cooperative breeder with all group members participating in nesting season activities and territorial defense (Lennartz et al. 1987). Groups can consist of a solitary bird, usually a male, or a potential breeding pair with zero to four helpers. Helpers are usually male offspring from previous breeding seasons; however, female helpers are common in some populations (DeLotelle and Epting 1992). Walters et al. (1988a) studied the demographics and sociobiology of a large population in the North Carolina sandhills. Their model is appropriate to North Carolina and perhaps other populations, but some populations in different parts of the species' range exhibit different levels of these demographic components. The following percentages are from the Walters et al. (1988a) North Carolina study; percentages are the annual change from one breeding season to the next.

Male fledglings may remain (27%) on their natal territory as helpers for several years waiting for the opportunity to inherit it, or occupy an adjacent territory, and subsequently acquire breeding status upon the death of the breeding male. Approximately 13% of male fledglings disperse and 39% of those become breeders on another territory. Another 31% of the dispersing males find a vacant territory or establish a new one and remain there as a solitary bird. These birds have trouble attracting mates and those that do are usually unsuccessful as breeders the first year (Daniels and Walters 2000a; Leonard et al. 2004). Dispersing males that do not establish or occupy a territory, but continue searching for one, are called "floaters"; they represent about 25% of the annual fledgling pool. Fledgling male annual mortality is high, about 57%. Female fledgling mortality is also high, estimated at 68%. Of the female fledglings that disperse from their natal territory and survive (about 31%), 92% become breeders their first year. Once established as breeders, females remain in that status (56%), disappear (31%), or move to another territory (12%) the following year. Young males and females typically disperse more frequently in the early fall (October) and the late spring (March) than other times of year (J. Walters, Virginia Polytechnic Institute, personal communication, and R. DeLotelle unpublished data). Dispersal patterns of both female and male subadults from their natal territories and their ability to breed their first year are behaviors that apparently make the species conducive to translocation.

Red-cockaded Woodpecker Dispersals and Metapopulation Theory

With the exception of a few large data sets for the North Carolina Sandhills red-cockaded woodpecker population (see Walters et al. 1988a; Daniels and Walters 2000b), rangewide knowledge regarding the frequency of both short- and long-distance dispersals of this species was limited (or unpublished) prior to the 1990s. With the increased emphasis on saving small populations and the associated need to band large segments of donor populations and all birds in recipient populations for the translocation program, the number of birds annually banded since the early 1990s has steadily increased. This existing and growing inventory of banded birds has proven invaluable to conservation efforts for the species.

Not only has our knowledge about numbers of dispersing birds and distances moved been considerably expanded in recent years, but also our understanding of how dispersal potentially affects demographic and genetic health of populations substantially increased. Lay and Swepston (1973), studying red-cockaded woodpeckers in east Texas, recorded the first long-distance movement of the species, 42 km by an adult male. It would be 15 years before Walters et al. (1988b) reported the next long-distance dispersal of a bird, 86 km by a breeding female in the North Carolina Sandhills. Subsequent long-distance dispersals ranging in distance from 27 to 275 km have been recorded in South Carolina (Jackson 1990), from Oklahoma to Arkansas (Montague and Bukenhofer 1994), from North Carolina to South Carolina (Ferral et al. 1997), in Texas (Conner et al. 1997), and in Florida (Lowery and Perkins 2002).

Schiegg et al. (2002) examined effects of fragmented populations on dispersal. Their results suggested that habitat fragmentation could affect dispersal success, thereby adversely affecting population dynamics in small (25 groups) populations. They suggested that the cooperative breeding system of the red-cockaded woodpecker "...may stabilize small aggregated populations, which may even act as populations sources." Schiegg et al. (2002) concluded that small populations likely have conservation value. Similarly, Haig et al. (1993:296) suggested that as her small study population (Savannah River Site, SC) increased in size, it "...could serve as a source of immigrants to numerous other local populations." Indeed, at least five birds have dispersed from this small population to distant populations: two to Fort Gordon, GA, and three to Fort Jackson, SC; four of five became breeders postdispersal. Daniels et al. (2000), investigating inbreeding, also concluded that small populations are important and suggested enhancing dispersal via retention and management of as many small populations as possible within a region. Additionally, they recommended "...linking disjunct inhabited areas..." (Haig et al. 1993:147) using recruitment clusters.

TABLE 2. Long-distance dispersal patterns for 110 marked red-cockaded woodpeckers.

	Total	Age		Breeder[a]		Mean distance dispersed	Population size[b]	
		Subadults	Adults	Yes	No		Small & medium	Large
Female	64 (58.2%)	38 (52.7%)	21 (65.6%)	41 (60.3%)	9 (52.9%)	31.4 ($n=59$)	28 (66.7%)	31 (53.5%)
Male	46 (41.8%)	34 (47.3%)	11 (34.4%)	27 (39.7%)	8 (47.1%)	27.8 ($n=44$)	14 (33.3%)	28 (46.5%)
Total	110	72	32	68	17	29.6 ($n=103$)	42	59

[a] Indicates breeding status for 85 of the 110 birds where status was documented.
[b] Size of population bird dispersed to: small (40 potential breeding groups; PBG), medium (41–249 PBG), large (>249 PBG).

DeLotelle et al. (2004) summarized 110 dispersals (see Table 2) ranging in distance from 7 to 325 km; 60% exceeded 17 km. More dispersers were females (58.2%) than males and subadults (69.2%) than adults. Adult females dispersed more frequently (65.6%) than adult males. The majority (80%) of birds dispersing became breeders. Overall, females and males dispersed about the same average distance (about 30 km); however, subadults dispersed farther (30 km) than adults (22 km). Finally, 41.6% of dispersals were within or to small and medium populations from a larger population, e.g., large to medium or medium to small. DeLotelle et al. (2004), based on these data, years of research in central Florida, and a broader (see Harrison 1994; Gutierrez and Harrison 1996) interpretation of "classic" metapopulation theory (see Levins 1969), suggested that red-cockaded woodpeckers may currently function as metapopulations in certain parts of their range.

The metapopulation definition that we choose is less stringent than the conservative definition of "demographic rescue" (Hanski and Simberloff 1997; DeLotelle et al. 2004). We believe that present-day aggregations of small red-cockaded woodpecker populations in relatively close proximity to one another more appropriately fit the "demographic recovery" definition. In this sense, demographic recovery involves movement among red-cockaded woodpecker populations sufficient to assist one another in demographic stability (see DeLotelle et al. 2004). Further, a recent book (*Genetics, Demography, and Viability of Fragmented Populations*, edited by A.G. Young, G.M. Clarke, M.L. Gosling, G. Cowlishaw, R. Woodroffe, and J. Gittleman, 2000; chapter entitled "The Metapopulation Paradigm: A Fragmented View of Conservation Biology") states that "[t]he ever-growing diversity of empirical and theoretical studies that demonstrate the importance of spatial structure in deterring ecological and evolutionary trajectories also indicates that long-term conservation programs need to focus on regional rather than local within-population dynamics." This is the approach we advocate.

In the context of metapopulation theory, all of the above findings and recommendations represent important conservation issues. With few exceptions, all researchers and managers who have investigated red-cockaded woodpecker dispersal as it relates to genetic, and in some cases demographic, maintenance of small populations concluded that saving such populations has conservation value. Translocations provide the opportunity, both short and long term, to save, expand, and maintain isolated and insular small populations. Additionally, translocations, along with improvements in habitat linkages via "steppingstones" (Stangel et al. 1992) or "island corridors" (Costa and Edwards 1997), can be potentially used to restore or maintain structure and function of metapopulations.

The relatively high and improving success rate of red-cockaded woodpecker translocations is not surprising when the bird's sociobiology, likely historic and documented present dispersal patterns, genetic and demographic metapopulation structure, and population demographics are understood (Walters et al. 1988a; Haig et al. 1993, 1996; Letcher et al. 1998; DeLotelle et al. 2004).

That is, translocations replicate probable historic short- and long-range dispersal patterns that would have occurred in large contiguous areas of forest, and in naturally fragmented landscapes where birds likely functioned as a metapopulation, e.g., central and south Florida. That red-cockaded woodpeckers disperse within and among today's highly fragmented populations and landscapes is well documented (Walters et al. 1988a; Daniels and Walters 2000a,b; DeLotelle et al. 2004). Dispersal is a primary behavior of the species, directly related to an individual's fitness, i.e., the ability of most females and some males to obtain breeding status. Translocations simulate subadults' natural behavior to disperse. Given the very high, annual natural mortality rates (57% male, 68% female) of subadults (Walters et al. 1988a), and high retention rates (percent of birds remaining in their recipient population through at least one breeding season) of translocated subadults (65%, Hedman et al. 2004; 72%, Hagan et al. 2004a; 82%, Stober and Jack 2004), the probability of survival and breeding for individual subadult birds, on average, is actually increased if they are translocated from their natal territory to a new population.

Translocation Success

Definitions of Success

Translocation success of individual red-cockaded woodpeckers has been defined in various ways (Allen et al. 1993; Costa and Kennedy 1994; Hess and Costa 1995; Carrie et al. 1999; Franzreb 1999; Edwards and Costa 2004). Therefore, meaningful interpretations and comparisons of existing data sets remain difficult. Translocation success at the population scale is relatively easy to quantify, being measured in number of potential breeding groups. Instituted in 1998, the U.S. Fish and Wildlife Service's Annual Red-cockaded Woodpecker Population Data Report now provides a comprehensive and systematic reporting procedure to help track and document success of all translocated birds. Intensive monitoring at recipient populations is crucial to understanding successes and failures of any translocated animals. Without such information adaptive management will not be possible.

In this section, we primarily summarize overall success of translocations at the population scale. That is, we measure success by evaluating changes of basic population parameters, including number of occupied clusters, birds, and potential breeding groups. All of our study populations harbored fewer than 23 potential breeding groups when they began a translocation program. Therefore, by definition, all were considered "small," i.e., fewer than 30 potential breeding groups. Indeed, 27 (90%) of the 30 populations harbored fewer than 14 potential breeding groups; 17 (57%) had fewer than 5. Populations exceeding 30 potential breeding groups are not eligible to be translocation recipients (U.S. Fish and Wildlife Service 2003). Such populations are expected to expand on their own via use of recruitment clusters, prescribed burning, and, as needed, intrapopulation translocation.

The growth of the Marine Corps Base Camp Lejeune red-cockaded woodpecker population (34 potential breeding groups in 1992 to 67 groups in 2002) is an excellent example of how populations larger than 30 potential breeding groups can be significantly increased in a relatively short time, without translocation, by using appropriate habitat improvement and management techniques (Walters 2004). Similarly, recipient populations must also have rigorous habitat programs in place to receive birds, including four suitable cavities per recruitment cluster, two recruitment clusters available for each pair of birds received, and an aggressive prescribed burning program (U.S. Fish and Wildlife Service 2003). Given that quality habitat conditions were required and all recipient populations were essentially consistent in this regard, we believe the rapid growth of our study populations was directly related to translocation. Furthermore, we have no examples of populations similar in size than our study populations that have substantially increased in size without translocations.

We strongly recommend identification of success criteria for all longleaf ecosystem fauna translocations prior to program implementation. However, more than one definition may be necessary depending on the programs' objectives. Additionally, it is imperative that development and implementation of comprehensive postrelease monitoring protocols for all translocation programs be established. Without clearly defined and understood measures of success and comprehensive monitoring to confirm translocation results, program improvements will be questioned and support for the initiative will be at risk.

Translocation Results

Here we offer a preliminary summary of the overall population level successes of red-cockaded woodpecker translocations. From 1989 to 2002, 1014 birds were translocated: 941 came from 11 major donor populations and 73 came from 19 minor donor populations. Birds were translocated to 30 primary recipient populations and 19 secondary recipient populations. During the 14 years, the mean number of birds translocated from the 19 minor donors was 4, ranging from 1 to 8. Most of these birds went to primary recipient populations. Of the 19 secondary recipient populations, 17 (89%) received fewer than 10 birds (range 1–14). These birds came from both major and minor donor populations. Because the majority of translocations involving minor donor populations and secondary recipient populations involved fewer than 10 birds during the entire period they are not included in the analyses below.

Overall, numbers of occupied clusters, total birds, and potential breeding groups more than doubled in the 30 primary recipient populations from initiation of translocation to the nesting season of 2002 or 2003; the latest years available for the corresponding data (Table 3). On average, by population, occupied clusters increased from 7.5 to 15.8, number of birds from 17.7 to 43.6, and potential breeding groups from 6.0 to 13.9. Populations received birds for 2 to 14 years, average of 6.4. The number of birds received annually ranged from 1.4 to 10.8 and averaged 5.0. From initiation of translocation to 2002 or 2003, each population received, in total, an average of 29 birds; range of 3–130. The majority (70%) of populations received more than 20 birds and 43% received 30 or more. These results demonstrate that, on average, small populations receiving five birds per year doubled in size in approximately 6 years.

In the 11 major donor populations, number of pre-translocation occupied clusters and potential breeding groups increased 8.7% and 17.3%, respectively, during this period (Table 4). Over the 14-year period, 941 birds were translocated from the 11 donor populations, averaging 85.5 (range 12–253) birds per

TABLE 3. Pretranslocation program initiation and 2002 or 2003 population demography (number of active clusters, adult birds, and potential breeding groups) for 30 primary recipients of translocated red-cockaded woodpeckers, and total number of birds translocated to the properties from 1989 to 2002.

		Pretranslocation			# birds translocated[d]	# years[e]	# birds per year[f]	2002 or 2003		
	# populations	# clusters[a]	# birds[b]	# PBG[c]				# clusters[g]	# birds[h]	# PBG[i]
Total	30	225	532	180	866		148.5	475	1307	416
Mean		7.5	17.7	6.0	28.9	6.4	4.95	15.8	43.6	13.9

[a] Number of active clusters prior to translocation.
[b] Number of adult birds prior to translocation.
[c] Number of potential breeding groups prior to translocation.
[d] Total number of birds translocated to the populations from different populations (i.e., intrapopulation moves are not included).
[e] Mean number of years populations have received birds; most populations received some birds each year.
[f] Mean number of birds received annually.
[g] Number of active clusters pre-nesting season in 2002 or 2003.
[h] Number of adult birds pre-nesting season in 2002 or 2003.
[i] Number of potential breeding groups pre-nesting season in 2002 or 2003.

TABLE 4. Pretranslocation program initiation and 2002 or 2003 population demography (number of active clusters and potential breeding groups) for the 11 primary donors of translocated red-cockaded woodpeckers and total number of birds translocated from the properties from 1989 to 2002.

	# populations	Pretranslocation		# birds translocated[c]	# years[d]	# birds per year[e]	2002 or 2003	
		# clusters[a]	# PBG[b]				# clusters[f]	# PBG[g]
Total	11	1958	1626	941		97.8	2145	1968
Mean		178.0	148.0	67 birds per year	8.3	8.9	195.0	179.0

[a] Number of active clusters prior to translocation.
[b] Number of potential breeding groups prior to translocation.
[c] Total number of birds translocated from all populations from 1989 to 2002.
[d] Mean number of years populations donated birds.
[e] Number of birds per year donated.
[f] Number of active clusters pre-nesting season in 2002 or 2003.
[g] Number of potential breeding groups pre-nesting season in 2002 or 2003.

population and 67 birds per year. On average, populations donated birds for 8.3 (range 2–14) years and translocated 8.9 (range 3–18.1) birds annually. Although, in total, donor populations increased in size, 6 of the 11 populations only remained stable during their donor years. Population increases in the other 5 populations were directly attributable to habitat management programs, including prescribed burning and artificial cavity installation in occupied and recruitment clusters. Without such population and habitat management and evaluation procedures, managers may falsely conclude that removal of birds from the population may contribute to their inability to increase their population. Conversely, a declining population with an aggressive habitat management and translocation program in place may indicate an adverse effect from translocation. However, no such examples exist, and furthermore U.S. Fish and Wildlife Service donor population translocation guidelines (see U.S. Fish and Wildlife Service 2003) would very likely prevent this scenario from occurring.

Factors Potentially Related to Success

Various factors, working independently or cumulatively, affect the short- and long-term success of individual bird translocations. Similarly, these factors in combination with population-level variables affect the long-term translocation success at the population scale. Because more than 1000 red-cockaded woodpeckers have been translocated to 49 different populations since 1986, we have a good understanding of what affects success at both the individual bird and population level for this species. Notably, most studies prior to those published in *Red-cockaded Woodpecker: Road to Recovery* (Costa and Daniels 2004) focused on factors related to success of individual bird translocations, e.g., weather, forest type, recruitment cluster conditions, age, status, and sex of bird, distance moved, time of year, and method of release. However, numerous small population translocation studies published in Costa and Daniels (2004) focused on factors that, in the authors' opinions, affected overall success of their population-level growth, e.g., kleptoparasite control, numbers of birds translocated, and condition of forest structure. Using all of these "success" studies, we review factors important for red-cockaded woodpecker translocations and suggest how these and perhaps other factors would affect reintroduction of other fauna into longleaf pine forests.

Factors affecting translocation success of red-cockaded woodpeckers and other fauna may be broadly categorized as follows: environmental, e.g., weather, forest type, physiographic province (equivalent to ecoregions and recovery units), habitat conditions (e.g., cavity suitability and forest structure), and kleptoparasites (predators and competitors for other species); demographic, e.g., age, status (breeder, helper, fledgling), and sex; logistics, e.g., distance translocated; temporal, e.g., season of year and time of day moved; translocation type, e.g., single bird to an established solitary bird or multiple, unrelated subadult pairs; release method, e.g., from cavity, aviary, or free-release, and program methodology, e.g., duration of translocations and numbers of animals translocated. Although no red-cockaded woodpecker research team has incorporated, or controlled for all of these variables in their translocation analyses, one or more authors have examined these factors for their effect at the individual bird or population level.

Environmental Factors

Weather, Forest Type, Physiographic Province

Only one study has examined effects of weather conditions, forest type, and physiographic province at both capture and release sites on translocation success of individual birds. Edwards and Costa (2004), by establishing the conservative success criteria of "an individual remaining at the release cluster, followed by pairing and breeding," were able to make inferences regarding effects of selected environmental variables on translocation success of about 158 individual birds from 1989 to 1995. Overall, 27% of translocations were successful. Success was not related to weather, forest type, or physiographic province. Haig et al. (1996) suggested forest habitat type does not limit dispersal of red-cockaded woodpeckers. This may, in part, explain why Edwards and Costa (2004) found no significant differences in success of translocated birds from one forest type to another. Although Edwards and Costa (2004) found no differences in success related to physiographic province, in consideration of minimizing the potential for genetic disruption of "locally adapted populations," Stangel et al. (1992) and Haig et al. (1994, 1996) suggested using local or geographically proximate populations as donors. Typically, these donors would be in the same physiographic province as the recipient.

Published reviews of translocations (e.g., Griffith et al. 1989; Wolf et al. 1996) do not consider effects of weather, forest type, or physiographic province on translocation success of any animal species. Effects of these variables on red-cockaded woodpecker translocations were not significant. Likely, in part, the nonsignificance of weather was directly related to temperature and precipitation guidelines established for red-cockaded woodpecker translocations to avoid potential adverse effects (see U.S. Fish and Wildlife Service 2003). We suggest that similar guidelines be developed for any animal species being considered for translocation. Additionally, we recommend that for other species being considered for reintroduction into longleaf pine forests, caution be exercised in choosing forest type and physiographic province or ecoregion of the donor population; near is better than distant. We base this recommendation on the need to minimize disruption of locally adapted populations.

Habitat Conditions

Cavity Suitability and Quantity. Hess and Costa (1995) attributed several failed translocations in 1989/1990 to poor condition and quality of release cavities. Fortunately, all of these situations could be remedied post-1990/1991 due to development of artificial cavity technology (see Copeyon 1990; Allen 1991). They also recommended a minimum of three suitable cavities be available when a female is being released in a solitary male cluster. Similarly, numerous managers have recommended that alternative roost cavities, for example, at least four per cluster, should be available at translocation recipient clusters (Hagan et al. 2004a; Lohr 2004; Stober and Jack 2004). Others recommended providing extra cavities to minimize impacts of kleptoparasites and ensure all potential fledglings had a roost cavity (Marston and Morrow 2004).

After a decade of artificial cavity use for translocations, several researchers have presented data suggesting cavity age and cavity maintenance may affect translocation success of individual birds. Saenz et al. (2004) suggested that the newest artificial (inserts) cavities, those less than 1 year old, positively influenced translocation success when compared to translocations to older (greater than 1 year old) artificial cavities. However, due to confounding variables in their data set they acknowledged, "... it is possible that the number of successful translocations is independent of the condition [age] of the inserts...." However, Marston and Marrow (2004) also suggested that their population increase was in part attributed to maintaining cavity inserts in a "newer-like condition." Neal and Montague (2000), strong proponents of prolonging insert suitability via regular interior cleaning and exterior maintenance, have substantially increased their red-cockaded woodpecker population using artificial cavities and translocations. Many of their cavities have been in use for numerous (5–10+) years.

Forest Structure. Most managers and researchers agree that habitat quality of the release cluster and surrounding territory is important to translocation success; however, its impact has never been measured. It is, and will remain for the near future, difficult to impossible to evaluate effects of some factors thought to affect translocation success. Primarily, this is because investment in individual birds is high and managers cannot afford (ecologically, politically, socially, or economically) to use translocated birds in experiments to test, for example, their response to less-than-optimal habitat. Additionally, sample sizes would always be small, i.e., fewer than 30 potential breeding groups, for both control and treatment groups, leaving correlative or experimental results to be questioned.

Red-cockaded woodpeckers evolved in an open, parklike ecosystem and have subsequently been extirpated from forests where these conditions were not present, e.g., forests with a dense hardwood midstory (see Costa and Escano 1989). We believe translocation success is improved when forest structure replicates historic conditions to the maximum extent possible (see Platt et al. 1988a). Although untested, and, therefore, speculative, professional opinions among managers and researchers responsible for expanding small populations is that their population-level translocation successes are largely a result of "best possible condition" of cluster habitat (Hess and Costa 1995), "high-quality" habitat (Hagan et al. 2004a), "high quality of the habitat's structure" (Stober and Jack 2004), and plentiful habitat in "excellent condition" (Hedman et al. 2004). Maintaining the open, midstory-free forest structure required for participation in red-cockaded woodpecker translocation programs is best accomplished with prescribed fire (see U.S. Fish and Wildlife Service 2003).

Habitat condition at recipient sites for all animal reintroductions is of paramount importance (Griffith et al. 1989, 1990; Kleiman 1990; IUCN/SSC RSG 1995; Wolf et al. 1996). It is critical that all limiting habitat factors be identified, addressed, and mitigated. All of the target species' required habitat components, including suitable area for translocations and natural population growth, and appropriate structure and composition, must be in place prior to program initiation. Adequate amounts of high-quality habitat, including microhabitat components, like nest structures, must be the foundation of any longleaf pine ecosystem fauna translocation program.

Kleptoparasites

Relationships between kleptoparasites (Kappes 1997) and translocated red-cockaded woodpeckers are difficult to study due to the significant value of individual donor birds and sample size (see above). For example, managers would not intentionally want to expose their translocated birds to predation or cavity kleptoparasitism by not controlling target predators or kleptoparasites if they believed control was necessary. Although some associations between red-cockaded woodpeckers and red-bellied woodpeckers have been investigated, our overall knowledge of kleptoparasite impacts remains very limited, particularly for small populations (Kappes and Harris 1995). For example, little to no research exists on relationships between other avian cavity usurpers, e.g., red-headed woodpeckers (*Melanerpes erythrocephalus*) and northern flickers (*Colaptes auratus*), and red-cockaded woodpeckers.

Similarly, only one study examined effects of southern flying squirrels on red-cockaded woodpecker nesting success in a small population (Conner et al. 1996). However, relationships between translocated birds and southern flying squirrels were not investigated in that study. Additionally, the study was correlative, not experimental, and was based on observational evidence of a small sample size. Conner et al. (1996) found no negative effect of southern flying squirrels on red-cockaded woodpeckers and concluded that any effect on a "healthy" red-cockaded woodpecker population is likely minimal; we believe impacts may not be minimal if set within an existing metapopulation structure. By definition, small populations receiving birds via translocations cannot be considered healthy. They are constantly at risk from potential stochastic demographic and environmental threats. Indeed, Conner et al. (1996) acknowledged that when control is considered beneficial, it should only occur while the red-cockaded woodpecker population is "small and vulnerable to extirpation." Small and vulnerable describes translocation recipient populations.

A few additional studies, both controlled experiments, researched southern flying squirrel and red-cockaded woodpecker relationships (Laves and Loeb 1999; Mitchell et al. 1999). However, both of these studies were conducted in large (greater than 100 potential breeding groups) populations and neither involved translocated birds. Laves and Loeb (1999) found that southern flying squirrels' use of red-cockaded woodpecker cavities in the breeding season had a significant negative effect on red-cockaded woodpecker reproduction. Southern flying squirrel removal resulted in significantly more fledglings produced in treatment (removal) clusters versus control (no removal) clusters. They concluded that in very small or declining red-cockaded woodpecker populations (e.g., translocation recipient populations), where any nest loss could result in extirpation, southern flying squirrel control might be necessary. In contrast, Mitchell et al. (1999) found no increase in the number of fledglings in groups where southern flying squirrels were removed. They suggested removal of southern flying squirrels is unnecessary in "healthy" populations of red-cockaded woodpeckers.

Given the current absence of experimental data regarding interactions between southern flying squirrels and red-cockaded woodpeckers in small populations involved with translocations, we rely, for now, on professional opinion whether controlling southern flying squirrels increases population-level translocation success. Numerous managers and researchers involved in growing very small populations to larger sizes, e.g., from fewer than 3 to 8 to 10 occupied clusters, within a relatively short period of time (4 to 5 years), suggested that controlling southern flying squirrels was a contributing factor in their translocation success. In some cases, large numbers of squirrels were removed, e.g., 26–108 per month for 5 years (Allen et al. 1993) and 2304 over 9 years (Franzreb 1997b), while red-cockaded woodpecker populations increased substantially from 1 to 6 pairs and 1 to 19 pairs, respectively.

Managers of many other small population expansion programs, dependent on translocations, have agreed that southern flying squirrel control is an important management activity. Marston and Marrow (2004) suggested that southern flying squirrel removal contributed to: (1) increased nesting and fledgling success, (2) occupation of recruitment clusters by dispersing fledglings, and (3) quality and longevity of cavities. All of these observations related directly or indirectly to translocation success. Other studies have been more direct about benefits of southern flying squirrel removal on translocation success. Stober and Jack (2004) stated, "flying squirrel removal is an essential management action when establishing new red-cockaded woodpecker populations." Hedman et al. (2004) believed that southern flying squirrels were "...exerting considerable pressure..." on translocated birds (presumably for cavities, although this was not stated), and concluded that controlling southern flying squirrels had a "major role" in the growth of their red-cockaded woodpecker population. Similarly, Poirier et al. (2004) believed that the 14-fold red-cockaded woodpecker population increase would not have occurred without a southern flying squirrel removal program. Lohr (2004), studying a very small population, suggested southern flying squirrels posed a threat to efforts to expand that population. In the first red-cockaded woodpecker reintroduction (see Hagan and Costa 2001), Hagan et al. (2004a) believed southern flying squirrels were potentially "limiting population expansion"; therefore, they were controlled for the first several years of the reintroduction program.

In conclusion, many managers and researchers agree that controlling southern flying squirrels is/may be necessary in small, at-risk red-cockaded woodpecker populations. We do not advocate control as a long-term population maintenance activity in medium to large populations, but as a potential short-term population establishment or small population stabilization strategy. Kappes (2004) concluded that kleptoparasites might affect red-cockaded woodpecker population size (number of groups) through their impacts on reproduction and exacerbating cavity limitation (see Poirier et al. 2004). For these reasons and others discussed above, it may/will be necessary, on a case-by-case basis, to control kleptoparasites, particularly southern flying squirrels, in translocation recipient populations.

Translocation success has been positively associated with introducing animals into areas without congeneric competitors, morphologically similar species, or species with similar life history traits (Griffith et al. 1989, 1990). Results discussed above, regarding southern flying squirrels, support the premise that competitors (kleptoparasites) with at least one similar, but critical, life history trait, i.e., cavity use, can potentially affect success of a translocation program. This effect will likely be particularly severe if the shared trait involves the primary limiting factor for the introduced species, for example roosting and nesting cavities for red-cockaded woodpeckers. Reintroduction programs for other longleaf pine ecosystem fauna should identify potential competitors, their presence and abundance, and need for and method of control at the recipient location. Control, if required, should be carefully monitored and designed in conjunction with the increasing size of the translocated population to be a relatively short-term program. If, based on preliminary analyses, it is determined control will be necessary for years (10+) even as the translocated population increases in size, the overall reintroduction program should be reevaluated.

Demographic Factors
Age, Status, and Sex

Early (1986 to 1995) research and population rescue translocations of red-cockaded woodpeckers used birds of various ages and status, e.g., breeder, helper, or subadult, of both sexes (Allen et al. 1993; Hess and Costa 1995). Documented successes of these translocations were somewhat inconsistent, in part due to differences in success definitions, small sample sizes, and inadequate monitoring (Rudolph et al. 1992; Costa and Kennedy 1994). In spite of these inconsistent results, several findings were noteworthy. Franzreb (1999) found no significant differences in success between translocation of younger (5–7 months old) and older (8–12 months old) subadult females. Similarly, success was not related to subadult age (4 to 8 months) for either males or females in another study (Edwards and Costa 2004). Most studies did not recommend translocating helper or adult males due to their homing tendency or poor success rates (Rudolph et al. 1992; Carrie et al. 1996; Franzreb 1999; see Table 5). Conversely, most studies recommended translocating either subadult or adult females; success ranged from moderate to high (Allen et al. 1993; Costa and Kennedy 1994; Hess and Costa 1995; Franzreb 1999; Edwards and Costa 2004; see Table 5).

Based on past research and translocation results, we have an excellent understanding of which age class and sex ratios to translocate to maximize translocation success for red-cockaded woodpecker populations. It is imperative for other translocation candidates

TABLE 5. Red-cockaded woodpecker translocation success[a] reported from 1994 to 2004.[b]

Study	Translocation type			Success criteria
	Female to male	Male to female	Potential pair	
Allen et al. (1993)	40% (4 of 10)	25%	33% (2 of 6)	Remained in the population and successfully bred
Costa and Kennedy (1994)	62% (48 of 77)	38%	33% (18 of 54)	Ranged from "interacted well" to "fledged young"
Hess and Costa (1995)	61% (11 of 18)	—	—	Remaining at release cluster through subsequent breeding season
Carrie et al. (1999)	—	—	65% (11 of 17)[c]	Remained in the population and successfully bred
Franzreb (1999)	82% (18 of 22)	33%	40% (4 of 10)	Remained in the vicinity of release cluster for ≥30 days
Edwards and Costa (2004)	42% (36 of 86)	15%	13% (10 of 79)	Remaining at the release cluster, followed by pairing and nesting

[a] Success was measured in various ways. Success applies to individual or multiple translocation "events" over time involving individual birds, pairs of birds, or multiple pairs of birds. Success is not, in this table, referring to the success of growing target populations.
[b] Source (with slight modifications): Edwards and Costa (2004).
[c] Included translocation of 5 potential pairs to inactive clusters and later 6 additional single birds (3M, 2F) to solitary individuals, and 1 solitary male was released to determine whether he would remain at the release site.

that appropriate age class, breeding status, and sex ratios are determined. This is best accomplished basing decisions on species' life history, ecology gained from studies of wild populations, and by following adaptive management principles based on field trials with rigorous monitoring. We recommend following these procedures with all translocation programs.

Logistic Factors

Distance Translocated

Several studies have evaluated effects of distance moved on red-cockaded woodpecker translocation success. Franzreb (1999) found that less than 8% of birds moved 18 to 464 km returned home, while almost 42% of birds moved less than 6 km returned home. Edwards and Costa (2004), summarizing 159 translocations, found no significant differences among distances moved. Their mean distance for successful moves was 350 km (range of 2 to 630) and for unsuccessful moves was 405 km (range of 8 to 493). In another study, four adult males moved less than 19 km returned to their capture clusters (Allen et al. 1993). Others have also documented birds, particularly males, returning to their capture clusters when moved short to moderate distances, e.g., 3 km (Rudolph et al. 1992), 35 km (Reinman 1984) and 16, 26, and 29 km (same bird moved three times) (Carrie et al. 1996). Following the U.S. Fish and Wildlife Service's translocation bird allocation guidelines and given the paucity of donor populations and dispersion of recipient populations throughout the Southeast, few red-cockaded woodpeckers are now translocated less than 32 km, except for intrapopulation moves.

We found no "distance moved" related summaries for other species' translocations. For longleaf pine ecosystem fauna translocations, we recommend that unless there is some compelling reason to move animals long distances, for example, their ability to home, they be moved as short a distance as practical. This strategy may minimize costs, stress related to time in captivity, and potential for disease transmission. Additionally, doing so may help avoid disruption of local gene complexes and ensure that probable local morphological adaptations are not compromised in a different environment.

Temporal Factors

Season of Year

Edwards and Costa (2004) found that success rates for birds translocated in fall (September–December) and spring (January–April) were not different. Since birds are not moved until the fall, it must be that maturity of young birds has been accounted for by the time they reach 5–6 months of age. These translocation periods are consistent with dispersal patterns for both young males and females in natural situations.

Time of Day

Edwards and Costa (2004) found no significant differences in success whether birds were moved during the day or night. Day moves involved trapping the bird at dawn, transporting and feeding it all day, and placing it in its release cavity at dusk to be released the following morning. Night moves involved trapping the bird at dusk, transporting it to its recipient cluster, typically by midnight, placing it in its release cavity, and releasing it at dawn. Although "time in captivity" data were not collected, typically day moves involved considerably more time in captivity than night moves. Additionally, day moves required handling birds six to eight times for feeding (at 45-minute intervals), a potentially stressful situation. However, on average, a bird moved in the day spent 4 to 6 more hours in its release cavity, for a total of about 12 hours, than a bird moved in the night. This not only provided more potential rest compared to birds moved at night, which only spent 6 to 8 total hours in their release cavity, but also allowed for additional cavity and site acclimation time. However, the contribution, if any, of these factors to success is questionable, given the lack of significant differences between day and night moves.

Based on our red-cockaded woodpecker data set we offer the following recommendations for other fauna being considered for reintroduction in longleaf pine forests. First, we suggest the time of year animals are translocated should be based on their dispersal patterns or response to breeding seasons. Additionally, weather, ability to obtain food, and ability to avoid, or at least minimize, predation and competition must be considered. Finally, translocations should attempt to minimize social disruption of, and maximize the probability of integration into, the extant population. Regarding time of day, and with the exception of the possible need for daylight to acclimate to specific habitat needs for some species, we believe it to be a minor or unimportant factor for most species' translocations. If these or other temporal factors are thought to affect the potential for translocation success, they should be addressed prior to initiation of the program.

Translocation Type

Single Bird to Established Solitary Bird

Success of translocating single birds to established solitary birds has varied depending on age and sex of both the bird moved and the recipient bird (see Table 5). Allen et al. (1993) found that success of translocating females to solitary males had mixed results, primarily depending on the male's age. Females moved to adult males all bred, whereas no females moved to subadult males bred. Costa and Kennedy (1994) documented that 62% of females translocated to solitary males were successful. In contrast, only 25% of males moved to solitary females were successful. Hess and Costa (1995) found that 61% of females translocated to solitary males were successful. Franzreb (1999) recorded 82% success for females moved to resident males. Finally, Edwards and Costa (2004) documented success rates of 42% for females moved to solitary males.

Multiple, Unrelated Subadult Pairs to Recruitment Clusters

Although early translocations (1986 to 1995) of unrelated, subadult pairs to establish potential breeding groups were minimally successful (see Table 5), most managers and researchers continued to believe the technique had promise (Costa and Kennedy 1994; Franzreb 1999; Edwards and Costa 2004). Importantly, none of these early attempts involved multiple pairs of birds, a technique that would not be instituted as a management practice until the late 1990s (see Rudolph et al. 1992).

In 1994/1995, Carrie et al. (1999) further investigated the pair translocation technique, but used multiple pairs, although not simultaneously. They achieved a 65% success rate, defined as birds remained in the population and bred. Based on this success, and the relatively poor success of single-pair translocations, the U.S. Fish and Wildlife Service adopted a policy of not, except in extenuating circumstances, translocating single pairs. Instead, pair translocations after about 1997 involved multiple pairs, typically a minimum of two to three, and not uncommonly four or five, and were conducted on the same day.

It was decided to thoroughly evaluate the multiple-pair, simultaneous release concept while concurrently attempting for the first time to reintroduce a red-cockaded woodpecker population *de novo*, i.e., in the absence of a founder population. The Turner Endangered Species Fund and the U.S. Fish and Wildlife Service embarked on a 5-year research project (Hagan and Costa 2001). The study proposal was fully implemented in 1998 and results were remarkably successful. By 2002, the population harbored 15 occupied clusters, 15 potential breeding groups, and 58 birds (Hagan et al. 2004a). The success of simultaneously releasing multiple pairs of birds is directly related to the principle that success increases with increasing numbers of animals released.

All small population growth success stories from the mid to late 1990s through the early 2000s employed the multiple-pair (typically two to five pairs), same-day release strategy (Hagan and Costa 2001) rigorously implemented by Hagan et al. (2004a) (Brown and Simpkins 2004; Hedman et al. 2004; Lohr 2004; Marston and Marrow 2004; Morris and Werner 2004; Stober and Jack 2004). This model is the new and successful paradigm for expanding small, at-risk red-cockaded woodpecker populations, and reintroducing new populations.

We recommend the structure and composition of fauna translocation cohorts that are reintroduced into an area be based on the species sociobiology. The number of animals to introduce with each translocation attempt, by age class, sex, breeding or social status, and other demographic factors, should be initially determined based on knowledge of wild populations' demographics and sociobiology. If a thorough understanding of species sociobiology is lacking, then research should be conducted to answer critical questions.

Release Method

From Cavity

With very few exceptions, all red-cockaded woodpecker translocations have involved the release method described by DeFazio et al. (1987). This usually involves trapping the bird at dusk with a small mist net or mosquito bag net attached to a telescoping pole (Jackson 1977), placing it in a wooden/wire box, transporting it to the release cluster, placing it in the appropriate cavity, placing wire screen over the cavity entrance with a string attached that reaches the ground, and pulling the screen off at dawn. When a bird is released in an established solitary bird group the translocated bird is not released until the resident bird exits its cavity. When unrelated subadults are released in a recruitment cluster, they are released simultaneously after each bird appears at its screen. Although this release method has proven acceptable for red-cockaded woodpeckers, different methods had been used for other bird species (see Ellis et al. 1978; Barclay and Cade 1983).

From Aviary

Scott and Carpenter (1987) suggested that one technique for translocating birds is to do a "soft" release, i.e., holding them in captivity for some designated time at the release site. They hypothesized this process may help birds become acclimated, and perhaps imprinted, to their new environment. Franzreb (1997a) designed and tested for durability a mobile, metal-framed aviary for the eventual holding and release of red-cockaded woodpeckers. Although proving durable, Edwards et al. (1999) found this design's takedown and reassembly to be complex and, in combination with its weight, essentially nonmobile. Therefore, Edwards et al. (1999) modified the design and constructed an aviary using PVC plastic pipe for the frame. Their final design met their objectives; it was lightweight and easy and fast (less than a day) to assemble. The efficacy of this design was used in two different experiments.

Franzreb (2004) confined 22 red-cockaded woodpeckers in aviaries. Thirteen (59%) birds completed the experiment, remaining in the aviary for 9 to 14 days. The other 9 birds either died in the aviary ($n = 2$) or were released ($n = 7$) for various reasons prior to 9 days. Of the 13 birds remaining in the aviary, 61% were successful upon release; success was defined as remaining at, or in the vicinity of, the release cluster for 30 days or more. Therefore, overall success rate was 36% (59% × 61%). The 61% of the birds successfully completing the experiment were not statistically different from "hard" release (see DeFazio et al. 1987) results (63%) reported by Franzreb (1999) for translocations at the same location. However, the 36% overall success is considerably poorer than hard release results. Edwards et al. (2004), using the same success criteria as Franzreb (2004), conducted six aviary translocations (birds held 10 to 14 days), paired with six concurrent hard releases. Only one soft-release bird (17%) remained for greater than 30 days (256+ days); another stayed 5 days and the other four dispersed the day of release. Based on results of these two studies, which had worse or no better success rates than hard releases, managers are understandably unwilling to use aviaries for red-cockaded woodpecker translocations, given equipment and construction costs when compared to hard releases, bird-maintenance requirements, and logistics.

Free-Release

Walters et al. (2004) reported success rates for free-released and cavity released birds. Free-release involved capturing birds at dawn, transporting them to their release cluster, and immediately releasing them. Of the free-release birds, 36% remained at their release cluster for at least 30 days, while 14% of cavity released birds did so. Although retention at the release clusters was relatively low, the majority (71%) of both the free-released and cavity released (76%) birds stayed within the study area (see Walters et al. 2004). In addition, six free-releases in central Florida were successful while two failed (R. DeLotelle unpublished data). Two of the six successful translocations included males to solitary females. If further studies verify free-release success rates are at least equivalent to cavity release rates, it may be a preferred method for short-distance moves. All of the free-releases of Walters et al. (2004) were intrapopulation and, therefore, birds were at their release cluster in several hours or less, providing adequate time for locating cavities and acclimating to the site prior to dark. Free-releases made later in the day would minimize or eliminate these potential benefits and may decrease success rates.

We believe that for the majority of potential longleaf pine ecosystem reintroduction candidates, some variation of cavity (e.g., nest box, burrow) or free-release method will prove practicable and successful. With some exceptions, e.g., red wolf, captive propagation and holding wild animals in captivity prior to release is not likely to become a major program in translocations. These methods can be prohibitively expensive and logistically challenging and, therefore, with the exception of animals at high risk (e.g., whooping cranes and Florida panthers), not preferred as long as wild, donor stock is available.

Translocation Methodology

Number and Duration of Translocations and Numbers of Animals Translocated

Griffith et al. (1990) found that success rates of threatened, endangered, and sensitive species translocations improved with increasing numbers of animals released and years of releases. Wolf et al. (1996) also found numbers of animals released to be associated with success; however, number and duration of releases were not important. Although our red-cockaded woodpecker data did not provide comparisons between populations for numbers released and duration and number of releases, we did find that, on average, small red-cockaded woodpecker populations could be doubled in size in approximately 6 years by translocating about 30 birds at a rate of 5 per year. Based on our results, and those of others, we recommend that to the maximum extent practicable translocation of other fauna into longleaf pine forests involve relatively large numbers of individuals released annually over multiple years. We believe such a strategy would increase both the survival probability of individual animals during the short term and the probability of successfully augmenting an existing, or reintroducing a new, population over the long term.

Translocation Concerns

Effects of Translocation on Donor Populations

A fundamental premise of red-cockaded woodpecker translocation programs is that donor populations must remain stable, if they have attained their population goal, or a minimum size or be increasing if they have not (U.S. Fish and Wildlife Service 2003). Expanding the size of larger populations has proven relatively straightforward when proper habitat and population management techniques are applied. Numerous examples of the population growth potential (5–10% annually, on average) of medium to large populations exist, including populations at Fort Bragg, NC (Britcher and Patten 2004), Fort Stewart, GA (Carlile et al. 2004), Fort Benning, GA (Doresky et al. 2004), Elgin Air Force Base, FL (Petrick and Hagedorn 2004), Marine Corps Base Camp Lejeune, NC (Walters 2004), and Carolina Sandhills National Wildlife Refuge, SC (U.S. Fish and Wildlife Service unpublished data). The common attributes of their recovery programs included aggressive recruitment cluster establishment, an artificial cavity management program, prescribed burning, and well-planned and implemented silvicultural practices. Except for Marine Corps Base Camp Lejeune, all of these sites are translocation donor populations.

Several studies have investigated effects of removing subadult birds from donor populations. Walters et al. (2004) documented a 4% increase in their "donor plot" (a subset of the entire population) as five to eight birds were removed annually. While unable to specifically identify why removal had no effect, they suggested the location of donor clusters in the center of the population and rigorous adherence to U.S. Fish and Wildlife Service guidelines specifically designed to minimize group- and population-level impacts, as probable reasons. However, Walters et al. (2004) did report a reduction of group size in their donor plot. Hagan et al. (2004b) found no significant differences in number of fledglings, mean group size, and percent change in occupied clusters between donor and nondonor groups. Acknowledging an experimental design problem, they still concluded that overall, translocations did not appear to harm the donor population.

In addition to these studies, it is encouraging that several primary donor populations continued to significantly increase in size, while birds were removed (see above). These and other examples provide critical documentation supporting the premise that large populations can function as a valuable source for translocated birds while simultaneously increasing their own size. Smaller populations can also serve as donors as long as rigorous group and population monitoring is conducted, translocation guidelines are followed, and birds are removed during years of surplus (DeLotelle et al. 1995; U.S. Fish and Wildlife Service 2003).

Because of the paucity of literature on effects of removing animals from wild donor populations, we strongly recommend that wild populations proposed as donors in longleaf pine fauna translocation programs be monitored annually. Monitoring must be adequate to ensure that potential adverse impacts from removals can be detected and distinguished from normal demographic variation. Monitoring is also required to determine availability, i.e., numbers and frequency, of animals for translocation. Understanding demography, reproductive potential, current size, and overall trend of donor populations is critical not only to determine availability of animals but also to evaluate potential impacts of removal.

Diseases and Parasites

Prior to the 1990s, only one study had been conducted on blood parasites of red-cockaded woodpeckers (Love et al. 1953). In recent studies several blood parasites have been found and some have been identified (Luttrell et al. 1995; Pung et al. 2000). Based on their findings, Pung et al. (2000) suggested caution if subadult, unexposed birds are translocated to populations where risk of infection is elevated by other woodpeckers or higher numbers of dipteran vectors. However, examination of individual birds for blood parasites is not practical given the numbers and timing of birds translocated. Importantly, no instances of disease-related mortality have been documented from the many translocations conducted to date. Nor do we believe the risk is significant, primarily based on the species' natural dispersal behavior.

Pung et al. (2000) found a "rich community" of arthropods in cavities, a cosmopolitan mite, *Androlaelaps casalis*, being the most common blood-feeding arthropod. This mite was found in 76% of the cavities at an average density of 51 (range 1–233) per cavity. However, they concluded that *A. casalis* had no effect on red-cockaded woodpecker fitness. Overall, they found occurrence and density of blood-feeding insects low, even in cavities that had been in constant use for 20 years.

There is potential to transmit diseases, endo- and ectoparasites, and infectious or contagious pathogens via translocation programs for longleaf pine forest fauna. For species that would otherwise not emigrate from one population to another, e.g., reptiles, amphibians, and small mammals, potential for adverse effects at both the recipient population and for translocated animals may be elevated. That is, the recipient population may never have been exposed to certain diseases, parasites, or contagious pathogens and, therefore, have no acquired immunity to them if they arrive with translocated animals (IUCN/SSC RSG 1995). For migratory and wide-ranging species and those species prone to dispersals, e.g., red-cockaded woodpeckers, between geographically isolated populations, the potential for transmission of diseases, parasites, and other pathogens may be minimal. However, translocating large numbers of animals over relatively short time periods may be a concern. At relatively low rates of natural immigration, recipient populations may have time to adapt and develop normal levels of resistance and immunity to diseases, parasites, and other pathogens. Under high rates of immigration, as mimicked by an intensive translocation program, adaptation may be difficult to impossible and result in deleterious effects to recipient populations. Caution must also be exercised to minimize risk of exposing translocated animals to vectors of disease agents that may be present at release sites, but not donor sites, for which they have no acquired immunity (IUCN/SSC RSG 1995).

The IUCN/SSC RSG (1995), Kleiman (1990), and many others have cautioned against the possibility of transmitting diseases via a translocation program. We share these concerns. However, given the frequency of long-distance dispersals (DeLotelle et al. 2004) and research results to date, we do not believe that disease or transmissions of other pathogens via translocations is a serious concern for red-cockaded woodpeckers. Given the level of current knowledge about high disease level in gopher tortoises (*Gopherus polyphemus*) and lack of information on other potential longleaf pine ecosystem translocation candidates, we are not as confident suggesting that transmissions of disease and other pathogenic agents are not a serious consideration for other species. We recommend a thorough review of available literature on each species, and field and laboratory analyses for diseases, parasites, bacteria, and other potential pathogens on a sample of both donor and recipient populations. These measures and precautions will help guide final decisions on whether or not to proceed with translocations. Additionally, based on findings of literature reviews and field and laboratory studies, if potential pathogens are discovered but a decision is made to proceed with a translocation program, specific guidelines, procedures, and protocols must be developed and implemented to avoid or minimize impacts on recipient populations and translocated animals.

Geographic Variation

Mengel and Jackson (1977) measured multiple morphological characteristics of 430 red-cockaded woodpecker museum specimens from throughout their range. They found latitudinal differences in wing and tail length, both being longer farther north. Additionally, culmens were relatively shorter in interior and northern populations, suggesting morphological differences can also occur along habitat gradients from Coastal Plain to Interior forest populations.

Although small and inconclusive from a management perspective, the body of knowledge regarding diseases, parasites, bacteria, and geographic variation suggests that for all longleaf pine ecosystem fauna translocations, the distance translocated should be limited as much as possible. Limiting distance translocated may help minimize transmission of pathogens and parasites because closer populations may be locally adapted to potential threats. Similarly, minimizing latitudinal distance moved may improve an animal's probability of survival given presumed adaptive significant of known, or assumed, morphological differences seen geographically or by habitat gradients.

Translocation Strategies

Regional Scale

In 1995, a multistate (Arkansas, Louisiana, Oklahoma, and Texas) partnership was formed to allocate annually the limited number of red-cockaded woodpeckers available from the only 2 donor populations to the approximately 20 recipient populations (Saenz et al. 2002). This partnership is known as the Western Range Translocation Cooperative (WRTC). With some modifications, this model has been replicated for remaining portions of the birds' range. In 2003, the Southern Range Translocation Cooperative (SRTC) (Alabama, Florida, Georgia, and Mississippi) had 7 donor populations and approximately 30 recipient populations, while the Northern Range Translocation Cooperative (NRTC) (Kentucky [formerly], North Carolina, South Carolina, and Virginia) had 3 donor populations and 3 recipient populations. Within cooperatives, birds are primarily allocated to eligible populations based on population size, availability of suitable habitat with artificial cavities installed, and success of past translocations. With release of the red-cockaded woodpecker recovery plan revision, it is expected that a population's role in recovery, e.g., a "primary core" versus a "significant support" population, will be increasingly important in its recipient status in the bird allocation process (U.S. Fish and Wildlife Service 2003).

The allocation process is essentially one of informed consent, whereby participants at annual translocation cooperative meetings determine recipient population priorities (see Saenz et al. 2002). Importantly, respective state wildlife agencies and the U.S. Fish and Wildlife Service have necessary veto power if they believe a particular recipient or donor is not eligible to receive or provide birds, respectively. For example, a recipient may be determined ineligible if release cluster habitat is not in excellent condition. A donor population would likely be denied interpopulation translocation authority if its population trend were not stable or increasing.

With few donors and many recipients in the WRTC and SRTC, the annual supply of available birds never meets the demand. This fact, along with the informed consent process, leaves managers unsure, on a year-to-year basis, whether they will be receiving birds. Saenz et al. (2002) recognized this issue and designed and tested various potential regional translocation models. Overall, they concluded that a partitioning strategy that annually allocated all available birds to the six largest recipient populations, called the "elitist" model, was best at meeting conservation and translocation goals, defined as management objectives. Importantly, the "random" model, representing the informed consent system, performed well, achieving rates of population growth only slightly lower, and with fewer population extirpations, than the elitist model.

Although Saenz et al. (2002) suggested using a single strategy to minimize costs and simplify the program, they also recognized flexibility is important for other reasons, including ensuring genetic diversity, addressing fluctuating budgets, meeting previous commitments from prior years, and satisfying diverse goals of managers. For the foreseeable future, the informed consent system will likely continue as the basic red-cockaded woodpecker allocation process in the WRTC and SRTC. In the NRTC, with only three recipients and three donors, long-term agreements have been established between donor and recipient pairs, eliminating the need for annual allocation meetings. However, numerous additional populations in North and South Carolina harbor fewer than 30 potential breeding groups and could be participating in translocation programs if they desired, thereby likely creating a need for annual meetings.

Population Scale

At the population scale, red-cockaded woodpecker (and most other longleaf pine forest species) translocations have four potential applications: (1) prevent extirpations, (2) reduce isolation of groups or subpopulations, (3) genetic management, and (4) reintroduction (U.S. Fish and Wildlife Service 2003). Red-cockaded woodpecker translocations have been used since the late 1980s and early 1990s to achieve goals 1 and 2, and since the mid-1990s to explore goal 4. Genetic management is likely occurring as other goals are being achieved, but only one translocation program specifically designed to achieve genetic management has been established (Haig et al. 1993). After more than a decade (1989–2002) of pursuing these goals, we understand how to stabilize, expand, and establish red-cockaded woodpecker populations using translocations.

The basic strategy for any small red-cockaded woodpecker population translocation program involves the following: (1) augment solitary bird groups, typically male, (2) build from edges of the population's core, (3) connect demographic subpopulations by building from their proximate borders, (4) accomplish 2 and 3 above by simultaneously translocating multiple (at least three), unrelated subadult pairs annually or as frequently as possible until the population goal is reached, (5) provide one additional recruitment cluster for each pair released, and (6) arrange recruitment clusters in clumped arrays as much as possible. Following these guidelines will help ensure that population-specific translocation strategy goals of: maximizing retention and survival of individual birds, minimizing cost, and expanding the population to its potential size of 10–30 potential breeding groups, will be achieved as rapidly as possible (U.S. Fish and Wildlife Service 2003).

Overall, regional and population-level strategies for reintroducing fauna in longleaf pine ecosystems will vary by species. Types and scales of strategies employed will be dependent upon number and types of cooperators involved, number of animals available, resources available, logistical considerations, number of populations at risk of extirpation, calculated extirpation risk to each population, and species life history and ecology. Developing and testing species-specific strategies provides a foundation for adaptive management and increasing success.

The Future

The red-cockaded woodpecker translocation program has been remarkably successful. However, various challenges remain for maintaining and improving the program. Continued success hinges on filling key information gaps and increasing the number of available birds to meet population needs.

Additional research on rates of natural dispersal and new research on factors promoting juvenile dispersal between populations are needed (U.S. Fish and Wildlife Service 2003; DeLotelle et al. 2004). Investigating these issues will help us evaluate genetic threats and determine whether translocations will be necessary for genetic management of small- to medium-size populations over the long term. The paucity of information on diseases, parasites, and other potential pathogens, e.g., West Nile virus, within translocation donor and recipient populations suggests additional studies on this aspect of red-cockaded woodpecker biology should be increased.

Although translocation success increased substantially from the initial efforts in the late 1980s/early 1990s to the late 1990s/early 2000s, additional research is needed to improve methods. Specifically, more information on time of year of release, release method, number of pairs simultaneously released, effects of southern flying squirrels, and different regional translocation strategies would be valuable. Additionally, more research is needed on effects of subadult removal on donor populations. The limited research to date on this issue has suffered from small sample size and inadequate study design. Data on this subject have been gathered from monitoring programs rather than well-designed research projects.

Several approaches may increase the annual supply of birds available for translocation. However, fully implementing these options is challenging, given budget constraints, logistical challenges, and unknown potential impacts to donor populations. As an example, increasing output of birds from donor populations would likely require establishment of recipient-driven compensation systems for donors, or sharing resources to minimize economic burden on donors. The argument could be made that resources used by donors to help recipients could be used to increase the rate of growth of donors' own population via habitat improvement initiatives. Ironically, as donor populations increase in size they could provide additional subadult birds for translocation if managers could increase their nesting season monitoring. In the NRTC, the three recipient populations all pay their respective donors approximately $1000 for the opportunity to receive each bird. Donors use these funds to maintain the level of monitoring necessary to translocate subadult birds.

It may be possible to increase the number of birds available for translocation in the short-term by allowing smaller populations to be donors. This strategy may be particularly applicable when the only "official" donor populations (see U.S. Fish and Wildlife Service 2003) are far from the recipient population and economic, research, and logistical opportunities for intensive monitoring are in place at these small populations. The South/Central Florida Recovery Unit may provide such an opportunity. It consists of many (about 16), small (3–50 occupied clusters) populations. Most are intensively monitored, i.e., totally marked, and several have long-term research programs established. This level of knowledge may make it ecologically possible and logistically practical to move carefully selected birds between these populations and among potential metapopulations. Assessing the potential donor population level impact would be a critical and required component of any program involving removing birds from small populations.

There are numerous, small red-cockaded woodpecker populations across the Southeast well below their potential population size and yet not participating in translocation programs. Reasons for nonparticipation vary but include inadequate funding to monitor their population and prepare habitat, unqualified staff, or lack of commitment. Overcoming these barriers is challenging and ultimately requires leaders willing to secure funding, hire and train staff, and identify red-cockaded woodpecker recovery as a management priority. It is hoped that as more donor populations become available, current recipient populations leave the translocation program (by reaching the threshold of 30 potential breeding groups), and public awareness increases regarding longleaf pine ecosystem restoration, more small populations will participate in translocations.

Other Potential Species for Reintroduction and Translocation in Longleaf Pine Forests

Will Translocations Work for Other Longleaf Pine Ecosystem Fauna?

Understanding, and then designing the translocation program for red-cockaded woodpeckers to capitalize on the life history and ecological behaviors of the species has been largely responsible for the documented successes. Before considering translocation programs for other longleaf pine fauna, biologists must ask several important questions. Does the species exhibit any life history traits that make it "naturally" conducive to translocations, like dispersal behavior? Conversely, does its ecology preclude likely success if translocated, e.g., a species with a strong homing instinct? Are there age- or status-specific (e.g., reproductive or social structure) characteristics that enhance or detract from the probable success of translocating specific individuals? What is the species' annual reproductive potential, that is, are few offspring or multiple broods common? Answers to these and other questions directly related to the species' ecology, as it relates to being receptive to translocations, are critical. These questions must be addressed and evaluated to not only increase success rates for appropriate species, but also to identify those fauna that are poor candidates for translocation or are candidates for other kinds of enhancement efforts that do not involve translocation.

Numerous other rare vertebrate fauna, besides red-cockaded woodpeckers, are potential candidates for reintroduction into longleaf pine forests (Table 1). Most have ongoing recovery efforts including some form of reintroduction method (Table 6). Table 6 includes five species each of mammals and reptiles, four bird species (not including red-cockaded woodpeckers), and two amphibian species. Five species are listed as endangered, seven are listed as threatened, and five have critically small populations requiring the growth of existing populations or the establishment of additional populations. Disease appears to be a major problem for several species; therefore, any risk of disease via translocation efforts must be minimized (Cunningham 1996; Trenham and Marsh 2002; L. LaClaire, U.S. Fish and Wildlife Service, personal communication 2003). Monitoring efforts must be conducted to evaluate success and potential problems with translocations.

Species Accounts

Below we provide species accounts for 10 vertebrates listed in Table 6 that have, at some level, been involved in translocation programs. Although the remaining six species (not including the red-cockaded woodpecker) may have potential for translocation programs, available literature on translocation for these species is inadequate to provide meaningful species accounts.

Amphibians

Two listed amphibian species are dependent on longleaf or slash pine forests that surround isolated, ephemeral wetlands. The Mississippi gopher frog (*Rana capito sevosa*) is listed as endangered in Mississippi (66 *Federal Register* 62993, 2001) and the flatwoods salamander (*Ambystoma cingulatum*) is listed as threatened in Florida, Georgia, and South Carolina. Several populations of flatwoods salamander recently have been documented in the above states (64 *Federal Register* 15691, 1999); however, only one population of Mississippi gopher frogs is known to exist. Primary habitat for adults of both species includes upland sandy areas of longleaf or slash pine forest surrounding ephemeral wetlands. Small adult population size, vulnerability to environmental changes or conditions, and loss of breeding sites threaten the flatwoods salamander and Mississippi gopher frog.

Both species breed in flooded, ephemeral grassy ponds or cypress domes, and other isolated wetlands that lack predatory fish populations (Mount 1975). The grassy nature of the

TABLE 6. Population trend and reintroduction methods for listed and rare vertebrates found in longleaf pine forests.

Common name	Population	Trend	Method of increase
Amphibians			
Mississippi gopher frog	1 population in southern MS	Stable, vulnerable	Translocation
Flatwoods salamander	51 populations with 71% in Florida; 122 breeding sites.	Stable, vulnerable	Translocation?
Reptiles			
Eastern Indigo snake	Mostly FL and south GA; possible in AL, MS, and SC	Declining, particularly on private lands	Translocation questionable
Gopher tortoise	Southwestern AL, northeastern panhandle of LA, and southeastern MS	Probably declining	Translocation
Blue-tailed mole skink	Limited to southern portion of Lake Wales Ridge in FL	Probably declining	Unknown
Sand skink	Ocala National Forest and southern portion of Lake Wales Ridge in FL	Probably declining	Translocation?
Louisiana pine snake	Small population in western LA and eastern TX	Unknown	Some releases, program on hold
Birds			
Mississippi sandhill crane	100 adults	Stable	Captive release
Red-cockaded woodpecker	5800 groups	Increasing throughout their range	Translocation/ artificial cavities
Southern bald eagle	~1000 territories in FL	Increasing throughout their range	Translocation/ artificial nest sites
Southeastern American kestrel	Southern portion of LA, MS, AL, GA, and SC; throughout FL highlands	Increasing on protected lands	Nest boxes
Bachman's sparrow	Common in appropriate habitat	Declining due to habitat loss	No information
Mammals			
Red wolf	1 small population in eastern NC	Stable, vulnerable	Captive breeding and release
Florida panther	1 population with 2 subpopulations; 30–50 adults	Declining/stable, vulnerable	Captive breeding and release, possibly
Florida black bear	4 primary populations	Declining/stable	Unknown
Louisiana black bear	40 to 60 adults	Declining	Unknown
Sherman's fox squirrel	Widely distributed in FL	Declining, habitat loss and degradation	Translocation? see Delmarva fox squirrel literature

entire pond or surrounding grassy transitions of wetlands are particularly important to survival of tadpoles or larvae. Loss of this microhabitat has occurred through detrimental timber practices, other development activities, and anthropogenic changes to drainage patterns. The single remaining breeding pond used by Mississippi gopher frogs is on a topographic high within a small drainage basin. Nearby wells are used to maintain appropriate water levels during years of low rainfall (L. LaClaire, U.S. Fish and Wildlife Service, personal communication 2003). Ponds used by flatwoods salamanders are usually down slope from surrounding uplands and may be seasonally connected to other wetlands. Hollow stumps and occupied and abandoned gopher tortoise burrows are underground refugia important to adult Mississippi gopher frogs, which historically occurred in areas with gopher tortoises. The remaining Mississippi gopher frog population occurs in an area where gopher tortoises have been extirpated, although efforts are underway to reestablish a resident gopher tortoise population (L. LaClaire, U.S. Fish and Wildlife Service, personal communication 2003).

Translocations of these species have not been conducted and using translocations as a means of enhancing or reestablishing amphibians is controversial (Marsh and Trenham 2001, Seigal and Dodd 2002). Dodd and Seigal (1991) conducted a through review of the benefits and costs of amphibian and reptile relocations, repatriations, and translocations. The possibility of disease transmission appears to be the most serious potential negative aspect of translocation (Cunningham 1996; Daszak et al. 1999; L. LaClaire, U.S. Fish and Wildlife Service, personal communication 2003). For example, Seigal and Dodd (2002) point out that one potential site for translocation of Mississippi gopher frog tadpoles or eggs contained an undescribed fungus that decimated all ranid tadpoles within a matter of weeks. Additionally, there appear to be several diseases to which frogs are susceptible, particularly under stressful conditions (L. LaClaire, U.S. Fish and Wildlife Service, personal communication 2003).

Reptiles

The gopher tortoise is a resident of high sandy areas and dry microhabitats within wet pine flatwoods distributed throughout the Lower Coastal Plain and Sandhills. The species is listed as federally threatened in the westernmost portion of its range, including southwestern Alabama, the northeastern panhandle of Louisiana, and southeastern Mississippi (U.S. Fish and Wildlife Service 1990a). The listed population occurs within the historic range of longleaf pine. Threats include habitat alteration, predation by humans, road mortality, and isolation of mature animals from one another (Lohoefener and Lohmeier 1981; U.S. Fish and Wildlife Service 1990a). In Florida, translocation of gopher tortoises is often conducted as part of development mitigation. However, there are significant problems with this process related to the social system of the animal and the potential to spread upper respiratory tract disease, as well as other diseases, to uninfected populations (Cox et al. 1994, J. Barish, Florida Fish and Wildlife Conservation Commission, personal communication 2003). Currently, all donor populations must be screened for disease prior to any movement of individuals.

The sand skink (*Neoseps reynoldsi*) is listed as threatened by the U.S. Fish and Wildlife Service and inhabits sandy ridges with underlying moist soil (Christman 1992). This species may be particularly abundant in ecotonal areas between pine savannas and oak scrub. It also occurs in rosemary scrub, turkey oak barrens, and sandy areas of high pine. Within their known distribution all suitable habitats were probably inhabited (P. Molar, Florida Fish and Wildlife Conservation Commission, personal communication 2003). The preferred habitat within its range is extremely limited. Areas free of abundant roots, with scattered shrubby vegetation and patches of bare sand seem to be preferred habitats; however, surveys have found them in a few areas with an extensive tree canopy. Habitat loss, both historic and current, is the major threat to this species. Much of its habitat along Florida's Lake Wales Ridge has been converted to citrus groves and human development. Within apparently continuous habitat, small drainages likely result in genetic segregation and this genetic structuring has implications for conservation of all scrub-associated lizards (Branch et al. 2003, P. Molar, Florida Fish and Wildlife Conservation Commission, personal communication 2003). Any strategy for population enhancement by means of translocation or relocation should consider genetic implications and require a comprehensive plan to conserve genetic diversity (Branch et al. 2003).

To date, the only translocation effort for sand skinks involved complete removal and transportation of the top 6 inches of soil from the project area along with the resident sand skink population to an abandoned citrus grove. The recipient site was believed to be suitable and occupied by sand skinks prior to its conversion to citrus. The skinks have persisted, but apparently are not breeding (P. Molar, Florida Fish and Wildlife Conservation Commission, personal communication 2003). If additional relocations are attempted, recipient sites must include appropriate edaphic conditions and be restored to native conditions. Diseases do not appear to be a problem for this species.

The eastern indigo snake (*Drymarchon corais*) occurs in southeast Georgia and peninsular Florida with disjunct populations in the Florida Panhandle and perhaps Alabama (Conant 1975). The species inhabits a variety of habitats in southern Florida; however, in the northern parts of its range it is more restricted to sand ridge habitats of pine interspersed with wetlands (Speake et al. 1978). Fire exclusion, habitat loss, and extensive (historical) trapping for the pet trade are responsible for their decline. In northern parts of their range, loss of gopher tortoises and their burrows also are limiting factors (Speake and Mount 1973). Males are territorial and cannibalistic and may cover up to 160 ha (P. Molar, Florida Fish and Wildlife Conservation Commission, personal communication 2003). To date, the only translocation efforts included releases of snakes by Speake, from a variety of sources, on St. Marys and St. Vincent Islands in Georgia and southern Alabama (P. Molar, Florida Fish and Wildlife Conservation Commission, personal communication 2003). Feral hogs occurred in all release sites and apparently killed most snakes during high water conditions. Regular monitoring was apparently not conducted. If reintroduction programs are implemented, release sites should contain 4000 to 4800 ha of good habitat and be free of feral and domestic predators (P. Molar, Florida Fish and Wildlife Conservation Commission, personal communication 2003).

Birds

There are three nonmigratory species of sandhill cranes in the Southeast. The Mississippi population (*Grus canadensis pulla*) is federally listed as endangered, occurring only on the Mississippi Sandhill Crane National Wildlife Refuge in southeastern Mississippi (U.S. Fish and Wildlife Service 1991). This population was greatly reduced by hunting prior to the twentieth century. Currently the population is limited by suitable habitat, primarily longleaf pine savannahs. Lower levels of genetic heterozygosity may result in poor hatching success and some debility in captive chicks (U.S. Fish and Wildlife Service 1991). Over the past several years the population has remained relatively stable at approximately 100 adults, with slightly more than 20 nests during 2003. Although apparently stable, predation of nests from a variety of mammals is problematic (L. LaClaire, U.S. Fish and Wildlife Service, personal communication 2003). Coyotes, foxes, and bobcats are the most significant predators and are controlled through trapping programs. Coupled with habitat restoration, an ongoing program of releasing captive-bred cranes helps stabilize the single extant population.

The southeastern American kestrel (*Falco sparverius paulus*) is resident in the lower southeastern Coastal Plain and primarily uses open habitats such as pastures, longleaf pine/turkey oak sandhills, and grasslands (Johnsgard 1990; Stys 1993). The largest population occurs in Florida and is associated with upland ridges in the northern two-thirds of the peninsula (Stys 1993). Loss of habitat from logging or fire suppression appears to be the primary cause of the species' decline. Additionally, the loss of red-cockaded woodpeckers and their cavities may have contributed to declines in southeastern American kestrels. Nest boxes installed in formerly occupied habitat, where adjacent extant populations still exist, have been successful in reestablishing breeding populations. For example, at Fort Jackson in South Carolina installation of nest boxes over the past 8 years has resulted in a large breeding population (T. Marston, U.S. Army, personal communication

2003). Nest boxes are repaired as needed, breeding is monitored, and nestlings are banded. Over the past 6 years an average of 17 nests produced an average clutch size of 4.7 eggs with about 34% of eggs failing. On average, these nests produced about 2.9 nestlings at time of nestling banding (T. Marston, U.S. Army, personal communication 2003). Similar success has occurred in other areas of the species range, for example, Fort Gordon, GA.

The southern bald eagle (*Haliaeetus leucocephalus leucocephalus*) is federally listed as endangered (Curnutt 1996). Historically distributed throughout North America, pesticide poisoning and habitat loss eliminated the species from most of its former range. Banning DDT in 1972 resulted in a dramatic increase in Florida populations while the 10 other southeastern states only had 187 breeding pairs in 1991 (U.S. Fish and Wildlife Service 1987; Jenkins and Sherrod 1993). Bald eagles occupy a variety of habitats, although proximity to large bodies of open water with large trees (especially pine) for nest sites, are important requirements. Nests are usually in live, tall trees within several miles of a large lake or river.

Florida currently has one of the largest eagle populations in the lower 48 states and it continues to expand. Eggs from Florida populations have been used to reintroduce eagles into several other states (Wood 1982). Typically, eggs were collected after 2 weeks of natural incubation, transported to the Sutton Avian Research Center in Oklahoma, incubated by Bantam hens (*Gallus* sp.) or artificially (Nesbitt et al. 1998), and young released at 11–12 weeks of age at recipient sites (Sherrod et al. 1987; Nesbitt et al. 1998). Florida populations remained stable or increased during the period of egg removal.

An alternate strategy for population enhancement or establishment includes the construction of artificial nesting platforms to encourage the return of eagles to former breeding territories (Hunter et al. 1997). The nest platforms complete with nest material are constructed in tops of well-formed trees similar in structure to those normally used by eagles. From 1994 to 1998, three successful artificial nests accounted for 15.4% of the nesting attempts in a southern Ontario population (Hunter et al. 1997). Because of the DDT ban and, to a lesser extent, reintroduction programs, populations are growing in most areas of the Southeast and the species has been proposed for delisting by the U.S. Fish and Wildlife Service.

Mammals

Two large, predatory, federally endangered mammals with similar ecological and social constraints to their reintroduction are the Florida panther (*Felis concolor coryi*) and red wolf (*Canis rufus*). Historically, these mammals occurred over much of the southeastern United States and ranged over a variety of habitats including longleaf pine forests (Anderson 1983; U.S. Fish and Wildlife Service 1990b). Translocation of red wolves and the Florida panthers' close relative, mountain lions (*Felis concolor stanleyana*) have resulted in individuals surviving and breeding in the wild (U.S. Fish and Wildlife Service 1990b; Belden and McCown 1996).

For red wolves, interbreeding with coyotes was a major problem in the last remaining population and potentially a problem for any reintroduction efforts. During the 1980s, the remaining red wolves persisted in low numbers in a small area of southeastern Texas and southwestern Louisiana. From this area, over 400 canids were captured and screened to obtain 43 red wolf-types for the captive breeding program. Of these, only 15 adults survived and produced offspring that fit the red wolf-type for reintroduction.

Reintroduction of red wolves has been considered or attempted in the Land between the Lakes (LBL) region in western Tennessee and Kentucky and Cades Cove (Great Smoky Mountains National Park) in eastern Tennessee. The LBL project failed prior to the release of the first animal because of public opposition (U.S. Fish and Wildlife Service 1990b). Captive-bred wolves were successfully introduced at Cades Cove and survived and bred in the wild. However, because of poor pup survival, inability of adults to establish territories, and low availability of prey, the wolves were removed from the Park. Since these initial attempts, red wolves have been successfully reintroduced to the remote Alligator River National Wildlife Refuge in eastern North Carolina (Parker 1987; Phillips 1990). Releases usually included family units of adults and offspring. Mortality of adult females was high; however, the population has survived and is producing young.

In the late 1990s, it was estimated that 30 to 50 adult Florida panthers remained in the Big Cypress National Preserve and Everglades National Park physiographic regions in south Florida (Belden and McCown 1996). In 2003, 87 adult Florida panthers were known to exist. For recovery it is proposed that this population be managed and that panthers be reintroduced to three additional areas (Belden and Hagedorn 1993). To that end, and as a feasibility study, 19 mountain lions were released as Florida panther surrogates into northern Florida. These releases included 11 females and 8 vasectomized males. These lions, including captive-raised and wild-caught animals, established 15 home ranges. Unfortunately, the captive-raised animals (particularly the males) were involved in a number of interactions with humans and resulted in bad publicity (Belden and McCown 1996). Based on this study, Belden and Hagedorn (1993) concluded that reintroductions of a panther population in northern Florida or elsewhere were biologically feasible. However, issues of funding, gaining public support, and identifying suitable landscape-scale areas need to be resolved before such reintroductions can be attempted.

Planning for Future Reintroduction of Longleaf Pine Ecosystem Fauna

Based on our literature review, case study of red-cockaded woodpeckers, and what we have learned about reintroduction of other longleaf pine ecosystem fauna, we have developed a translocation success ranked response model (see Table 7). Prior to discussing its application, one observation about the model

TABLE 7. Translocation success ranked response model for longleaf pine ecosystem potential reintroduction fauna candidates.

Factors potentially influencing translocation success	Probability of success		
	High	Medium	Low
Ownership of recipient property	Federal	State	Private
Public support	Substantial	Likely	Questionable
Availability of funding	Good	Fair	Poor
Cost per animal translocated	<$1500	$1500–$5000	$5000+
Logistical challenges	Few	Manageable	Numerous
Source of donor stock	Wild (hard release)	Wild (soft release)	Captive bred
Number of animals available annually	Many	Some	Few
Number of years required to reach population goal	5	5 to 10	10+
Spatial scale (number of acres required per territory)	<100	100 to 1000	1000 to 10,000+
Species status	Federally listed	State listed/federal candidate	Rare/sensitive/special concern
Species ecological classification	Generalist	Adaptable	Specialist
Number of species unique habitat requirements	Few (1 or 2)	Manageable (3 to 4)	Numerous (5 or more)
Number of significant, non-habitat-related, threats	Few (1)	Manageable (2 to 3)	Numerous (4 or more)
Target population size (number of breeding adults)	Small (20 to 50)	Medium (50 to 200)	Large (200+)
Species reproductive potential	High	Moderate	Low
Species dietary habit	Omnivorous	Herbivorous	Carnivorous

is noteworthy. The ranking factors fall in various categories, including administrative, economic, social/cultural (i.e., human dimensions), ecological, and management-oriented. This is significant because it highlights the need to approach each potential translocation program with a multidisciplinary coalition of partners. The partners, individually and collectively, must not only gain knowledge about each factor, but also develop an adequate understanding of the interrelationships among factors to assess, design, and ultimately implement a translocation program. Typically, partners may include governmental natural resource agencies (federal and state), nongovernmental conservation organizations, private industry, private landowners, and universities.

The model has multiple applications. First, it can be used to highlight areas of program vulnerability by identifying those factors for a particular species with a "low" ranking. While nothing can be done for some factors with a low ranking, for example species is an ecological specialist or has low reproductive potential, other factors can often be addressed. For example, if the reintroduced species will be exposed to numerous threats from competitors, resources and management emphasis can be directed toward mitigating this factor, thereby minimizing its impact and reducing the overall vulnerability of the project's success. In other words, the model can be used to focus limited resources on management of key limiting factors.

Second, the model can also direct biologists and researchers to those factors for which more information is required about their target species to make informed decisions. By doing so, research, monitoring, and management programs, capable of filling in the blanks in the model, can be designed and implemented. Again, this helps focus limited resources on those factors that may affect program success.

Probable success of any fauna translocation program will ultimately be directly related to the ability of managers and biologists to implement actions necessary to reach the target population size. Therefore, many factors listed in the ranked response model could also be discussed and evaluated in the context of "probability of program implementation." For example, implementation will be more difficult without public support and adequate amounts of funding and donor stock. Even if the above factors are adequately addressed, implementation will still be challenging if many unique habitat requirements must be satisfied and numerous threats, e.g., predators and competitors, must be managed. Additionally, if a species has low reproductive potential, specific dietary requirements, or requires a vast territory, implementation will also be very challenging. Therefore, from an implementation perspective, the model can be used to help quantify and qualify both the annual and long-term probability that the program will be successful. With this knowledge, administrators and managers can project and plan for future needs, while simultaneously calculating expected program benchmarks given start-up and probable short-term needs for each appropriate factor.

In addition to the above applications, the model has also been designed to assist managers, biologists, and administrators predict the overall probability of success of translocation programs for vertebrate, terrestrial fauna of longleaf pine ecosystems. The model contains 16 factors that we consider critical for assessing probable success of translocations. Initially, ranking each factor as high, medium, or low could assess the probability of a successful translocation program. Determination of success probability after this initial analysis may be possible if a clear majority (e.g., 75%; 12 or more) of factors fall within one rank. However, using this system and lacking a clear majority, it would be necessary to "score" or "interpret" the results to assign an overall probability rank to the program. For example, if six factors were assigned "high,", six factors "medium," and four factors "low," one could logically conclude at least a medium probability of success for the project. A scoring system provides challenges when no clear pattern develops, therefore, a factor weighting system may be necessary to help interpret results. Once weighted, some factors for any number of reasons, e.g., economic, administrative, social, or ecological, become more important to program success than other factors. Under this scenario, a minority of weighted factors in one rank (e.g., four in "low") may "outweigh" the influence of the ranks with more variables listed, e.g., five in "medium" and seven in "high." Ultimately, the inability to satisfy an individual, but critical, factor, e.g., availability of funding or number of animals available annually, could make program implementation impossible.

Although we considered weighting each factor, at this time we do not believe adequate information exists to do so with reasonable accuracy for the entire suite of potential longleaf pine fauna translocation candidates (see Table 1). That is, the multiple and complex relationships that exist between and among these various factors, in the context of all longleaf pine fauna as one group, are not well known or understood. However, at the single species scale, knowledgeable individuals may be able to weight factors for which adequate information exists. The red-cockaded woodpecker may provide such an opportunity.

We encourage managers to use and build on our model but we caution that it has limitations related to its generic coverage. Simultaneously, we encourage biologists who develop species-specific models to consider a weighting system. In conclusion, we recommend that this model be used, in conjunction with other appropriate decision-making tools, and likely more species-specific variables, to guide officials responsible for listed and rare species conservation.

Conclusion

Conservation of biodiversity in longleaf pine forests via fauna translocations will be challenging. Once life history and ecology of each species are well understood and it has been determined to be a viable candidate for translocation, complex operational issues must still be considered. Program implementation-related issues include: availability of donor stock, landscapes large enough to accommodate viable populations, adequate funding to initiate and sustain the program, including postrelease monitoring, ability, via technology and habitat management programs, to restore and maintain the forest's structure and composition, and public support. Many of these issues are interrelated and must be addressed in a comprehensive fashion. Additionally, some programs such as prescribed burning, while essential for long-term management of most longleaf pine ecosystem species, are not necessarily controlled or administered by the fauna reintroduction team. Therefore, reintroduction efforts also require close coordination and cooperative partnerships with habitat managers and others to ensure success.

We believe significant potential exists to further the cause of longleaf pine ecosystem restoration via establishing new and augmenting existing small, at-risk populations of listed and rare vertebrate fauna. We encourage responsible officials, knowledgeable scientists, and interested conservationists to explore the possibilities, identify opportunities, and implement well-designed programs. Each success will be an important contribution to our legacy of restoring the biodiversity of the endangered longleaf pine ecosystem.

Acknowledgments

We appreciate the editorial reviews received from Dr. George Tanner, University of Florida, and an anonymous reviewer; their comments and suggestions improved our chapter considerably. A special thank you to Dr. Jeff Walters, Virginia Polytechnic Institute, for his detailed review of an earlier draft of our manuscript. His insights and knowledge regarding red-cockaded woodpeckers focused his review and comments on issues most relevant to a thorough understanding of our topic. Addressing his thoughtful comments significantly improved our chapter. We also thank Ms. Nickie Nichols, U.S. Fish and Wildlife Service student intern and Clemson University graduate student, working on red-cockaded woodpeckers, for formatting and constructing our tables. Thanks to Ms. Nicole Edwards, U.S. Fish and Wildlife Service computer assistant, Clemson University, for assembling our literature cited section. Finally, we thank our editors, Dr. Shibu Jose, Dr. Eric Jokela, and Dr. Debbie Miller, for inviting us to prepare this chapter and, thereby, contribute to this much-needed text on the longleaf pine ecosystem.

References

Abrahamson, W.G., and Hartnett, D.C. 1990. Pine flatwoods and dry prairies. In *Ecosystems of Florida*, eds. R.L. Myers and J. J. Ewel, pp. 103–149. Orlando: University of Central Florida Press.

Allen, D.H. 1991. An insert technique for constructing artificial red-cockaded woodpecker cavities. U.S. Forest Service General Technical Report SE-73.

Allen, D.H., Franzreb, K.E., and Escano, R.F. 1993. Efficacy of translocation strategies for red-cockaded woodpeckers. *Wildl Soc Bull* 21:155–159.

Anderson, A.E. 1983. A critical review of literature on puma (*Felis concolor*). Colorado Division of Wildlife Special Report 54.

Barclay, J.H., and Cade, T.J. 1983. Restoration of the Peregrine Falcon in the eastern United States. In *Bird Conservation*, ed. S.A. Temple, pp. 3–40. Madison: University of Wisconsin Press.

Belden, R. C., and Hagedorn, B. W. 1993. Feasibility of translocating panthers into northern Florida. *J Wildl Manage* 57:388–397.

Belden, R.C., and McCown, J.W. 1996. Florida panther reintroduction feasibility study. Florida Game and Fresh Water Fish Commission, Bureau of Wildlife Resources. Study No. 7507, Final Report. Tallahassee, FL.

Branch, L.C., Clark, A.M., Moler, P.E., and Bowen, B.W. 2003. Fragmented landscapes, habitat specificity, and conservation genetics of the three lizards in Florida scrub. *Conserv Genet* 4:199–212.

Britcher, J.J., and Patten, J.M. 2004. Red-cockaded woodpecker management on Fort Bragg: Then and now. In *Red-cockaded Woodpecker: Road to Recovery*, eds. R. Costa and S. J. Daniels. Blaine, WA: Hancock House Publishers. In press.

Brown, C., and Simpkins, S. 2004. The Chickasawhay Ranger District story: Saving a small population from extirpation. In *Red-cockaded Woodpecker: Road to Recovery*, eds. R. Costa and S. J. Daniels. Blaine, WA: Hancock House Publishers. In press.

Carlile, L.D., Ten Brink, C., Mitchell, C.L.R., Puder, S.E., Spadgenske, E. W., and Beaty, T. A. 2004. An intensively managed and increasing red-cockaded woodpecker population at Fort Stewart, Georgia. In *Red-cockaded Woodpecker: Road to Recovery*, eds. R. Costa and S. J. Daniels. Blaine, WA: Hancock House Publishers. In Press.

Carrie, N.R., Moore, K.R., Stephens, S.A., and Keith, E.L. 1996. Long-distance homing of a translocated red-cockaded woodpecker. *Wildl Soc Bull* 24:607–609.

Carrie, N.R., Conner, R.N., Rudolph, D.C., and Carrie, D.K. 1999. Reintroduction and postrelease movements of red-cockaded woodpecker groups in eastern Texas. *J Wildl Manage* 63:824–832.

Christman, S.P. 1992. Threatened: sand skink, *Neoseps reynoldsi* (Stejneger). In *Rare and Endangered Biota of Florida*, ed. P.E. Moler. Gainesville: University Press of Florida.

Conant, R. 1975. *A Field Guide to Reptiles and Amphibians of Eastern and Central North America*. Boston: Houghton Mifflin Co.

Conner, R.N., and Rudolph, D.C. 1991. Forest habitat loss, fragmentation, and red-cockaded woodpecker populations. *Wilson Bull* 103:446–457.

Conner, R.N., Rudolph, D.C., Saenz, D., and Schaefer, R.R. 1996. Red-cockaded woodpecker nesting success, forest structure, and southern flying squirrels in Texas. *Wilson Bull* 108:697–711.

Conner, R.N., Rudolph, D.C., Schaefer, R.R., and Saenz, D. 1997. Long-distance dispersal of red-cockaded woodpeckers in Texas. *Wilson Bull* 109:157–160.

Copeyon, C.K. 1990. A technique for constructing cavities for the red-cockaded woodpecker. *Wildl Soc Bull* 18:303–311.

Costa, R. 2001. Red-cockaded woodpecker. In *Wildlife of Southern Forests: Habitat and Management*, ed. J.G. Dickson, pp. 309–321. Blaine, WA: Hancock House Publishers.

Costa, R., and Daniels, S.J., eds. 2004. *Red-cockaded Woodpecker: Road to Recovery*. Blaine, WA: Hancock House Publishers. In Press.

Costa, R., and Edwards, J.W. 1997. Cooperative conservation agreements for managing red-cockaded woodpeckers on industrial forest lands: What are the motivations? In *Proceedings of the Symposium on the Economics of Wildlife Resources on Private Lands*, ed. R. Johnson, pp. 111–124. Auburn, AL: Auburn University.

Costa, R., and Escano, R. 1989. Red-cockaded woodpecker: Status and management in the southern region in 1986. U.S. Forest Service Technical Publication R8-TP12.

Costa, R., and Jordan, N. 2003. Biological opinion on all section 10(a)(1)(A) management, monitoring and research permits issued to all private, state and federal agencies and individuals involved with management, conservation and recovery of the red-cockaded woodpecker throughout the range of the species. U.S. Fish and Wildlife Service, Clemson, SC.

Costa, R., and Kennedy, E. 1994. Red-cockaded woodpecker translocations 1989–1994: State-of-our-knowledge. *Annual Proceedings of the American Zoo and Aquarium Association*, pp. 74–81. Atlanta, GA: Zoo Atlanta.

Cox, J., Kautz, R., MacLaughlin, M., and Gilbert, T. 1994. Closing the Gaps in Florida's Wildlife Habitat Conservation System. Florida Game and Fresh Water Fish Commission. Tallahassee, FL.

Crowder, L.B., Priddy, J.A., and Walters, J.R. 1998. Demographic isolation of red-cockaded woodpecker groups: A model analysis. Project Final Report, prepared for U.S. Fish and Wildlife Service.

Cunningham, A.D. 1996. Disease risks of wildlife translocation. *Conserv Biol* 10:349–353.

Curnutt, J.L. 1996. Southern bald eagle. In *Rare and Endangered Biota of Florida, Volume IV. Birds*, eds. J.A. Rodgers, Jr., H.W. Kale II, and A. T. Smith, pp. 179–187. Orlando: University Press of Florida.

Daniels, S.J., and Walters, J.R. 2000a. Inbreeding depression and its effects on natal dispersal in red-cockaded woodpeckers. *Condor* 102:482–491.

Daniels, S.J., and Walters, J.R. 2000b. Between-year breeding dispersal in red-cockaded woodpeckers: Multiple causes and estimated cost. *Ecology* 81:2473–2484.

Daniels, S.J., Priddy, J.A, and Walters, J.R. 2000. Inbreeding in small populations of red-cockaded woodpeckers: Insights from a spatially explicit individual-based model. In *Genetics, Demography and Viability of Fragmented Populations*, eds. A.G. Young and G.M. Clarke, pp. 129–147. London: Cambridge University Press.

Daszak, P., Berger, L., Cunningham, A.A., Hyatt, A.D., Green, D.E., and Speare, R. 1999. Emerging Infectious diseases and amphibian population declines. *Emerg Infect Dis* 5:735–748.

DeFazio, J.T., Hunnicutt, M.A., Lennartz, M.R., Chapman, G.L., and Jackson, J.A. 1987. Red-cockaded woodpecker translocation experiments in South Carolina. *Proc Southeast Assoc Fish Wildl Agenc* 41:311–317.

DeLotelle, R.S., and Epting, R.J. 1992. Reproduction of the red-cockaded woodpecker in central Florida. *Wilson Bull* 104:285–294.

DeLotelle, R.S., Epting, R.J., and Newman, J.R. 1987. Habitat use and territory characteristics of red-cockaded woodpeckers in central Florida. *Wilson Bull* 99:202–217.

DeLotelle, R.S., Epting, R.J., and Demuth, G. 1995. A 12-year study of red-cockaded woodpeckers in central Florida. In *Red-cockaded Woodpecker: Recovery, Ecology and Management*, eds. D.L. Kulhavy, R.G. Hooper, and R. Costa, pp. 259–269. Center for Applied Studies in Forestry. College of Forestry, Stephen F. Austin State University, Nacogdoches, TX.

DeLotelle, R.S., Epting, R.J., Leonard, D.L., Jr., and Costa, R. 2004. Management strategies for recovery of red-cockaded woodpeckers populations: A metapopulation proposal. In *Red-cockaded Woodpecker: Road to Recovery*, eds. R. Costa and S.J. Daniels. Blaine, WA: Hancock House Publishers. In Press.

Dodd, C.K., Jr., and Seigal, R.A. 1991. Relocation, repatriation, and translocation of Amphibians: Proven Management Method or Experimental Technique? 16:552–554.

Doresky, J., Barron, M., and Swiderek, P. 2004. Landscape scale restoration and red-cockaded woodpecker recovery? In *Red-cockaded Woodpecker: Road to Recovery*, eds. R. Costa and S.J. Daniels. Blaine, WA: Hancock House Publishers. In press.

Drumm, R.L., Boyd, K., and Camp, S.N. 2004. Red-cockaded woodpecker management on Fort Gordon, Georgia. In *Red-cockaded Woodpecker: Road to Recovery*, eds. R. Costa and S.J. Daniels. Blaine, WA: Hancock House Publishers. In Press.

Edwards, J.W., and Costa, R. 2004. Rangewide success of red-cockaded woodpecker translocations. In *Red-cockaded Woodpecker: Road to Recovery*, eds. R. Costa and S.J. Daniels. Blaine, WA: Hancock House Publishers. In Press.

Edwards, J.W., Dachelet, C.A., and Smathers, W.M. 1999. A mobile aviary to enhance translocation success of red-cockaded woodpeckers. In *Fish and Wildlife Research and Management: Applying Emerging Technologies*, pp. 48–53. Proceedings of the 37th Annual Meeting of the Canadian Society of Environmental Biologists. Edmonton, Alberta, Canada.

Edwards, J.W., Mari, Y., and Smathers, W. 2004. Evaluation of a mobile aviary to enhance translocation success of red-cockaded woodpeckers. In *Red-cockaded Woodpecker: Road to Recovery*, eds. R. Costa and S.J. Daniels. Blaine, WA: Hancock House Publishers. In Press.

Ellis, D.H., Dobrott, S.J., and Goodwin, J.L., Jr. 1978. Reintroduction techniques for masked bobwhite. In *Endangered Birds: Management Techniques for Preserving Threatened Species*, ed. S.A. Temple, pp. 345–354. Madison: University of Wisconsin Press.

Engstrom, T., Crawford, R.I., and Baker, W.W. 1984. Breeding bird populations in relation to changing forest structure following fire exclusion: A 15-year study. *Wilson Bull* 96:437–450.

Ferral, P., Edwards, J.W., and Armstrong, A.E. 1997. Long-distance dispersal in red-cockaded woodpeckers. *Wilson Bull* 109:277–284.

Franzreb, K.E. 1997a. A mobile aviary design to allow the soft release of cavity nesting birds. U.S. Forest Service Research Note SRS-5.

Franzreb, K.E. 1997b. Success of intensive management of a critically imperiled population of red-cockaded woodpeckers in South Carolina. *J Field Ornithol* 68:458–470.

Franzreb, K.E. 1999. Factors that influence translocation success in the red-cockaded woodpecker. *Wilson Bull* 111:38–45.

Franzreb, K.E. 2004. Effects of a mobile aviary in translocation success of red-cockaded woodpeckers. In *Red-cockaded Woodpecker: Road to Recovery*, eds. R. Costa and S.J. Daniels. Blaine, WA: Hancock House Publishers. In press.

Gordon, D.R. 1994. Translocation of species into conservation areas: A key for natural resource managers. *Nat Areas J* 14:31–37.

Griffith, B., Scott, J.M., Carpenter, J.W., and Reed, C. 1989. Translocation as a species conservation tool: Status and strategy. *Science* 245:477–480.

Griffith, B.J., Scott, J.M., Carpenter, J.W., and Reed, C. 1990. Translocations of captive-reared terrestrial vertebrates, 1973–1986. *Endangered Species Update* 8(1):10–13.

Gutierrez, R.J., and Harrison, S. 1996. Applications of metapopulation theory to spotted owl management: A history and critique. In *Metapopulations and Wildlife Conservation Management*, ed. D. McCullough, pp. 167–185. Covelo, CA: Island Press.

Hagan, G., and Costa, R. 2001. Rare woodpeckers reintroduced to North Florida. *Endangered Species Bulletin* 26:30–31.

Hagan, G., Costa, R. and Phillips, M.K. 2004a. The first reintroduction of red-cockaded woodpeckers into unoccupied habitat: A private land and conservation success story. In *Red-cockaded Woodpecker: Road to Recovery*, eds. R. Costa and S.J. Daniels. Blaine, WA: Hancock House Publishers. In press.

Hagan, G. Engstrom, R.T., Cox, J., and Spivey, P. 2004b. Effects of translocation on a large red-cockaded woodpecker population. In *Red-cockaded Woodpecker: Road to Recovery*, eds. R. Costa and S.J. Daniels. Blaine, WA: Hancock House Publishers. In press.

Haig, S.M., and Nordstrom, L.H. 1991. Genetic management of small populations. In *Westview Special Studies in Natural Resources and Energy Management: Challenges in the Conservation of Biological Resources: A Practitioner's Guide*, pp. 119–138. Boulder, CO: Westview Press.

Haig, S.M., Belthoff, J.R., and Allen, D.H. 1993. Population viability analysis for a small population of red-cockaded woodpeckers and an evaluation of enhancement stragegies. *Conserv Biol* 7:289–301.

Haig, S.M., Walters, J.R., and Plissner, J.H. 1994. Genetic evidence for monogamy in the red-cockaded woodpecker, a cooperative breeder. *Behav Ecol Sociobiol* 34:295–303.

Haig, S.M., Bowman, R., and Mullins, T.D. 1996. Population structure of red-cockaded woodpeckers in south Florida: RAPDs revisited. Mol Ecol 5:725–734.

Hanski, I., and Simberloff, D. 1997. The metapopulation approach, its history, conceptual domain, and application. In *Metapopulation Biology, Ecology, Genetics, and Evolution*, eds. I. Hanski and M.E. Gilpin, pp. 5–26. San Diego, CA: Academic Press.

Harris, L.D., and Vickers, C.R. 1984. Some faunal community characteristics of cypress ponds and the changes induced by perturbation. In *Cypress Swamps*, eds. K.C. Ewel and H.T. Odum, pp. 171–185. Gainesville: University Press of Florida.

Harrison, S. 1994. Metapopulations and conservation. In *Large-scale Ecology and Conservation Biology*, eds. P.J. Edwards, N.R. Webb, and R.M. May, pp. 111–128. Oxford: Blackwell.

Hedman, C.W., Poirier, J.R., Durfield, P.E., and Register, M.A. 2004. International Paper's habitat conservation plan for the red-cockaded woodpecker: Implementation and early success. In *Red-cockaded Woodpecker: Road to Recovery*, eds. R. Costa and S. J. Daniels. Blaine, WA: Hancock House Publishers. In press.

Hess, C., and Costa, R. 1995. Augmentation from the Apalachicola National Forest: The development of a new management technique. In *Red-cockaded Woodpecker: Recovery, Ecology and Management*, eds. D.L. Kulhavy, R.G. Hooper, and R. Costa, pp. 385–388. Center for Applied Studies in Forestry, College of Forestry, Stephen F. Austin State University, Nacogdoches, TX.

Hooper, R.G., and McAdie, C.J. 1995. Hurricanes and the long-term management of the red-cockaded woodpecker. In *Red-cockaded Woodpecker: Recovery, Ecology and Management*, eds. D.L. Kulhavy, R.G. Hooper, and R. Costa, pp. 148–166.

Center for Applied Studies in Forestry, College of Forestry, Stephen F. Austin State University, Nacogdoches, TX.

Hooper, R.G., Niles, L.J., Harlow, R.F., and Wood, G.W. 1982. Home ranges of red-cockaded woodpeckers in coastal South Carolina. *Auk* 99:675–682.

Hooper, R.G., Watson, J.C., and Escano, R.E.F. 1990. Hurricane Hugo's initial effects on red-cockaded woodpeckers in the Francis Marion National Forest. *Trans North Am Wildl Nat Resourc Conf* 55:220–224.

Hunter, P., Mahony, N.A., Ewins, P.J., and Field, M. 1997. Artificial nesting platforms for bald eagles in southern Ontario, Canada. *J Raptor Res* 31(4):321–326.

International Union for Conservation of Nature and Natural Resources/Species Survival Commission, Re-introduction Specialist Group (IUCN/SSC RSG). 1995. IUCN/SSG RSG guidelines for re-introductions. African Wildlife Foundation, Nairobi, Kenya, Africa.

Jackson, J.A. 1977. A device for capturing tree cavity roosting birds. *North Am Bird Bander* 2:14–15.

Jackson, J.A. 1990. Intercolony movements of red-cockaded woodpeckers in South Carolina. *J Field Ornithol* 61:149–155.

Jenkins, M.A., and Sherrod, S.K. 1993. Recent bald eagle nest records in Oklahoma. *Bull Okla Ornithol Soc* 26:25–28.

Johnsgard, P.A. 1990. *Hawks, Eagles, Falcons of North America*. Washington, DC: Smithsonian Institution Press.

Jones, S.R. ed., 1990. Captive propagation and reintroduction: A strategy for preserving endangered species? *Endangered Species Update* 8(1):3–88.

Kappes, J.J., Jr. 1997. Defining cavity-associated interactions between red-cockaded woodpeckers and other cavity-dependent species: Interspecific competition or cavity kleptoparasitism? *Auk* 114:778–780.

Kappes, J.J., Jr. 2004. Community interactions associated with red-cockaded woodpecker cavities. In *Red-cockaded Woodpecker: Road to Recovery*, eds. R. Costa and S. J. Daniels. Blaine, WA: Hancock House Publishers. In press.

Kappes, J.J., Jr., and Harris, L.D. 1995. Interspecific competition for red-cockaded woodpecker cavities in the Apalachicola National Forest. In *Red-cockaded Woodpecker: Recovery, Ecology and Management*, eds. D.L. Kulhavy, R.G. Hooper, and R. Costa, pp. 389–393. Center for Applied Studies in Forestry, College of Forestry, Stephen F. Austin State University, Nacogdoches, TX.

Kleiman, D.G. 1990. Decision-making about a reintroduction: Do appropriate conditions exist? *Endangered Species Update* 8(1):18–19.

Lacy, R.C. 1987. Loss of genetic diversity from managed populations: Interacting effects of drift, mutation, immigration, selection, and population subdivision. *Conserv Biol* 1:143–158.

Laves, K.S., and Loeb, S.C. 1999. Effects of southern flying squirrels *Glaucomys volans* on red-cockaded *Picoides borealis* reproductive success. *Anim Conserv* 2:295–303.

Lay, D.W., and Swepston, D.A. 1973. Red-cockaded woodpecker study. Job Completion Report, Project W-80-R-16, Job 10. Texas Parks and Wildlife Department.

Lennartz, M.R., Hooper, R.G., and Harlow, R.F. 1987. Sociality and cooperative breeding of red-cockaded woodpeckers (*Picoides borealis*). *Behav Ecol Sociobiol* 20:77–88.

Leonard, D.L., Jr., DeLotelle, R.S., and Epting, R.J. 2004. Factors contributing to variation in fledgling production in central Florida red-cockaded woodpeckers. In *Red-cockaded Woodpecker: Road to Recovery*, eds. R. Costa and S.J. Daniels. Blaine, WA: Hancock House Publishers. In press.

Letcher, B.H., Priddy, J.A., Walters, J.R., and Crowder, L.B. 1998. An individual-based, spatially-explicit simulation model of the population dynamics of the endangered red-cockaded woodpecker, *Picoides borealis*. *Biol Conserv* 86:1–14.

Levins, R. 1969. Some demographic and genetic consequences of environmental heterogeneity for biological control. *Bull Entomol Soc Am* 15:237–240.

Lohefener, R., and Lohmeier, L. 1981. Comparison of gopher tortoise (*Gopherus polyphemus*) habitats in young slash pine and old longleaf pine areas of southern Mississippi. *J Herpetol* 15:239–242.

Lohr, S.M. 2004. Red-cockaded woodpecker recovery efforts in an isolated and small South Carolina Sandhills population. In *Red-cockaded Woodpecker: Road to Recovery*, eds. R. Costa and S.J. Daniels. Blaine, WA: Hancock House Publishers. In press.

Love, G.J., Wilkin, S.A., and Goodwin, M.H. 1953. Incidence of blood parasites in birds collected in southwestern Georgia. *J Parasitol* 38:52–57.

Lowery, L., and Perkins, J. 2002. Long dispersal of a red-cockaded woodpecker in central Florida. *Fla Field Nat* 30:42–43.

Luttrell, M.P., Stangel, P.W., and Bartlett, C.M. 1995. A survey for blood parasites in red-cockaded woodpeckers. In *Red-cockaded Woodpecker: Recovery, Ecology and Management*, eds. D.L. Kulhavy, R.G.

Hooper, and R. Costa, pp. 332–334. Center for Applied Studies in Forestry, College of Forestry, Stephen F. Austin State University, Nacogdoches, TX.

Marsh, D.M., and Trenham, P.C. 2001. Metapopulation dynamics and amphibian conservation. *Conserv Biol* 15:40–49.

Marston, T.G., and Morrow, D.M. 2004. Red-cockaded woodpecker conservation on Fort Jackson military installation: A small population's response to intensive management in the Sandhills of South Carolina. In *Red-cockaded Woodpecker: Road to Recovery*, eds. R. Costa and S.J. Daniels. Blaine, WA: Hancock House Publishers. In press.

Mengel, R.M., and Jackson, J.A. 1977. Geographic variation of the red-cockaded woodpecker. *Condor* 79:349–355.

Mills, L.M., Feltner, K.J., and Reed, T.O. 2004. The rise and fall of the red-cockaded woodpecker population in Kentucky: A chronology of events preceding extirpation. In *Red-cockaded Woodpecker: Road to Recovery*, eds. R. Costa and S.J. Daniels. Blaine, WA: Hancock House Publishers. In press.

Mitchell, L.R., Carlile, L.D., and Chandler, C.R. 1999. Effects of southern flying squirrels on nest success of red-cockaded woodpeckers. *J Wildl Manage* 63:538–545.

Montague, W.G., and Bukenhofer, G.A. 1994. Long range dispersal of a red-cockaded woodpecker. *Proc Arkansas Acad Sci* 48:259–260.

Morris, V., and Werner, C. 2004. Withlacoochee State Forest Croom Tract red-cockaded woodpecker recovery effort. In *Red-cockaded Woodpecker: Road to Recovery*, eds. R. Costa and S.J. Daniels. Blaine, WA: Hancock House Publishers. In press.

Mount, R. 1975. *The Reptiles and Amphibians of Alabama*. Auburn, AL: Auburn Printing Company.

Neal, J.C., and Montague, W.G. 2000. Serviceability of RCW cavities. U.S. Forest Service, Ouachita National Forest, Waldron, AR.

Nesbitt, S.A., Jenkins, A., Sherrod, S.K., Wood, D.A., Beske A., White, J.H., Schulz, P.A., and Schwikert, S.T. 1998. Recent status of Florida's bald eagle population and its role in eagle reestablishment efforts in the southeastern United States. *Proc Annu Conf Southeast Assoc Fish Wildl Agenc* 52:377–3883.

Parker, W.T. 1987. A plan for reestablishing the red wolf on Alligator River National Wildlife Refuge, North Carolina. Red Wolf Management Series: Technical Report No. 1. U.S. Fish and Wildlife Service, Atlanta, GA.

Petrick, C.J., and Hagedorn, B.W. 2004. Population status and trend of red-cockaded woodpeckers on Eglin Air Force Base, Florida. In *Red-cockaded Woodpecker: Road to Recovery*, eds. R. Costa and S.J. Daniels. Blaine, WA: Hancock House Publishers. In press.

Phillips, M.K. 1990. Red wolf: Recovery of an endangered species. *Endangered Species Update* 8(1):79–81.

Platt, W.J., Evans, G.W., and Rathbun, S.L. 1988a. The population dynamics of a long-lived conifer (Pinus palustris). *Am Nat* 131:491–525.

Platt, W.J., Evans, G.W., and Rathbun, S.L. 1988b. Effects of fire season on flowering of forbs and shrubs in longleaf pine forests. *Oecologia* 76:353–363.

Poirier, J., Hedman, C., and Register, M. 2004. Southern flying squirrel use of artificial red-cockaded woodpecker cavities at International Papers Southlands Forest: The past, present, and future. In *Red-cockaded Woodpecker: Road to Recovery*, eds. R. Costa and S.J. Daniels. Blaine, WA: Hancock House Publishers. In press.

Pung, O.J., Carlile, L.D., Whitlock, J., Vives, S.P., Durden, L.A., and Spadgenske, E. 2000. Survey and host fitness effects of red-cockaded woodpecker blood parasites and nest cavity arthropods. *J Parasitol* 86:506–510.

Ralls, K., Ballou, J.D., and Templeton, A. 1988. Estimates of lethal equivalents and the cost of inbreeding in mammals. *Conserv Biol* 2:185–193.

Reinman, J.P. 1984. Woodpeckers receive second chance. *Fla Nat* 57:17.

Robbins, L.E., and Meyers, R.L. 1992. Seasonal effects of prescribed burning in Florida: A review. Miscellaneous Publication No. 8. Tall Timbers Research Station, Tallahassee, FL.

Rudolph, D.C., Conner, R.N., Carrie, D.K., and Schaefer, R.R. 1992. Experimental reintroduction of red-cockaded woodpeckers. *Auk* 109:914–916.

Saenz, D., Baum, K.A., Connor, R.N., Rudolph, D.C., and Costa, R. 2002. Large-scale translocation strategies for reintroduced red-cockaded woodpeckers. *J Wildl Manage* 66:212–221.

Saenz, D., Schaefer, R.R., Conner, R.N., Best, M.S., Rudolph, D.C., and Carrie, D.K. 2004. Influence of artificial cavity age on red-cockaded woodpecker translocation success. In *Red-cockaded Woodpecker: Road to Recovery*, eds. R. Costa and S. J. Daniels. Blaine, WA: Hancock House Publishers. In press.

Schiegg, K., Walters, J.R., and Priddy, J.A. 2002. The consequences of disrupted dispersal in fragmented red-cockaded woodpecker *Picoides borealis* populations. *J Anim Ecol* 71:710–721.

Scott, J.M., and Carpenter, J.W. 1987. Release of captive-reared or translocated endangered birds: what do we need to know? *Auk* 104(3):544–545.

Seigal, R.A., and Dodd, C.K., Jr. 2002. Translocations of amphibians: Proven management method or experimental technique? *Conserv Biol* 16(2):552–554.

Selender, R.K. 1983. Evolutionary consequences of inbreeding. In *Genetics and Conservation: A Reference for Managing Wild Animal and Plant Populations*, eds. C.M. Shonewald-Cox, S.M. Chambers, B. MacBryde, and L. Thomas, pp. 201–215. Menlow Park, CA: Benjamin/Cummings.

Shaffer, M.L. 1981. Minimum population sizes for species conservation. *Bioscience* 31:131–134.

Shaffer, M.L. 1987. Minimum viable populations: Coping with uncertainty. In *Viable Populations for Conservation*, ed. M.E. Soule, pp. 69–86. London: Cambridge University Press.

Sherrod, S.K. Jenkins, M.A., McKee, G., Tatum, S., and Wolfe, D. 1987. Using wild eggs for production of bald eagles for reintroduction into southeastern United States. In *Proceedings Third Southeast Nongame and Endangered Wildlife Symposium*, eds. R.R. Odum, K.A. Riddleberger, and J. C. Ozier, pp. 14–20. Georgia Department of Natural Resources.

Speake, D.W., and Mount, R.H. 1973. Some possible ecological effects of rattlesnake roundups in the southeastern Coastal Plain. *Proc 27th Annu Conf Southeast Assoc Game Fish Comm* 27:267–277.

Speake, D. W., McGlincy, J.A., and Colvin, T.A. 1978. Ecology and management of the eastern indigo snake in Georgia: A progress report. In *Proceedings of the Rare and Endangered Wildlife Symposium*, Technological Bulletin WL 4, eds. R.R. Odom and L. Landers, pp. 64–73. Game and Fish Division, Georgia Department of Natural Resources.

Stangel, P.W., Lennartz, M.R., and Smith, M.H. 1992. Genetic variation and population structure of red-cockaded woodpeckers. *Conserv Biol* 6:283–292.

Stober, J.M., and Jack, S.B. 2004. Down for the count? Red-cockaded woodpecker restoration on Ichauway. In *Red-cockaded Woodpecker: Road to Recovery*, eds. R. Costa and S.J. Daniels. Blaine, WA: Hancock House Publishers. In press.

Stout, I.J., and Marion, W.R. 1993. Pine flatwoods and xeric pine forests of the southern (Lower) Coastal Plain. In *Biodiversity of the Southeastern United States: Lowland Terrestrial Communities*, eds. W. H. Martin, S. G. Boyce, and A. C. Echternacht, pp. 373–446. New York: John Wiley and Sons.

Stys, B. 1993. Ecology and habitat protection needs of southeastern American kestrel (*Falco sparverius paulus*) on large-scale development sites in Florida. Florida Game and Fresh Water Fish Commission, Nongame Wildlife Program Technical Report No. 13. Tallahassee, FL.

Trenham, P.C., and Marsh, D.M. 2002. Amphibian translocation programs: Reply to Seigel and Dodd. *Conserv Biol* 16(2):555–556.

U.S. Fish and Wildlife Service. 1987. Habitat management guidelines for the bald eagle in the southeast region. U.S. Fish & Wildlife Service, Washington, DC.

U.S. Fish and Wildlife Service. 1990a. Gopher tortoise (*Gopherus polyphemus*) recovery plan. U.S. Fish & Wildlife Service, Atlanta, GA.

U.S. Fish and Wildlife Service. 1990b. Red wolf recovery plan. U.S. Fish & Wildlife Service, Atlanta, GA.

U.S. Fish and Wildlife Service. 1991. Mississippi sandhill crane recovery plan. U.S. Fish & Wildlife Service, Atlanta, GA.

U.S. Fish and Wildlife Service. 2003. Recovery plan for the red-cockaded woodpecker (*Picoides borealis*): second revision. U.S. Fish & Wildlife Service, Atlanta, GA.

Vickers, C.R., Harris, L.D., and Swindel, B.F. 1985. Changes in herpetofauna resulting from ditching of cypress ponds in coastal plains flatwoods. *For Ecol Manage* 11:17–29.

Walker, J.L., and Peet, R.K. 1983. Composition and species diversity of pine–wiregrass savannahs of the Green Swamp, North Carolina. *Vegetatio* 55:163–179.

Walters, J.R. 1991. Application of ecological principles to the management of endangered species: The case of the red-cockaded woodpecker. *Annu Rev Ecol Syst* 22:505–523.

Walters, J.R. 2004. Unusual dynamics in a rapidly increasing population of red-cockaded woodpeckers at Camp Lejeune, North Carolina. In *Red-cockaded Woodpecker: Road to Recovery*, eds. R. Costa and S. J. Daniels. Blaine, WA: Hancock House Publishers. In press.

Walters, J.R., Doerr, P.D., and Carter, J.H., III. 1988a. The cooperative breeding system of the red-cockaded woodpecker. *Ethology* 78:275–305.

Walters, J.R., Hansen, S.K., Carter, J.H., III, Manor, P.D., and Blue, R.J. 1988b. Long distance dispersal of an adult red-cockaded woodpecker. *Wilson Bull* 100:494–496.

Walters, J.R., Crowder, L.B., and Priddy, J.A. 2002. Population viability analysis for red-cockaded woodpeckers using an individual-based model. *Ecol Appl* 12:249–260.

Walters, J.R., Gault, K.E., Hagedorn, B., Petrick, C.J., Phillips, L.F., Jr., Tomcho, J., Jr., and Butler, A. 2004. Effectiveness of recruitment clusters and intrapopulation translocation in promoting growth of the red-cockaded woodpecker population on Eglin Air Force Base, Florida. In *Red-cockaded Woodpecker: Road to Recovery*, eds. R. Costa and S.J. Daniels. Blaine, WA: Hancock House Publishers. In press.

Wolf, C.D., Griffith, B., Reed, C., and Temple, S.A. 1996. Avian and mammalian translocations: Update reanalysis of 1987 survey data. *Conserv Biol* 10:1142–1154.

Wood, D.A. 1982. Florida's bald eagle. *Fla Wildl* 35:42.

Chapter 12

Spatial Ecology and Restoration of the Longleaf Pine Evosystem

Thomas S. Hoctor, Reed F. Noss, Larry D. Harris, and K. A. Whitney

Introduction

Trees of the genus *Pinus* have been dominant species in the southeastern Coastal Plain (SECP) landscape since the Pleistocene (c. 2 million years). When Spanish conquistadors arrived 500 years ago, longleaf pine (*Pinus palustris*) forest covered millions of hectares, and was the dominant land cover type throughout the SECP from southeastern Virginia to eastern Texas. Longleaf-dominated natural communities were also a significant component of some Piedmont and southern Appalachian landscapes up to 600 m above mean sea level (Varner et al. 2003a). From the wet flatwoods of Louisiana to the sandhills of southeastern Virginia, and from the ancient sand dune ridges and flatwoods of south-central Florida to the Ridge and Valley, Cumberland Plateau, and Blue Ridge physiographic provinces of northern Georgia and Alabama, longleaf pine has played a leading role in the ecology and evolution of southeastern biota. Over much of its former extent, longleaf pine communities served as the matrix surrounding many different wetland and smaller upland communities. In these landscapes, interactions between fire and hydrology were key ecological processes facilitated and mediated largely by the fire ecology of longleaf pine. These processes combined with variation in climate, topography, and soils resulted in spatial heterogeneity across spatial scales ranging from local sites and ecotones to regional landscapes. This heterogeneity supported a vast array of flora and fauna, unsurpassed virtually anywhere in the temperate zone. Species that evolved in this ecological context were frequently dependent on vast longleaf pine landscapes maintained by spatially extensive disturbance regimes (Means and Grow 1985; Noss 1988; Myers 1990; Means 1996; Harris et al. 1996a).

Not surprisingly, then, the wholesale reduction of longleaf pine landscapes to small remnant patches has led to widespread ecological dysfunction and endangerment of many species (Noss et al. 1995). Remnant patches are incapable of supporting the functional ecological processes needed to maintain ecological integrity or the spatial extensiveness and ecological juxtapositions necessary to maintain viable populations of many species that evolved in longleaf pine landscapes (Fig. 1).

Thomas S. Hoctor • Department of Landscape Architecture, University of Florida, Gainesville, Florida 32611. **Reed F. Noss** • Department of Biology, University of Central Florida, Orlando, Florida 32816. **Larry D. Harris** • Department of Wildlife Ecology and Conservation, University of Florida, Gainesville, Florida 32611. **K. A. Whitney** • Department of Urban and Regional Planning, University of Florida, Gainesville, Florida 32611.

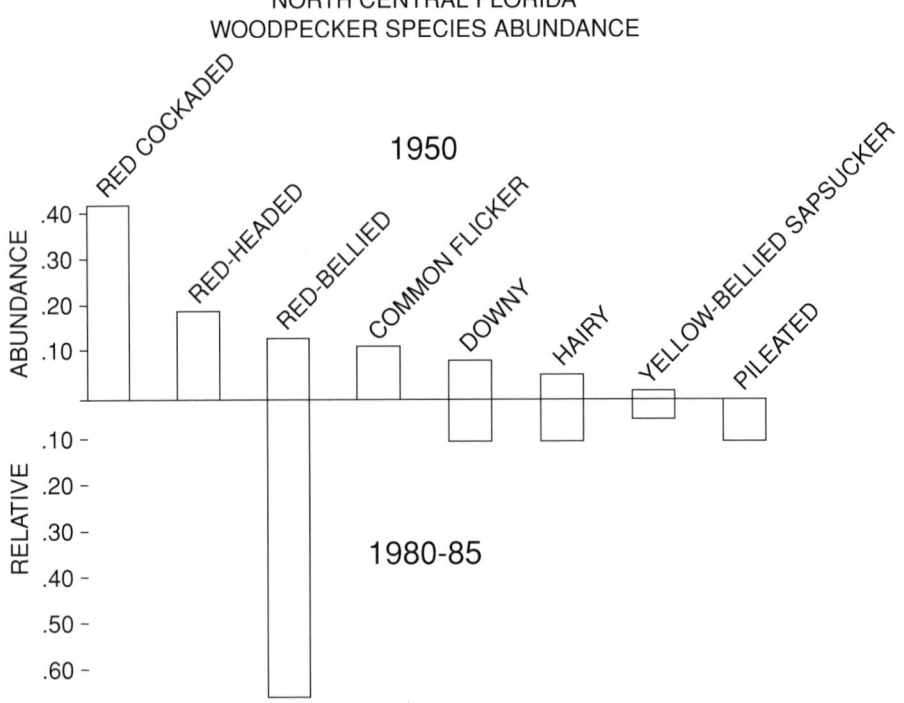

FIGURE 1. Dramatic changes in wildlife communities are one result of the demise of longleaf pine forests and ecological dysfunction. As longleaf pine communities have become vastly diminished in north-central Florida, the woodpecker guild has been altered dramatically.

The demise of longleaf pine began soon after European colonization, with commercial timbering and agricultural conversion decimating some longleaf pine landscapes well before 1900. These changes have left an indelible mark on our understanding of longleaf pine ecology, which may cloud future attempts at restoration.

Given the distinctive ecology and current condition of longleaf pine communities, an understanding of landscape ecology and regional reserve design principles is crucial for guiding restoration. A regional-scale strategy that integrates the principles and techniques of landscape ecology, conservation biology, and restoration ecology is required to restore the diversity of longleaf pine communities. Such a strategy should include restoration and maintenance of all variants of longleaf pine communities across their historic range. Conservation at this scale requires attention to between-stand diversity and function, spatial ecological processes, ecological context, interactions between natural communities, sufficient aerial extent (both of individual patches and networks of patches), and functional connectivity. The restoration of longleaf pine landscapes is required to reestablish the dominant role of this species in the regional ecology of the southeastern United States.

We use the term "evosystem" in this chapter to emphasize the critical influence of a regionally dominant ecosystem on the evolution of associated species (Hutchinson 1965; Harris et al. 1996a). As the predominant matrix vegetation in much of the southeastern United States, longleaf pine provided the context within which a diverse flora and fauna evolved (Means 1996; Platt 1999). This context has been disrupted, and longleaf pine no longer exerts the ecological or evolutionary influence it once did. The massive decline (greater than 97% from the pre-European distribution, by virtually all estimates; e.g., Ware et al. 1993) of a regional matrix vegetation has ramifications on the persistence and future evolution of far more species—and on the operation of fundamental ecological processes—than would the loss of unique natural communities that were never common (Noss et al. 1995). Yet, with its emphasis on rarity, conservation strategy has more often focused on the latter than the former.

Landscape Ecology and Historic Longleaf Pine Communities

In its pristine condition with millions of trees measuring a yard or more in basal diameter, the *Pinus palustris* consocies unquestionably presented one of the most wonderful forests in the world.... The complete destruction of this forest constitutes one of the major social crimes of American history. (Wells and Shunk 1931)

In presettlement times, longleaf pine forests covered approximately two-thirds (more than 37 million hectares) of the upland surface area in the SECP (Wahlenberg 1946; Noss 1988; Ware et al. 1993; Means 1996; Platt 1999; Frost this volume). The concept of matrix communities or dominant land cover is integral in landscape ecology. Landscape ecology has become increasingly relevant in efforts to understand, protect, restore, and manage biodiversity because it focuses on the interactions between spatial patterns and ecological processes (Turner 1989; Forman 1995). These processes and patterns generate and control the distribution of biodiversity. The concept of landscape, either implicitly or explicitly, incorporates the notion that two or more identifiably different ecosystems (or natural communities), and their associated species, interact in a spatially patterned mosaic (Forman and Godron 1986; Harris et al. 1996a).

Landscape ecology increasingly has recognized the significance of spatial heterogeneity over a range of scales from regional to local (Forman 1995; Turner et al. 1995; Poiani et al. 2000). Longleaf pine communities exhibited striking spatial heterogeneity across these scales, occurring on extremely well-drained sandhills (e.g., "desert in the rain," Wells and Shunk 1931), in seasonally flooded flatwoods throughout the SECP, on mountain slopes in Georgia and Alabama, and on relict sand dune ridges in central and south-central Florida (Noss 1988; Myers 1990; Platt 1999). Across this ecological spectrum, longleaf pine communities varied from open savannas with often stunted trees on drier, less productive sites to massive old-growth forests with trees at least 450 years old on more productive clay soils or adjacent to riparian communities (Noss 1988; Platt et al. 1988; Means 1996; Harris 1999). Although there are commonalities among these communities, associated herbaceous species and associated natural communities show a broad range of variation (Platt 1999). In the SECP, even extremely subtle differences in elevation can have profound effects on species composition and vegetation structure (Noss 1988; Noss and Harris 1990; Platt 1999). For instance, a change in elevation of less than a few meters can result in a gradient of communities ranging from extremely well-drained and desertlike sandhills to wet flatwoods and swamps. Along seepage slopes and other ecotones between uplands and wetlands, pronounced changes in plant species composition and structure can occur over an elevation range of only a few centimeters (Noss and Harris 1990; Means 1996; Platt 1999). As a fire-adapted, matrix land cover, longleaf pine communities play a keystone role in maintaining between-habitat diversity throughout the SECP (Means 1996). Both gradual and abrupt ecotones are prominent in longleaf pine-dominated landscapes and play significant roles in the propagation and influence of spatial ecological processes such as fire. High plant diversity, primarily in the herbaceous layer, is a prominent characteristic of most intact longleaf pine communities (Clewell 1981; Walker and Peet 1983; Peet and Allard 1993), and many of these plant species achieve their greatest abundance within ecotones (Noss 1988). Hence, traditional plant community classifications, which avoid ecotones and sample the "characteristic" interior areas of communities (Noss 1987a), miss much of the diversity in landscapes such as those dominated by longleaf pine. Due to the complexity of environmental gradients at multiple spatial scales, several hundred plant species can occur in areas as small as 1 ha. Beta (between-habitat) and gamma (regional) diversity are correspondingly high (Platt 1999).

We consider landscapes to include two or more, spatially and compositionally distinct community types that interact through a variety of spatial and temporal processes (Noss 1987a; Turner 1989; Forman 1995; Harris et al. 1996a; Poiani et al. 2000). Adjacency and/or connectivity are important aspects of landscape pattern that facilitate interactions among landscape components through the flow of organisms, materials, water, and energy (Forman 1995; Harris et al. 1996a). Such interactions and functions were commonplace before wholesale destruction and degradation of longleaf pine landscapes. Across much of the range of longleaf pine, ecological processes such as fire and the myriad species completely or partially associated with longleaf pine strongly influenced the ecological structure and function of the SECP. Wiregrass (*Aristida stricta* and *A. beyrichiana*), for example, along with other grasses and the duff (fallen needles) of longleaf pine, promoted the spread of frequent, low-intensity ground fires across the landscape (Platt 1988; Noss 1989), which in turn shaped vegetation structure and spatial patterns at multiple scales. Interactions between longleaf pine-dominated uplands, scrub (mostly in Florida, with a somewhat lower fire frequency), hardwood hammocks on sites topographically protected from fire such as islands, peninsulas, and karst grottos (Harper 1911; Platt and Schwartz 1990; Means 1996), and various types of riparian wetlands were critical for generating and maintaining the region's biodiversity. Importantly, the archetypal longleaf pine ecosystem—the open, park-like forests of the Coastal Plain maintained by very high fire frequencies—does not apply to all longleaf pine communities. In particular, the montane longleaf pine communities of northern Georgia and Alabama have been poorly studied but seem to differ in disturbance regime, species composition, and physiognomy. Dominance in the overstory is shared with such species as blackjack oak (*Quercus marilandica*) and sand hickory (*Carya pallida*), while ice and snowstorms join the relatively less frequent fire in maintaining longleaf pine (Varner et al. 2003a,b). On other sites in the lower Piedmont and upper Coastal Plain, tree co-dominants include southern red oak (*Quercus falcata*), flowering dogwood (*Cornus florida*), post oak (*Quercus stellata*), loblolly pine (*Pinus taeda*), shortleaf pine (*Pinus echinata*), and sweetgum (*Liquidambar styraciflua*) (Beckett and Golden 1982; Peet and Allard 1993).

Longleaf Pine Exploitation and the Deconstruction of Longleaf Pine Ecology

It is not our purpose here to review the long history of longleaf pine utilization and exploitation, but we have concerns about the usual accounts of longleaf pine history and their effects on perceptions of its ecology. It is well accepted that, since European settlement, the extent of longleaf pine has declined to less than 2 or 3% of its former area (Means and Grow 1985; Noss 1988, 1989; Ware et al. 1993; Means 1996). However, it is important to recognize the great length of time and increasing intensification of use and destruction of longleaf pine resources since European exploitation began. By the year 2108 Florida will have been part of the United States for the same length of time that it was under Spanish rule. Thus, human impact on southeastern forest resources has occurred over a substantial period of time and has taken place under the influence of several cultural groups (Harris 1980; Myers 1990; Frost 1993, this volume). Review of the historical literature regarding the use and exploitation of forest resources in the SECP indicates that significant impacts have been ongoing for most of the period since Europeans arrived. Although human impact is not a particularly modern influence on this ecosystem, the intensive site preparation and other practices of modern industrial silviculture may indeed have more lasting effects on the ecology of longleaf stands and landscapes than did more primitive forms of exploitation (Noss 1988).

Although the types of forest products extracted and forms of exploitation changed over the centuries, the trend of resource use has been one of continually increasing intensity, at least until very recently when clearcutting of longleaf pine on public lands diminished (Elliott 1912; Greeley 1920; Tyson 1956; Adams 1976; Frost 1993; Stout and Marion 1993; Harris 1999). Both the forest and wildlife populations reflect these changes.

The long history of exploitation in longleaf pine forests in the SECP is an essential consideration for restoring the ecological function of this natural community. Although the usual definition of longleaf pine forests as open-canopy "savannas" certainly is supported by anecdotal accounts and the ecology of remaining longleaf pine ecosystems on less productive sites within the SECP, the distinctive characteristics of longleaf pine forests on more productive soils and outside the SECP have been obscured by the trends and pattern of historical exploitation of this species. Longleaf pines on bottomland and other productive soils were frequently cleared first because of easier access and conversion to agricultural uses. And fire suppression has resulted in hardwood dominance in alluvial and other productive sites that once supported impressive forests of longleaf pine (Harris 1999). Hence, few examples of these communities remain. This pattern is not exceptional. Across the United States and in other countries, the proportion of protected areas is lowest on low-elevation lands with highly productive soils (Harris 1984; Scott et al. 2001). In other cases, longleaf pine on high-productivity sites was likely selectively cut, so that the density and average size of trees declined over time (Fig. 2). Because longleaf pine communities occurred on a variety of sites and have been affected by numerous forms of exploitation over a long period of time, consideration of the landscape ecology of longleaf pine communities on regional, landscape, and local scales is critical for developing restoration and management strategies. With this in mind, in the following sections we focus on the primary aspects of the spatial ecology of longleaf pine and on the need for restoration that accounts for this spatial heterogeneity and landscape-level processes.

The Spatial Ecology of Longleaf Pine Natural Communities: From Basic Biology to Biogeography

Relevant Basic Longleaf Pine Biology

Longleaf pine and wiregrass (and other herbaceous species in some cases) are the ecologically pivotal species of these pyrophilic systems, and frequent, low-intensity fire is the single most critical ecological process necessary for their maintenance (Noss 1988; Harris et al. 1996a; Whitney 1999). Longleaf pines convert lightning strikes to ground fires (Platt et al. 1988). Fallen pine needles, with their volatile oils and resins, combined with highly flammable grasses promote frequent fires that eliminate most hardwoods and other potentially competing plants (Noss and Harris 1990; Harris et al. 1996a; Platt 1999). This situation supports the hypothesis of Mutch (1970), who suggested that the flammable properties of some plant species have been favored by natural selection because they reduce competition with other plant species. In addition, longleaf pine's entire life cycle is linked to fire, and seed germination and seedling establishment are enhanced if the seed falls on sparsely vegetated or bare mineral soil (Myers 1990). Volatile resins produced by longleaf pine not only promote fire but also deter pest outbreaks that are more common in less fire-adapted yellow pine species (Harris et al. 1996a; Whitney 1999).

Lighterwood in the form of standing and prostrate longleaf pine logs existed in large volumes in the original forest (Harris 1999). Settlers could use this readily available "down wood" for many years before they had to do actual logging. Indeed, the turpentine industry got its start by using down literwood and

Before logging

After logging

FIGURE 2. Low-density stands of longleaf pine have now become so commonplace that many writers consider them to be characteristic of original conditions. Rich sites, such as those that were settled first, and alluvial sites that have now become dominated by hardwoods once supported dense stands of large longleaf pines. Source: Mohr, C. 1896. Timber pines of the southern United States. Bulletin No. 13, USDA Forest Service, Washington, DC.

dragging it into piles on the ground and simply letting the various grades of tars flow into the barrels (Butler 1998). Such wood played a key role in virgin longleaf pine forests by providing unique niches for various plants and animals in the ground cover (Hermann 1993).

Nutrient cycling is also an important issue on sandy sites where nutrients quickly leach downward from surface layers. Gopher tortoise (*Gopherus polyphemus*), pocket gopher (*Geomys pinetis*), and scarab beetles (*Peltotrupes* sp.) all help reverse the direction of nutrients leaching downward (Kalisz and Stone 1984; Myers 1990). An endemic scarab beetle from north-central Florida (*Peltotrupes youngi*) digs 360 cm or deeper and can transport as much as 8 mg of subsoil to the surface per hectare per year (Kalisz and Stone 1984).

Heterogeneity: Between- and among-Stand Considerations, Ecotones, and Mosaics

Spatial heterogeneity among stands of longleaf pine and other natural communities embedded within the longleaf pine matrix was a key characteristic of these landscapes. Tree form, density, and related characteristics govern the within-stand heterogeneity, while the size, shape, and arrangement of the stands in the larger forest management unit affect landscape heterogeneity (Harris 1980, 1984). Within stands of longleaf pine, elevation, soils, and disturbance history all influenced stand structure. Various disturbances, including hurricanes, tornadoes, floods, and diseases, also strongly influenced stand structure (Harris et al. 1996a). These patterns and processes are essential for creating and maintaining the required environmental conditions for many species. In addition, because many species require different community types during different seasons or life stages, and must move between communities on a regular basis, the arrangement of community types within the landscape is of critical importance (Noss 1987a; Harris 1989; Forman 1995; Harris et al. 1996a).

Gradual ecotones and abrupt edges between longleaf pine and other natural communities were both common in intact landscapes (Barry 1980; Noss and Harris 1990; Means 1996; Platt 1999). Ecotones reflect changes in edaphic conditions and interactions between processes, especially fire and flooding, in longleaf pine landscapes. Ecotones included what could be considered unique natural communities that were dependent on more productive soils or wetter soils along with frequent, or somewhat frequent, fires. In Marion and Alachua counties in north-central Florida a prime example is the southernmost extension of upland pine forest (sometimes referred to as "red oak woods") found on richer soils including on the edges of ravines, sinkholes, or other relatively steep slopes (Harper 1915; Duever 1983; Myers 1990). These communities had higher productivity and harbored some plant species [including canopy co-dominants southern red oak and mockernut hickory (*Carya alba*)] that were absent or less common in adjacent communities both up- and downslope. The greater production of food resources was likely significant for herbivorous species such as fox squirrels (*Sciurus niger*), Florida mice (*Podomys floridanus*), and gopher tortoises (*Gopherus polyphemus*), and for carnivores including the eastern indigo snake (*Drymarchon corais couperi*) that would benefit from increased prey densities.

Fires in longleaf pine communities are also important for maintaining fire-adapted plant communities such as pitcher plant bogs. Frequent fires prune back wetland shrubs that would otherwise encroach upslope, maintaining an open herbaceous bog community in some cases (Fig. 3; Noss and Harris 1990; Means 1996). This is particularly pronounced on seepage slopes in the Florida Panhandle and southern Alabama. Ecotones in longleaf pine landscapes tend to be very high in plant diversity, and their degradation or outright destruction results in significant erosion of biodiversity and ecological integrity. Fires moving downhill from longleaf pine communities also help maintain graminaceous vegetation within ephemeral wetlands needed or preferred by flatwoods salamanders (*Ambystoma cingulatum*) and other amphibians, many of which are imperiled or declining (Huffman and Blanchard 1990; Dodd and LeClaire 1995; Palis 1996; Dodd and Cade 1998; Cox and Kautz 2000; Kautz and Cox 2001).

In Florida, longleaf pine sandhills are frequently sharply juxtaposed with scrub communities. These communities are dependent on starkly different fire regimes, with scrub, which is dominated by short-stature oaks or sand pine (*Pinus clausa*), typically burning catastrophically on 15- to 80-year intervals

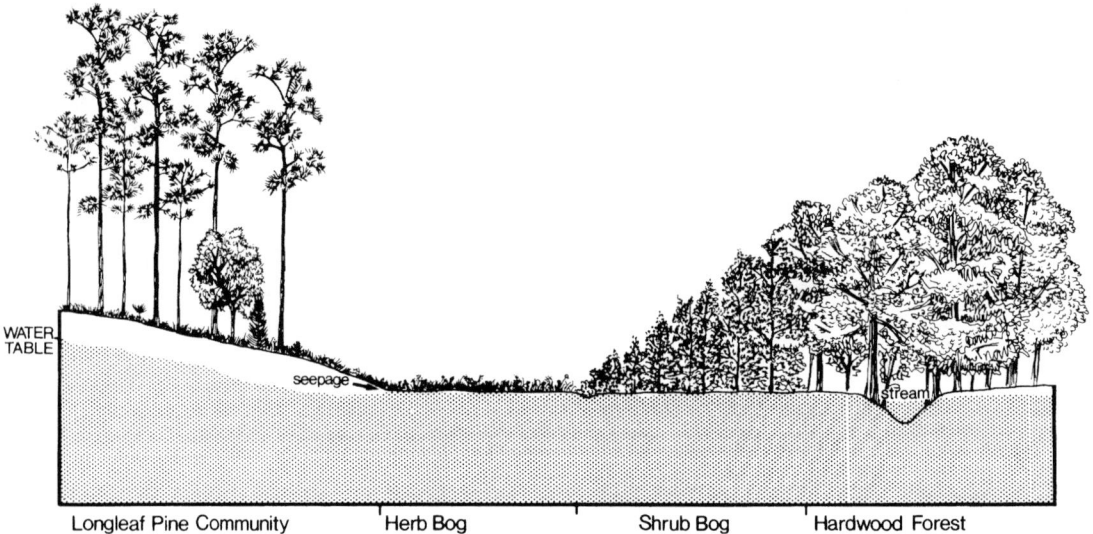

FIGURE 3. Fire moving downslope from longleaf pine communities maintains herbaceous wetland communities along ecotones that would otherwise be dominated by shrubby or forested wetlands. Such "ecotone communities maintained by a combination of wetter or richer soils (including herb bogs and red oak woods) are often the first to disappear through fire suppression. Source: Means (1996).

(Myers 1990). The transition between these two communities is often sharply demarcated and is thought to represent changes in historic fire regimes that then tend to be perpetuated by the vastly different vegetative physiognomy and flammability (Myers 1990).

Spatial mosaics are an integral part of functional landscapes (Forman 1995). The discussion regarding ecotones touches on this issue, but the interactions between different patches within a longleaf pine landscape transcend edge or ecotonal phenomena. Forman's (1995) concept of the "ecosystem cluster" is relevant. An ecosystem cluster "is recognized as a spatial level of hierarchical organization between the local ecosystem and the landscape. It describes a group of spatial elements connected by a significant exchange of energy or matter" (Forman 1995:287). This is similar to the "functional landscape" concept of Noss (1987a) where multiscalar interactions between patches are considered essential for maintaining ecological integrity and biodiversity (Harris 1984; Noss and Harris 1986; Noss 1987a). The concept of catena, used to describe connected sequences of soils from ridge top to valley bottom, is also relevant (Milne 1935, 1947; Woodmansee 1990; Forman 1995). Though most energy and matter-flows in a catena may be downslope, processes such as flooding, fire, movement of nutrients, animal movements, and windflows can move upslope as well (Boerner and Kooser 1989; Woodmansee 1990; Forman 1995; Harris et al. 1996a). Therefore, ecosystem clusters and catenas can be seen as networks connected by ecological processes flowing across landscapes.

Spatial mosaics have a strong influence on fire patterns within longleaf pine landscapes. Linear riparian networks, wetland patches, sinkholes, and ravines all can serve as significant barriers that create fire shadows of various sizes (Harper 1911; Noss and Harris 1990; Platt and Schwartz 1990; Means 1996). Fire shadows, from scales ranging from individual down lighterwood (Hermann 1993) to wetland patches and riparian strips, include significant spatial heterogeneity and often support natural communities intolerant of fire, or frequent fire, and in turn provide habitat for additional species and produce additional resources

FIGURE 4. The often interdigitated spatial relationship between longleaf pine forest and adjacent riparian forests and wetland communities is a critical ecological characteristic of intact longleaf pine landscapes. Many species require both upland and wetland communities to meet their life history needs, interactions between fire and flooding along these juxtapositions maintain unique natural communities, and riparian swaths provide corridors for many species. Reprinted from Harris, L.D., Hoctor, T. S., and Gergel, S. E. 1996a, with permission from the University of Chicago Press.

 Hardwood Forest

 Longleaf Pine Upland Forest

such as mast (e.g., acorns) for wildlife. Fire also affects longleaf pine landscape mosaics beyond the boundaries of fire-adapted communities. For example, *Torreya taxifolia*, an endangered tree found only along bluffs of the Apalachicola River in the Florida Panhandle, may be imperiled because, due to fire suppression, smoke from adjacent longleaf pine communities no longer controls fungal diseases (Schwartz et al. 1995).

The significance of the relationship between longleaf pine uplands and wetlands extends beyond interactions between flooding and fire. In many landscapes formerly dominated by longleaf pine, riparian strips bisected uplands in interdigitated patterns that resulted in firebreaks, facilitated functional juxtapositions between natural communities, and provided corridors for wetland species (Fig. 4; Delcourt and Delcourt 1977; Harris 1988; Noss and Harris 1990; Delcourt et al. 1993; Harris et al. 1996a):

The role of riparian, bottomland forests in the landscape is also critical. Between-community qualities include their high contrast with conifer-dominated uplands, and a phenology of mast and other resource production that is non-synchronous with upland communities (Harris 1989).

... Finally, the landscape-level functions of riparian forests include dispersal of plants and animals along both terrestrial and aquatic pathways. The linear, dendritic pattern of riparian forests makes them ideal collectors and transporters of wildlife across regional landscapes. (Noss and Harris 1990: 133)

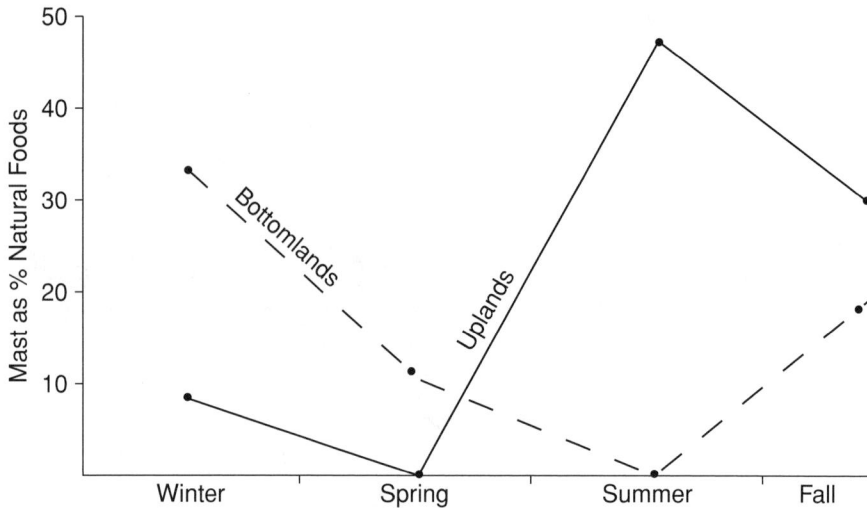

FIGURE 5. Florida black bear (*Ursus americanus floridanus*) in longleaf pine landscapes require both uplands that produce saw palmetto fruit (*Serenoa repens*) and other soft mast and riparian hardwood forests that produce acorns and other hard mast.

Spatial mosaics are also essential for maintaining viable populations of many species found in longleaf pine landscapes. For example, Weigl et al. (1989) argued that management prescriptions for the red-cockaded woodpecker (*Picoides borealis*) calling for extensive, frequent burning and removal of almost all hardwoods throughout longleaf pine forests would not necessarily provide ideal habitat for fox squirrels, which are also imperiled:

any action that removes all or most of the larger oaks (or hickories) from among the pines of the preferred pine-oak habitat will have a devastating effect on the food supply and nest cavity availability for the fox squirrel and many other kinds of wildlife. The continuous thinning of dense small hardwoods by fire or other means would be highly beneficial in most forests, but the management goal should be an open stand of large pines and scattered 30+ year-old oaks or oak groves ... [W]hile the mature pine-oak forest represents the major habitat of the fox squirrel, this is not the only vegetation type which this animal can use, nor does such prime habitat have to occur in single large units. A mosaic of habitats with substantial pine-oak representation, large areas of edge, some open land, and access to bottomland seems to support squirrels as well as larger pine tracts. (Weigl et al. 1989: 79–80)

Other species found in longleaf pine landscapes are also dependent on functional spatial mosaics (Fig. 5). Without downplaying the significance of frequently burned, open-canopy longleaf pine stands, less frequently burned communities are critical components of the broader landscape mosaic for many species. For example, sandhills and other open pinelands are considered only secondary habitat for the Florida black bear (*Ursus americanus floridanus*), and frequent fires that remove too much shrubby cover can reduce habitat quality (Cox et al. 1994; Maehr 1997; Maehr et al. 2001a; Hoctor 2003). In particular, saw palmetto fruit (*Serenoa repens*) is an exceptionally important food resource, and fruit production is highest in palmetto stands that are at least 5 years old (Hilmon 1969; Maehr 1997; Maehr et al. 2001a). To facilitate mosaics that could provide high-quality bear habitat while meeting the habitat needs of species requiring more frequent fire, Maehr et al. (2001a: 43) recommended:

Ecosystem approaches to management on public lands may dictate prescribed fire regimes that are not always optimal for bears (i.e. some public lands include timber harvesting as a primary objective, and others are very close to urban areas). In these cases, a multiple use module (MUM) approach could be applied to create natural, heterogeneous landscapes that also provide high quality habitat nodes, such as old-growth forest for high quality den sites, and altered fire regimes to provide dense stands of infrequently burned saw palmetto for food and cover.

Other species that either favor or require habitat mosaics in longleaf pine landscapes include the eastern indigo snake, which prefers upland/wetland mosaics and uses gopher tortoise burrows as refugia (Moler 1992), and various amphibians that require ephemeral ponds for breeding but live in natural pinelands as adults (Huffman and Blanchard 1990; Dodd and LeClaire 1995; Palis 1996; Dodd and Cade 1998; Cox and Kautz 2000; Kautz and Cox 2001).

Area and Connectivity

The former extensivity and connectivity of longleaf pine landscapes suggests that myriad functional processes are dependent on these characteristics. Within natural longleaf pine forests, large swaths of connected uplands were necessary for maintaining the high-frequency, low-intensity fire regime needed to perpetuate the system (Table 1):

Theoretically, a few ignitions in each state would be sufficient to burn over most of the longleaf pine landscape. That this happened as recently as the 1880s was reported by Hough (1882), who said that fires burned for weeks at a time over several counties (when counties were much larger than at present). (Means 1996: 213)

Under current conditions, use of prescribed fire to restore and manage longleaf pine is easier in larger, rural landscapes than in smaller patches surrounded by intensive development (Harris et al. 1996a). Red-cockaded woodpeckers, fox squirrels, and indigo snakes all need large areas to support viable populations (Cox et al. 1994; Cox and Kautz 2000). In addition, amphibians that breed in ephemeral ponds also require extensive, intact uplands around suitable breeding ponds. Dispersal distances from natal ponds are frequently in excess of 2000 m for various species including gopher frog (*Rana capito*), striped newt (*Notophthalmus perstriatus*), and flatwoods salamander (Ashton 1992; Means et al. 1996; Palis 1996; Dodd and Cade 1998; Kautz and Cox 2001; Semlitsch 2000).

Connectivity is a ubiquitous attribute of natural landscapes. It occurs at many spatial and temporal scales and allows for flows of disturbances, water, energy, and nutrients, as well as organisms and their genes (Harris 1984, 1985; Forman and Godron 1986; Noss and

TABLE 1. The probability of fire and the effects of fire are scale dependent.[a]

Scale of focus	Probability of lightning ignition in a decade	Significance of occurrence
Individual tree (0.001 acre)	Very small	Lethal
Remnant patch (0.01–1 acre)	Modest	Serious
Isolated tract (10–100 acres)	High	Important
Forest (10,000–100,000 acres)	Certainty	Essential

[a] Habitat fragmentation has resulted in much smaller patches of longleaf pine forest that are much less likely to harbor lightning ignitions. Large, connected areas are much more likely to experience the fires needed to maintain functional longleaf pine landscapes.

Harris 1986; Noss 1987b; Noss and Harris 1990; Harris and Scheck 1991; Noss and Cooperrider 1994; Harris et al. 1996b). Connectivity in the sense of intact ecotones and functional juxtapositions between natural communities within longleaf pine landscapes has been discussed above. Nevertheless, regional-scale reserve design and conservation planning must also consider the functional connectivity between landscapes to maintain viable metapopulations of sensitive, wide-ranging species and to provide opportunities for species to respond to climate change. Long-term survival for many species will depend on interconnected landscapes that are sufficiently integrated to allow functional dynamics in both time and space (Peters and Darling 1985; Hunter et al. 1988; Peters and Lovejoy 1992; Noss and Cooperrider 1994; Harris et al. 1996a,b). Important, narrower corridors include riparian networks that serve as dwelling habitat and travel routes for many species (Harris 1988, 1989) and pinelands that may allow the dispersal of red-cockaded woodpeckers (Walters et al. 1988), fox squirrels (Weigl et al. 1989), and other species. Larger landscape linkages would ideally be wide enough to encompass a diversity of habitat types, including complete topographic gradients, and to allow sensitive wildlife species to travel undisturbed by human activities (Harris and Gallagher 1989; Noss and Harris 1990; Harris and Atkins 1991; Harris and Scheck 1991; Noss 1992, 1993; Noss and Cooperrider 1994; Harris et al. 1996b).

Landscape-Level Strategies, Opportunities, and Challenges

One of the primary lessons of landscape ecology is that spatial context matters (Harris 1984; Harris et al. 1996a). Harris and Kangas (1988:141–142) recognized the significance of contextual challenges for restoring and managing biodiversity:

As humans modify land use—and thus the composition of landscapes—the matrix and context of individual habitats are changed. Regardless of whether the content of the surviving habitat fragments is directly altered, changes in the contextual setting will inevitably lead to indirect changes in the structural content of these habitat fragments ... [O]nly habitat assessments that evaluate the internal characteristics of these very large habitats in the context of their location and surrounding environments will lead to reasonable conclusions about their adequacy.

Hence, natural resource conservation and land-use planning must consider the effects of actions within their largest spatial and temporal contexts (Forman 1987, 1995; Harris et al. 1996a). Landscape ecologists and conservation biologists recognize habitat loss and fragmentation as the primary threats to biodiversity and many ecological processes and services (Wilcox and Murphy 1985; Harris and Silva-Lopez 1992; Meffe and Carroll 1997). Fragmentation occurs at multiple spatial scales, from the regional-scale separation of major blocks of longleaf pine by extensive agricultural and urban development; to landscape-scale fragmentation by agriculture, major roads, clearcuts, and residential/commercial development; to site-level fragmentation by forest roads, trails, and plowed fire lines. The understanding and maintenance of critical landscape functions, including interactions between natural communities in time and space, are central considerations in landscape ecology and are essential for long-term monitoring and conservation of biodiversity. Fragmentation ultimately leads to landscape dysfunction and the erosion of biodiversity; thus, strategies to foster landscape heterogeneity and connectivity, and maintain and/or restore critical landscape processes, are essential to both landscape viability and biodiversity conservation (Harris 1984; Weigl et al. 1989; Turner et al. 1995; Harris et al. 1996a; Gordon et al. 1997; Poiani et al. 2000). Such comprehensive landscape planning could maximize extensivity and connectivity by linking reserves and multiple-use conservation lands into functional networks. The goal is to protect and manage a landscape to preserve and/or mimic natural processes and evolutionary forces. Only in such a landscape can we expect to conserve the remaining biodiversity and functional processes that will allow further evolution and adaptation (Harris et al. 1996a). Its spatial ecology demands a restoration and conservation strategy that will return longleaf pine communities to their role as the matrix ecosystem. Such a strategy must include efforts to restore and maintain intact ecotones, functional juxtapositions of uplands and wetlands needed by many species, and the spatial extent and connectivity needed by wide-ranging and other species adapted to longleaf pine forests. Harris et al. (1996a: 341) describe a vision for longleaf pine landscape restoration that is relevant here:

[S]imply saving some islands of old-growth longleaf pine is not the issue; saving the red-cockaded woodpecker is not the issue; and providing an interconnected habitat system that protects both the longleaf pine and the red-cockaded woodpecker should not be the issue. Rather, we see the issue as being that of restoring and maintaining a spatially integrated longleaf pine ecosystem that can and will maintain the full suite of landscape ecological processes including fire that is ignited in one place but allowed to disperse across the system; a system that can withstand the effects of major hurricanes and still remain viable and resilient because of its extensive nature; a system that is capable of sustaining natural outbreaks of beetles and fungi; and a system that is interdigitated with other community types that provide seasonally important services for the longleaf pine community and vice versa.

Landscape Conservation Strategies

Given the current status of longleaf pine communities as remnant islands in a sea of intensive forestry, hardwood stands that developed after fire suppression, agriculture, and, increasingly, urban areas, opportunities to restore and manage at landscape scales are rare and probably will be fleeting. However, awareness of longleaf pine biology combined with planning guidelines from conservation biology and landscape ecology can be used as a sound basis for restoring ecological processes in longleaf pine landscapes that will support native biodiversity. Overall, regional-scale restoration and management of longleaf pine landscapes must include reestablishing large patches of old growth, connectivity among patches, intact ecotones, and functional environmental and community gradients.

The restoration and maintenance of viable metapopulations of longleaf-associated animal species (e.g., fox squirrel, red-cockaded woodpecker) within connected networks of core protected areas on public conservation lands, augmented by multiple-use public and private lands managed to contribute to ecological integrity, is an overarching goal (Harris 1984; Noss and Harris 1986; Noss 1987b; Noss and Cooperrider 1994; Harris et al. 1996b; Hoctor et al. 2000). Such networks provide the opportunity to reestablish landscapes where fragmentation is minimized and natural connectivity, habitat juxtapositions, ecological gradients, and spatial and temporal heterogeneity are enhanced (Harris 1984; Noss and Harris 1986; Noss and Cooperrider 1994; Harris et al. 1996a; Soulé and Terborgh 1999; Margules and Pressey 2000; Poiani et al. 2000).

How do we restore longleaf pine landscapes starting from the current situation of primarily small, isolated tracts of longleaf, virtually all of which are degraded to some degree? Although urbanization is increasing rapidly in parts of the SECP, rural land uses and forestry still dominate. Even in Florida, 80% of the state is currently classified as rural. Therefore, the protection and restoration of functional networks of longleaf pine and associated communities is still feasible. Significant opportunities to restore and maintain longleaf pine landscape-types exist on large public lands, including national forests and military lands. In fact, large-scale restoration to longleaf pine and increased use of prescribed fire is underway on various public conservation lands and military installations, including Eglin Air Force Base in northwest Florida (Gordon et al. 1997). Even the national forests, where longleaf pines were clear cut and replaced with plantations of slash, loblolly, and sand pines, have reduced large-scale logging, have moved toward more uneven-aged management, and are now replanting longleaf or allowing natural regeneration on many sites.

Restoration of functional longleaf pine landscapes must include consideration of between-stand characteristics, including management for old-growth longleaf within forests also managed for timber production (Harris 1984). Reestablishment of longleaf pine diversity within multiple-use landscapes should include rehabilitation of sandhills and clayhills using prescribed fire, replanting of longleaf on sites that originally supported longleaf, and restoration of longleaf on other selected sites, including bottomland ecotones.

The importance of considering ecological context for restoring and maintaining functional landscapes cannot be overstated. Ecology has long suffered from a focus on within-system attributes and functions (Harris et al. 1996a). Landscape ecology has made clear that multiscale spatial patterns and processes across heterogeneous landscapes control key ecological functions. Another crucial design principle for longleaf pine landscapes is the restoration of functional ecotones and habitat juxtapositions between uplands dominated by longleaf and associated smaller patch upland communities with wetlands including bogs, marshes, shrub swamps, bottomland forests, and forested swamps. Fires should be allowed to burn into wetland edges to reestablish ecotonal communities shaped by the interaction of fire and flood and to restore breeding habitat for the myriad plant, invertebrate, and amphibian species characteristic of intact longleaf pine landscapes (Noss 1988; Huffman and Blanchard 1990; Noss and Harris 1990; Means 1996; Palis 1996; Dodd and Cade 1998; Platt 1999; Cox and Kautz 2000; Kautz and Cox 2001). Interactions, juxtaposition, and interdigitation with forested wetlands are essential characteristics of longleaf pine landscapes, and wetlands provide key functions including mast production, drought refugia, and breeding habitat (Clewell et al. 1982; Hart 1984; Harris and Vickers 1984; Vickers et al. 1985; Gross 1987; Harris 1988, 1989; Forman 1995).

Restoration and management of ecological context require spatially extensive approaches. We again emphasize that, in the case of longleaf pine, restoration of context means reestablishing the role of longleaf pine as the landscape matrix for other communities. This will often require restoration on the scale of tens to hundreds of thousands of hectares, in addition to the more typical site-level restoration projects. Although ecological function might be restored on single, very large public land holdings, regional landscape restoration requires design and management across ownerships, which in turn will require multiagency and public–private cooperation (Harris 1984; Noss and Harris 1986; Noss and Cooperrider 1994). Context management also requires consideration of adjacent land uses and the utilization of buffer-zone principles to encourage compatible land uses that can provide habitat for wide-ranging species, facilitate critical ecological fluxes, minimize external threats and negative edge effects, and potentially provide functional connectivity among landscapes. Such context management also is critical for restoring landscapes that can support more natural fire regimes and other landscape processes:

The prospect of restoring an interconnected landscape system of conservation areas where ecological processes such as naturally ignited and naturally propagated fire exist (that is, a LET-BURN POLICY) appears imminent but problematic in at least certain states. These problems may be overcome by managing entire landscapes to emphasize compatibility of adjacent land uses in a manner so that conservation lands will be shielded from the negative impacts of adjacent intensive development, and developed areas will be shielded from the potential negative impacts of conservation land management. (Harris et al. 1996a:334)

Therefore, protection and restoration coordinated with private partners will be essential. Red-cockaded woodpeckers are the flagship species needing inclusion of, and incentive for, private landowners to restore longleaf pine landscapes. One project, discussed below, involves efforts to reestablish red-cockaded woodpeckers on lands around military bases to increase the viability of regional populations and reduce the management burden that currently lies almost exclusively within military installations.

Opportunities for restoring connectivity among widely separated longleaf pine landscapes are particularly challenging.

Nevertheless, current land uses could facilitate the restoration of longleaf pine across larger areas over time.

Opportunities

A key fact is that there are still millions of hectares of forested land in the SECP. Although much of this forest has been converted to pine plantation or otherwise degraded by fire suppression and other significant changes in ecological processes (Ware et al. 1993), emerging trends toward landscape and regional-scale conservation planning, protection, and restoration provide significant opportunities to reestablish longleaf pine as the dominant matrix vegetation. The following examples of conservation initiatives include landscape and regional-scale projects and other programs that may facilitate restoration of longleaf pine landscapes:

Cross-Florida Greenway

This conservation area is an approximately 190-km-long corridor of more than 40,000 ha that runs from the Florida Gulf Coast north of Tampa to the St. Johns River in northeastern Florida. Conservation efforts include the restoration of remaining longleaf pine forests degraded by fire suppression and replanting of longleaf pine on former agricultural lands. Ecotonal communities including degraded red oak woodlands are also being restored. One of the primary goals of this project is to re-create a continuous corridor of longleaf pine and related natural communities from the Ocala National Forest to the Gulf Coast and to protect and restore populations of Sherman's fox squirrel (*Sciurus niger shermani*), gopher tortoises, indigo snakes, red-cockaded woodpeckers, and many other species dependent on longleaf pine landscapes (Harris and Hoctor 1992).

Department of Defense Projects

U.S. Department of Defense (DOD) installations contain some of the most significant longleaf pine forest remaining (Means 1996; Gordon et al. 1997). The DOD has developed integrated natural resource management plans that include restoration of longleaf communities. One major effort involves Eglin Air Force Base in the Florida Panhandle, which seeks to restore extensive longleaf pine forests through prescribed fire and other efforts "utilizing integrated natural resources management and principles of ecosystem management to ensure ecosystem viability and biodiversity while providing compatible multiple uses" (Means 1996: 219).

The DOD has also been working with partners in the North Carolina sandhills around Fort Bragg and other project areas to protect additional habitat for federally listed species (especially the red-cockaded woodpecker), connect existing conservation lands, and reduce the potential for further urban encroachment around bases that could interfere with military training operations (Goodison and Hoctor 2003). These efforts will now be facilitated by recent authorization for DOD to work with partners to protect lands near military bases (Robert Lozar personal communication).

An additional project involves cooperation between DOD, the state of Florida, The Nature Conservancy, the University of Florida, and private landowners (e.g., the Nokuse Plantation) to protect a 90-km landscape linkage between the Apalachicola National Forest and Eglin Air Force Base in northwest Florida. This landscape linkage will protect existing longleaf pine forests and provide opportunities to restore longleaf pine while protecting an essential flight-training corridor for DOD. This project illustrates the concept of "sharing the burden" in the management of area-sensitive species such as the red-cockaded woodpecker and, more importantly, in the restoration of intact landscapes that will require public–private cooperation on a regional scale.

The Nature Conservancy's Ecoregional Planning

The Nature Conservancy has developed ecoregion-based biodiversity plans across the range of longleaf pine. Ecoregional planning is a biodiversity assessment effort attempting to comprehensively identify all sites needed to conserve biodiversity within all ecoregions in the United States (Groves et al. 2000, 2002). The plans for ecoregions in the southeastern United States identify various areas for protecting existing longleaf pine sites and landscape-scale restoration opportunities to restore longleaf pine as a matrix community. In the Peninsular Florida Ecoregion, sites capable of supporting potential matrix-quality longleaf pine landscapes were defined as having at least 2000 ha of sandhill or flatwoods cover. Twenty-seven such sites were identified for flatwoods and eight for sandhills within the ecoregion (Hilsenbeck et al. 2001; Hoctor 2003).

The Florida Ecological Network

The Florida Ecological Network is part of the Florida Greenways Program administered by the Florida Department of Environmental Protection. The Florida Ecological Network was developed using a GIS-based regional landscape analysis to delineate the best opportunities to protect large, connected landscapes across the state (Hoctor et al. 2000; Hoctor 2003). Existing longleaf pine communities and focal species dependent on longleaf pine were included as priorities in the analysis (Hoctor et al. 2001, 2002b; Kautz and Cox 2001). Other areas of degraded pineland cover that could contribute to protecting large landscapes also were included, and such areas may provide significant opportunities for longleaf pine restoration in the future. An implementation plan has been developed for the Florida Greenways Program that includes the identification of Critical Linkages within the Florida Ecological Network (Hoctor et al. 2002b). These Critical Linkages are now the primary focus of landscape protection opportunities in Florida and include projects relevant to longleaf pine conservation, including the Apalachicola National Forest–Eglin Air Force Base project mentioned above and a 60,000-ha landscape linkage between the Camp Blanding military training site and Osceola National Forest in northeast Florida.

EPA Southeastern Ecological Framework

The EPA Southeastern Ecological Framework is a cooperative effort between EPA Region 4 and the University of Florida that was delineated using a GIS-based analysis to identify priority land-conservation areas across an eight-state region including several states within the range of longleaf pine: North Carolina, South Carolina, Georgia, Florida, Alabama, and Mississippi. The framework is meant to encourage federal and state interagency cooperation and coordinate protecting and restoring landscapes efficiently (Hoctor et al. 2002a). As with the Florida Ecological Network, millions of hectares of pineland are included, which could provide significant opportunities for longleaf pine restoration (Fig. 6).

Forest Certification/Landowner Programs

Forest certification is a developing, incentives-based strategy to facilitate ecologically sound management of forest resources. Forest certification involves an independent, third-party assessment of field-level forest management practices against specified social, ecological, and economic standards (http://www.sfrc.ufl.edu/Extension/ffws/fc.htm#ctcs). In the Southeast, forest certification is beginning to be used to restore longleaf pine on private lands that have been converted to other uses or other tree species. The success of certification will be based on consumer demand for wood products that are certified as being produced using management practices that promote healthy forests. If successful, forest certification could provide significant economic incentives for private landowners to restore longleaf pine forests. However, whether sufficient demand exists is unknown. Also, restoration of functional longleaf pine landscapes will require the restoration and better management of very large areas. Whether forest certification will produce sufficient incentive to restore hundreds of thousands of hectares or more remains to be seen.

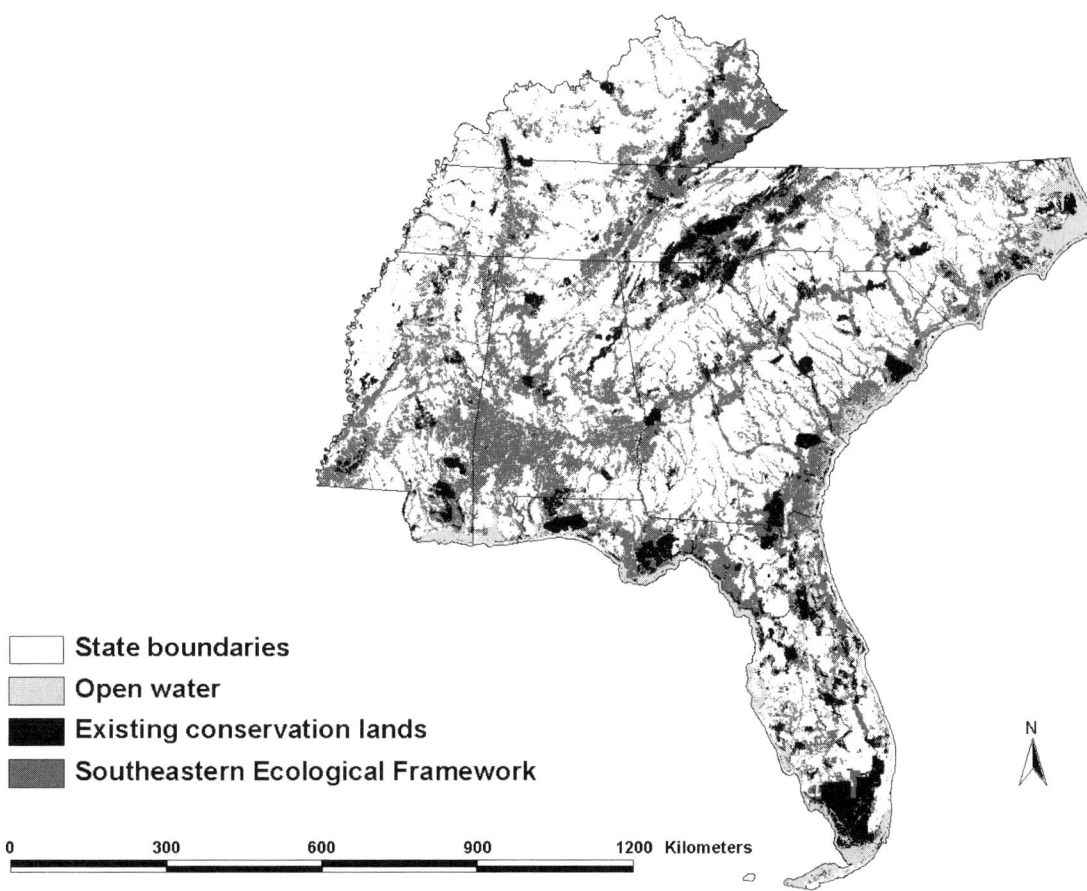

FIGURE 6. The EPA Southeastern Ecological Framework (SEF) covers almost half of EPA Region 4. Over 80% of the SEF is forested and approximately 25% is within existing conservation lands or open water. Although much of the forest cover on private land is intensively managed, the SEF suggests that there are still very good opportunities to restore large longleaf pine-dominated landscapes across much of its historic range.

Challenges

Climate Change

Although connected networks of public and private conservation lands may provide some opportunity for species to respond to climate change, the rate and magnitude of climate change combined with the current level of landscape-scale habitat fragmentation may likely exceed the capacity of many species to adapt (Peters and Lovejoy 1992). In coastal areas, especially in very low elevations such as Florida, sea-level rise could inundate hundreds of thousands of hectares. In the future, conservationists will likely face the need to transplant many species to new sites in desperate efforts to avoid extinctions. Such efforts, including the reestablishment of functional natural communities and landscapes, will obviously be difficult (Peters and Lovejoy 1992).

Water Mining and Aquifer Impacts

In Florida most water for drinking, agriculture, and industry comes from the Floridan aquifer. Greatly increased demand combined with increasing drought have resulted in significantly lower aquifer levels in recent years (Bacchus 2000). Though longleaf pine is adapted to tolerate xeric conditions, the lack of recruitment and the death of mature trees during prolonged droughts may be a significant concern (Richard Franz personal communication). In addition, aquifer drawdowns can result in significantly reduced hydroperiods (or conversion to uplands) of ephemeral ponds and other wetlands (Bacchus 2000). Therefore, continued lowering of aquifer levels especially in combination with drought could seriously impact the ecological integrity of remaining longleaf pine landscapes and significantly hinder restoration.

Funding for Land Conservation, Growth Management, and Regional Conservation Planning

Florida has spent over 4 billion dollars to acquire conservation lands and establish conservation easements on ecologically significant private lands over the past 14 years. Nevertheless, even with Florida's existing commitment, millions of unprotected hectares important for conservation remain (Hoctor et al. 2000; Hoctor 2003). Though land acquisition for conservation purposes is generally popular, and various local and state governments and the federal government are spending significant amounts on land conservation, needs greatly exceed available funds (Hoctor et al. 2004). In Florida and other sunbelt states, development induced by rapidly growing human populations will continue to erode the available private land base that could support the restoration of longleaf pine landscapes.

Florida enacted growth management legislation in the 1970s and 1980s in response to rapid development. This legislation created state oversight for local future land-use plans and large developments with regional impacts. Drafters of the legislation felt that state oversight would expose the link between growth and infrastructure costs and consequently might control growth in a manner that could be accommodated by infrastructure improvements and direct growth to areas most suitable (Nicolas and Steiner 2000). Good comprehensive plans would limit sprawling development and would give land acquisition efforts more time to protect lands critical for biodiversity (Hoctor 2003). Finally, good planning also is needed to contribute to the core area-buffer model of reserve design, where intensive development is separated from biodiversity reserves by rural lands including silviculture, agriculture, and other uses more compatible with conservation objectives (Harris 1984; Noss and Harris 1986; Soulé 1991). However, it is now widely acknowledged that growth management in Florida has largely failed to stop sprawling development (Nicolas and Steiner 2000; Hoctor 2003). With significant private property rights issues and political leanings that currently impede regional conservation planning (Hocter et al. 2004), efforts to enact or enforce effective growth management will be challenging.

Nevertheless, there is a growing recognition of the essential ecological services that natural landscapes provide to humans such as clean drinking water, storm water management, flood control, particulate matter removal, and carbon sequestration, as well as food and shelter for native species (Daily 1997, 2000; Benedict and McMahon 2001). One possibility for restoring longleaf pine landscapes is the development of a "landscape utility" concept that recognizes the services provided by intact ecosystems and compensates private landowners for ecologically sound management that provided such services or purchases land from them for public use (Maynard Hiss personal communication). Florida's water management districts already purchase lands or acquire conservation easements to ensure that critical hydrological functions are protected or restored. Broader representation of ecosystem services (including carbon sequestration) that could be funded both through mitigation funds and utility taxes (Maynard Hiss personal communication) would be challenging but could provide a significant tool for restoring and managing longleaf pine landscapes.

Wildland–Urban Interface

Related to issues regarding increasing development and urban/suburban sprawl is the concomitant increase in the wildland–urban interface, which can be defined as "an area where increased human influence and land-use conversion are changing natural resource goods, services, and management" (Macie and Hermansen 2002:2). Although the threat of wild fire ideally may be used as a tool for promoting prescribed burning to restore and manage longleaf pine communities in the urban–wildland interface, efforts to use prescribed fire, especially in the growing season, are hindered by concerns about smoke drift and associated declines in air quality and highway safety. As the public becomes increasingly urban and unfamiliar with the natural role of fire in southeastern ecosystems, the difficulty of implementing prescribed burning is bound to increase. Other important issues within the wildland–urban interface include continued habitat loss and increasing habitat fragmentation and increased demands for resource-based recreation activities (Macie and Hermansen 2002).

Making Longleaf Pine a Competitive Economic Benefit on Private Lands

Efforts to restore and maintain longleaf pine landscapes would clearly benefit if longleaf was considered an economic resource competitive with other uses of land, or at least if tax benefits or other economic returns were provided (Macie and Hermansen 2002). Forest certification may promote this situation, especially in large rural landscapes. However, whether longleaf pine communities can compete with the economic gains from development is questionable, especially as land values continue to rise. Paying landowners for providing ecological services such as carbon sequestration and aquifer recharge could help increase the value of longleaf pine beyond selective timber harvesting, hunting rights, or other traditional economic uses.

The Inverse of Ecological Networks: Road Networks

Highway and other linear transportation projects (such as proposed high-speed train routes) are a prominent threat to biodiversity (Trombulak and Frissell 2000). Existing highways need to be retrofitted to facilitate the movement of focal species and to reduce roadkills (Hoctor et al. 2000; Hoctor 2003; Smith 2003). Furthermore, systemwide planning should be conducted to avoid ecologically sensitive areas (especially large, intact landscapes and corridors) when planning new transportation routes, and to enhance efforts to minimize and mitigate impacts that may be unavoidable. Currently, such efforts are usually limited to individual projects, where politics, road design constraints, and limited budgets are frequently invoked to avoid spending sufficient money and time to truly mitigate ecological impacts. Systemwide planning should ensure that an appropriate budget is developed to minimize and mitigate impacts. State departments of transportation typically have very large budgets (the Florida Department of Transportation's budget for infrastructure improvements between 2001 and 2020 is $108 billion; Florida Department of Transportation 2001). Even if only 5% of these budgets could be used to counteract the ecological costs of roads by building bridge spans, widening bridges, building wildlife underpasses and overpasses, and protecting large areas of habitat, biodiversity could be protected while transportation capacity is increased (Smith 1999; Hoctor et al. 2000; Hoctor 2003; Smith 2003).

Conclusions

Restoring the great longleaf pine forest of the Southeast is, essentially, an effort to re-create G. Evelyn Hutchinson's ecological theater and evolutionary play (Hutchinson 1965) in one of the most biologically outstanding regions of the world. Saving a few patches of longleaf pine and focusing attention on a few legally protected species—the course we have been on in conservation planning for several decades in the region—is not sufficient. As of 1993, there were 27 federally listed species and 99 candidates for listing that were closely associated with or dependent on longleaf pine and related (i.e., wiregrass) ecosystems (Noss et al. 1995). Recently, many of the candidate species have been removed from the list for essentially political reasons (i.e., misplacement of the burden of proof; Noss et al. 1997), but the status of biologically imperiled species generally continues to decline.

What is needed is a multifaceted and comprehensive program to protect, restore, and manage longleaf pine landscapes. Such a program has several essential components:

- A detailed gap analysis of all the variants of longleaf pine communities across the range of the species should be performed to assess how well each type is represented in various categories of conservation areas. This assessment should include the associations described under the Longleaf Pine Alliance in the National Vegetation Classification System (Anderson et al. 1998, Grossman et al. 1998). However, for a more comprehensive assessment of representation needs, a rangewide map of existing longleaf pine cover should be overlaid on a map of abiotic habitats, i.e., defined by soils, elevation, geologic substrate, and so on.
- An interorganizational recovery team, involving government and private partners, should be created to develop a Longleaf Pine Ecosystem Recovery Plan. Such a plan must incorporate goals to restore the full range of longleaf pine community variation (i.e., using the results of the gap analysis described above), along with fire and other natural disturbances within historic ranges of variability in space, time, and other attributes. The Ecosystem Recovery Plan would not replace the need for more detailed, species-level plans; in fact, there is evidence that ecosystem plans produced to date provide less assurance of species recovery than single-species plans (Clark et al. 2002). To the extent feasible, the Ecosystem Recovery Plan should go beyond the minimalist goals of existing species recovery plans toward the goal of restoring viable populations and metapopulations of longleaf-associated species across their historic distributions. In most cases, this involves protecting and restoring large tracts of old-growth longleaf pine that are well connected at local, landscape, and regional scales.
- Longleaf-associated species that are not currently listed under the U.S. Endangered Species Act, but which are declining and vulnerable to existing and future threats, should be identified and used as focal species (e.g., Lambeck 1997) for conservation planning. Modeling of habitat and population viability of focal species addresses specific issues of habitat area, distribution, and configuration (including connectivity) that are poorly addressed by conventional conservation assessments. For example, "area-limited species" are those that occur in low densities or require large areas, and "dispersal-limited species" include those that are sensitive to barriers or sources of mortality, such as roads, when attempting to move across the landscape. Lambeck (1997) hypothesized that the most sensitive species in these groups may serve as effective umbrellas for other species with similar vulnerabilities. A suite of such species, intelligently selected, can serve as focal species for conservation programs on landscape to regional scales (Lambeck 1997; Carroll et al. 2001; Noss et al. 2002). An example of a potential focal species closely associated with longleaf pine is the fox squirrel. In general, the southeastern subspecies of fox squirrels appear to be long-lived, with low adult mortality and few, small litters each year. These life history characteristics may explain why most populations of southeastern fox squirrel are declining and have failed to recover even after preservation of potential habitats (Tappe and Guynn 1998). Other potential focal species that currently receive no protection or planning under the Endangered Species Act include several upland snakes such as the eastern diamondback rattlesnake (*Crotalus adamanteus*), pine snake (*Pituophis*

melanoleucus), and kingsnake (*Lampropeltis getula*).
- The longleaf pine ecosystem must be considered together with other vegetation types in the region. As noted previously, ecotones are sites of high diversity in the longleaf pine landscape, and these extend from the montane peaks in the extreme north of the region to the low swamps and coastal marshes. Ideally, the entire complex of environmental gradients would be restored to intactness across much of the region. When planning across vegetation types and other gradients, wide-ranging species that utilize multiple habitats, for example the Florida panther/eastern cougar and black bear, are ideal focal species (Maehr et al. 2001a,b). Importantly, we urge the active reestablishment of Florida panther populations—and their connection into functional, self-sustaining metapopulations—across the historic distribution of the subspecies in the Southeast.
- We emphasize the importance of "sharing the burden" in the management of area-limited species such as the red-cockaded woodpecker and fox squirrel, as well as in the restoration and management of longleaf pine communities generally. A fully restored and intact landscape requires public–private cooperation across counties and states.

References

Adams, A.L. 1976. The Fourth Quarter, autobiography by same (as told to Tom Dunkin). Published privately 379 pp.

Anderson, M., Bourgeron, P., Bryer, M.T., Crawford, R., Engelking, L., Faber-Langendoen, D., Gallyoun, M., Goodin, K., Grossman, D.H., Landaal, S., Metzler, K., Patterson, K.D., Pyne, M., Reid, M., Sneddon, L., and Weakley, A.S. 1998. *International Classification of Ecological Communities: Terrestrial Vegetation of the United States. Volume I. The National Vegetation Classification System: List of Types*. Arlington, VA: The Nature Conservancy.

Anon. 1987. *Pensacola, the Lumbering Era (1887)*. Pensacola, FL: The John Appleyard Agency, Inc.

Ashton, R.E., Jr. 1992. Flatwoods salamander. In *Rare and Endangered Biota of Florida, Vol. 3, Amphibians and Reptiles, 2nd Edition*, ed. P.E. Moler, pp. 39–43. Gainesville: University Press of Florida.

Bacchus, S. 2000. Uncalculated impacts of unsustainable aquifer yield including evidence of subsurface interbasin flow. *J Am Water Resour Assoc* 36(3):457–481.

Barry, J.M. 1980. *Natural Vegetation of South Carolina*. Columbia: University of South Carolina Press.

Beckett, S., and Golden, M.S. 1982. Forest vegetation and vascular flora of Reed Brake Research Natural Area, Alabama. *Castanea* 47(4):386–392.

Benedict, M.A., and McMahon, E.T. 2001. Green infrastructure: Smart conservation for the 21st century. Sprawl Watch Clearinghouse Monograph Series, Washington, DC.

Boerner, R.E.J., and Kooser, J.G. 1989. Leaf litter redistribution among forest patches within an Allegheny Plateau watershed. *Landscape Ecol* 2:81–92.

Butler, C. 1998. *Treasures of the Longleaf Pines: Naval Stores*. Shalimar, FL: Tarkel Publishing.

Carroll, C., Noss, R.F., and Paquet, P.C. 2001. Carnivores as focal species for conservation planning in the Rocky Mountain region. *Ecol Appl* 11:961–980.

Clark, J.A., Hoekstra, J.M., Boersma, P.D., and Kareiva, P. 2002. Improving U.S. Endangered Species Act recovery plans: Key findings and recommendations of the SCB Recovery Plan Project. *Conserv Biol* 16:1510–1519.

Clewell, A.F. 1981. Natural setting and vegetation of the Florida Panhandle. USACE, Contract No. DACWO1-77-C-0104. Mobile, AL.

Clewell, A.F., Goolsby, J.A., and Shuey, A.G. 1982. Riverine forests of the South Prong Alafia River System, Florida. *Wetlands* 2:21–72.

Cox, J., and Kautz, R. 2000. Habitat conservation needs of rare and imperiled wildlife in Florida. Office of Environmental Services, Florida Fish and Wildlife Conservation Commission, Tallahassee.

Cox, J., Kautz, R., MacLaughlin, M., and Gilbert, T. 1994. *Closing the Gaps in Florida's Wildlife Habitat Conservation System*. Tallahassee: Florida Game and Freshwater Fish Commission.

Daily, G.C., ed. 1997. Introduction: What are ecosystem services? *Nature's Services: Societal Dependence on Natural Ecosystems*. Washington, DC: Island Press.

Daily, G.C. 2000. Management objectives for the protection of ecosystem services. *Environ Sci Policy* 3(6):333–339.

Delcourt, H., and Delcourt, P. 1977. Presettlement magnolia-beech climax of the Gulf Coastal Plain: Quantitative evidence from the Apalachicola River Bluffs, north-central Florida. *Ecology* 58:1085–1093.

Delcourt, P.A., Delcourt, H.R., Morse, D.F., and Morse, P.A. 1993. History, evolution, and organization of vegetation and human culture. In *Biodiversity of the Southeastern United States: Lowland Terrestrial Communities*, eds. W.H. Martin, S.G. Boyce, and A.C. Echternach, pp. 47–79. New York: John Wiley & Sons.

Dodd, C.K., Jr., and Cade, B.S. 1998. Movement patterns and the conservation of amphibians breeding in small, temporary wetlands. *Conserv Biol* 12:331–339.

Dodd, C.K., Jr., and LeClaire, L.V. 1995. Biogeography and status of the striped newt (*Notophthalmus perstriatus*) in Georgia, USA. *Herpetol Nat Hist* 3:37–46.

Duever, L.C. 1983. Natural communities of Florida's inland sand ridges. Palmetto. Winter Park, FL: Florida Native Plant Society 3(3):1–3, 10.

Elliott, S.B. 1912. *The Important Timber Trees of the United States*. Boston: Houghton Mifflin.

Florida Department of Transportation. 2001. 2020 Florida transportation plan. Florida Department of Transportation, Tallahassee.

Forman, R.T.T. 1987. The ethics of isolation, the spread of disturbance, and landscape ecology. In *Landscape Heterogeneity and Disturbance*, ed. M.G. Turner, pp. 213–229. New York: Springer-Verlag.

Forman, R.T.T. 1995. *Land Mosaics: The Ecology of Landscapes and Regions*. London: Cambridge University Press.

Forman, R.T.T., and Godron, M. 1986. *Landscape Ecology*. New York: John Wiley & Sons.

Frost, C. 1993. Four centuries of changing landscape patterns in the longleaf pine ecosystem. *Proc Tall Timbers Fire Ecol Conf* 18:17–44.

Goodison, C., and Hoctor, T.S. 2003. Defining the Southeastern Ecological Framework for Military Installations. Report to the U.S. Army Corps of Engineers Construction Engineering Research Laboratory, Champaign, IL.

Gordon, D.R., Provencher, L., and Hardesty, J.L. 1997. Measurement scales and ecosystem management. In *The Ecological Basis of Conservation: Heterogeneity, Ecosystems, and Biodiversity*, eds. S.T. A. Pickett, R.S. Ostfeld, M. Shachak, and G.E. Likens, pp. 262–273. New York: Chapman & Hall.

Greeley, W.B. 1920. Forester. USDA report to U.S. Senate.

Gross, F.E.H. 1987. Characteristics of small stream floodplain ecosystems in north and central Florida. Master's thesis, University of Florida, Gainesville.

Grossman, D.H., Faber-Langendoen, D., Weakley, A.S., Anderson, M., Bourgeron, P., Crawford, R., Goodin, K., Landaal, S., Metzler, K., Patterson, K. D., Pyne, M., Reid, M., and Sneddon, L. 1998. *International Classification of Ecological Communities: Terrestrial Vegetation of the United States. Volume II. The National Vegetation Classification System: Development, Status, and Applications*. Arlington, VA: The Nature Conservancy.

Groves, C., Valutis, L., Vosick, D., Neely, B., Wheaton, K., Touval, J., and Runnels, B. 2000. *Designing a Geography of Hope: A Practitioner's Handbook for Ecoregional Conservation Planning*. Arlington, VA: The Nature Conservancy.

Groves, C.R., Jensen, D.B., Valutis, L.L., Redford, K.H., Shaffer, M.L., Scott, J.M., Baumgartner, J.F., Higgins, J. V., Beck, M.W., and Anderson, M.G. 2002. Planning for biodiversity conservation: Putting conservation science into practice. *BioScience* 52:499–512.

Harper, R.M. 1911. The relation of climax vegetation to islands and peninsulas. *Bull Torrey Bot Club*. 38:515–525.

Harper, R.M. 1915. Vegetation types. In Natural Resources of an Area in Central Florida. Florida Geological Survey, 13th Annual Report, eds. E.H. Sellards, R.M. Harper, E.N. Mooney, W.J. Latimer, H. Gunter, and E. Gunter, pp. 135–188.

Harris, L.D. 1980. Forest and wildlife dynamics in the Southeast. *Trans North Am Wildl Nat Resourc Conf*. 45:307–322.

Harris, L.D. 1984. *The Fragmented Forest: Island Biogeography Theory and the Preservation of Biotic Diversity*. Chicago: University of Chicago Press.

Harris, L.D. 1985. Conservation corridors: A highway system for wildlife. ENFO Report 85-5, Florida Conservation Foundation, Winter Park.

Harris, L.D. 1988. The nature of cumulative impacts on biotic diversity of wetland vertebrates. *Environ Manage* 12:675–693.

Harris, L.D. 1989. The faunal significance of fragmentation of southeastern bottomland forests. In *Proceedings of the Symposium: The Forested Wetlands of the Southern United States, Orlando, FL*. USDA Forest Service, General Technical Report SE-50, pp. 126–134. Asheville, NC.

Harris, L.D. 1999. Remembering Florida's ancient forests: Old Florida in words and pictures. Florida Defenders of the Environment Special Bulletin, Gainesville, FL, Summer 1999.

Harris, L.D., and Atkins, K. 1991. Faunal movement corridors in Florida. In *Landscape Linkages and Biodiversity*, ed. W. E. Hudson, pp. 117–134. Defenders of Wildlife. Washington, DC: Island Press.

Harris, L.D., and Gallagher, P.B. 1989. New initiatives for wildlife conservation: The need for movement corridors. In *Defense of Wildlife: Preserving Communities and Corridors*, ed. G. Macintosh, pp. 12–34. Washington, DC: Defenders of Wildlife.

Harris, L.D., and Hoctor, T.S. 1992. Cross Florida Greenbelt State Recreation and Conservation Area Management Plan. Volume IV: Report on Biological Issues. University of Florida, Gainesville.

Harris, L.D., and Kangas, P. 1988. Reconsideration of the Habitat Concept. *Trans North Am Wildl Nat Resour Conf* 53:137–144.

Harris, L.D., and Scheck, J. 1991. From implications to applications: The dispersal corridor approach to the conservation of biological diversity. In *Nature Conservation 2: The Role of Corridors*, eds. D.A. Saunders and R. J. Hobbs, pp. 189–220. Chipping Norton, New South Wales, Australia: Surrey Beatty and Sons.

Harris, L.D., and Silva-Lopez, G. 1992. Forest fragmentation and the conservation of biological diversity. In *Conservation Biology: The Theory and Practice of Nature Conservation*, eds. P. Fielder and S. Jain, pp. 197–237. New York: Chapman & Hall.

Harris, L.D., and Vickers, C.R. 1984. Some faunal community characteristics of cypress ponds and changes induced by perturbation. In *Cypress Swamps*, eds. K. C. Ewel and H. T. Odum, pp. 171–185. Gainesville: University Press of Florida.

Harris, L.D., Hoctor, T.S., and Gergel, S. E. 1996a. Landscape processes and their significance to biodiversity conservation. In *Population Dynamics in Ecological Space and Time*, eds. O. E. Rhodes, Jr., K. Chesser, and M.H. Smith, pp. 319–347. Chicago: University of Chicago Press.

Harris, L.D., Hoctor, T.S., Maehr, D., and Sanderson, J. 1996b. The role of networks and corridors in enhancing the value and protection of parks and equivalent areas. In *National Parks and Protected Areas: Their Role in Environmental Areas*, ed. R. G. Wright, pp. 173–198. Cambridge, MA: Blackwell Science.

Hart, R. 1984. Evaluation of methods for sampling vegetation and delineating wetlands transition zones in coastal west-central Florida. Technical Report Y-84-2, U.S. Army Engineer Waterways Experiment Station, Vicksburg, MS. NTIS No. AD A144 677.

Hermann, S.M. 1993. Small-scale disturbances in longleaf pine forests. *Proc Tall Timbers Fire Ecol Conf* 18:265–274.

Hilmon, J.B. 1969. Autecology of saw palmetto (*Serenoa repens* (Batr.) Small). Ph.D. dissertation, Duke University, Durham, NC.

Hilsenbeck, R., Hoctor, T., Goodison, C., Hernandez, P., Caster, W., and Hernandez, J. 2001. The Peninsular Florida Ecoregional Plan. The Florida Chapter of The Nature Conservancy, Tallahassee.

Hoctor, T.S. 2003. Regional landscape analysis and reserve design to conserve Florida's biodiversity. Ph.D. dissertation, University of Florida, Gainesville.

Hoctor, T.S., Carr, M.H., and Zwick, P.D. 2000. Identifying a linked reserve system using a regional landscape approach: The Florida ecological network. *Conserv Biol* 14:984–1000.

Hoctor, T.S., Teisinger, J., Carr, M.H., and Zwick, P.D. 2001. Ecological Greenways Network Prioritization for the State of Florida. Final Report. Office of Greenways and Trails, Florida Department of Environmental Protection, Tallahassee.

Hoctor, T.S., Carr, M.H., Goodison, C., Zwick, P.D., Green, J., Hernandez, P., McCain, C., Whitney, K., and Teisinger, J. 2002a. Final report: Southeastern Ecological Framework. Environmental Protection Agency Region 4, Atlanta, GA.

Hoctor, T.S., Teisinger, J., Carr, M.H., and Zwick, P.D. 2002b. Identification of Critical Linkages Within the Florida Ecological Greenways Network. Final Report. Office of Greenways and Trails, Florida Department of Environmental Protection, Tallahassee.

Hoctor, T.S., Carr, M., Zwick, P., and Maehr, D.S. 2004. Identification and realization of a Florida Ecological Network: the Florida Statewide Greenways Project and its political context. In *Ecological Networks and Greenways: Concept, Design, Implementation*, eds. R.H.G. Jongman and G. Pungetti, pp. 222–250. London: Cambridge University Press.

Hough, F.B. 1882. Report on forestry, submitted to Congress by the Commissioner of Agriculture. U.S. Department of Agriculture, Washington, DC.

Huffman, J.M., and Blanchard, S.W. 1990. Changes in woody vegetation in Florida dry prairie and wetlands during a period of fire exclusion, and after dry-growing-season fire. In *Fire and the Environment. Ecological and Cultural Perspectives*, pp. 75–83. General Technical Report. SE-69. USDA Forest Service, SE Forest Experiment Station, Asheville, NC.

Hunter, M.L., Jr., Jacobson, G.L., Jr., and Webb, T., III, 1988. Paleoecology and the coarse-filter approach to maintaining biological diversity. *Conserv Biol* 2: 375–384.

Hutchinson, G.E. 1965. *The Ecological Theater and the Evolutionary Play*. New Haven, CT: Yale University Press.

Kalisz, P.J., and Stone, E.L. 1984. Soil mixing by scarab beetles and pocket gophers in north-central Florida. *Soil Sci Soc Am J* 48(1):169–172.

Kautz, R.S., and Cox, J.A. 2001. Strategic habitats for biodiversity conservation in Florida. *Conserv Biol* 15:55–77.

Lambeck, R.J. 1997. Focal species: A multi-species umbrella for nature conservation. *Conserv Biol* 11:849–856.

Macie, E.A., and Hermansen, L.A., eds. 2002. Human influences on forest ecosystems: The southern wildland–urban interface assessment. General Technical Report SRS-55. USDA Forest Service, Southern Research Station, Asheville, NC.

Maehr, D.S. 1997. The comparative ecology of bobcat, black bear, and Florida panther in south Florida. *Bull Fla Mus Nat Hist* 40:1–176.

Maehr, D.S., Hoctor, T.S., Quinn, L.J., and Smith, J.S. 2001a. Black bear habitat management guidelines for Florida. Florida Fish and Wildlife Conservation Commission, Tallahassee, FL.

Maehr, D.S., Hoctor, T.S., and Harris, L.D. 2001b. The Florida panther: A flagship for regional restoration. In *Large Mammal Restoration: Ecological and Sociological Challenges in the 21st Century*, eds. D.S. Maehr, R.F. Noss, and J. L. Larkin, pp. 293–312. Washington, DC: Island Press.

Margules, C.R., and Pressey, R.L., 2000. Systematic conservation planning. *Nature* 405:243–253.

Means, D.B. 1987. Impacts on diversity of the 1985 Land and Resource Management Plan for National Forests in Florida. Unpublished report to the Wilderness Society.

Means, D.B. 1996. Longleaf pine forest: going, going, . . . In *Eastern Old-growth Forest: Prospects for Rediscovery and Recovery*, ed. M. Davis, pp. 210–229. Washington, DC: Island Press.

Means, D.B., and Grow, G. 1985. The endangered longleaf pine community. ENFO (Florida Conservation Foundation) Sept: 1–12.

Means, D.B., Palis, J.G., and Baggett, M. 1996. Effects of slash pine silviculture on a Florida population of flatwoods salamander. *Conserv Biol* 10:426–437.

Meffe, G.K., Carroll, C.R., and contributors. 1997. *Principles of Conservation Biology*, 2nd edition. Sunderland, MA: Sinauer Associates.

Milne, G. 1935. Some suggested units of classification and mapping for East African soils. *Soil Res* 4:183–198.

Milne, G. 1947. A soil reconnaissance journey through parts of Tanganyika Territory, December 1935 to February 1936. *J Ecol* 35:192–265.

Mohr, C. 1896. Timber pines of the southern United States. Bulletin No. 13, USDA Forest Service, Washington, DC.

Moler, P. 1992. Eastern indigo snake. In *Rare and Endangered Biota of Florida, Volume III: Amphibians and Reptiles*, ed. P. Molar, pp. 181–186. Gainesville: University Press of Florida.

Mutch, R. W. 1970. Wildland fires and ecosystems—A hypothesis. *Ecology* 51:1046–1051.

Myers, R.L. 1990. Scrub and high pine. In *Ecosystems of Florida*, eds. R. Myers and J. Ewel, pp. 150–193. Orlando: University of Central Florida Press.

Nicolas, J. C., and Steiner, R.L. 2000. Growth management and smart growth in Florida. *Wake Forest Law Rev* 35(3):645–670.

Noss, R.F. 1987a. From plant communities to landscapes in conservation inventories: A look at The Nature Conservancy (USA). *Biol Conserv* 41:11–37.

Noss, R.F. 1987b. Protecting natural areas in fragmented landscapes. *Nat Areas J* 7:2–13.

Noss, R.F. 1988. The longleaf pine landscape of the Southeast: Almost gone and almost forgotten. *Endangered Species Update* 5 (5):1–8.

Noss, R.F. 1989. Longleaf pine and wiregrass: Keystone components of an endangered ecosystem. *Nat Areaes J* 9:211–213.

Noss, R.F. 1992. The Wildlands Project: Land conservation strategy. *Wild Earth* (Special Issue):10–25.

Noss, R.F. 1993. Wildlife corridors. In *Ecology of Greenways*, eds. D.S. Smith and P.C. Hellmund, pp. 43–68. Minneapolis: University of Minnesota Press.

Noss, R.F., and Cooperrider, A. 1994. *Saving Nature's Legacy: Protecting and Restoring Biodiversity*. Washington, DC: Defenders of Wildlife and Island Press.

Noss, R.F., and Harris, L.D. 1986. Nodes, networks, and MUMs: Preserving diversity at all scales. *Environ Manage* 10:299–309.

Noss, R.F., and Harris, L.D. 1990. Connectivity and conservation of biological diversity: Florida as a case history. In *Proceedings of the 1989 Society of American Foresters National Convention, Spokane, WA, September 24–27*, pp. 131–135. Society of American Foresters, Bethesda, MD.

Noss, R.F., LaRoe, E.T., and Scott, J.M. 1995. *Endangered Ecosystems of the United States: A Preliminary Assessment of Loss and Degradation*. Biological Report 28. Washington, DC: USDI National Biological Service.

Noss, R.F., O'Connell, M.A., and Murphy, D. D. 1997. *The Science of Conservation Planning: Habitat*

Conservation under the Endangered Species Act. Washington, DC: Island Press.

Noss, R.F., Carroll, C., Vance-Borland, K., and Wuerthner, G. 2002. A multicriteria assessment of the irreplaceability and vulnerability of sites in the Greater Yellowstone Ecosystem. *Conserv Biol* 16:895–908.

Palis, J.G. 1996. Element stewardship abstract: flatwoods salamander (*Ambystoma cingulatum* Cope). *Nat Areas J* 16:49–54.

Peet, R.K., and Allard, D.J. 1993. Longleaf pine vegetation of the southern Atlantic and eastern Gulf Coast regions: A preliminary classification. *Proc Tall Timbers Fire Ecol Conf* 18:45–81.

Peters, R.L., and Darling, J.D.S. 1985. The greenhouse effect and nature reserves. *BioScience* 35:707–717.

Peters, R.L., and Lovejoy, T.E., eds. 1992. *Global Warming and Biological Diversity*. New Haven, CT: Yale University Press.

Platt, W.J. 1999. Southeastern pine savannas. In *Savannas, Barrens, and Rock Outcrop Plant Communities of North America*, eds. R. C. Anderson, J.S. Fralish, and J.M. Basin, pp. 23–51. London: Cambridge University Press.

Platt, W.J., and Schwartz, M.W. 1990. Temperate hardwood forests. In *Ecosystems of Florida*, eds. R. Myers and J. Ewel, pp. 194–229. Orlando: University of Central Florida Press.

Platt, W.J. , Glitzenstein, J.S., and Rathbun, S. K. 1988. The population dynamics of a long-lived conifer (Pinus palustris). *Am Nat* 131:491–525.

Poiani, K.A., Richter, B.D., Anderson, M.G., and Richter, H.E. 2000. Biodiversity conservation at multiple scales: Functional sites, landscapes, and networks. *BioScience* 50:133–146.

Rerick, R. H. 1902. *Memoirs of Florida*, 2 vols. Atlanta.

Schwartz, M.W., Hermann, S.M., and Vogel, C.S. 1995. The catastrophic loss of *Torreya taxifolia*: Assessing environmental induction of disease hypotheses. *Ecol Appl* 5:501–516.

Scott, J.M., Davis, F.W., McGhie, G., and Groves, C. 2001. Nature reserves: Do they capture the full range of America's biological diversity? *Ecol Appl* 11:999–1007.

Semlitsch, R. D. 2000. Principles for management of aquatic-breeding amphibians. *J. Wildl Manage* 64:615–631.

Smith, D.J. 1999. Highway–wildlife relationships (Development of a decision-based wildlife underpass road project prioritization model on GIS with statewide application). Technical report, Florida Department of Transportation, Tallahassee.

Smith, D.J. 2003. The ecological effects of roads: Theory, analysis, management, and planning considerations. Ph.D. dissertation, University of Florida, Gainesville.

Soulé, M.E. 1991. Land use planning and wildlife maintenance: Guidelines for conserving wildlife in an urban landscape. *J Am Plann Assoc* 57:313–323.

Soulé, M.E., and Terborgh, J., eds. 1999. *Continental Conservation: Scientific Foundations of Regional Reserve Networks*. Washington, DC: Island Press.

Stout, I.J., and Marion, W.R. 1993. Pine flatwoods and xeric pine forests of the southern (lower) coastal plain. In *Biodiversity of the Southeastern United States: Lowland Terrestrial Communities*, eds. W.H. Martin, S.G. Boyce, and A.C. Echternach, pp. 373–446. New York: John Wiley & Sons.

Tappe, P.A., and Guynn, D.C., Jr. 1998. Southeastern fox squirrels: R- or K-selected? Implications for management. In *Ecology and Evolutionary Biology of Tree Squirrels*, eds. M.A. Steele, J. F. Merritt, and D. A. Zegers, pp. 239–247. Virginia Museum of Natural History Special Publication 6. Martinsville, VA.

Trombulak, S.C., and Frissell, C.A. 2000. Review of ecological effects of roads on terrestrial and aquatic communities. *Conserv Biol* 14:18–30.

Turner, M.G. 1989. Landscape ecology: The effect of pattern on process. *Annu Rev Ecol Syst* 20:171–197.

Turner, M.G., Gardner, R.H., and O'Neill, R.V. 1995. Ecological dynamics at broad scales: Ecosystems and landscapes. *BioScience Suppl* S29–S35.

Tyson, W.K. 1956. History of the utilization of longleaf pine (Pinus palustris Mill.) in Florida from 1513 until the twentieth century. Master's thesis, University of Florida, Gainesville.

Varner, J.M., III, Kush, J.S., and Meldahl, R.S. 2003a. Vegetation of frequently burned old-growth longleaf pine (*Pinus palustris* Mill.) savannas on Choccolocco Mountain, Alabama, USA. *Nat Areas J* 23:43–52.

Varner, J.M., III, Kush, J.S., and Meldahl, R.S. 2003b. Structural characteristics of frequently-burned old-growth longleaf pine stands in the mountains of Alabama. *Castanea* 68:211–221.

Vickers, C.R., Harris, L.D., and Swindel, B.F. 1985. Changes in herpetofauna resulting from ditching of cypress ponds in coastal plains flatwoods. *For Ecol Manage* 11:17–29.

Wahlenberg, W.G. 1946. *Longleaf Pine: Its Use, Ecology, Regeneration, Protection, Growth and Management*. Washington, DC: Charles Lathrop Pack Forestry Foundation.

Walker, J., and Peet, R.K. 1983. Composition and species diversity of pine-wiregrass savannas of the Green Swamp, North Carolina. *Vegetatio* 55:163–179.

Walters, J.R., Hansen, S.K., Carter, J.H., III, Manor, P.D., and Blue, R.J. 1988. Long-distance dispersal of an adult red-cockaded woodpecker. *Wilson Bull* 100:494–496.

Ware, S., Frost, C., and Doerr, P.D. 1993. Southern mixed hardwood forest: The former longleaf pine forest. In *Biodiversity of the Southeastern United States: Lowland Terrestrial Communities*, eds. W.H. Martin, S.G. Boyce, and A.C. Echternacht, pp. 447–493. New York: John Wiley & Sons.

Weigl, P.D., Steele, M.A., Sherman, L. J., Ha, J., and Sharpe, T. 1989. The ecology of the fox squirrel (Sciurus niger) in North Carolina: Implications for survival in the southeast. *Tall Timbers Bull* 24.

Wells, B.W., and Shunk, I. V. 1931. The vegetation and habitat factors of the coarser sands of the North Carolina plain: An ecological study. *Ecol Monogr* 1:465–521.

Whitney, K. 1999. Spatial structure affects landscape ecology function. Master's thesis, University of Florida, Gainesville.

Wilcox, B.A., and Murphy, D. D. 1985. Conservation strategy: The effects of fragmentation on extinction. *Am Nat* 125:879–887.

Woodmansee, R. G. 1990. Biogeochemical cycles and ecological hierarchies. In *Changing Landscapes: An Ecological Perspective*, eds. I. S. Zonneveld and R. T. T. Forman, pp. 57–71. New York: Springer-Verlag.

Chapter 13

Longleaf Pine Restoration

Economics and Policy

Janaki R. R. Alavalapati, G. Andrew Stainback, and Jagannadha R. Matta

Introduction

Public preference for native forest ecosystems is on the rise throughout the world because of their valuable market outputs, i.e., timber and nontimber products, and nonmarket outputs such as biodiversity, ecological services, and aesthetics. As a result, restoration of native forest ecosystems has become an important component of sustainable forest management (Stainback and Alavalapati 2004). Longleaf pine (*Pinus palustris*) forests are one of the most biologically diverse native ecosystems in North America, supporting hundreds of plant and animal species. When Europeans first colonized North America, forests dominated by longleaf pine covered vast areas of the southeastern Coastal Plain. At that time longleaf pine forests may have existed on close to 36 million hectares (Landers et al. 1995). Due to landscape changes brought on by colonization, agricultural expansion, and population growth over the past several centuries, longleaf pine today covers only a small fraction of its historical range.

This chapter focuses on the economics and policy aspects of longleaf pine restoration on private lands. In particular, we discuss the policy and economic opportunities and challenges associated with the restoration of this ecosystem and make an important observation that private landowners are key because a large percentage of land in the Southeast lies in their hands. Next, we review some of the major policy initiatives that have been either tried or suggested toward restoring longleaf pine on private lands. Finally, we provide some conclusions along with future directions for policy research and development.

Longleaf Pine: Past and Present

The history of longleaf pine forests has been described in detail by Frost (this volume). During colonial times longleaf pine was harvested mostly for its valuable wood. However, large-scale logging occurred in the nineteenth and early twentieth centuries. By 1935 approximately 8 million hectares of the original 36 million hectares of longleaf pine forests remained (Landers et al. 1995). These forests were further reduced during the 1950s when timber and pulp industries started to convert

Janaki R. R. Alavalapati, G. Andrew Stainback, and Jagannadha R. Matta • School of Forest Resources and Conservation, University of Florida, Gainesville, Florida 32611.

land that previously supported longleaf pine to the fast-growing loblolly pine (*Pinus taeda*) and slash pine (*Pinus elliottii*). Lack of reforestation and government policies to exclude fire also contributed to the decline of longleaf pine forests. Finally, the conversion of forests to agriculture led to further reductions. Today, virgin longleaf stands exist only in a few isolated areas (Abrahamson and Hartnett 1990).

Over two-thirds of remaining longleaf pine forests in the United States is found on private lands, with most of this on nonindustrial private land. The only exception is the state of Florida where the majority of longleaf pine is in public ownership (Kelly and Bechtold 1983). Most longleaf pine stands are natural in origin as historically very little was planted on cutover sites. Longleaf pine forests support a diverse collection of plant and animal species (see chapters by Peet and Means in this volume) many of which have declined with the loss of their habitat. More than 30 species associated with this forest are listed as endangered or threatened under the Endangered Species Act, with the red-cockaded woodpecker (*Picoides borealis*) being the most notable example.

The red-cockaded woodpecker (RCW) requires large tracts of mature pine stands, preferably longleaf, with a relatively open understory free of midstory vegetation. RCWs live in groups that consist of a breeding pair, the current year's offspring, and sometimes helpers that usually are the male offspring from the previous breeding season (Kennedy et al. 1996). The RCW excavates cavities in mature living pine trees for its nest and roosts. It also feeds on the insects that live under the bark of the trees. A group of cavity trees that are used by a group of RCWs is referred to as a "cluster."

Currently there are fewer than 5000 RCW groups scattered throughout the southeastern United States and populations are still being lost (Kennedy et al. 1996). The largest populations occur on public lands where extensive tracts of mature longleaf pine are actively maintained. In 1973 Congress passed the Endangered Species Act (ESA) to prevent the loss of vulnerable species, including the RCW. The ESA prohibits harm or taking of a listed species. When an RCW is on private land, management activities can be significantly restricted. These restrictions can result in a significant loss of timber revenue to the landowner. Thus, landowners have a strong incentive to avoid managing land in a way that may attract RCWs. This includes restoring longleaf pine. Thus, many landowners and resource professionals believe that economic incentives that encourage landowners to plant longleaf pine and actively manage land to produce habitat for the RCW and other species may be more effective than relying solely on command and control regulations.

Economics of Longleaf Pine Restoration

Much of the old-growth longleaf pine forests that exist today are on public land. This includes National Forest land, military bases such as Eglin Air Force Base, state-owned land and to a lesser extent land owned by the Department of Interior. As part of its stated objective to protect and enhance biodiversity on national forests the U.S. Forest Service has made conserving the remaining longleaf on its land a top priority. In addition, the U.S. Forest Service has planted tens of thousands of hectares of longleaf pine forests. The military has made it a priority to maintain existing stands on its land, and the Department of the Interior has set a goal of maintaining and restoring longleaf pine. States and local governments have also recognized the value of longleaf and implemented policies for its restoration. For example, the Blackwater State Forest in Florida maintains extensive longleaf pine stands and longleaf pine forests on private lands have been purchased by Alachua County, Florida, as part of an effort to protect sensitive ecosystems. All of these efforts can make important and unique contributions to the restoration of longleaf pine and its associated environmental benefits. However, in the traditional range of longleaf pine only 10% of the land base is in public ownership (Johnson and Gjerstad 1999). The remaining lands are privately owned.

Because of the large amount of land in private ownership, private landowners will play a crucial role in restoring longleaf forests. There are several advantages that longleaf pine offers landowners. Longleaf is generally more resistant to fire, hurricane damage, and pine bark beetle attacks than other common commercial pine species in the region. It grows relatively straight and firm, making it a good choice for producing sawtimber and poles. Longleaf pine needles make an excellent straw, which can provide periodic income for landowners. It can produce excellent habitat for game such as bobwhite quail and is generally perceived as more aesthetically pleasing than other pine species. Despite these advantages, except for very xeric sites, longleaf has generally proved to be less profitable than other timber species in the region.

Investigating various site indices, thinning schedules, rotation ages, and discount rates, Yanquoi and Flick (1994) found longleaf pine to be less profitable than loblolly pine. Alavalapati et al. (2002) found longleaf pine to be less profitable than slash pine for producing sawtimber and pulpwood from unthinned even-aged plantations. Stainback and Alavalapati (2004) found that longleaf pine in the form of silvopasture (growing longleaf pine along with pasture) can be more profitable than growing pure longleaf pine. There are several reasons for the lower returns of longleaf pine. First, longleaf pine has historically been difficult to regenerate. In addition, it is not a prolific seed producer and its seeds are relatively large, resulting in lower dispersal rates. Second, longleaf pine is generally intolerant of competition (Boyer and Peterson 1983). Thus, natural regeneration is typically more difficult than with other common pine species of the region. Third, because of the early grass stage, longleaf pine typically exhibits slower growth at young stand ages (Boyer 1996). Finally, managing understory in longleaf pine forests is a major hurdle that landowners face. Maintaining sunny and open areas is required in order to ensure native plant and animal diversity in longleaf pine forests. A healthy longleaf pine community will have two principal vegetation layers: the pine canopy, and a rich ground cover of grasses, legumes, and other flowering herbs. The understory species of longleaf pine forests are fire-adapted and they will resprout quickly after fire. Reduction in fire frequency to intervals longer than 5 years leads to elimination of the herb layer, which provides seeds and legumes for small game species such as rabbits and quail, as well as hundreds of nongame birds and small mammals like mice and voles which, in turn, support hawks, owls, and mammalian predators (http://www.forestry.auburn.edu/sfnmc/class/longleaf.html). Much of this complex food web collapses when fire is eliminated from woodlands. Direct costs and liability issues associated with managing fire-dependent longleaf pine forest ecosystem are additional hurdles that discourage landowners from restoring longleaf pine.

There have been some recent improvements in establishing stands of longleaf pine. Artificial regeneration is being advanced by container seedlings and herbicide control of competing vegetation. For instance, harvesting and regeneration using the shelterwood method, where mature pines are left during initial harvest to provide a seed source, are well suited for longleaf pine (Demers and Long 2003). There is also an increasing interest in providing incentives to private landowners to plant longleaf pine. Many environmental benefits such as biodiversity, carbon sequestration, recreational and aesthetic benefits that are commonly associated with longleaf pine are at least partly external to the landowner. Thus, although these attributes of longleaf may provide significant benefits to society, the landowner rarely receives financial compensation for producing them. If public preferences for hunting in longleaf pine ecosystems, longleaf pine straw, and longleaf sawtimber reflect in the market, in the form of price premiums for example, competitiveness of longleaf relative to slash or loblolly might improve.

Currently, longleaf pine restoration and its associated benefits may be underproduced from a social perspective. Several policy mechanisms have been suggested and some have been implemented to internalize the environmental benefits associated with longleaf pine management. The most notable to date have been centered around inducing landowners to create or improve habitat for the federally endangered RCW. These policies include Habitat Conservation Plans (HCPs), Safe Harbor programs, and marketable Transferable Endangered Species Certificates (TESCs). Other policies include cost sharing programs, efforts to disseminate information to landowners about the potential benefits of longleaf pine and conservation easements purchased by governments or nonprofit organizations. For example, the Florida Fish and Wildlife Conservation Commission encourages management and enhancement of RCW foraging habitat and nesting sites through its Landowner Incentive Program (LIP). Similarly, USDA's Natural Resources Conservation Service provides both technical assistance and up to 75% cost-share assistance to establish and improve wildlife habitat under its Wildlife Habitat Incentives Program (WHIP). Longleaf pine areas have been accorded a priority ecological site status for this program in Florida. Potential policies for restoring longleaf pine also include setting up markets that allow landowners to sell credits for the additional carbon that would be sequestered as a result of switching to longleaf pine from other species that have shorter rotations. These credits could potentially be sold to governments or private companies if a cap and trade system for CO_2 emissions were implemented similar to the one set up for sulfur dioxide emissions and acid rain. Changes can be made to the Conservation Reserve Program that would provide greater incentives for landowners to plant and maintain biologically important ecosystems such as longleaf pine forests. Finally, research and information about alternative land uses that may make longleaf pine more attractive to landowners can be disseminated. For example, using longleaf pine in agroforestry systems such as silvopasture or selling pine straw can generate an annual cash flow to help defray the disadvantage of the long time periods involved in producing sawtimber and poles (Stainback and Alavalapati 2004). In the remainder of this paper we will discuss these policy options in greater detail.

Policy Options to Restore Longleaf Pine on Private Land

Endangered Species Act and Incentives to Protect RCW

The ESA, which was enacted by Congress in 1973, is to protect and recover species that are close to extinction. The Act prohibits any individual to "take" a species, where "take" is defined as "to harass, harm, pursue, hunt, shoot, wound, kill, trap, capture, collect, or attempt to engage in any such conduct." Under this Act, the U.S. Fish and Wildlife Service oversees species that live on land, while the National Marine Fisheries Service focuses on marine species. The RCW was one of the first species listed as endangered under the ESA. In 1995 in a landmark case, *Sweet Home v. Babbitt*, the U.S. Supreme Court ruled that it is permissible for the U.S. Fish and Wildlife Service to define "take" to include any modification of habitat of a listed species that would impair behavior essential for the survival of the species. Such behavior potentially includes mating, feeding, sheltering, or any other behavior essential to survival.

The ESA can require public agencies, such as the U.S. Forest Service, not only to prevent taking a listed species but also to implement plans to recover the species. However, private landowners have a responsibility only to prevent a taking of a listed species. This is an important distinction, as over 80% of the species listed under the Act have at least some habitat on private land (Zhang and Mehmood 2002). As mentioned before, more than 90% of the potential longleaf pine (preferred RCW habitat) occurs on private land. Thus, implementing polices that effectively restore longleaf pine on private land is essential for the recovery of the RCWs.

The benefits of biological diversity and endangered species conservation may be large but are generally diffuse throughout society. Conversely, the cost of protecting biodiversity on private land is concentrated on a relatively small number of landowners. In addition, the benefits of land management aimed at maximizing private goods are received mostly by the individuals who own the land, whereas the costs associated with the degradation of public goods such as RCW habitat are diffuse. Thus, the interests of the private landowner often are not the same as those of the public when it comes to biodiversity. The ESA restrictions on private lands are largely aimed at mitigating the latter type of market failure. However, addressing the private cost of protecting biodiversity is also very critical.

The restrictions placed on private land use through the ESA to protect RCWs can be costly to the landowner. For example, around each cavity tree at least 60 acres of foraging habitat must be maintained within a half-mile of the tree. In addition, there are restrictions placed on the harvesting of large pines, pesticide use, and road construction within a 200-foot radius of a cavity tree. These restrictions can significantly reduce the value of the land to the landowner. For example, Lancia et al. (1989) found that the opportunity cost of providing foraging habitat for RCWs can be as high as $155 per acre. This cost is an incentive for landowners to manage their land in a way that prevents RCWs from inhabiting their property. A well-publicized example often touted by those seeking reform is Ben Cone, a private landowner in North Carolina.

Ben Cone owns large tracks of longleaf pine in North Carolina. He managed his land using longer harvesting rotations and frequent low-intensity fires. This management regime gave him a stream of income from quail hunting leases and timber harvests as well as creating habitat for RCWs. In 1991 the U.S. Fish and Wildlife Service prevented Cone from harvesting timber on more than 1,500 acres of his land in order to protect 29 RCWs that resided on his property. Cone now clearcuts around the restricted RCW habitat not allowing any trees to mature to an age that would attract more woodpeckers. He claims that he is motivated to prevent further RCWs from inhabiting his land to prevent further regulation of his property. Thus, while the ESA is efficient in protecting the existing RCWs on private property, it is also creating a perverse incentive discouraging the production of more habitat.

(Based on testimony of Benjamin Cone, Jr. before the U.S. Congress, March 31, 1997)

In 1982 the ESA was amended in order to lessen the economic burden of the Act on private landowners. The amendments allow a landowner to incidentally take a listed species if the take is appropriately mitigated. "Incidental" is defined as an action that is not intended to take a species but nonetheless results in a take. Residential development or timber harvests are examples of actions that can be considered incidental. Appropriate mitigation involves the development of a Habitat Conservation Plan (HCP) that may include relocating populations or creating new habitat elsewhere. These provisions allow much greater flexibility for landowners to comply with the act. HCPs involving many landowners or statewide HCPs allow for further reduction in costs. The flexibility of HCPs can be enhanced to provide financial incentives for landowners to enhance and expand habitat for listed species.

One example of this involving longleaf pine is the Safe Harbor program developed by the U.S. Fish and Wildlife service in the late 1990s. In this program landowners sign a contract obligating them to protect a baseline population of RCWs. In addition, landowners in this program must agree to provide additional habitat by planting more trees, removing understory, and providing artificial nesting cavities (Zhang and Mehmood 2002). In exchange the landowner will not be subject to further land use regulations if more RCWs inhabit their land. The landowner thus eliminates the risk that the property will be further regulated under the ESA. The Safe Harbor program started in North and South Carolina but it is expanding in other states. Landowners are generally supportive of the Safe Harbor program. Zhang and Mehmood (2002) found through surveys that participants in Safe Harbor are pleased overall and favor the program over a strict command and control regime.[1] However, the authors conclude that the expansion of the program could be greatly helped by tying the program with more direct financial incentives to the landowners.

Along these lines, Kennedy et al. (1996) proposed developing marketable Transferable Endangered Species Certificates (TESCs) for RCWs. In this program a landowner would receive TESCs for engaging in activities outlined in a statewide HCP. The landowner could then use these certificates to incidentally take RCWs or sell them to other landowners. The market created in RCW habitat would have several distinct advantages. First, allowing TESCs to be traded would lower the overall cost of mitigation activities. Landowners who have a low cost for the mitigation required by the HCP would have an incentive to acquire these certificates and sell them to landowners with higher costs. Landowners who have no RCWs on their property can also participate. The HCP baseline for such a landowner would be zero. Any additional RCW habitat produced would be extra and the landowners could sell any TESCs they received. Such a program could provide a strong incentive for landowners to produce RCW habitat because it would now have a marketable value. If the program required more than 1 hectare of RCW habitat to be created for every permit to destroy 0.4 hectare, it would result in a net increase in RCW habitat.

Similar programs have been implemented on small scales to protect other endangered species and wetlands on private land. For example, in 1995 the Bank of America created a 73-hectare conservation bank in southern California to preserve threatened and endangered species. The bank then received credits and sold those credits to developers who needed them to develop other lands. The Florida Wetlands' Bank has restored 140 hectares of wetlands and sold credits for $40,000 each under provisions of the Clean Water Act (Shogren 1998). In order for these programs to work effectively they need careful government oversight to ensure no net loss or even require a gain in habitat or wetlands. However, by providing an economic value for ecological services they can provide incentives for landowners to more explicitly consider environmental benefits in land management decisions.

Carbon Credits, Pasture (Cattle), and Longleaf Pine

Global climate change, induced by combustion of fossil fuels and land use change, is a growing concern among many policymakers and environmentalists. Forests are thought to play a crucial role in the global carbon cycle by sequestering carbon dioxide as they grow and emitting it when biomass decays or burns. As such, planting trees on marginal agricultural land, growing trees on ranchlands (silvopasture), and changing forest management regimes to increase biomass production may be desirable options to sequester additional carbon. This potential influenced participants in the Kyoto protocol to allow countries to count carbon sequestered in forests toward obligations under the protocol. The United States has been a strong proponent of this idea. Thus, even though the United States has pulled out of Kyoto, President Bush's alternative proposal includes over $3 billion for agricultural and forestry carbon sequestration activities (Bush 2002). Research indicates that forestry can be a cost-effective option for reducing atmospheric carbon dioxide concentrations. Dixon (1997), for example, estimated carbon could be sequestered in U.S. forests for between $2 and $56 per ton.

Research focusing on sequestering carbon in southeastern U.S. pine forests suggests that private landowners could be induced to sequester additional carbon for relatively modest prices. Stainback and Alavalapati (2002) investigated the impact of internalizing carbon benefits onto private forest landowners on the optimal management of slash pine plantations. The results indicate that carbon prices of less than $50 per metric ton can induce landowners to lengthen their rotations to sequester additional carbon. Huang and Kronrad (2001) found that private landowners would be willing to shift from the financially optimal timber rotation to a rotation that maximizes sequestered carbon for prices of carbon less than $70 per metric ton in loblolly plantations.

Stainback and Alavalapati (2004) conducted an in-depth economic analysis of restoring

TABLE 1. Land expectation values ($/hectare) under different management regimes and carbon prices ($/metric ton) for longleaf pine in the southeastern United States.[a,b]

Carbon price	Management regime						
	p	tf	sp	tf (rot = 60)	tf (rot = 80)	sp (rot = 60)	sp (rot = 80)
0	1700.18	2088.04	2804.64	1507.34	1186.10	2471.05	1927.42
10	1700.18	3088.81	3805.42	2644.02	2372.21	3558.31	3014.68
20	1700.18	4151.36	4843.26	3756.00	3558.31	4695.00	4200.79
30	1700.18	5201.56	5893.45	4867.97	4744.42	5806.97	5386.89
40	1700.18	6276.47	6956.01	5979.94	5905.81	6918.94	6572.99
50	1700.18	7376.08	8055.62	7116.62	7091.91	8043.27	7882.65

[a] Reprinted from Stainback and Alavalapati (2004) with permission from Elsevier.
[b] p, tf, and sp represent traditional pasture, traditional forestry, and silvopasture, respectively, "rot" stands for a fixed rotation at either 60 or 80 years.

longleaf pine. They developed a stand-level economic model of a silvopasture (pasture and longleaf pine) and compared the profitability of silvopasture with traditional pasture (no trees) and traditional forestry (only trees). Within the silvopasture and traditional forestry regimes they estimated the initial tree density and rotation age that maximize land value. They then incorporated carbon benefits into the model and investigated how payments to the landowner for sequestering carbon in trees impact the profitability of silvopasture and traditional forestry with longleaf pine. Specifically, they modeled that a landowner is paid for sequestering carbon as the stand grows and the landowner pays for carbon emissions from the decay of sawtimber, pulpwood, and wood waste left at harvest. They incorporated revenues associated with pine straw. They simulated the model to estimate the opportunity cost of fixing the rotation age at 60 and 80 years to provide habitat for the RCW for both silvopasture and traditional forestry with and without carbon payments.

Stainback and Alavalapati's (2004) results suggest that silvopasture is more profitable than both traditional forestry and traditional pasture with and without carbon payments (Table 1). As expected, land values for silvopasture and traditional forestry increase with increasing carbon prices. If the price of carbon is $0 and the rotation age were to be fixed to 60 and 80 years, respectively, to create foraging and nesting habitat for RCW, the optimal initial tree density for silvopasture would be 1236 and 988 trees per hectare, respectively. Increasing the rotation ages to create foraging and nesting RCW habitat was found to decrease the land value under both traditional forestry and silvopasture. The opportunity cost of extending the rotation for traditional forestry and silvopasture is shown in Fig. 1. The opportunity cost of extending rotation to 60 years is smaller than that of traditional longleaf pine forests. Results suggest that the opportunity cost of extending the rotation, however, declines with an increase in carbon price. In the face of carbon payments and policy support to ensure longer rotations, longleaf pine restoration through silvopasture can be a viable option. With more than 2 million hectares of ranchlands, Florida provides unique opportunities to restore longleaf pine through silvopasture.

There may be other payment mechanisms that are easier to administer and have lower transaction cost. For instance, a lump sum could be paid to landowners to plant longleaf pine on marginal pasture or other land with no trees with the payer receiving the carbon credits. Alternatively, payments could be made on the basis of temporarily sequestering carbon. In this type of program, a landowner could be paid a smaller amount to sequester carbon for a year. As long as the carbon remains sequestered, the landowner continues to receive this rental payment. Even though these mechanisms exclude the carbon that is potentially

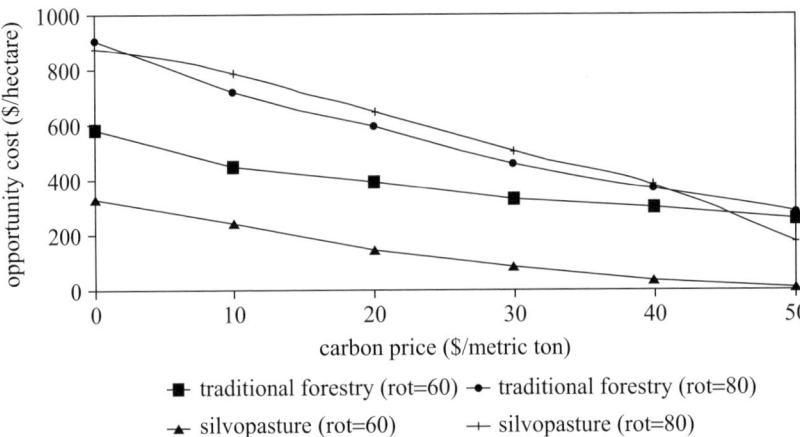

FIGURE 1. Opportunity cost of fixing the rotation age at 60 and 80 years under traditional forestry and silvopasture with longleaf pine in the southeastern United States. Reprinted from Stainback and Alavalapati (2004) with permission from Elsevier.

sequestered in end products, they reduce the need to have reliable estimates of product decay and resulting carbon emissions, thus making it easier to deal with risk and changing market conditions.

Other Potential Policy Mechanisms

The Conservation Reserve Program (CRP) has been responsible for a significant portion of the longleaf pine that has been planted on private lands to date. This can be furthered by expanding this program or by combining it with other programs. The CRP emerged from the Soil Bank Act of the 1950s. It was originally conceived as a means to take highly erodible agricultural land out of production. Usually land managed under the program is planted in trees, although sometimes other vegetative cover can be used. There has been a trend in recent years to expand the mission of the CRP to address other environmental concerns as well. For instance, basing on an Environmental Benefits Index (EBI), landowners make offers to enroll their land through a competitive process that chooses land with the highest EBI. The area enrolled each year has traditionally been capped around 14.5 million hectares. In 2002, the CRP was reauthorized to expand the program to over 15.6 million hectares and to place an even greater emphasis on environmentally sensitive lands. Even though erosion is still the top priority, other environmental benefits such as generation of wildlife habitat receive greater attention in calculating the new EBI. Landowners who receive contracts are given annual payments of 10 to 15 years and cost sharing assistance to establish a permanent ground cover such as trees.

Because of its high environmental value and ability to grow on marginal pastureland, longleaf pine is a good candidate for CRP participation. In 1998 a longleaf pine national CRP priority area was established to help restore longleaf pine on private lands in the Southeast. This approach has proved effective. For instance, since 1998 over 10,000 hectares of longleaf pine have been planted in South Carolina as a result of this program (USDA 2003). Expanding the acreage allowed under the CRP and possibly combining it with other incentive programs such as those discussed earlier to encourage RCW habitat improvement may be very effective in restoring longleaf on private land. Programs such as LIP and WHIP are particularly worth exploring.

In addition to expanding the CRP, efforts can be made to disseminate information regarding nontimber products that can be marketed from longleaf pine. These include pine straw, hunting leases, and silvopasture. Roise et al. (1991)

found that pine straw from a longleaf pine plantation can yield $32 to $64 per hectare per year. This translates into an increase in land value of more than $80 per hectare.[2] Longleaf forests provide good habitat for quail and turkey and are compatible with grazing in a silvopasture system. These nontimber products not only provide an additional source of income but also provide intermediate income while waiting for a stand of trees to mature. This can be especially important for longleaf pine because of the longer rotations associated with sawtimber and pole production as well as RCW habitat production.

Conclusions

Longleaf pine is regarded by many as one of the most treasured forest resources in the southern United States. Yet, because of past policies and the fact that many of the benefits of longleaf pine are diffuse, this forest has dramatically declined since European colonization. Recent efforts of landowners, resource specialists, and governments have made some progress in restoring this once dominant forest on private lands. However, more efforts are needed to ensure the perpetuation of longleaf pine and all of its benefits into the future.

The increased interest in harnessing market forces to help solve environmental problems may be particularly useful in the restoration of longleaf pine. The vast majority of potential longleaf pine sites are located on private property. Thus, traditional command and control mechanisms may not be conducive for its restoration. Further, focusing on just one of the many environmental services associated with longleaf pine may not provide a strong enough incentive for its restoration. For example, Alavalapati et al. (2002) found that carbon benefits alone were not enough to make longleaf pine as profitable as slash pine. A combination of subsidizing landowners for carbon sequestration, RCW habitat, and other amenities was necessary. Landowners growing longleaf pine in the future may be able to participate in several markets such as a Transferable Endangered Species Certificate market for the RCW and a carbon market as well as participating in the Conservation Reserve Program. This type of co-benefit approach is increasingly recognized as a necessity for making market mechanisms work for the conservation of ecologically valuable lands (Daily and Ellison 2002). Furthermore, an increase in straw price, demand for quail hunting, and premiums for graded lumber from longer rotation might stimulate landowners to restore longleaf pine. However, it is doubtful that market mechanisms are going to prove to be a panacea for longleaf pine restoration. Instead, a diversity of approaches is needed. This includes restoration of longleaf on public lands, public purchase of private lands with longleaf pine ecosystems, especially those under a threat of conversion to other uses, the use of conservation easements and appropriate regulation of private land, as well as market incentives for landowners. With a growing interest in longleaf pine, and innovative policies, this species may once again be a substantial component of our southern forests.

Endnotes

1. Overreliance on command and control policies to regulate private lands may be politically unsustainable. At the time of writing, a bill (HR 4840) is being considered by Congress that would allow landowners to participate more effectively and provide input in developing recovery plans.
2. Reported figures on pine straw revenues may not account for additional fertilization that is needed to offset nutrient removal associated with straw collection. This means additional cost to the landowner and lower profits.

References

Abrahamson, W.G., and Hartnett, D.C. 1990. Pine flatwoods and dry prairies. In *Ecosystems of Florida*, eds. R.L. Myers and J.J. Ewel, pp. 103–149. Orlando: University of Central Florida Press.

Alavalapati, J.R.R., Stainback, G.A., and Carter, D.R. 2002. Restoration of the longleaf pine ecosystem on private lands in the U.S. South: An ecological economic analysis. *Ecol Econ* 40:411–419.

Boyer, W.D. 1996. Longleaf pine can catch up. In *Proceedings of the First Longleaf Alliance*

Conference—*Longleaf Pine: A Regional Perspective on Challenges and Opportunities*, ed. J. S. Kush, pp. 28–29. Mobile, AL, September 17–19.

Boyer, W.D., and Peterson, D.W. 1983. Longleaf pine. In *Silvicultural Systems for the Major Forest Types of the United States*. Agricultural Handbook No. 445, ed. R.M. Burns, pp. 153–156. Washington, DC: USDA Forest Service.

Bush, G.W. 2002. Global Climate Change Policy Book. The White House, Washington, DC.

Daily, G.C., and Ellison, K. 2002. *The New Economy of Nature: The Quest to Make Conservation Profitable*. Washington DC: Island Press.

Demers, C., and Long, A. 2003. Longleaf pine regeneration. Retrieved 11 April 2003 from http://edis.ifas.ufl.edu/BODY_FR064.

Dixon, R.K. 1997. Silvicultural options to conserve and sequester carbon in forest systems: Preliminary economic assessment. *Crit Rev Environ Sci Technol* 27:139–149.

Huang, C.H., and Kronrad, G.D. 2001. The cost of sequestering carbon on private forest lands. *For Policy Econ* 2:133–142.

Johnson, R., and Gjerstad, D. 1999. Restoring the longleaf pine forest ecosystem. *Alabama's Treasured Forests* Fall:18–19.

Kelly, J.F., and Bechtold, W.A. 1983. The longleaf pine resource. In *Proceedings of the Symposium on the Management of Longleaf Pine*, ed. R.M. Farrar, Jr., pp. 11–22. Long Beach, MS, April 4–6.

Kennedy, E.T., Costa, R., and Smathers, W.M., Jr. 1996. New direction for red-cockaded woodpecker habitat conservation: Economic incentives. *J For* 94:22–28.

Lancia, R.A., Roise, J.P., Adams, D.A., and Lennartz, M.R. 1989. Opportunity costs of red-cockaded woodpecker foraging habitat. *South J Appl For* 13:81–85.

Landers, J.L., Van, L., David, H., and Boyer, W.D. 1995. The longleaf pine forest of the Southeast: Requiem or renaissance? *J For* 93:39–43.

Roise, J.P., Chung, J., and Lancia, R. 1991. Red-cockaded woodpecker habitat management and longleaf pine straw production: An economic analysis. *South J Appl For* 15:88–92.

Shogren, J.F. 1998. A political economy in an ecological web. *Environ Resour Econ* 11:557–570.

Stainback, G.A., and Alavalapati, J.R.R. 2002. Economic analysis of slash pine forest carbon sequestration in the southern U.S. *J For Econ* 8:105–117.

Stainback, G.A., and Alavalapati, J.R.R. 2004. Restoring longleaf pine through silvopasture practices: An economic analysis. *For Policy Econ* 6(3–4): 379–390.

USDA. 2003. Farmers save environmentally-sensitive farmland with CRP. Retrieved 15 April 2003 from http://www.scda.state.sc.us/AgLinks/ANRCESITE/fsa_enviro_sens_CRP.html.

Yanquoi, R., and Flick, W.A. 1994. Economics of longleaf pine management. In *Forest Economics on the Edge: Proceedings of the 24th Annual Southern Forest Economics Workshop*, eds. D.H. Newman and M.E. Aronow, p. 89. Savannah, GA, March 27–29.

Zhang, D., and Mehmood, S.R. 2002. Safe harbor for the red-cockaded woodpecker: Private forest landowners share their views. *J For* 100:24–29.

Chapter 14

Role of Public–Private Partnership in Restoration

A Case Study

Vernon Compton, J. Bachant Brown, M. Hicks, and P. Penniman

Introduction

With today's increasing challenges in restoring the longleaf pine ecosystem, land managers, both public and private, need innovative management solutions. Since most challenges are shared across the landscape and desired end results are similar for land managers, one innovative approach that is proving effective is working in partnership with multiple organizations, agencies, and stakeholders. Within a partnership, members share the risks and the challenges of managing the longleaf pine ecosystem, as well as the benefits, such as healthier, more functional ecosystems. Focus and emphasis on collaboration, cooperation, and consensual goals provide the foundation for positive and productive partnership actions, which usually result in successful attainment of partnership and member goals and objectives.

The Gulf Coastal Plain Ecosystem Partnership (GCPEP) is an example of a partnership that has been able to frequently attain challenging and ambitious landscape-scale conservation goals and objectives through positive, result-oriented action and collaboration. GCPEP was formed because several landowners and managers shared concerns and challenges regarding the decline of the longleaf pine ecosystem in northwest Florida and south Alabama. In 1996, seven public and private landowners formed a partnership to address common land- and water-conservation concerns and challenges, and to utilize the opportunity to act collaboratively and cooperatively. Currently, there are ten partners in GCPEP that share landscape-scale conservation goals in the region.

This chapter will describe how the framework and function of GCPEP may provide a "blueprint" for other partnerships, and will explain the ecological rationale behind the creation of GCPEP. In addition, some of the early and current successes as well as challenges the partnership has experienced will be discussed. The chapter will also examine how the partnership maintains a common focus on, and kinetic progress toward, conservation goals through planning and prioritization methods.

The chapter is approached in sequence beginning with the concepts that lead to the formation of the partnership, including landscape-scale conservation and the advantages of ecosystem management through partnerships. The following sections explain in

---

Vernon Compton, J. Bachant Brown, M. Hicks, and P. Penniman • The Nature Conservancy, Jay Florida Office, The Gulf Coastal Plain Ecosystem Partnership of Jay, Florida 32565.

detail the various aspects of GCPEP, including inception, discussion about each individual partner, and many of the conservation projects that have been identified as priorities by the partners.

Landscape-Scale Conservation

Landscape-scale conservation served as the primary purpose for establishing GCPEP. Successful landscape-scale conservation usually involves actions that affect large and numerous parcels of land, typically owned by multiple persons or organizations. Conserving functional landscapes improves the likelihood of achieving sustainable conservation of biodiversity. According to Low (1999), emphasis on conserving functional landscapes dramatically improves efficiency and effectiveness for the following reasons:

- Conservation actions that simultaneously affect ecological systems, communities, and species at multiple scales within a single intact landscape provide a more ecologically integrated conservation strategy that better protects functional landscapes and biodiversity.
- Functional landscapes typically include private and public lands both of which are frequently needed to protect and restore ecological processes.
- Landscape-scale conservation requires an ecosystem approach involving multiple strategies to abate critical threats driven by incompatible human uses of the lands and waters.
- Landscape-scale conservation focuses on restoration of conservation targets.

Ecologically important natural systems and resources are typically embedded within a large working landscape, which includes the people who live and work in these places. Except for isolated wilderness areas, threats to conservation targets often involve incompatible human uses and economic development. Solutions invariably require working with local landowners, community leaders, and governments. Long-term conservation of these places will only happen through support of and participation in conservation planning and implementation by the local community.

Partnerships in Conservation

When forming a landscape-scale partnership, consideration of many different factors is essential. Partnerships require a clear understanding of the purpose of the individual organizations interested in becoming enrolled, as well as the manner in which the coalition of organizations will operate. A successful partnership will become an entity of its own that ideally will be greater than the sum of its parts. This best occurs when each organization is well established and committed to remaining involved in the partnership for the long term.

Partnerships are often guided by a Steering Committee, the method used by GCPEP. Ideally, the Steering Committee has agreed-upon operating guidelines to ensure efficient operation of the partnership. During Steering Committee meetings, and day-to-day operations and interactions, it is important to approach all topics and issues with the utmost respect for members and their respective organizations, as well as to minimize preconceived expectations and conceptions. Negotiations are most successful when all partners view one another as equal. When the playing field is level for everyone involved, it provides an effective environment for cooperation, communication, and understanding. Greatest potential for success is realized when goals for far-reaching cooperative restoration projects are shared and involvement for partners is maximized. The end results of such an approach can be extremely positive and may produce widespread benefits that may never have been imagined when initially planning meetings and projects.

GCPEP: An Example of an Effective Partnership

The Gulf Coastal Plain Ecosystem Partnership is a successful collaboration among ten public and private organizations that collectively

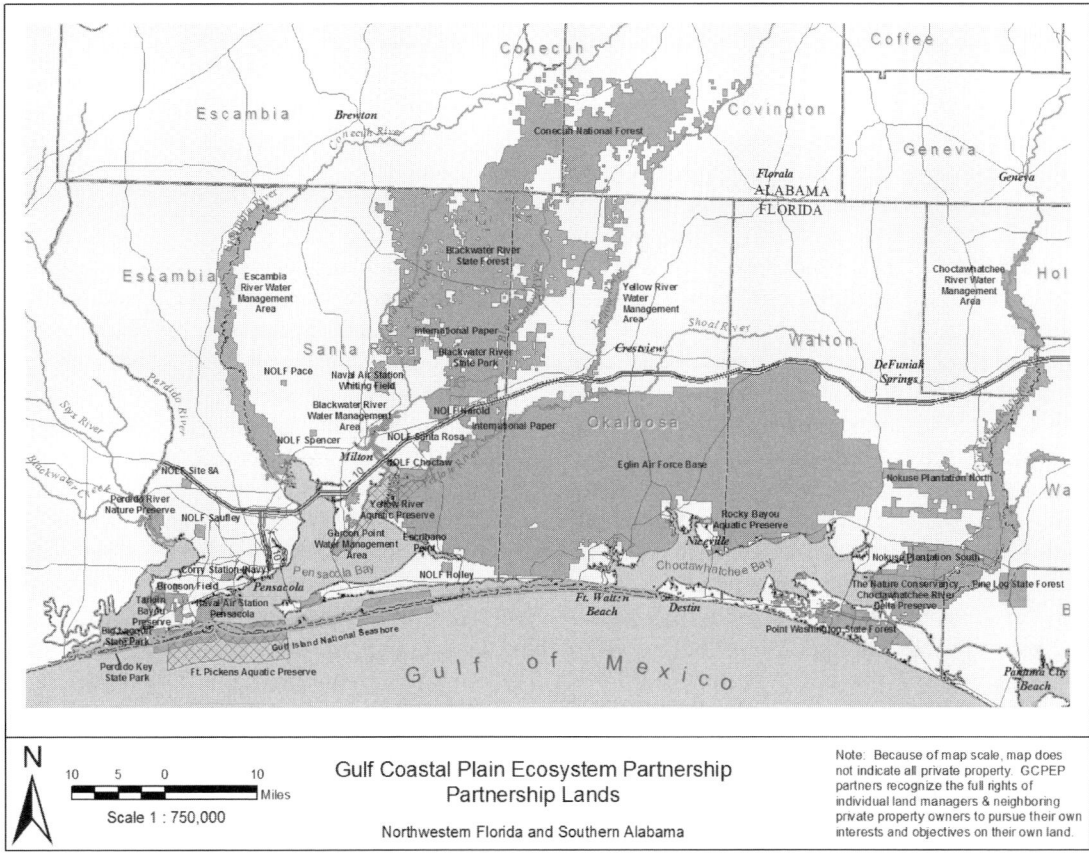

FIGURE 1. Gulf Coastal Plain Ecosystem Partnership lands and surrounding landscape in northwest Florida and south Alabama.

manage more than 425,859 ha of land in one of the most biologically significant regions in North America (Fig. 1). The GCPEP landscape has the vast majority of the world's remaining old-growth longleaf pine ecosystems, containing some longleaf pine trees that are over 500 years old. GCPEP partners include the Departments of Defense, Florida Department of Environmental Protection, Florida Division of Forestry, Florida Fish and Wildlife Conservation Commission, International Paper, National Forests in Alabama, National Park Service, Nokuse Plantation, Northwest Florida Water Management District, and The Nature Conservancy (Compton et al. 2002a and The Nature Conservancy 2005).

Explaining the inception of GCPEP may provide guidelines for initiating a partnership. An understanding of the partnership framework and projects may offer measures of success for cooperative restoration methods, which have been successful for GCPEP. The partnership has proven to be more effective and productive than expected, achieving goals that no one organization could individually accomplish.

How GCPEP Began

The GCPEP began with an idea. One agency contacted another to discuss the possibility of combining efforts to create a contiguous landscape for recovery efforts for the federally endangered red-cockaded woodpecker (RCW; *Picoides borealis*). By reconnecting the longleaf pine ecosystem, northwest Florida and south Alabama lands could provide enough contiguous forest to aid in the recovery of the RCW and other rare species, such as Florida black bears (*Ursus americanus floridanus*). The original

GCPEP landscape consisted of connected lands that were primarily undeveloped, but became fragmented by roads and increasing development. Reconnecting these lands through sharing resources, cooperating on management activities, and protecting important conservation lands would potentially restore the landscape to establish a more functional metapopulation of RCWs and other species requiring broad and largely intact longleaf pine and associated ecosystems.

Since GCPEP originally formed there have been several changes. New partners have joined GCPEP, while existing partners have enrolled additional lands into the partnership landscape. Steering Committee representatives have changed due to shifting responsibilities, relocations, and retirements. The GCPEP staff, which is explained later in more detail, plays an important role of providing continuity over time as changes occur both within and surrounding the partnership. Additional GCPEP staff has been added to support strategies and actions set by the Steering Committee.

The GCPEP Framework

It was decided that a Steering Committee would allow the GCPEP to function best because each partner would have equal representation and decision-making power. The GCPEP is guided by the Steering Committee, which is composed of two representatives from each of the partner organizations. Each partner organization chooses the representatives, which include one primary and one alternate contact. Representation at the Steering Committee meetings by one of the representatives from each partner organization is encouraged. Occasionally when a representative is unable to attend the meeting a designee chosen by the primary contact may represent the organization. The GCPEP Steering Committee, which meets biannually, has established guidelines to ensure efficient operation of the partnership.

Consensus is desired in reaching agreements among the partners during the Steering Committee meetings to ensure an equal voice for all. If there is minority dissent, then the majority is charged with finding an alternative solution acceptable to all. The goal is to always maintain productivity while keeping the consensus process efficient. Decisions are based upon Steering Committee voices only—the GCPEP staff does not vote. The Steering Committee functions best when everyone participates and ensures input from their respective organizations in all decisions.

The Steering Committee established GCPEP's mission: to develop a set of long-term strategies to abate the critical threats and to improve regional ecosystem health; to recover listed species of plants and animals and avoid new listings; to restore and protect large, connected, functional examples of native ecosystems; and to provide ecosystem goods and services compatible with the above to surrounding communities.

At each GCPEP Steering Committee meeting, a research, scientific, and general information manual that highlights all the partners' progress since the last meeting is distributed by the GCPEP staff to each representative (Compton et al. 2002b). The manual is then disseminated to the widest audience possible, particularly within and to the partners' agencies and supervisors. The Steering Committee recognizes the importance of exporting the lessons learned from the partnership to other landowners, organizations, community leaders, and the general public. Scientific research and knowledge gained remains limited in value if not shared with either those who manage land or influence the management of land.

Early Successes of GCPEP

GCPEP Memorandum of Understanding

The wording of the GCPEP Memorandum of Understanding (MOU) was established through a series of meetings to discuss the elements that each agency could agree upon, which would also fit within legal and interagency requirements. The MOU recognizes that the individual public and private agencies have legitimate and varied management goals. The MOU is in no way intended to limit or constrain the individual goals and missions of each partner's organization.

The purpose of the GCPEP MOU is to develop and implement a voluntary and cooperative stewardship strategy to sustain the long-term viability of native plants and animals, the integrity of ecosystems, the production of commodities and ecosystem services, and the human communities that depend upon them.

The goals of the GCPEP MOU are to assist, share information, and coordinate efforts with the member partners in fulfilling the purposes of the MOU; to provide a model for local, state, federal, and private entities working together to fulfill the purposes of the MOU; and to communicate to the public the success in meeting both individual and common goals related to the MOU.

Conservation Area Planning

Conservation Area Planning, originally known as Site Conservation Planning, which is discussed in further detail later in this chapter, represents a tremendous partnership accomplishment by going beyond thinking within individual boundaries to thinking at a landscape level. The completion of a Conservation Area Plan allows for more effective management and restoration across large landscape areas, according to Compton et al. (2002a).

GCPEP Challenges

GCPEP has encountered many unexpected challenges, some of which required extensive cooperation to reach solutions. An initial challenge for the partnership was bringing together a committed, well-established nucleus of organizations. The establishment early on of the GCPEP Steering Committee to set overall goals and priorities was a challenge but led to a stronger partnership and much faster success on the ground.

When working through any challenging process it is important that each partner be cognizant of language to remain positive and solution-oriented in conversations, written documents, and while communicating with the media. Continuous and careful planning assists with the challenge of allotting the amount of time, staff, and resources necessary to manage required tasks, while maintaining and balancing the prioritization of crucial conservation opportunities that may be lost if not promptly addressed.

Challenges are experienced during the process of receiving approval and submitting funding proposals with numerous partner organizations. Clear and constant communication with each department involved is required to complete proposal submissions. Ensuring each partner involved in the agreement has reviewed and approved the proposal typically requires additional time. Close attention must be paid to tracking the progress and the reporting requirements for each project that is awarded funding to ensure the deliverables stated in the agreement are routed and received in a timely manner.

GCPEP Benefits Individual Partners

In addition to collective accomplishments, each GCPEP partner has achieved outstanding individual conservation successes. The partnership has played an important role in projects by providing assistance, scientific expertise, funding, in-kind donations, and public education support. Important contributions include facilitation of projects using unconventional methods such as cross-boundaries projects like RCW translocation between forests. Additional in-kind contributions to assist the partners include sharing supplies, equipment, and personnel required for support of landscape-scale conservation, such as office space and staff, sharing GIS data, burn prioritization modeling, endangered-species management, and road maintenance.

The following section highlights each individual GCPEP partner and briefly describes the lands they manage in the partnership.

Department of Defense

At 187,548 ha, Eglin Air Force Base holds the largest amount of land of all partners in the

GCPEP. Undeveloped lands serving as buffers for military operations contain old-growth longleaf pine forests and red-cockaded woodpecker clusters, along with other unique natural communities and species. Eglin projects include biodiversity restoration, native plant demonstration areas, and native plantings along roads and streams for erosion control. Eglin has led the way with developing a burn prioritization model and assisting with exporting it to other partner lands.

Naval Air Station Pensacola manages 3409 ha of forest, wetlands, shoreline, and outdoor recreation areas in the GCPEP. Naval Air Station Pensacola leadership highlights include maintaining the regional osprey population with 20 new fledglings produced annually, honeybee relocation programs, sea oat plantings for shoreline stabilization, International Coastal Cleanups, and Tree City USA designation on the base.

Naval Air Station Whiting Field manages 3795 ha in the partnership and natural resource efforts include gopher tortoise (*Gopherus polyphemus*) and flatwoods salamander (*Ambystoma cingulatum*) protection, public nature trails, Tree City USA designation, agricultural and timber projects, and regional support for conservation land purchases.

Florida Department of Environmental Protection

The Florida Department of Environmental Protection (DEP) manages 23,176 ha in GCPEP. The Coastal and Aquatic Managed Areas, a Division of DEP, manages four aquatic preserves. Beneficial efforts include Gulf sturgeon (*Acipenser oxyrinchus desotoi*) studies, shoreline vegetation restoration, and coastal cleanups. The Blackwater River State Park and the Yellow River Marsh Preserve State Park, also managed by DEP, maintains and restores lands and waters that provide a variety of recreational opportunities including swimming, canoeing, hiking, birding, botanizing, and camping.

Big Lagoon, Tarkiln Bayou Preserve, and Perdido Key State Parks bring coastal and barrier island habitats to the partnership, which are surrounded by urban development posing significant challenges with prescribed burning and roads leading into the properties. Ongoing projects in the parks include bird counts, protection of bird nesting areas, and beach mouse habitat restoration.

Florida Division of Forestry

Blackwater River State Forest is one of the largest state forest in Florida with 78,779 ha managed in the partnership. Working with GCPEP, the Blackwater River State Forest has increased erosion control efforts by using native plants to protect the entire Blackwater River watershed. The Division of Forestry has improved road management programs that construct stream crossings to protect water quality and aquatic habitat. Successful red-cockaded woodpecker recovery programs have also been implemented such as translocation of birds and installation of cavity inserts.

Pine Log State Forest with 2797 ha is the oldest state forest in Florida containing sandhills, flatwoods, cypress swamps, and titi forests located on the eastern border of the GCPEP. Point Washington State Forest with 6170 ha borders an area of rapid residential and commercial growth along spectacular, fragile coastal areas on the Gulf of Mexico and contains rare species such as white-topped pitcher plants (*Sarracenia leucophylla*).

Florida Fish and Wildlife Conservation Commission

The Florida Fish and Wildlife Conservation Commission (FWCC) manages the parcel of land and habitat between the Northwest Florida Water Management District and Eglin Air Force Base known as Escribano Point. Escribano Point is comprised of 472 ha, which is a mosaic of habitats including pine and scrubby flatwoods, inshore marine habitat, oak hammocks, wet prairies, and wetlands. Escribano Point also includes high-quality submerged plant communities with many rare plant species. The FWCC has technical knowledge of endangered species, animal population records, bear information, game and nongame ecology, wildlife expertise, and prescribed fire, and provides assistance to landowners as well as operational support for the partnership.

International Paper

International Paper is a private timber and paper products company that has dedicated 9819 ha of crucial conservation lands to GCPEP, including an important connector parcel linking Eglin Air Force Base and the Blackwater River State Forest. This parcel serves as a critical wildlife corridor for wide-ranging species such as the Florida black bear and provides important habitat for the rare Florida bog frog (*Rana okaloosae*) and flatwoods salamander. With GCPEP support, the company implemented a cooperative gully restoration project that helped to protect Florida bog frog habitat. Additionally, International Paper has included other important conservation lands within the Blackwater River watershed to GCPEP.

National Forests in Alabama

Conecuh National Forest in south Alabama is composed of 33,909 ha of longleaf pine habitat. Conecuh has received recognition for continually meeting annual prescribed burning goals. Conecuh sets examples through successful longleaf pine restoration and monitoring projects and the use of native grasses for road maintenance and erosion control. The forest also protects crucial Gulf sturgeon spawning areas and red-cockaded woodpecker nesting habitats.

National Park Service

More than 80 percent of Gulf Islands National Seashore is under water, but the barrier islands are the most outstanding features to those who visit. The Seashore stretches 170 km from Cat Island in Mississippi to the eastern tip of Santa Rosa Island in Florida, but only the portions within the Florida Panhandle are enrolled within GCPEP which consists of 10,034 ha. There are snowy-white beaches, sparkling blue waters, fertile coastal marshes, and maritime forests all of which are important GCPEP conservation targets. This is the most highly visited GCPEP natural area and the National Park Service uses the opportunity to focus on resources interpretation and education.

Nokuse Plantation

Nokuse Plantation is an ambitious and exciting project by a private conservation buyer. The project includes 21,448 ha east of Eglin Air Force Base that was chosen for the biological significance and the importance of connectivity to GCPEP as a wildlife corridor and to restore highly degraded lands. The objective of the visionary project is to protect regionally significant areas that may serve as a critical wildlife habitat for species such as the Florida black bear and to restore the historical longleaf pine ecosystems.

Northwest Florida Water Management District

One of the five water management districts in Florida, the Northwest Florida Water Management District (NWFWMD) is charged with protecting watersheds, providing drought control, and maintaining drinking water supplies in the Florida Panhandle. The NWFWMD manages 45,715 ha in the GCPEP area, which serve to protect rivers, associated floodplains, estuarine systems, and wildlife habitat. With GCPEP assistance the NWFWMD has added to its landholdings, conducted important prescribed burns on wetland savannas, and constructed trail systems for public use and education.

The Nature Conservancy

The Nature Conservancy manages 2056 ha with its Perdido River Nature Preserve and Choctawhatchee River Delta Preserve. Participation in GCPEP has helped The Nature Conservancy advance its mission to conserve biodiversity in northwest Florida and south Alabama through community involvement, landscape-scale conservation and restoration, and land acquisitions.

New Partners

Other important conservation lands may be added to the GCPEP with unanimous agreement from the Steering Committee and the landowners. The Steering Committee has established the following criteria to admit new partners to the GCPEP:

1. Understands and supports the purposes of the GCPEP and can clearly articulate both what their organization has to gain from and what they plan to contribute to the partnership.
2. Meets one or both of the following criteria:
   (a) Manages or owns significant land or water holdings in the GCPEP geographic area with strong preference given to those sharing a border with one or more existing GCPEP partners, or
   (b) Can offer significant expertise in one or more of the following management or conservation disciplines: forestry, water and watersheds, wildlife, biodiversity, prescribed fire, endangered species, or nature-based recreation.
3. Commits to appointing and sending at least one, and preferably two, representatives to all GCPEP Steering Committee meetings and other functions as needed.
4. Agrees to lead or co-lead one or more cooperative GCPEP projects per year.
5. Agrees in principle to provide financial or operational support to the GCPEP, either as direct funds or as in-kind support, and agrees to seek additional resources to support cooperative projects.
6. Understands and agrees to adhere to the GCPEP operating guidelines.
7. Agrees to keep all appropriate people within their organization informed and knowledgeable about the GCPEP purposes and activities.

Partnership Staff

Multiple partners contribute the necessary staff to facilitate the GCPEP. The GCPEP staff provides assistance, support, coordination, and information to the Steering Committee and their organizations. The GCPEP staff does not vote on any topics. The Conservation Area Plan, which was completed by the Steering Committee, provides guidance for the staff.

To enable adequate operations, a partnership staff is recommended whose primary focus is the overall partnership. These positions may include a project director to lead important meetings and coordinate multiple projects; scientists to lead restoration, research, and monitoring; and a program manager to facilitate a wide range of administrative and financial tasks. This office provides a central location for functional communication among the different partners.

Approaching all interactions with flexibility and accommodating such a diverse group enables the staff to take advantage of vast opportunities. Remaining focused on priorities that have been identified, while at the same time incorporating unexpected changes, provides a dynamic forum for coordinating multiple projects. While attending to the requests of one individual partner, it is also essential to ensure that all of the partners receive a timely response when requesting assistance. Depending on the size of the landscape there may be a considerable amount of travel involved to address all of the partners' needs.

Low (1999) stated that the local project staff, particularly the project director, is the single most important element of success and the local partnership staff is possibly the most important factor that determines the success of a partnership for landscape-scale conservation. An ability to multitask and attend to numerous issues with various degrees of prioritization is essential. Some of the qualities include commitment to the future; ability to handle risk and uncertainty; ability to form constructive relationships with all kinds of people; and aptitude for problem solving.

Support Staff

Good project support for the GCPEP has been critical and extensive. According to Low (1999), no local partnership, particularly in the early stages of development, should be an island. Each project needs high-level assistance from a support team. A local project needs to be able to call upon experienced ecosystem conservation practitioners to serve as sounding boards for ideas, to provide advice and counsel, to provide contacts with outside sources of assistance, and to provide hands-on help. The local GCPEP staff has received tremendous additional support from the partners' regional offices, providing assistance in numerous areas including conservation science, land protection, government relations, communications, and operations.

Conservation Area Planning

Once a partnership is established, use of a planning framework tool is highly recommended to help maintain focus, assist with prioritizing, and make the best use of limited time and funding. Any type of conservation planning must overcome many challenges, one of which is the need to simultaneously accommodate many different, sometimes competing, goals, only one of which may be conserving biodiversity. This planning tool also needs to be readily available, reasonably fast, and cost-effective. The GCPEP utilized The Nature Conservancy's Conservation Area Plan process identified by Low (1999). With this planning framework tool, the partners were able to determine local threats to the long-term persistence of conservation targets, which include specific focal species and natural communities at a site, and to identify the most important management actions needed to conserve selected conservation targets.

The Conservation Area Plan approach explicitly recognizes that humans are part of ecosystems, ecosystems are complex moving targets, ecosystem structure and composition are controlled by processes operating at many different spatiotemporal scales simultaneously, and the scientific community has little understanding of the structure and function or life history needs of most of the ecosystems and species that they seek to conserve. Thus, all knowledge is treated as provisional, and the planning process becomes as important as the information used in planning (The Nature Conservancy 1998).

The Conservation Area Plan is broken down into a Five-S Framework (The Nature Conservancy 2000). The Five S's are:

- Systems: the conservation targets at a site and the natural processes that maintain them
- Stresses: the causes of destruction, degradation, or impairment of the systems at a site
- Sources: the agents or activities generating the stresses
- Strategies: the types of conservation activities deployed to abate sources of stress (threat abatement) and enhance or restore the system (restoration)
- Success: measures that monitor the effectiveness of implemented strategies often involving tracking of biodiversity health and threat abatement at a site.

For the purposes of this chapter, the planning process will be briefly described. In order to gain a thorough understanding of the process, it is highly recommended to refer to The Nature Conservancy's "Landscape-Scale, Community-Based Conservation: A Practitioner's Handbook" (Low 1999).

The first "S," Systems, captures the conservation targets at a site. Conservation targets include significant and possibly unique ecosystems, biological communities, and species. Identifying the appropriate targets is the single most important step, since it lays the foundation for all subsequent steps in the planning process. The goal is to choose conservation targets that represent multiple levels of biological organization, have different life history requirements, depend on different ecological processes, and encompass a variety of different spatial scales. In effect, planning targets act as conservation umbrellas or surrogates, however imperfectly, for all other target species and natural communities occurring in the geographic area. Thus, targets, whether community- or species-level, are used to cumulatively address the ecological requirements for all species and communities occurring at a site.

After the selection of conservation targets, key ecological attributes associated with each of the targets are identified and defined. This important step allows for accurate assessment of target viability, threats to the targets, and subsequent strategies identified to abate the threats. It also allows the planners to better understand and identify the data gaps and uncertainties associated with the targets and their respective key ecological attributes.

The next two "S's" in the framework, Stresses and Sources, combine to determine the threats to a system. In order to develop effective conservation strategies, one must understand both the stresses affecting the conservation targets and their key ecological attributes, and the sources of stress. In this stage of the planning process, after identifying the major stresses to the targets, the stresses are then ranked based upon the severity and scope of damage. For each stress there may be one or more sources of stress. After the major sources of stress are identified, they are ranked according to specific guidelines. The source is what managers must focus on for threat-abatement strategies.

After having identified and ranked what the primary critical threats are to the conservation targets, conservation strategies are identified based on their ability to abate the threats to key ecological attributes of the conservation targets and ultimately improve and/or maintain target viability or health. Strategies can be either threat abatement, which focuses on preventing, diminishing, or removing one or more sources of stress, or restoration, which directly enhances or restores the viability of the conservation target. When identifying and developing strategies, it is important to first consider an array of strategic approaches and then formulate a suite of potential strategies. Next, evaluate and rank the potential strategies as to their impact and feasibility in order to identify the top priorities for immediate action.

The last "S" in the framework is Success. Measuring conservation success is an important step in order to monitor whether or not actions or implemented strategies are having the desired and anticipated outcome. Success can be defined as making substantial progress toward the long-term abatement of critical threats and the sustained maintenance or enhancement of biodiversity health at sites. Commonly, it takes a long time for implementation of a conservation strategy to manifest in the actual improvement or maintenance of target viability and health, signified by desired performance of indicators of biodiversity health. Therefore, indicators are needed and used to account for incremental short-term success. These indicators can reflect the capacity to implement strategies. The three key factors that can account for early success within a project such as a partnership are: project leadership and support, strategic approach, and adequate funding.

As a planning tool, the Conservation Area Plan should be adaptive in order to make accommodations for individual circumstances. For instance, GCPEP used the Five-S Framework as a foundation, which was modified and built upon to develop the overall GCPEP Conservation Area Plan. When the partners first established the partnership, one of the first agenda items for planning was to share individual land management conservation objectives. From these individual objectives, the partners then collectively prioritized overall GCPEP objectives. It was these conservation objectives and the process used in identifying them, as well as the identification of common challenges and conservation issues that laid the foundation for the Conservation Area Plan process.

The agreed-upon GCPEP objectives in priority order were:

1. Conserve viable populations of target species
2. Introduce relatively natural fire regimes and protect key ecotypes
3. Protect urban interface and reduce fragmentation by use of conservation easements
4. Control erosion in ecologically sensitive areas
5. Manage recreation and public access to maximize compatibility with conservation objectives

14. Public–Private Partnership in Restoration

6. Increase communication, interaction, and training among partners
7. Increase inventory and monitoring to further adaptive management
8. Increase public education and stakeholder involvement
9. Share resources on priority projects
10. Secure outside funding and support
11. Inventory and control exotic species
12. Protect aquatic resources
13. Increase understanding of successful economic management of longleaf pine
14. Restore and manage the longleaf pine ecosystem
15. Recover the red-cockaded woodpecker
16. Manage populations of game species
17. Conserve functional community types

From these objectives, a list of potential conservation planning targets was identified. Then, using the partners' knowledge of the GCPEP area and the ecological analysis done by Hardesty and Moranz (1999), the partners selected, by consensus, a subset of targets. These 18 primary conservation-planning targets included 8 species and 10 natural communities. The 8 species were chosen because they were declining across their range, they had large area requirements (relative to their body size), they were found on the majority of GCPEP lands or waters, and they would not necessarily be well protected through appropriate management of natural community-level targets. The 10 communities were chosen because they are important for facilitating functional ecological processes and each included many rare, threatened, endangered, and/or ecologically significant species.

The GCPEP conservation targets identified in the Conservation Area Plan are:

- Alluvial rivers/floodplains
- Barrier island complex
- Blackwater rivers/floodplains
- Depression wetlands
- Estuarine systems
- Fish/mussel complex
- Flatwoods salamander (T)
- Florida black bear (t)
- Florida bog frog (e)
- Gulf sturgeon (T)
- Longleaf pine sandhill matrix
- Mainland sand pine scrub
- Okaloosa darter (E, e)
- Pine flatwoods matrix
- Red-cockaded woodpecker (E)
- Seepage slopes
- Steephead stream/slope systems
- Upland game birds

T = federally threatened
t = state threatened
E = federally endangered
e = endemic to GCPEP lands

In choosing these 18 primary conservation-planning targets, GCPEP deviated from the recommendations in the Five-S Framework. The handbook recommends that no more than eight focal targets be chosen; however, due to the large land area that was enrolled in the partnership and the large number of varying partner needs, a larger number of targets was deemed necessary. Also, based upon the partners' needs that were identified through the objectives, a species target (upland game birds) was chosen that might not have apparent "ecological" significance. Game birds were chosen because several partners identified this species group as one for which they needed assistance and guidance with regard to population management, in order to meet particular land management objectives.

Once the targets were identified, GCPEP staff met with each partner individually to conduct threats analyses for the targets that occurred on the lands they manage. During these sessions, which included the partners' scientists and managers, partner comments regarding specific targets were also incorporated. The GCPEP staff combined each of the individual partner target threat analyses into an overall GCPEP target threat analysis. This allowed the partners to gain a sense of the threats per target across the landscape as well as on individual properties. The final step was to compile overall stresses and sources (overall threats) for all of the GCPEP conservation targets. This part of the planning process is a work in progress since it is still being determined how to obtain a more accurate picture by weighting the rankings of the stresses and sources based on the number of conservation targets and the number of partners affected.

Many threats were identified as being directly or indirectly related to the burgeoning growth of residential and commercial land uses in the region over the last decade. This growth has been forecasted to increase even more so over the next decade. With this growth have come increased land and water supply demand, recreational pressures, water quality degradation, strain on infrastructure, and other pressures on public and natural resources.

According to Hiers et al. (2002), from 1990 to 2000, the population of the seven-county GCPEP area increased by 18.5%. In Florida, the populations of Escambia, Okaloosa, Santa Rosa, and Walton counties increased by 12, 19, 44, and 46, respectively. Walton and Santa Rosa counties rank in the top ten fastest growing counties in Florida between 1990 and 2000.

The threats analysis process identified a number of primary threats that endanger terrestrial and aquatic targets. These threats can be considered "killer threats" to several of the targets. The biggest terrestrial "killer threat" identified by the partners was altered fire regime (stress) due to inadequate or incompatible fire management (source). These sources included, but were not limited to, the partners' ability to burn, the seasonality of burns, and the placement of plow lines. Another terrestrial "killer threat" identified was decreased reproductive fitness (stress) due to demographic isolation (source).

The identified aquatic "killer threat" concerned the hydrological and ecological impacts (stress) that a proposed dam (source) would have within one of the five watersheds within GCPEP. Another aquatic "killer threat" was alteration to the natural hydrologic, chemical, and physical characteristics of aquatic systems and subsequent degradation of aquatic ecological community and species integrity (stress) due to incompatible land use practices in agriculture, recreation, road construction and maintenance, forestry, and urban development (source).

The GCPEP staff then selected ten strategies considering all of the partners' conservation objectives, issues, and challenges, and their ability to abate threats to the identified 18 conservation targets as explained through the threats analyses. The following were identified for each strategy: the overall goal, the partner contacts for whom GCPEP staff will work, the potential expected accomplishments, and the conservation targets addressed by the strategy.

The issues that the ten GCPEP strategies address are as follows:

- Inadequate/incompatible fire management
- Incompatible development
- Inadequate/incompatible dirt roads, utility corridors, culverts, or clay pits management
- Surveying, mapping, and monitoring of conservation targets
- Incompatible recreation
- Invasive and native species management
- Inadequate/incompatible agriculture management
- Inadequate/incompatible forestry management
- Internal and community GCPEP communications and education
- Illegal trash dumps

After the strategies were selected, specific action items were identified that would accomplish the overall goals of the strategies. For each of the strategies, the partners prioritized the actions, which served as the basis for current and future GCPEP projects and activities.

As was mentioned earlier for the last "S," Success, early success can be shown with leadership and support, a strategic approach, and adequate funding. Early on, the GCPEP partnership concept appeared to be potentially successful. Since then, the partnership has proven effective in minimizing and eliminating critical threats due to the commitment of the partners to accomplish the top action priorities chosen in the Conservation Area Plan. A few of the top action project categories include prescribed fire, endangered species management, and land protection. They are described below. Projects can range from assisting individual partners at specific sites, to landscape-scale in scope, involving coordination with multiple partners.

Prescribed Fire

Longleaf-pine-dominated sandhills and flatwoods provide the matrix within which many other communities, such as seepage slopes and depression wetlands, are embedded. These embedded communities require the same prescribed fire treatments as surrounding sandhills and flatwoods. Others, such as baygalls, have a less frequent fire return interval, approximately every 50–100 years. The exceptional diversity of animals and plants in the GCPEP landscape is a result of frequent fire. For instance, the federally endangered red-cockaded woodpecker depends on fire-maintained longleaf pine sandhills and flatwoods for foraging within the understory. The federally threatened flatwoods salamander also depends on fire to maintain the necessary ecotone of the depression wetlands where they breed (Hardesty et al. 1999). According to Provencher et al. (2000), plant diversity on fire-maintained ecosystems in the GCPEP is very rich: as many as 45 plant species have been found in 400-m^2 plots and at least 293 species of plants have been identified within sandhills on Eglin Air Force Base lands. Understory species richness and cover have been positively correlated with insect species abundance and biomass. Fire-adapted understory plant species also play an important role in this ecosystem by carrying the fire that limits the invasion of competing hardwoods and sand pines.

Significant partnership support for fire management exists, as evidenced by: prescribed burning on public lands and cooperative GCPEP prescribed burns, annual smoke management meetings, completion of a peer-reviewed landscape-disturbance model, partner involvement in a fire council, and the start-up of an Ecosystem Support Team that will provide prescribed burning assistance across GCPEP lands.

A priority conservation objective identified in the GCPEP Conservation Area Plan is the reintroduction of natural fire regimes to protect key ecosystems, embedded communities, and species. The challenges that led to incompatible and inadequate fire management being identified as a "killer threat" included insufficient amount of area burned, insufficient return of fire intervals, and resistance to growing season burning due to public misconceptions.

Collaborative work at Eglin led to the development of an innovative landscape disturbance computer model that simulates management of longleaf pine habitats in modeled landscapes. The landscape disturbance model creates "movies" of expected landscape changes over time resulting from different management scenarios. The model identified the need to burn on a shorter return interval than previously planned. Eglin and the GCPEP are collaborating to continue the development of a spatially explicit model that uses GIS data layers to evaluate ecological condition of upland longleaf pine ecosystems, which will ultimately help to prioritize management actions across the landscape. Another effort is the development of a spatial model that will help prioritize limited prescribed fire resources to areas where fire is most needed.

Development pressures are intense across the GCPEP landscape and have led to increasing wildland–urban interface challenges such as lack of prescribed fire near urban areas due to public misconceptions and concerns about fires being conducted close to neighborhoods. The wildland–urban interface challenge was highlighted when in 1998 the wildfire season proved to be devastating in Florida: nearly 2300 wildfires burned almost 202,500 ha throughout the state, and more than 300 homes and 30 businesses were damaged. As a result, greater statewide emphasis has been placed on managing the wildland–urban interface. The Division of Forestry, along with the GCPEP has created several fire teams to be proactive in the prescribed fire management of these wildland–urban interface areas in order to decrease the high fuel loads.

Endangered Species Management

Partners are working together to improve habitat and recover populations of several rare and endangered species, including the flatwoods salamander, the Florida bog frog, the Gulf sturgeon, the Okaloosa darter (*Etheostoma okaloosae*), and the red-cockaded woodpecker. The red-cockaded woodpecker, a medium-sized woodpecker that inhabits open, mature pine or pine-oak woodlands, was federally listed as endangered under the Endangered Species Act due to dramatic declines that had occurred across their range. During the previous decade the population had also declined in the three main population centers within the GCPEP landscape: Eglin Air Force Base, Blackwater River State Forest, and Conecuh National Forest. These rare woodpeckers have often been labeled "indicators" of a healthy ecosystem. They depend upon southern pine forests managed well with prescribed fire, midstory management, and stand density control.

Several research studies suggest that RCW productivity is directly related to the diversity and quality of the understory plant–insect community. One study at the Savannah River Ecology Laboratory in South Carolina by Hanula and Franzreb (1998) observed that up to 70% of the prey captured by red-cockaded woodpeckers below the canopy was mainly from the soil/litter layer. In addition to healthy ground cover maintained by regular prescribed burning, the woodpeckers also require old pines for nest cavity construction. Very few old-growth pine trees remain, another limiting factor across the range of the RCW. RCW family groups also need large habitat areas, defending home ranges of 61–202 ha.

The local recovery effort for the RCW has been a GCPEP success story. Several of the partners have worked cooperatively to reverse the RCW's decline. Across the GCPEP landscape the RCW population is increasing due to a cooperative and intensive habitat improvement program ranging from increased lightning season prescribed burning, installation of cavity inserts, and supplementing the population with females from other southeastern population strongholds. In addition, partners share equipment and training opportunities.

The RCW population at Eglin Air Force Base has increased significantly over the last 10 years (Moranz and Hardesty 1998). In the late 1980s Eglin lost a military test mission due to a jeopardy opinion from the U.S. Fish and Wildlife Service. The main reason for the jeopardy opinion was the lack of information about the population of two endangered species: the RCW and the Okaloosa darter. This loss of a mission sparked the development of a management program that has produced the fastest growing large population of RCWs. Populations on Eglin Air Force Base lands have grown from 217 active clusters of cavity trees in 1994 to 308 active clusters currently.

A complete systematic survey of RCW habitat was reported in 1993 and a monitoring and banding program was established in 1992. Along with the continued survey and monitoring program, Eglin has developed an intensive management program, which has included constructing over 800 artificial cavities, translocating over 40 birds, conducting growing-season fires, and protecting cavity trees. Eglin has also completed the first landscape-level research program to determine the best combination of management techniques to increase the population.

At Blackwater River State Forest currently all RCWs are banded using color leg bands for identifying and monitoring the birds' activity. All tree clusters are surveyed for activity yearly. All nestlings and immigrating adults are color-banded yearly. During the 2001 breeding season, 26 of 27 clusters had successful nests. Of the 21 nests where chicks were banded, 42 nestlings were produced. Of this total, 19 nests fledged 34 young, with 13 males and 21 females. The artificial cavity insert program has also proven to be successful, with 67 out of 130 installed inserts currently occupied by RCWs. The population has also been augmented with 24 juvenile RCWs translocated from other populations, including Apalachicola National Forest and Eglin Air Force Base. These intensive management efforts have succeeded in stopping the decline of the RCW population on Blackwater. In a period of 4 years the RCW population has increased from 13 active groups and 6 single males, a total of 19 clusters in 1998, to 33 active groups and no single males, a total of 33 clusters.

Conecuh National Forest, in Covington and Escambia counties of Alabama, manages 22 active RCW clusters. Although a smaller population compared to that of its neighbors to the south, it has increased steadily from the 14 clusters that remained after Hurricane Opal in 1995. In keeping with the RCW Recovery Plan, the U.S. Forest Service maintains an inventory of both active and inactive nesting sites, monitors nesting activity, bands fledglings each spring, provides "recruitment" habitat by installing artificial cavities, and maintains existing habitat by prescribed burning approximately 10,000 ha each year. In addition, the agency is developing future habitat for the RCW by actively working to restore the native longleaf pine ecosystem.

Land Protection

Land protection was chosen as a high-priority strategy by the GCPEP Steering Committee due to the large number of inholdings, buffers, and connectors needed to protect the biological diversity of the GCPEP landscape over the long term. In addition, the partners recognized the benefit these lands would provide concerning prescribed burning, especially by reducing smoke management concerns and urban–wildland interface issues.

The Florida Forever Program is the state's blueprint for conservation of the unique natural resources and is the largest program of its kind in the United States. The program encompasses a wide range of goals, including: restoration of damaged environmental systems, water resource development and supply, increased public access, public lands management and maintenance, and increased protection of land by acquisition of conservation easements.

The Nature Conservancy, Florida Division of Forestry, and International Paper have worked closely on numerous Florida Forever projects, one of which is Yellow River Ravines, a 6500-ha project. The purchase of this important parcel would connect the two largest landholdings in GCPEP, Eglin Air Force Base and Blackwater River State Forest, and also includes a 1600-ha inholding in the state forest. The GCPEP Steering Committee has long recognized the importance of the Eglin–Blackwater connector parcel as a significant conservation land and as a buffer for Eglin Air Force Base. The property is a Stage 1 Priority Site identified in The Nature Conservancy's East Gulf Coastal Plain Core Team (1999). The Florida Game and Freshwater Fish Commission has also identified it as an important conservation area in the report "Closing the Gaps in Florida's Wildlife Habitat Conservation System" by Cox et al. (1994). The property includes important tributaries of the Yellow River that protect water quality and species diversity and which provide habitat for several rare species. Other important parcels in the project area buffer Blackwater River State Forest and Naval Air Station (NAS) Whiting Field.

The following are significant reasons to protect this and other GCPEP area lands long term:

1. Military Mission—Protecting the military mission is dependent upon ensuring adequate acreage for military training. Encroachment along the boundaries of a military base can have negative impacts on mission capacity through restrictions on low-level flights over developed areas or noise concerns from neighbors. The connector parcel is adjacent to an important Ranger training area along the Yellow River and an outlying field for NAS Whiting Field. Protecting this land would also provide long-term habitat for several rare species. Increasingly, as habitat around

bases is developed, more habitat demands fall upon the bases themselves. It then becomes more and more difficult to meet the military mission while supporting the rare species displaced from surrounding developed habitats.
2. Conservation Significance—The Yellow River Ravines land would serve as a critical wildlife corridor between Eglin Air Force Base and Blackwater River State Forest. Rare species include the Florida bog frog, flatwoods salamander, and Florida black bear. Natural communities of significance include steephead stream/slope systems and depression wetlands. The rare species and communities found on the connector parcel are also being managed for recovery on Eglin Air Force Base.
3. Water Quality/Quantity—Several creeks in the corridor, including Weaver, Garnier, and Julian Mill creeks, feed the Yellow River. This area serves as a water recharge area for Santa Rosa County. Planning for water recharge is important in an area that is dependent upon water from the shallow sand and gravel aquifer. Protecting water quality and quantity is important for biodiversity protection, providing water supply, and protecting military training that is dependent upon adequate water flow in the Yellow River watershed.
4. Recreation and Hunting—As the region continues to grow and develop, recreational space will become more limited and demands on remaining space will increase. The Yellow River Ravines connector parcel, located in Santa Rosa County, could decrease future recreational demands on Eglin and provide increased recreational opportunities for area residents and visitors alike.

The Yellow River Ravines project was approved by the State of Florida as an "A-ranked" Florida Forever Project, assuring funding for the project. Additional projects, which protect and buffer important conservation lands, were also approved by the State of Florida including the Northwest Florida Greenway Project. This 100-mile-long and 5- to 10-mile-wide conservation corridor will link Eglin Air Force Base to Apalachicola National Forest and the Gulf of Mexico. All of these projects were strongly supported by GCPEP, other state and federal agencies, the local county commission, environmental and recreational organizations, and the general public.

GCPEP has been instrumental in moving land and water management and land protection from being controversial community issues, historically, to the present, being more strongly supported issues. This has, in part, been due to a tremendous education effort aimed at community leaders and politicians. The GCPEP staff has served as an important communication and support link between the partnership and the surrounding communities.

Conclusion

Large-scale restoration of the longleaf pine ecosystem may be more effective if public and private landowners choose to work together in landscape-level partnerships. When partners who restore and manage longleaf pine use science-based planning as a common goal, partnerships can succeed though individual partner missions may vary widely. Completion of a Conservation Area Plan allows for more effective and efficient management and restoration across large landscape areas. Given limited funding for personnel, equipment, and projects, this method of planning is recognized by many to leverage strategies to accomplish short- and long-range goals.

The success of any partnership depends on respect and cooperation and may operate more efficiently with a staff dedicated to the effort. More may be accomplished when combining expertise and resources to effectively manage individual lands, while at the same time meeting the challenges of sustaining larger ecosystems. By doing this, GCPEP serves as an example of how organizations can work together to achieve common and important goals such as restoring and maintaining the longleaf pine ecosystem. This chapter may provide a "blueprint" for partnerships to set conservation and restoration objectives and priorities for both the individual partners and the collective partnership.

Acknowledgments

We extend gratitude to each of the GCPEP partners' Steering Committee Representatives and to their organizations for their commitment to, and support of, the GCPEP. The time dedicated by each of the partners is vital to the strength and success of the partnership. We are also grateful to the many individuals, the governmental and nongovernmental agencies, and the other essential organizations, which contribute generous amounts of time, expertise, and resources to make the GCPEP an effective conservation tool.

References

Compton, V. 2002. Gulf coastal plain ecosystem partnership—Achieving results through cooperation. *Florida Forests* 6 Issue 3:22–25.

Compton, V., Bachant, J., Hiers, S., and Penniman, P. 2002a. Gulf Coastal Plain Ecosystem Partnership: Site conservation plan. Jay, FL: The Nature Conservancy.

Compton, V., Bachant, J., and Penniman, P. 2002b. Gulf Coastal Plain Ecosystem Partnership: Steering committee manuals. Jay, FL: The Nature Conservancy.

Cox, J., Kautz, R., MacLaughlin, M., and Gilbert, T. 1994. Closing the gaps in Florida's wildlife habitat conservation system. Florida Game and Fresh Water Fish Commission, Tallahassee, FL.

East Gulf Coastal Plain Core Team. 1999. East Gulf Coastal Plain Ecoregional Plan. Chapel Hill, NC: The Nature Conservancy.

Hanula, J.L., and Franzreb, K.E. 1998. Source, distribution and abundance of macroarthropods on the bark of longleaf pine: Potential prey of the red-cockaded woodpecker. *For Ecol Manage* 102:89–102.

Hardesty, J.L., and Moranz, R.A. 1999. Longleaf pine ecosystem restoration in northwest Florida sandhills: Issues and recommendations. Gainesville, FL: The Nature Conservancy.

Hardesty, J.L., Moranz, R.A., Woodward, S., and Compton, V. 1999. The Gulf Coastal Plain Ecosystem Partnership: An assessment of conservation opportunities. Gainesville, FL: The Nature Conservancy.

Hiers, S., Bachant, J., Compton, V., and Penniman, P. 2002. The Gulf Coastal Plain Ecosystem Partnership: Freshwater ecosystem demonstration. Jay, FL: The Nature Conservancy.

Low, G. 1999. Landscape-scale, community-based conservation: A practitioner's handbook. Arlington, VA: The Nature Conservancy.

Moranz, R.A., and Hardesty, J.L. 1998. Adaptive management of red-cockaded woodpeckers in northwest Florida: Progress and perspectives. Gainesville, FL: The Nature Conservancy.

Provencher, L., Litt, A., Gordon, D., and Tanner, G. 2000. Reference condition variability: Product to Eglin Air Force Base, Natural Resources Division. Gainesville, FL: The Nature Conservancy.

The Nature Conservancy. 1998. An approach for conserving biodiversity at portfolio sites: Site conservation planning. Arlington, VA: The Nature Conservancy.

The Nature Conservancy. 2000. The five-s framework for site conservation. Arlington, VA: The Nature Conservancy.

The Nature Conservancy. 2005. gulf coastal plain ecosystem partnership. Altamonte Springs, FL: The Nature Conservancy.

Index

2,4-D, 307, 309
3-in-1 plow, 273

Accord™, 284
Adaptive management, 8, 297, 320, 346, 353, 360, 423
Age classes, 124, 220, 221, 228, 242, 243, 246, 248, 278, 279, 281, 291
Alabama, 3, 9, 10, 23, 25, 26, 28, 31, 36, 54, 55, 78–80, 90, 95, 97, 102, 171, 172, 181, 199, 200, 251, 254, 257, 271, 277, 302, 328, 359, 364, 365, 377, 379, 380, 383, 392, 413, 415, 419, 427
Allelopathy, 136, 138
American bison, 179, 185, 198
Amphibians, 158–161, 166–169, 174, 175, 179, 183, 189, 193, 194, 196–199, 337, 338, 358, 362–364, 383, 387
  resident amphibians of longleaf pine savannas, 161
Andropogon mohrii, 68
Andropogon virginicus, 68, 73
Angelina National Forest, 198
Aquifer impacts, 393
Arboreal, 158, 166, 172, 174, 178, 191, 194
Aristida beyrichiana, 4, 15, 57, 60–68, 72–76, 78, 81–88, 135, 217, 307, 310, 315, 330
Arsenal™, 283, 284
Artificial regeneration, 111, 135, 272, 279, 289
  direct seeding, 111

Augmentation, 6, 336, 338
  definition, 336
Autecology, 4, 231

Bare-root nursery seedlings, 112
Bark beetles, 43, 153
Basal area, 104, 105, 109, 117, 120, 123, 141–150, 220, 222, 225–238, 243, 246, 247, 253–256, 262–264, 272, 277, 279, 282, 286, 304, 308
BD-q, 120–124, 225, 226, 230, 233, 235, 248, 279, 282
Bedding, 193, 196, 273, 274, 285
Beech, 11, 16
Ben Cone, 407
Biodiversity, 4–7, 51, 52, 190, 200, 300, 305, 306, 311, 339, 340, 369, 379, 380, 383, 384, 388–395, 403–407, 414, 418–422, 428
Biogeography, 381
Biological legacy, 302
Birds, 32, 33, 38, 109, 112, 118, 158, 159, 171–177, 179, 183–191, 197–199, 271, 272, 280, 289, 331, 336–365, 405, 418, 423, 426
  characteristic birds of longleaf pine savannas, 176
Bison bison, 179, 185, 198
Black bear, 180, 181, 184, 338, 363, 386, 397, 415, 419, 423, 428
Black Water River State Forest, 418, 419, 426–428
Blackgum, 70, 161
Blackjack oak, 71, 284, 380
Bluejack oak, 71, 284

Bluestem, 4, 68, 71, 98, 135, 138, 141, 298
  Broomsedge bluestem, 68
  Mohr's bluestem, 68
Bobwhite quail, 124, 193, 199, 217, 231, 237, 246, 271, 300, 405
Bottlebrush, 105
Broomsedge bluestem, 68
Brown spot needle blight, 276, 326
Bulldozing, 273
Burning, 15, 20, 34, 43, 44, 106, 110, 112, 114, 118, 122, 125, 139, 179, 187, 194, 196, 200, 233, 234, 243, 248, 274, 276, 280, 286, 287, 289, 290, 300, 303–319, 326–328, 330–332, 339, 346, 348, 357, 369, 383, 386, 395, 418, 419, 425–427

Cabbage palmetto, 71
Candle stage, 328
Canis rufus floridanus, 198
Canopy gaps, 98, 110, 111, 115, 120–123, 125–127, 139, 289
Carbon credits, 408, 409
Carolina Sandhills National Wildlife Refuge, 304, 316, 317, 357
Carrying capacity, 31–33
Carry-log, 27, 38
Carya tomentosa, 16
Castanea dentata, 9
Catastrophic events, 72, 340, 343
Catena, 384
Catkins, 99, 100
Cattle, 3, 18, 31–34, 107, 109, 186, 337, 408
Chapman oak, 70
Charcoal root rot, 283

Chestnut, 9, 71
Chestnut oak, 71
Chopper, 285, 303, 307
Classification, 51–91, 226, 367, 396
　ecological, 51–91
　　national vegetation, 52, 58, 66, 90, 396
Clayey and rocky uplands, 59
Clearcutting, 95, 115, 116, 124–219, 227, 290, 305, 381
　alternative-strip, 115
　progressive-strip, 115
Climate change, 187, 188, 388, 393, 408
Climax, 194, 303–306, 310, 337
Coastal Plain, 3–15, 21, 26, 31, 44, 51–56, 59–67, 72–90, 97, 101, 107, 108, 157–163, 166, 169, 171–176, 179, 181, 185, 187, 188, 190, 194, 195, 197, 198, 200, 218, 223, 228, 229, 238, 242, 246, 252, 256, 258, 298, 315, 330, 343, 359, 364, 365, 377, 380, 403, 413, 414, 415, 427
Cogon grass, 310, 317
Colinus virginianus, 124, 176, 217
Columbian mammoth, 185, 187
Common names, 68, 72, 91, 160
Community succession, 6, 116
Competition, 5, 32, 46, 95, 98–124, 135–154, 195, 221, 228, 233, 236, 238, 246, 251, 252, 261, 272–274, 276–278, 281, 283, 285, 286, 288, 303, 309, 311, 316, 319, 320, 331, 332, 354, 381, 405
　apparent, 136–138, 151
　asymmetrical, 136, 142
Composition, 4, 6, 7, 14, 46, 51, 52, 57, 59, 67, 80–87, 97, 115–117, 138–141, 158, 159, 178, 188, 221, 298, 299, 300, 303, 305, 306, 310–313, 327, 330, 331, 350, 355, 369, 379, 380, 388, 421
　altering species, 310
Cone production, 95, 100, 101, 108, 118, 232
Conecuh National Forest, 419, 426, 427
Connectivity, 341, 378, 380, 387–390, 396, 419
Conservation, 3, 4, 6, 23, 52, 58, 90, 115, 125, 127, 153, 200, 242, 245, 251, 305, 306, 311, 313, 335, 336, 338–340, 342, 344, 345, 359, 364, 365, 368, 369, 377–379, 388–396, 403–429
　area planning, 417, 421
　landscape-scale, 414
　partnerships in, 414
　strategies, 58, 335–337
Conservation biology, 345, 378, 389
Conservation Reserve Program (CRP), 153, 406, 410, 411
Containerized nursery seedlings, 113
Cornus florida, 62, 69, 77, 380
Cotton mouse, 165, 180
Cotyledons, 102
Cougar, 180, 182, 184, 397
Cronartium quercuum, 99
Crossett Experimental Forest, 223, 224, 226
Cruise, 288
Ctenium aromaticum, 57, 64, 65, 66, 69, 81, 85, 87, 88, 89, 315
Cutting cycle, 121–124, 219–223, 227–238, 277, 280

Darling National Wildlife Refuge, 198
Darlington oak, 71
DBH class, 224, 226
Demographic factors, 352, 355
Demographic isolation, 424
Dendroctonus spp., 153
Density management diagram, 263
Department of Defense, 391, 417
DeSoto, 17, 30
Dionaea muscipula, 15, 69, 81
Direct seeding, 111, 112, 272, 280, 305, 310, 311, 316, 318, 320
Disking, 273, 317
Dispersal, 95, 100, 102, 108, 112, 115, 119, 165, 174, 180, 189, 302, 304, 313, 331, 336, 340–346, 349, 354, 358, 360–362, 385, 387, 388, 396, 405
　seed, 95, 100, 108, 116, 186
Disturbance, 7, 95–99, 110, 119–127, 135, 138, 196, 199, 218–221, 242, 273–275, 284, 299, 304, 307, 377, 380, 383, 425
　natural, 124–127
　regimes, 95, 125, 135, 377
　silviculture that mimics, 124–127
Disturbance dynamics, 95, 97, 125
Diversity, 4, 5, 37, 38, 51, 52, 56, 57, 67, 73, 76, 80, 81, 84, 87, 90, 97–99, 122, 125, 126, 136, 142, 144, 145, 154, 157–159, 161, 163, 165–201, 218, 242–246, 252, 297–380, 383, 388, 390, 397, 405, 406, 411, 425–427

biodiversity, 4–7, 51, 52, 190, 200, 300, 306, 311, 339, 340, 379, 388, 391, 392, 394, 395, 403, 404, 418–422
　faunal, 5, 157–201, 337
　plant, 37, 136, 383
　understory, 37, 327
　vertebrate, 5
Donor population, 336, 342, 344, 347–349, 353, 357–359, 361, 364
Drum chopping, 273, 303, 306–308
Dwarf live oak, 71

Eastern Indigo snake, 168, 171, 185, 199, 338, 363, 365, 383, 387
Ecological functions, 125, 390
Economic benefit, 395
Economics and policy, 403
Ecoregions, 53–55, 59, 338, 340, 348, 392
　Atlantic Coastal Plain, 53
　Eastern Gulf Coastal Plain, 55
　Fall-line Sandhills, 53, 54
　of the longleaf ecosystem, 53
　Piedmont and Montane Uplands, 55
　Southern Coastal Plain, 54, 55
　West Gulf Coastal Plain, 55
Ecosystem cluster, 384
Ecotones, 195, 377, 379, 383, 384, 388–390, 397
Eglin Air Force Base, 73, 75, 306, 389, 391, 392, 404, 417–419, 425–428
Endangered Species Act, 6, 335, 338, 396, 404, 406, 426
Endangered species management, 424, 426
Endemic species, 90, 159, 162, 169, 298
Entisol, 56
Environmental benefits index (EBI), 410
Environmental stochasticity, 339–341
EPA Southeastern Ecological Framework, 392, 393
Escambia Experimental Forest, 102, 231, 234, 264, 277
European settlers, 13, 16
Evosystem, 377, 378
Exotic plants, 124

Facilitation, 135–153, 417
Fagus grandifolia, 16
Faunal diversity, 5, 157, 159, 161, 163–201, 337
Felling, 236, 256, 306, 307

Fence laws, 33
Feral hogs, 3, 31, 32, 96, 365
Feral livestock, 29
Fire, 4, 6, 10–16, 19, 23, 29–38, 44, 46, 52–313, 319, 320, 326–330, 337, 339, 350, 365, 377, 379–381, 383–391, 419, 420, 422, 424–426
  dormant season, 152, 153, 286, 287, 290, 303, 307, 308
  effects of, 13, 95, 105, 106, 159, 194, 196, 387
  exclusion, 4, 11, 32, 34–38, 97, 108, 195, 290, 300–309, 365
  frequency, 14, 15, 37, 52, 84, 158, 298, 300–304, 326, 380
  growing season, 13, 35, 196, 285–288, 290, 291, 303
  history, 178, 272, 288, 290–292, 303, 312
  prescribed, 102, 108, 109, 112, 118–124, 135, 141, 142, 153, 154, 196, 199, 200, 217, 231, 232, 234, 235, 236, 243, 246, 272, 277, 278, 280, 281, 285, 289, 305, 306, 307, 310, 319, 320, 327, 350, 386–391, 395, 419, 420, 424–426
  regime, 10
  suppression, 13, 29, 34, 37, 52, 59, 73, 80, 86, 97, 178, 196, 284, 300, 307, 337, 339, 365, 381, 385, 389, 391
Fire ant, 199
Flatwoods, 3, 15, 56–59, 64, 73, 75, 81–86, 97, 103, 104, 123, 157–166, 169, 171–174, 178–181, 189, 193, 196–199, 298, 307, 308, 310, 313, 327, 328, 337, 338, 362–364, 377, 379, 383, 387, 392, 418–428
Flatwoods salamander, 161–163, 189, 193, 196, 198, 338, 362, 363, 383, 387, 418–428
Florida, 3, 4, 13, 16, 17, 23, 25, 26, 28, 33, 36, 43, 46, 47, 51, 54–58, 62, 68–78, 81–84, 87–90, 105, 110, 114, 138, 157, 159, 163, 167–218, 242, 246, 251, 254, 298, 307, 310, 313, 317–320, 335, 338, 339, 342, 344–346, 356, 359, 361–369, 377–397, 403–409, 413–428
Florida Department of Environmental Protection, 415, 418
Florida Division of Forestry, 415, 418, 427
Florida Ecological Network, 392

Florida Fish and Wildlife Conservation Commission, 90
Florida mouse, 180, 181, 199
Florida red wolf, 198
Flowering dogwood, 69, 302, 380
Forest certification, 392
Forest structure, 117, 123, 242, 337, 348, 350
Form class, 258, 259
Fort Benning, 357
Fort Bragg, 80, 87, 357, 391
Fort Stewart, 74, 316, 357
Fossil record, 179, 183, 186
Fossorial, 160, 169–173, 181, 191–193, 198
Fox squirrel, 179–181, 199, 271, 338, 363, 383, 386–389, 391, 396, 397
Frogs, 5, 159–167, 172, 173, 190, 191, 193, 196, 198, 362–364
  list of frogs of longleaf pine savanna, 161
Fuel, 15, 27, 29, 59, 98, 109, 110, 114, 140, 194, 248, 277, 281–291, 297, 300, 307, 312, 326–331, 426
Functional landscape, 384, 390, 414

Gallberry, 69, 163, 166, 174, 195
Gap-phase regeneration, 110, 111, 120
Genetic considerations, 312, 341
Georgia, 4, 23–28, 32, 36, 37, 53–58, 67, 68, 70, 76–87, 95, 97, 114, 124, 157, 159, 171, 178, 181, 198–200, 237, 242, 246, 254, 257, 260, 298, 303, 310, 313, 316, 326, 328, 330, 332, 359, 362, 365, 377, 379, 380, 392
Georgia oak, 70
Germination, 4, 95, 102, 103, 106–108, 111–114, 140, 146, 153, 175, 186, 222, 272, 278, 280, 291, 311, 313, 315–319, 381
  factors affecting, 315
Girdling, 18, 153, 306, 307
Glyphosate, 142, 152, 275, 280, 284, 307, 309
Gopher frog, 161, 164–167, 189, 191, 338, 362–364, 387
Gopher tortoise, 5, 165, 168, 171–175, 181, 185, 186, 191–199, 246, 271, 331, 332, 338, 358, 363–365, 383, 387, 391, 418
Gopherus polyphemus, 168, 175, 271, 338, 358, 383, 418
Graminoid, 75, 309

Grass, 4, 5, 16, 29, 32, 38, 51, 57, 59, 69–77, 86, 90, 99, 102–110, 114, 115, 118, 135, 138–141, 146–151, 161, 165, 166, 174, 178, 187, 231, 274, 276, 277, 280, 288, 304, 310, 315, 317, 330, 331, 405
  Bahia, 274, 280, 305
  cogon grass, 310
  Indiangrass, 71
  lopsided Indiangrass, 71
  switchgrass, 70
  toothache grass, 69
  warm season, 261
  wiregrass, 4, 15, 34, 38, 57, 59, 67, 68, 73, 74, 76, 78, 80, 98, 135, 138, 141, 165, 166, 173, 174, 195, 200, 217, 284, 307–309, 313–320, 330–332, 380, 381, 396
Grass stage, 5, 29, 32, 99, 103–110, 114, 115, 118, 139, 146, 150, 151, 231, 276, 277, 288, 405
Greenwood plantation, 244, 245
Ground layer restoration, 297
Ground sloth, 184, 187
Group selection, 95, 111, 120–127, 141, 201, 221, 222, 225, 227, 230–238, 279, 280, 291
Growth and yield, 5, 124, 235, 236, 251–264
  of natural stands, 5, 253–256
  planted stands, 5, 256–259
Gulf Coastal Plain Ecosystem Partnership, 6, 413–415

Habitat conditions, 126, 231, 306, 330, 346, 348, 349
Habitat Conservation Plans (HCPs), 406–408
Habitat fragmentation, 197, 337, 344, 387, 393, 395
Haliaeetus leucocephalus, 338, 366
Heartwood, 18, 43, 48, 191, 192
Herbaceous understory, 109, 271, 285, 289, 291
Hexazinone, 142, 143, 152, 275, 284–286, 307–309
High-grading, 124
Hunting, 25, 124, 184, 187, 242, 245, 248, 330, 365, 395, 405, 407, 410, 411, 428
Hurricane damage, 405
Hybridization, 99
Hypocotyl, 102

Ilex coriacea, 64, 69
Ilex glabra, 64, 65, 69, 77, 83, 84, 90, 163

Ilex vomitoria, 63, 64, 66, 69, 84
Imazapyr, 142, 152, 275, 280, 284, 285
Imperata cylindrica, 310, 317
Indiangrass, 71
Indians, 14, 15, 18, 24, 25, 31
Indicators, 5, 7, 8, 72, 74, 178, 422, 426
Inkberry, 69
Interactions, 5, 98, 135–151, 300, 309, 336, 351, 367, 377–380, 383–385, 388, 390, 414, 420
 case studies, 142
 overstory, 137
 understory, 135
Interference, 136
International Paper, 415, 419, 427
Invasion, 34, 97, 169, 195, 199, 279, 425
Inventory, 13, 58, 76, 122, 124, 157, 223–226, 236, 237, 246, 275, 288, 289, 344, 423, 427
Ips spp., 153

Kalmia latifolia, 63, 69, 79
Keystone species, 5, 98, 135, 138, 175, 178, 198, 338–340
Kleptoparasites, 348–350, 352
Kudzu, 124

Laddering of fuels, 287
Land conservation, 394
 funding for, 394
Landowner Incentive Program (LIP), 406, 410
Landowner programs, 392
Landscape, 6–9, 13–18, 29–38, 51, 54–56, 76–81, 84, 86, 89, 96, 97, 115, 119, 120, 157, 160, 163, 166, 175, 189, 195, 197, 200, 242, 254, 271, 278, 299, 300, 312, 330–332, 337, 339, 367, 377–397, 403, 413–428
Landscape ecology, 6, 378, 379, 381, 388–390
 conservation strategies, 389–391
Large gallberry, 69
Liberation cutting, 228, 236
Lightwood, 18–20, 27, 44
Liquidambar styraciflua, 65, 70, 141, 163, 284, 380
Lizard, 159, 167–174, 190–194, 364
 characteristic lizards of longleaf pine savannas, 168
Loblolly pine, 16, 18, 29, 34–37, 70, 96, 99, 120, 123, 139, 140, 144, 153, 166, 169, 170, 176, 179, 194, 218, 228, 232, 247, 262, 284, 285, 291, 339, 380, 404, 405
Logistic factors, 353
Longleaf alliance, 275
Lopsided Indiangrass, 71
Louisiana, 22, 25, 26, 28, 30, 36, 51, 55, 75, 78, 79, 89, 101, 157, 169, 172, 187, 199, 232, 251, 257, 259, 261, 262, 328, 338, 359, 363, 364, 366, 377

Magnolia grandiflora, 70, 84
Magnolia virginiana, 65, 70
Mammals, 158, 159, 167, 172, 174, 179–191, 197, 199, 280, 289, 331, 336, 338, 358, 362, 363, 365, 366, 405
 characteristic mammals of longleaf pine savannas, 180
Marking rules, 230
Mastodon, 186
Mesic sites, 57, 74, 77, 110, 111, 284, 285, 307, 309
Metapopulation dynamics, 189, 197
Metapopulation theory, 344, 345
Microhabitat, 163, 165, 192, 350, 363
Midstory, 38, 108, 115, 136, 141, 178, 221, 237, 239, 243, 246–248, 284, 307, 308, 326–328, 331, 332, 339, 350, 404, 426
Military mission, 427, 428
Mississippi, 15, 17, 25, 28, 36, 52, 55–58, 73, 75–81, 84, 89, 159, 165, 172, 173, 193, 200, 254, 257, 261, 262, 309, 328, 338, 359, 362–365, 392, 419
Mockernut hickory, 16, 302, 383
Mohr's bluestem, 68
Monitoring, 7, 8, 108, 109, 200, 243, 278, 331, 336, 337, 346, 347, 352, 353, 357, 361, 362, 365, 368, 369, 388, 419, 420, 423, 424, 426
Mosaic, 13, 15, 16, 97, 115, 120, 122, 200, 316, 379, 386, 418
Mountain laurel, 69, 79
Mutualism, 138
Mycosphaerella dearnessii, 99, 276
Myrtle oak, 71

National Forests in Alabama, 415, 419
Native Americans, 13, 15
Natural Heritage Program, 58, 91, 200
Natural regeneration, 95, 97, 101, 108, 109, 111, 115–120, 126, 140, 217, 218, 227, 228, 231–234, 238, 273, 277, 288, 289, 389, 405
Nature conservancy, 91, 124, 200, 310, 320, 338, 391, 413, 415, 419, 421, 427
Naval stores industry, 3, 18, 19, 21, 43, 44, 46, 193, 339
Nokuse plantation, 391, 415, 419
North Carolina, 3, 4, 9, 13–37, 43, 51–53, 57, 58, 67, 80, 81, 85, 88, 90, 101, 139, 163, 169, 171, 173, 199, 200, 301, 311, 319, 343, 344, 359, 367, 391, 392, 407
Northern Bobwhite Quail, 193, 199, 237, 246
Northwest Florida Water Management District, 415, 418, 419
Nurse crop, 285
Nyssa sylvatica, 70, 81, 161

Oak, 15, 16, 18, 22, 34, 57, 59, 67, 70–73, 76, 83, 84, 90, 105, 139, 141, 159, 161, 163, 165, 166, 169–175, 181, 187, 193, 284, 290, 301, 307, 308, 327, 330, 337, 364, 365, 380, 383, 384, 386, 391, 418, 426
Odocoileus virginianus, 179, 180, 185
Old growth, 151, 253, 300, 389
Oust^{TM}, 283
Overstory, 4, 6, 59, 98, 104, 105, 107, 109, 111, 115–126, 135–154, 220, 229, 231–236, 246, 248, 271–292, 331, 332, 339, 380
 common names, 68–72
 scientific names, 68–72

Panicum virgatum, 62–66, 70, 76, 77, 80, 90
Partial restoration, 300
Pasture, 3, 11, 25, 26, 32, 33, 36, 193, 274, 280, 283, 305, 316, 317, 405, 408, 409
Payne's Prairie, 198
Pensacola, 17, 418
Peromyscus polionotus, 165, 193
Physiognomy, 51, 59, 337, 380, 384
Picloram, 275
Picoides borealis, 176, 178, 217, 335, 338, 386, 404, 415
Pin oak, 71
Pine straw, 34, 271, 301, 307, 312, 316, 317, 319, 405, 406, 409–411
Pinus echinata, 16, 63, 70, 76, 80, 99, 286, 339, 380

Index   435

Pinus elliottii, 64, 65, 70, 82, 87, 287, 339, 404
Pinus serotina, 16, 64–66, 70, 81
Pinus sondereggeri, 99
Pinus taeda, 60, 62, 66, 70, 96, 139, 284, 339, 380, 404
Pinus virginiana, 70, 81, 343
Pinus, 3, 15, 16, 43, 48, 60–67, 70, 74, 76, 79, 80–87, 95, 96, 99, 135, 138–140, 150, 157, 163, 176, 177, 186, 217, 246, 251, 284, 286, 287, 337, 339, 343, 377, 379, 380, 383, 403, 404
Pioneer species, 4, 285
Pitch kettles, 18
Plant diversity, 379, 383, 425
Plantation, 5, 11, 13, 17, 27, 36, 119, 135, 138, 139, 144, 154, 163, 165, 193, 196, 197, 237, 244, 245, 247, 257, 260, 264, 300, 301, 303, 304, 309, 310, 391, 411, 415, 419
  first pine, 36
  Greenwood, 244, 245
  Nokuse, 391, 415, 419
  pine, 4, 11, 13, 36, 136, 139, 142, 153, 154, 163, 165, 169, 191, 196, 197, 262, 263, 300, 304, 305, 309, 310, 391, 411
  Red Hills, 248
  restoration, 300
Planting stock, 114, 233
Pleistocene (Rancholabrean) vertebrates, 184
Pocket gopher, 107, 171, 173, 180–182, 191, 193, 198–200, 383
Policy, 6, 355, 390, 403, 405, 406, 409, 410
Policy options, 406
Pollination, 95, 99, 100, 312
Pond cypress, 161
Pond pine, 11, 16, 70
Population, 17, 31, 95, 117, 126, 159, 167, 171, 172, 174, 182, 188, 189, 193, 194, 197–200, 230, 245, 246, 289, 305, 311, 312, 320, 336, 339–368, 396, 403, 407, 418, 419, 423–427
  donor population, 336, 342, 344, 347–349, 353, 357–359, 361, 364
  expansion, 351, 352
  founder populations, 331, 332, 355
  metapopulation, 189, 197, 344–346, 351, 361, 388, 389, 396, 397, 416
  scale, 346, 348, 360, 361

Post oak, 16, 71, 139, 284, 327, 380
Postplanting management, 286
Prairies, 15, 79, 83, 84, 90, 158, 173, 179, 183, 187, 303, 418
Preparatory cut, 117–119, 289
Prescribed burning, 110, 114, 118, 122, 125, 139, 179, 194, 200, 233, 248, 306, 307, 326, 330, 331, 332, 339, 346, 348, 357, 369, 395, 418, 419, 425–427
  negative impacts, 328
  precautions, 328
  role of, 326
  schematic, 327
  techniques, 326
Prescribed fire, 102, 108, 109, 112, 118, 120, 122, 124, 135, 141, 142, 153, 154, 196, 199, 200, 217, 231–236, 243, 246, 272, 277, 278, 280, 281, 285, 289, 305–307, 310, 319, 320, 327, 350, 386–391, 395, 419, 420, 424–426
  dormant season, 286
  growing season, 118
Pyrogenic, 141, 194, 337

Q factor, 123
Quercus, 16, 57–65, 67, 70–84, 89, 139, 141, 159, 284, 285, 290, 307, 326, 380
Quercus chapmanii, 61, 64, 67, 70, 75, 84
Quercus coccinea, 63, 70, 81
Quercus geminata, 57, 73, 84
Quercus georgiana, 63, 70
Quercus hemisphaerica, 60, 61, 71, 73
Quercus incana, 16, 59–62, 72–76
Quercus inopina, 71
Quercus laevis, 16, 57, 59–62, 67, 71, 73–76, 159, 284, 290, 307, 326
Quercus margarettiae, 60, 61, 71, 77
Quercus marilandica, 16, 61–63, 71, 78–80, 139, 284, 380
Quercus minima, 61, 64, 71, 75, 83
Quercus myrtifolia, 60, 64, 71
Quercus palustris, 63, 71
Quercus prinus, 63, 71, 81
Quercus pumila, 61–65, 71, 73, 76, 82, 83, 89
Quercus stellata, 16, 61, 62, 71, 77, 139

Radicle, 102, 103
Recreation, 200, 395, 418–428
Red Hills plantations, 248
Red-cockaded woodpecker, 5, 6, 176, 178, 185, 195–201, 217, 218, 231, 235, 237, 238, 245, 246, 289, 306, 330, 335, 336, 338–369, 386–391, 397, 404, 415, 418, 419, 423–426
  demographic, 340–344
  diseases and parasites, 358, 359
  dispersals, 344, 345
  geographic variation, 359
  metapopulation theory, 344, 345
  sociobiology, 343, 344
  translocation, 339–361
Reference communities, 5, 38
Reference information, 299
Reference sites, 299
Reference systems, 6
Regeneration, 4, 29, 30, 32, 35, 95–127, 135, 140, 141, 153, 178, 187, 195, 217–248, 272, 273, 277–281, 284, 286–305, 332, 389, 405
  artificial, 111, 135, 272, 279, 289
  ecology, 95–127
  environment, 95, 98
  natural, 95, 97, 101, 108, 109, 111, 115, 116, 118, 120, 126, 140, 217, 218, 227, 228, 231–234, 238, 273, 277, 288, 289, 389, 405
  problems, 95
  requirements, 108
Regional conservation planning, 394
Regulation, 5, 120–124, 222–227, 230–238, 407, 411
  area-based, 227
  stand structure, 225–227
  volume, 223–225
Rehabilitation, 228, 390
  of understocked conditions, 228
Reineke stand density index, 262, 263
Reintroduction, 6, 37, 38, 285, 287, 328, 332, 335–369, 425
  definition of, 336
Release method, 337, 349, 355, 356, 361
Release treatments, 219, 220, 222, 276
Removal cut, 118, 119, 235
Reproduction method, 4, 5, 95, 115, 120, 124–127
  alternative strip, 115
  clearcutting method, 115, 116
  progressive strip, 118
  seed-tree, 116
  shelterwood, 116, 117
Reproductive biology, 95, 99, 231, 320
Reptiles, 158, 159, 167–169, 179, 183, 189, 191–194, 197–199, 271, 337, 338, 358, 362–364

Reptiles (*cont.*)
  characteristic reptiles resident in longleaf pine savannas, 168
Resettlement Virgin longleaf pine, 10, 24, 28, 193, 382
Resilience, 6, 343
Restoration, 3–7, 37, 38, 52, 58, 90, 95, 97, 99, 124, 136, 139, 151, 153, 269–429
  economics of, 404
  ground layer, 297–299, 306
  landscape approach, 6
  of overstory, 271–292
  of understory, 6, 153, 275, 284
  partial, 300
  policy aspects of, 6
  socioeconomics of, 6
  strategies, 309
  thresholds, 6
  trajectories, 6, 7
Restoration ecology, 6, 378
Reverse J, 223, 237, 239, 246, 278, 279
Riparian strips, 384, 385
Road networks, 395
Rosin yards, 24
Rotation age, 219, 227, 405, 409, 410
Ruderal species, 304, 305
Runner oak, 57, 71, 159
Running oak, 71, 83
Rust, 99

Sabal palmetto, 58, 65, 71, 87
Salamander, 5, 159–163, 166, 189, 193–198, 338, 362, 363, 383, 387, 418, 419, 423–428
  dwarf, 163, 195
  list of salamanders of longleaf pine savanna, 161
    Mabee's, 161, 163
    tiger, 161, 163, 189
Sand laurel oak, 71
Sand live oak, 70
Sandhill oak, 71
Sandhills, 3, 15, 19, 53–68, 73–76, 79, 80, 86, 87, 97, 105, 107, 110, 157–175, 179, 181, 182, 190, 198, 199, 298, 301, 304, 307–309, 313, 316, 317, 326, 328, 343, 344, 357, 364, 365, 377, 379, 383, 386, 390–392, 418, 425
Savanna, 6, 10, 11, 16, 18, 34, 56, 79, 81, 84–90, 159–201, 316, 330–332
Savannah River Site, 136, 142, 330, 344
Saw palmetto, 58, 71, 122, 124, 166, 174, 307, 328, 386, 387

Saw timber, 308
Scalping, 114, 274, 281, 283, 315
Scansorial, 158
Scarlet oak, 70
Scientific names, 68, 160
Sciurus niger, 179, 180, 338, 383, 391
Scottish Whiskey stills, 46
Scrub oaks, 16, 59, 75, 139, 174, 179, 284, 285, 326, 327
Seed, 4, 22, 29, 32, 33, 95, 96–102, 108–113, 116–125, 141, 146–154, 179, 186, 219, 221, 222, 227, 228, 231–235, 238, 252, 272, 273, 277–282, 285–291, 298, 302, 304, 311–321, 331, 332, 381, 405
  bed, 277, 280, 304
  cleaning, 313, 314, 315
  collection, 311, 313, 314
  dispersal, 95, 100, 108, 116, 186
  germination, 315
  handling, 313
  harvest, 313, 314, 315
  production, 98, 101, 102, 108, 141, 146, 148, 149, 154, 232, 298, 312
  storage, 313, 314, 315
Seed cut, 117–119, 231, 234
Seed tree, 101, 108, 153, 219, 234, 235, 252, 277, 288–291
Seedling, 111, 112, 135, 272, 280, 290, 305, 310, 311, 315, 316, 318–320
  bare-root, 112–115
  containerized, 113, 114
  development, 95, 102
  growth, 104, 105, 107–109, 138, 140, 233, 276, 279
  mortality, 106, 107
  plug, 114, 305, 319
Seedling exclusionary zone, 110
Seeps, 56, 59, 65, 84, 87, 89, 97
Selection method, 120, 122–223, 227–233, 237–239
  group, 120
  single tree, 120
Serenoa repens, 58, 60, 61, 64, 67, 71, 73, 75, 81–84, 123, 307, 328, 386
Shade-intolerant, 138, 221
Shear and pile, 273
Shelterwood, 95, 109, 116–120, 124, 125, 218, 219, 231–238, 277–279, 288–290, 405
  irregular, 119, 120, 125
  modified, 278, 279
  preparatory cut, 117–119, 289
  removal cut, 118, 119, 235

seed cut, 118, 119
uniform, 117, 119, 120, 125
Shortleaf pine, 15, 16, 29, 70, 99, 120, 123, 218, 223–231, 238, 239, 254, 259, 286, 287, 290, 339, 343, 380
Silty uplands, 56, 59, 62, 76–78
Silviculture, 5, 95, 111, 115, 119, 120, 124–127, 135, 139, 196, 197, 215–264, 381, 394
  even-aged, 99
  uneven-aged, 5, 119–120, 217–239
Silvicultural system, 115, 125, 218–221, 223
Single tree selection, 201, 279, 281
Site index, 255, 259–262
Site index curves, 260, 261
Site preparation, 114, 115, 196, 197, 219, 228, 233, 236, 251, 272, 273, 275, 279, 284–287, 302–304, 310, 311, 316, 380
  3-in-1 plow, 273
  bedding, 273
  bulldozing, 273
  burning, 274
  chemical, 272, 273
  disking, 273
  drum chopping, 273
  mechanical, 114, 115
  scalping, 274
  shear and pile, 273
  sub soiling, 274
Site quality, 101, 154, 253, 259, 261, 262
Slash pine, 11, 18, 30, 35–37, 43–46, 70, 88, 96, 99, 163, 181, 191, 196, 197, 262, 285, 287, 300, 301, 310, 339, 342, 362, 404, 405, 408, 411
Snags, 98, 125, 174, 179, 181, 190, 191
Snakes, 5, 158, 159, 167–174, 190–193, 198, 365, 387, 391, 396
  characteristic snakes of longleaf pine savannas, 168
Society for Ecological Restoration, 5, 6
Socioeconomic, 6, 336
Soil orders, 56
Soil productivity, 6, 273
Soils, 14, 16, 21, 25, 29, 52–59, 73, 76–90, 98, 103, 105, 107, 110, 112, 142, 153, 157, 165, 167, 169, 171, 173, 175, 181, 186, 194, 195, 199, 252, 261, 274, 283, 290, 304, 308, 330–332,

Index

343, 377, 379, 381, 383, 384, 396
Solenopsis geminata, 199
Sonderegger invicta, 199
Sonderegger pine, 99
Sorghastrum nutans, 62, 71, 75
Sorghastrum secundum, 60, 71–73, 75
South Carolina, 4, 13, 15, 21, 23, 25, 27, 28, 30, 33–37, 53, 54, 57, 58, 67, 73, 74, 76, 81, 85–87, 108, 165, 189, 190, 193, 199, 200, 297, 298, 307, 308, 313, 330, 331, 335, 343, 344, 359, 360, 362, 365, 392, 407, 410, 426
Southern bald eagle, 363, 366
Southern magnolia, 70
Southern red oak, 16, 70, 380, 383
Sowing, 112, 113, 152, 272, 316–319
Spatial ecology, 377, 379, 381, 383, 385, 387–397
Species richness, 5, 8, 56, 57, 78, 86, 89, 159, 175, 176, 196, 200, 274, 302, 303, 305, 307–309, 317, 425
Species selection, 311
Spodosol, 81
St. Augustine, 17
Stands, 3, 4, 5, 11, 13, 15, 16, 22, 26, 28, 34, 37, 38, 44, 52, 58, 63, 97–99, 101, 104, 107, 108, 110, 111, 115–123, 135, 138, 139, 145–153, 163, 178, 179, 191, 195–198, 217– 239, 246–248, 251–263, 272, 275–280, 286–292, 303–308, 328, 330, 331, 337, 339, 343, 381–383, 386–389, 404, 405, 409
  stand density, 46, 101, 108, 110, 118, 158, 222, 225, 243, 253, 260, 262, 263, 281, 426
Stand regulation, 120–124, 235
Stem exclusion stage, 138
Stoddard–Neel approach, 5, 237, 239, 244
Stoddard–Neel System, 246
Stratification, 112, 146, 315
Strip scalping, 114
Striped newt, 161–163, 166, 167, 189, 193, 198, 387
Strobili, 99, 100
Stumpholes, 165, 191, 193
Stumpwood, 48
Sub soiling, 274
Subxeric sandy uplands, 56, 57, 74, 75
Succession, 6, 34–36, 116, 119, 140, 176, 180, 181, 218, 331, 332
  community, 6, 116
  old field, 11

Sulfometuron methyl, 275
Sweet gum, 163
Sweetbay, 70
Switchgrass, 70

Tar Heelers, 24
Taxodium ascendans, 161
Temporal factors, 354
Temporary ponds, 164–166, 175, 189, 196
Texas, 3, 4, 9, 13, 17, 18, 21, 22, 25, 28, 36, 51, 57, 68, 97, 99, 157, 158, 163, 172, 198, 200, 257, 259, 328, 339, 344, 359, 366, 377
The Nature Conservancy, 200, 310, 320, 338, 391, 413, 415, 419, 421, 427
Thinning, 38, 117, 119, 123, 135, 139, 142–147, 154, 201, 219–221, 229, 234, 248, 252, 259, 260, 264, 271, 281, 287, 288, 304, 305, 310, 331, 332, 386, 405
Threatened, 3, 5, 6, 171, 198–201, 271, 298, 309, 335, 336, 338, 342, 356, 362, 364, 404, 408, 423, 425
Thresholds, 6, 7, 8, 260, 264
Titi, 69, 160, 418
Toothache grass, 69
Trajectory, 5–7
Translocation, 6, 335–337, 339–369, 417, 418
  concerns, 357
  demographic factors, 352
  factors affecting success, 348
  habitat conditions, 349
  logistic factors, 353
  methodology, 356
  need for, 339
  of Red–Cockaded woodpecker, 339, 343, 344, 353
  population scale, 360
  regional scale, 359
  strategies, 359
  success, 346
  temporal factors, 354
  type, 354
Triclopyr, 142, 152, 275, 307, 309
Turkey oak, 16, 71, 159, 163, 166, 169, 171–173, 175, 181, 284, 290, 301, 307, 364, 365
Turpentine, 3, 4, 9, 18, 20–26, 29, 34, 35, 43, 46–48, 193, 381
Turtles, 5, 159, 168, 169, 171, 172, 174, 175, 183, 184, 186, 189, 190

characteristic turtles of longleaf pine savannas, 168

U.S. Fish and Wildlife Service, 178, 199, 200, 335, 339, 340, 346, 348, 349, 350, 353, 355, 357, 359–366, 369, 406, 426
Ultisol, 76
Understory, 4, 6, 10, 13, 16, 37, 59, 73, 74, 75, 76, 79, 82–84, 86, 98, 105, 108, 109, 114, 115, 122, 125, 127, 135–145, 151–154, 178, 197, 220, 221, 229, 233, 237, 243–248, 262, 271, 275, 278, 283–292, 307, 309, 326–328, 331, 337, 404–407, 425, 426
  common names, 68–72
  scientific names, 68–72
Understory diversity, 327
Understory reinitiation stage, 138, 139
Uneven-aged silviculture, 5, 119, 120, 217–249
USDA Forest Service, 44–47, 95, 135, 217, 222, 223, 232–239, 251, 254, 297, 326, 330, 382

Vegetation classification, 52, 58, 66, 90, 396
  national, 52, 58, 66, 90, 396
  of longleaf pine communities, 58
Vegetation management, 135, 154
Venus flytrap, 15, 69, 81
Vertebrate diversity, 5
Virginia, 3, 9, 12, 13, 16, 18–21, 25–37, 51–53, 67, 70, 72, 81, 97, 157, 168, 180, 200, 339, 343, 344, 359, 369, 377
Virginia pine, 70, 343
Volume, 3–5, 18, 51, 52, 96, 97, 107, 113, 116, 119–123, 135, 158, 195, 217–225, 229–231, 235, 236, 242–245, 248, 251–259, 262, 264, 271, 283, 297, 298, 337, 379, 380, 403, 404
Volume-guiding diameter limit (V-GDL), 120–124

Wade Tract, 78, 246
Warm-season grass, 158
Water mining, 393
Water quality, 273, 281, 418, 424, 427, 428
Wetland, 157, 173, 174, 189, 195, 196, 331, 377, 383–387, 390, 419
White fringed beetle, 283

White-tailed deer, 179, 183
Wildland–urban interface, 395, 425, 426
Wildlife Habitat Incentives Program (WHIP), 406, 410

Wiregrass, 4, 15, 34, 38, 57, 59, 68, 73, 74, 76, 78, 80, 98, 135, 138, 141, 165, 166, 173, 174, 195, 200, 217, 307–309, 313–320, 330–332, 380, 381, 396

Xeric sand barrens, 56–60, 67
Xeric sites, 56, 67, 73, 111, 285, 298, 302, 316, 405

Yaupon, 69